Plant Peroxisomes

Plant Peroxisomes

Biochemistry, Cell Biology and Biotechnological Applications

Edited by

Alison Baker

University of Leeds,
Leeds, United Kingdom

and

Ian A. Graham

University of York,
York, United Kingdom

KLUWER ACADEMIC PUBLISHERS

DORDRECHT / BOSTON / LONDON

A C.I.P. Catalogue record for this book is available from the Library of Congress.

ISBN 978-90-481-6007-5

Published by Kluwer Academic Publishers,
P.O. Box 17, 3300 AA Dordrecht, The Netherlands.

Sold and distributed in North, Central and South America
by Kluwer Academic Publishers,
101 Philip Drive, Norwell, MA 02061, U.S.A.

In all other countries, sold and distributed
by Kluwer Academic Publishers,
P.O. Box 322, 3300 AH Dordrecht, The Netherlands.

Cover figure courtesy of Dr. Robert Mullen.

Printed on acid-free paper

Printed in the Netherlands.

CONTENTS

CONTRIBUTORS

Editors	Dr. A. Baker, Centre for Plant Science, University of Leeds, Leeds LS2 9JT, UK Prof. I.A. Graham, Centre for Novel Agricultural Products, Department of Biology, University of York, P.O. Box 373, York YO10 5YW, UK.
Chapter 1	Prof. H. Beevers, Biology Department, University of California, Santa Cruz, CA 93940 USA.
Chapter 2	Dr. M. Hooks, School of Biological Sciences, University of Wales Bangor, Gwynedd LL57 2UW, UK.
Chapter 3	Dr. J.E. Cornah and Dr. S.M. Smith, Institute of Cell and Molecular Biology, University of Edinburgh, The King's Buildings, Mayfield Road, Edinburgh EH9 3JH, UK.
Chapter 4	Dr. M. Heinze and Dr. B. Gerhardt, Institut fuer Botanik, University of Muenster, Schlossgarten 3, D-48148 Muenster, Germany.
Chapter 5	Dr. S. Reumann, Department of Plant Biochemistry, University of Goettingen, Untere Karspuele 2, D037073, Goettingen, Germany.
Chapter 6	Prof. D.P.S. Verma, Biotechnology Center, Ohio State University, Columbus OH 43210, USA.

Chapter 7 Prof. L.A. del Río *et al*,
Depto. Bioquímica, Biología Celular y Molecular de Plantas, Estación Experimental del Zaidín, CSIC, Apartado 419, E-18080 Granada, Spain.

Chapter 8 Prof. R.P. Donaldson,
Department of Biological Sciences, George Washington University,Lisner Hall, Room 340, 2023 G Street NW, Washington DC 20052, USA.

Chapter 9 Dr. M. Hayashi and Prof. M. Nishimura,
Department of Cell Biology, National Institute for Basic Biology, Okazaki 4448585, Japan.

Chapter 10 Prof. R.N. Trelease,
Department of Botany, Arizona State University, Tempe, Arizona AZ 85287-1601, USA.

Chapter 11 Dr. R. Mullen,
Department of Botany, Room 321 Axelrod Building, University of Guelph, Guelph, Ontario N1G 2W1, Canada.

Chapter 12 Dr. W. Charlton[1] and Dr. E. López-Huertas[2]
[1]Centre for Plant Science, University of Leeds, Leeds LS2 9JT, UK. [2]Puleva Biotech, 66 Camino de Purchil, Granada 18004, Spain.

Chapter 13 Dr. R. Grene Alscher,
Department of Plant Pathology, Physiology and Weed Science, Virginia Polytechnic Institute and State University, Blacksburg, VA 24061, USA.

Chapter 14 Dr. E. Rylott and Dr. T. Larson,
Centre for Novel Agricultural Products, Department of Biology, University of York, P.O. Box 373, York YO10 5YW, UK.

Chapter 15 Prof. Y. Poirier,
Institut d'Ecologie, Laboratoire de Biologie et Physiologie Végétales, Bâtiment de Biologie, Université de Lausanne, CH-1015 Lausanne, Switzerland

PREFACE

In the two decades since the last comprehensive work on plant peroxisomes appeared[1] the scientific approaches employed in the study of plant biology have changed beyond all recognition. Development of techniques such as plant transformation, confocal microscopy, PCR and high throughput DNA sequencing has led to an explosion of information and an opportunity to do experiments that were beyond the wildest dreams of the founders of our subject. A major investment by public funding bodies in Europe, the US and Japan, as well as commercial enterprises, has led to the production of research tools that were unthinkable even a few years ago. We have the complete genome sequence of a dicot (*Arabidopsis thaliana*) and a draft sequence of a monocot (rice). EST projects are well advanced for many other crop species. The development of various post genomic technologies such as proteomics, transcriptomics and metabolomics, and the relative ease of identifying mutants by forward and reverse genetics, raises the prospect of knowing the function of every gene in Arabidopsis in just a few years time, and being able to apply this knowledge to more economically important species. The accelerating pace of plant research is leading us to appreciate that peroxisomes have many more roles than the earliest described metabolic functions in lipid mobilisation and photorespiration, vital though these functions are. Many researchers are finding, and no doubt will continue to find, that their favourite protein, metabolic pathway or physiological response is in, or in some way connected to, peroxisomes. In animals defects in peroxisome formation and function results in multi-system defects and often death. It is becoming apparent that peroxisomes and peroxisome functions are no less important for normal plant development. It, therefore, seemed timely to produce a book which would seek to draw together the current state of knowledge of the field, as a convenient starting point for anyone who wants to know about plant peroxisomes.

The first section of the book surveys the functions of plant peroxisomes. The first chapter by Beevers provides a personal historical perspective of those pioneering early studies and provides a link between the past and the present. Hooks, Smith and Cornah, and Reumann provide comprehensive reviews of fatty acid oxidation, the glyoxylate cycle and the photorespiratory pathway,

[1] Huang, A.C., Trelease, R.N. and Moore, T.S. (ed) (1983) Plant Peroxisomes. New York:Academic Press

integrating older observations with new data arising from mutants, genome and EST sequences about gene families, expression and function. Verma addresses the important functions of root nodule peroxisomes in nitrogen fixation in tropical legumes. Heinze and Gerhardt lead us through the immense complexities of catalase function and biogenesis with multiple genes regulated at different levels giving rise to a wide range of isoforms to defend plants against oxidative stress. The role of peroxisomes as both a source of reactive oxygen species and ROS related signal molecules and as a first line of defence against these potentially harmful species is addressed by Del Rìo and co-workers. Donaldson surveys the composition and function of the peroxisome membrane. The compartmentalization of pathways such as photorespiration between organelles necessitates the intra-compartment movement of many molecular species. To what extent the peroxisome membrane is responsible for compartmentalization and to what extent supra-molecular assemblies of the pathway enzymes serves to confine intermediates (metabolite channelling) is still a matter of active debate. This question is addressed by Reumann in the context of the glycolate and glycerate pathways of photorespiration, and by Smith and Cornah in the context of the glyoxylate cycle, as well as more generally by Donaldson. Despite our comparative lack of knowledge of the peroxisome membrane it is clear that it has important functions in transport of metabolites and proteins as well as in ROS metabolism. The section ends with a chapter by Hayashi and Nishimura that describes several peroxisome-deficient mutants including those that are disrupted in specific aspects of peroxisome metabolism and peroxisome biogenesis. These studies highlight the contribution that the genetic approach is now making to our understanding of the physiological relevance of these processes and the mechanisms underlying peroxisome biogenesis and protein import.

The second section deals with peroxisome biogenesis and protein import in detail. Trelease provides a thoughtful discussion on the origin and differentiation of peroxisomes, still a topic of some controversy within the field, and then reviews our limited knowledge of the targeting and insertion of peroxisome membrane proteins. Mullen considers the targeting and import of peroxisome matrix proteins and Charlton and Lopez-Huertas conclude this section with a chapter on genes required for peroxisome biogenesis in other organisms and their counterparts in plants. This survey leads them to consider whether the components for peroxisome biogenesis are universal.

The final section considers the potential to exploit our growing knowledge of peroxisome biogenesis and function for biotechnological ends. Alscher considers the possibilities for engineering plant peroxisomes to increase plant resistance to stress. Rylott and Larson address the problem of futile cycling through β-oxidation as a potential barrier to increased yields of novel oils in transgenic plants. Poirier reviews use of peroxisomes as a compartment for the synthesis of biodegradable plastics and other novel biopolymers.

It has been a pleasure for us to work with colleagues from 9 countries and 3 continents during the preparation of this book. We as editors have learned an enormous amount from reading and discussing the various contributions with the authors whom we thank for contributing their time and enthusiasm to the project. We hope that you the reader will find it stimulating and informative.

ALISON BAKER AND IAN A. GRAHAM, DECEMBER 2001.

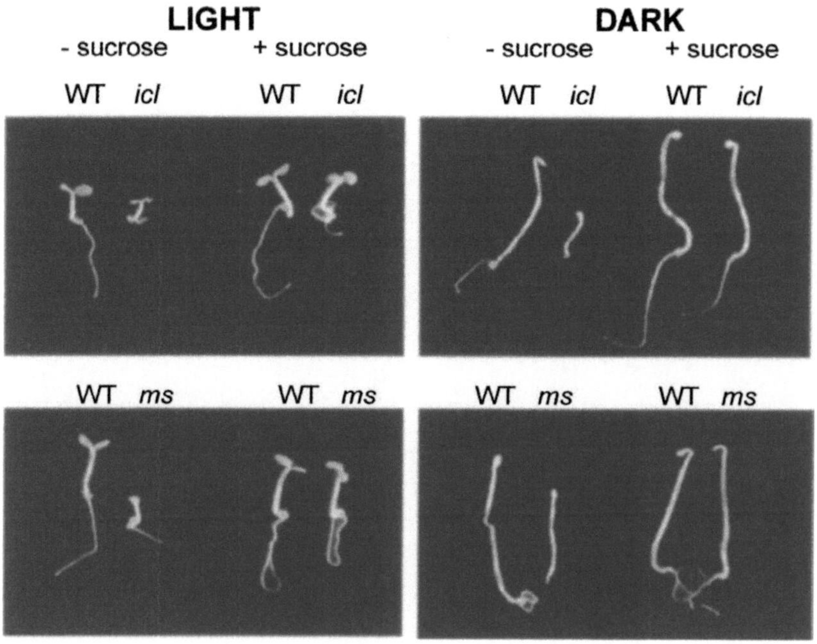

Growth of *icl* and *ms* mutant seedlings. See Chapter 3 Synthesis and Function of Glyoxylate Cycle Enzymes by Johanna E. Cornah and Steven M. Smith

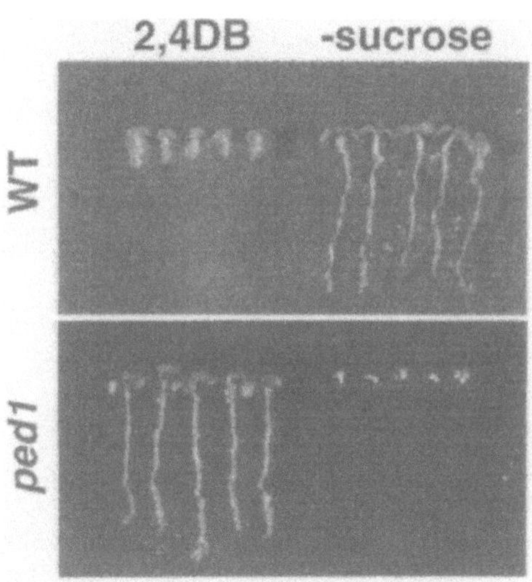

Effect of 2,4-DB and sucrose on the growth of wild type and *ped1* seedlings. See Chapter 9 Genetic Approaches to Understand Plant Peroxisomes by Makoto Hayashi and Mikio Nishimura.

ACKNOWLEDGEMENTS

We owe an enormous debt of gratitude to Sarah Pritchard for her efficient and thorough proof reading of all the chapters, and to Diane Baldwin, whose desk top publishing skills and careful attention to detail turned the various manuscripts into a book worthy of the name. It has been a pleasure working with them and without their patient and willing help the whole project would have been much more arduous.

1

EARLY RESEARCH ON PEROXISOMES IN PLANTS

Harry Beevers

University of California, Santa Cruz, USA

KEYWORDS

Microbodies; β-oxidation; glycolate pathway; two population hypothesis; single population hypothesis.

INTRODUCTION

The early research on peroxisomes in plants was centred mainly on three groups, my own at Purdue (later at University of California, Santa Cruz) and those of Tolbert at Michigan State University and Newcomb at Wisconsin. Some of the collaborators from these centres subsequently established their own research laboratories and, as interest in the field expanded, others independently joined the international association. Several authors in the present volume trace their lineage in peroxisomal research either directly or indirectly to these three groups, but progress has been such that the questions they now address are well beyond the vision of their ancestors.

The stage for the discovery of the functions of peroxisomes in plants was set in the early 1960s. Prior to 1965, distinctive organelles about 0.5μm in diameter, bounded by a single membrane and with an amorphous or granular interior had been recognized in electron micrographs of animal tissues and given the name 'microbodies' (Rhodin, 1958). Early cell fractionation studies on rat liver by Novikoff and others eventually led, in de Duve's laboratory, to the separation of a fraction, distinct from lysosomes and mitochondria,

A. Baker and I.A. Graham (eds.), Plant Peroxisomes, 1–17.

which was differentially enriched in urate oxidase and catalase. Examination of such fractions in the EM showed that organelles with the distinctive morphology of microbodies were the major component (Baudhuin *et al*, 1965). Since these organelles contained oxidases producing H_2O_2 and catalase, de Duve (1965) introduced the term peroxisomes to describe their biochemical capacity. They represent ~1% of the particulate protein in liver cells.

The identity of microbodies with peroxisomes was confirmed by enzyme cytochemistry when Novikoff and Goldfischer (1969) showed that catalase was specifically associated with the microbodies in electron micrographs of mammalian cells. Mollenhauer *et al* (1966) showed that organelles with the morphological characteristics of animal microbodies were present in electron micrographs of a variety of plant tissues and, later, these too were shown by enzyme cytochemistry to contain catalase (Vigil, 1970; Frederick *et al*, 1975).

The distinctive biochemistry of microbodies in plants became clear from two different lines of investigation. At Purdue, from the early 1950s, we had been working on the mechanism of conversion of fat to carbohydrate, which is a remarkably efficient metabolic process in young fatty seedlings (Beevers, 1961) and had shown that glycerol and acetate were each rapidly converted to sucrose in castor bean endosperm (Beevers, 1980). The elucidation of the glyoxylate cycle by Kornberg and Krebs in 1956 provided a mechanism for the net conversion of acetate to succinate and thus a plausible scheme for the overall process was as follows:

Fat → fatty acids → β-oxidation → acetyl CoA.
Acetyl Co A → (glyoxylate cycle) → succinate.
Succinate → oxalacetate → phosphoenolpyruvate + CO_2.
Phosphoenolpyruvate (reversed glycolysis) → sucrose.

With Kornberg, I was able to show that isocitrate lyase and malate synthase were present in the castor bean tissue (Kornberg and Beevers, 1957). Subsequently, the enzymology of the complete reaction sequence was firmly established at Purdue and labeling experiments proved that the sequence operated *in vivo* (Beevers, 1961, 1980). Since the glyoxylate cycle and the TCA cycle had three enzymes in common (citrate synthase, aconitase, malate dehydrogenase), it was expected that the glyoxylate cycle would be housed in the mitochondria and the particulate nature of some enzymes was suggested in earlier work. However, in 1966, when Breidenbach in my laboratory centrifuged a crude particulate fraction from castor bean endosperm on a sucrose density gradient, he made a surprising discovery. The mitochondria, with complete TCA cycle activity, were recovered from the gradient at equilibrium density 1.19g cm^{-3}, but did not show any activity of isocitrate lyase or malate synthase (Breidenbach and Beevers, 1967). The distinctive enzymes of the glyoxylate cycle were recovered instead in a clean peak at density 1.25g cm^{-3}, which contained ~25% of the protein within the gradient and was thus a major particulate fraction. EM examination showed that the fractions at density 1.19 were composed of mitochondria, while those at density 1.25 were dominated by ovate organelles ~1μm in diameter with amorphous interior and bounded by a single membrane (Breidenbach *et al*, 1968). Both organelle peaks contained malate dehydrogenase and citrate synthetase (later shown to be different iso-enzymes) and aconitase (Breidenbach *et al*, 1968, Cooper and Beevers, 1969a). The organelles containing the enzymes of the glyoxylate cycle were therefore named glyoxysomes. Soon afterwards, we found that they also contained catalase and glycolate oxidase (Breidenbach *et al*, 1968). Thus, the organelles had the morphological characteristics of microbodies and the definitive enzymes of peroxisomes. Vigil (1970) made a thorough investigation of castor bean endosperm tissue by electron microscopy. He showed that glyoxysomes and mitochondria were present in abundance when fat breakdown is occurring and that the glyoxysomes were the only organelle containing catalase. Glyoxysomes have since been isolated from a variety of fatty seedling tissues, including the aleurone and scutella of corn seeds, and are seen as a major organelle in such tissues when these are examined by electron microscopy. In this volume, the synthesis and function of glyoxylate cycle enzymes are reviewed in Chapter 3.

A further surprise was the finding by Cooper in my group that β-oxidation, which, at least in animal tissues, had been shown to be a mitochondrial function, was confined to the glyoxysomes in the castor bean preparations, and was independent of carnitine. We showed that acetyl CoA was produced from palmitoyl CoA, and this was accompanied by O_2-uptake and NAD^+ reduction (Cooper and Beevers, 1969b). In contrast to the first dehydrogenase step in the classical mitochondrial β-oxidation pathway, the first step in the glyoxysomal pathway was catalyzed by a flavin-linked oxidase, and H_2O_2 generated in this reaction was decomposed by endogenous catalase. Subsequent steps were analogous to the classical system, and thiolase (Cooper and Beevers, 1969b) and the fatty acid activating system (Cooper, 1971) were confined to the glyoxysomes. At about the same time, in Stumpf's laboratory, where lipid metabolism was a major industry, Hutton also showed that glyoxysomes were the site of β-oxidation activity (Hutton and Stumpf, 1969). Although, the acyl CoA oxidase was not recognized, the hydratase and β-OH fatty acyl CoA dehydrogenase were shown to be present (Hutton and Stumpf, 1969).

Thus, the following sequence of reactions was proposed for this glyoxysomal β-oxidation pathway:

(C_n) Fatty acid + ATP + CoA → Fatty acyl CoA + AMP + PP.
Fatty acyl CoA + O_2 (acyl CoA oxidase) → Enoyl CoA + H_2O_2.
H_2O_2 (catalase) → H_2O + 1/2 O_2.
Enoyl CoA + H_2O (Hydratase) → β-OH acyl CoA
β-OH acyl CoA + NAD^+ (dehydrogenase) → β-ketoacyl CoA + NADH
β-ketoacyl CoA + CoA (Thiolase) → (C_{n-2}) fatty acyl CoA + acetyl CoA

As shown, NADH accumulated during acyl CoA oxidation by the glyoxysomes; when mitochondria were also added, NADH was oxidized and the rate of O_2-uptake was doubled (Cooper and Beevers, 1969b). Gerhardt's review (1992) summarizes more recent work on β-oxidation by glyoxysomes, including the demonstration of complete conversion to acetyl-CoA. Nishimura's group has characterized both long-chain (Hayashi *et al*, 1998) and short-chain (de Bellis *et al*, 2000) fatty acyl CoA oxidases from pumpkin involved in this oxidation. Recent molecular and genetic studies on peroxisomal β-oxidation are described in Chapters 2 and 9.

Tolbert's laboratory at Michigan State University was the location of the second line of investigation leading to the biochemical definition of peroxisomes in plants. He had worked for many years on photosynthetic reactions in leaves, and particularly on the fate of glycolate which under some circumstances was a major photosynthetic metabolite. By a series of labelling experiments and enzymology, he had shown that, in the light, there was a sequence as follows:

2 glycolate → 2 glyoxylate → 2 glycine
2 glycine → serine + CO_2
serine → OH-pyruvate → glycerate
glycerate → P-glycerate → sugars

This is known as the glycolate or Tolbert pathway (Tolbert, 1971, 1980). The various enzymes in this pathway had been intensively studied, but their intracellular location was not revealed until Tolbert, following de Duve's demonstration of a hydroxyacid oxidase and catalase in animal peroxisomes and our own work on glycolate oxidase and catalase in glyoxysomes, showed first, that glycolate oxidase in leaves was at least partly particulate, and that organelles could be separated on stepped sucrose gradients of particulate preparations from spinach leaves, containing not only glycolate oxidase and catalase but OH-pyruvate reductase and various transaminases as well (Tolbert *et al*, 1969). The organelles from the gradient had the morphological features of microbodies and in deference to de Duve they were called peroxisomes. The leaf peroxisomes account for only a very small percentage of the protein of a crude particulate fraction from leaves which is dominated by chloroplast protein. It is now well established that peroxisomes in leaves play an essential role in the Tolbert pathway, along with the participation of mitochondria in the conversion of glycine to serine and the chloroplasts in the reactions following glycerate production in the leaf peroxisomes. Mutants defective in some of the enzymes of leaf peroxisomes are lethal under natural conditions (Somerville and Ogren, 1982).

The full significance of this pathway had to await the crucial discovery by Ogren and Bowes (1971) that the formation of glycolate in photosynthesis is a consequence of the fact that the enzyme responsible for the reaction of ribulose bis-phosphate with CO_2 to produce 2 moles of phosphoglycerate also reacts with O_2 to produce instead 1 mole of phosphoglycerate and 1 mole of phosphoglycolate. The glycolate pathway thus begins during photosynthesis in the light in the chloroplast, where the phosphoglycolate is hydrolyzed by a specific phosphatase (Tolbert, 1980). The overall pathway constitutes what is now known as photorespiration and is a normal accompaniment of photosynthesis, its magnitude depending on the relative concentrations of O_2 and CO_2 at the enzyme site. The photorespiratory pathway and transport of metabolites is reviewed in Chapter 5.

Both glyoxysomes and leaf peroxisomes co-operate with other organelles in the cells and during their functioning an inter-organelle traffic of metabolites occurs. In both types of organelle there is the necessity of transfer of reducing equivalents and specific transaminases in both are involved in this shuttling (Beevers, 1979; Huang *et al*, 1983; Tolbert, 1980; Mettler and Beevers, 1980). In glyoxysomes, the possibility of reoxidation of NADH by a membrane-bound electron transfer system has been suggested (Hicks and Donaldson, 1982). In the original isolations of glyoxysomes, it was recognized that the levels of aconitase were very low and this enzyme was apparently unstable (Cooper and Beevers, 1969a). More recent work has shown that there is no specific glyoxysomal aconitase (Coutois-Verniquet and Douce, 1993; de Bellis *et al*, 1993; Hayashi *et al*, 1995). This would require that a citrate-isocitrate shuttle occurs across the glyoxysome membrane, with the aconitase conversion occuring in the cytosol (Hayashi *et al*, 1995).

Newcomb's group at Wisconsin produced excellent EM pictures from a variety of leaves showing leaf peroxisomes (microbodies) in association with chloroplasts and mitochondria (Frederick *et al*, 1975). Some of these peroxisomes had crystalline inclusions, their morphological properties were clearly defined and they were shown to contain catalase by cytochemistry. These definitive studies were extended to include microbodies from a whole range of plant tissues, including fatty seedlings and root nodules. Although the number of microbodies per section (or per cell) varied widely, with fatty seedling tissues containing by far the most, it was established that they have a common morphology and are normal constituents of plant cells generally. It was noticed that in several sections there was a close association with endoplasmic reticulum.

In 1969, an important conference was held in New York, organized by de Duve of the Rockefeller University. It brought together not only workers on peroxisomes in mammalian tissues, but those in plants, fungi, and simple animals as well. The resulting report, The Nature and Function of Peroxisomes (Microbodies, Glyoxysomes) is a landmark in the history of the subject (Hogg, 1969). It was clear that, within the inclusive category of microbodies, there was a range of more or less specialized peroxisomes. At that time, the peroxisomes from rat liver were known to contain only a few enzymes, they represented a very minor fraction of the crude particulate material and no important metabolic role could be ascribed to them. By contrast, the leaf peroxisomes, with several enzymes of the glycolate pathway, played an essential role in photosynthetic metabolism, and the glyoxysomes, with more than a dozen enzymes associated with β-oxidation and the glyoxylate cycle, were the site of a major part of the metabolism of the tissue. They clearly belonged in the category of peroxisomes, but the defining enzymes of these latter organelles were almost incidental to glyoxysomal functioning and importance. During the following decade, research on microbodies (peroxisomes) in plants expanded as new approaches were developed and techniques became available. Several reviews appeared summarizing this work (Tolbert, 1971; Beevers, 1979; Kindl, 1982a) and a particularly valuable book was published by Gerhardt (1978) which provides a comprehensive account and bibliography concerning these organelles in all types of plant cells. In the following, only highlights of the progress in work with higher plants will be outlined.

The recognition that, though in widely different numbers, microbodies were present in essentially all plant cells (Mollenhauer *et al*, 1966; Frederick *et al*, 1975) was accompanied by their isolation from a variety of tissues other than leaves and fatty seedlings by applying similar techniques of sucrose gradient density separation. From nine different organs, including roots, tubers, petals, and coleoptiles, Huang and Beevers (1971) obtained a fraction which contained the marker enzymes catalase, glycolate oxidase and uricase, equilibrating together at density 1.20-1.25 and definitively separated from the mitochondria. Thus, these tissues yielded organelles with the (limited) enzyme markers of peroxisomes from animal tissues. Since, on this basis, their function remained in question, they were referred to as unspecialized microbodies or peroxisomes, in contrast to leaf peroxisomes and glyoxysomes.

Considerable progress was made during this period in the purification and characterization of the enzymes of leaf peroxisomes and glyoxysomes (Gerhardt, 1978; Kindl, 1982a, 1987, 1993; Tolbert, 1971, 1980, 1981). Different iso-forms of enzymes in particular organelles were recognized (Becker *et al*, 1978; Hock *et al*, 1987; Huang *et al*, 1974; Kagawa and Gonzalez, 1981) and antibodies prepared against purified enzymes. Frevert and Kindl (1980) showed that, in the glyoxysomal β-oxidation pathway, the activities of the hydratase and β-OH acyl CoA dehydrogenase were contained in a single protein, the multifunctional protein or MFP. A thorough review of the enzymes in this pathway was provided by Kindl (1987).

The question of the biogenesis of the peroxisomes was addressed. Since the organelles contained neither DNA nor ribosomes, the synthesis of their enzymes occurred elsewhere in the cell. In the first few days of growth in fatty seedlings there was a co-ordinated several-fold increase in the amounts of enzymes of the glyoxylate cycle and glyoxysomes (Gerhardt and Beevers, 1970). In leaves during greening in the light, the numbers of peroxisomes increased, although the increases in glycolate oxidase and catalase did not occur in parallel (Feierabend and Beevers, 1972a, b). Early speculation on glyoxysome ontogeny followed that of workers on mammalian cells and was based primarily on the observations from electron microscopy that there were frequent close associations and occasional connections with elements of the endoplasmic reticulum, and no dividing peroxisomes had been described. It was largely on this information that the origin of peroxisomes by vesiculation of the endoplasmic reticulum was suggested (Beevers, 1979; Gerhardt, 1978; Huang *et al*, 1983; Trelease, 1984). It was shown in castor bean endosperm that clean preparations of ER membranes and also ER vesicles with associated ribosomes (rough ER) could be obtained. The ER membranes were shown to be the site of synthesis of the various phospholipid constituents of glyoxysomal membranes (Kagawa *et al*, 1973; Moore *et al*, 1973; Beevers, 1979) and pulse labelling, for example with ^{14}C-choline, showed that the ER membranes were the first to become labelled *in vivo* and that this fraction lost radioactivity as it appeared in glyoxysomes (and mitochondria) (Kagawa *et al*, 1973). The fact that some glyoxysomal enzymes were apparently associated with their membranes and that early in growth they were also recovered in ER in proportionately greater amounts was in accord with the proposal that these enzymes were synthesized on rough ER and incorporated into glyoxysomes as they developed by vesiculation from ER (Beevers, 1979).

At least for matrix proteins, this particular model was rejected when it became clear from the work of Kindl (1982b), Becker *et al* (1978), and Hock *et al* (1987) that the glyoxysomal enzymes were synthesized in the cytosol on free ribosomes and not on rough ER. From other lines of work, it was becoming clear at this time that mechanisms existed whereby a variety of proteins destined for the interior of organelles could be transported across their membranes and, further, that the enzymes of individual organelle types were specifically targeted to their appropriate organelles (Blobel and Dobberstein, 1975).

During this period, there were accelerated efforts to elucidate the molecular biology of the early events in the synthesis of glyoxysomal proteins and their incorporation into glyoxysomes (Becker, *et al*, 1978; Hock, *et al*, 1987; Kindl, 1982b; Trelease 1984). The present understanding of these events is described by authors in the present volume (Chapters 10, 11 and 12); see also the review by Olsen and Harada (1995). Although substantial progress has been made for both matrix and membrane proteins, formidable problems remain concerning the way in which phospholipid and protein components of the membrane become incorporated during biogenesis. The ER is clearly responsible for phospholipid synthesis and recent work shows that it is also involved in the production of membrane proteins (Mullin *et al*, 1999; Nito *et al*, 2001). Overall, the consensus seems to be that peroxisomes in higher plants multiply by fission, but the act itself remains elusive. This area remains one of active debate, see Chapter 10.

The development of leaf peroxisomes has not been intensively studied, except in cucumber and other cotyledons. In seedlings of this type, the reserve fat is stored in the cotyledons (i.e. part of the embryo) which occupy the bulk of the seed volume. As fat is consumed during early growth by the mechanism described earlier, the cotyledons emerge above ground and in the light they expand, become green, and function as leaves. Glyoxysomes are present during fat breakdown and, after chloroplast development in the light, typical leaf peroxisomes are present. Thus, there is a transition during early growth, requiring light for its fulfillment; there is a replacement of one kind of microbody by another.

For both glyoxysomes and leaf peroxisomes in this organ, the problems of the origin of the bounding membrane remain. But, the question of the way in which the enzymes of the glyoxysome disappear and those of the leaf peroxisome succeed them has engaged considerable attention. The details of those endeavors are summarized by Beevers (1979), a balanced summary is given in Huang *et al* (1983), and a thoughtful analysis by Gerhardt and Betsche (1976). Two major proposals for the changeover were advanced. Kagawa *et al* (1973) showed that in watermelon cotyledons growing in darkness, the levels of enzymes of the glyoxylate cycle rose strikingly in early growth and declined to essentially zero after the fat had been consumed. The provision of light at any time resulted in a striking increase in the distinctive enzymes of leaf peroxisomes (glycolate oxidase and OH-pyruvate reductase) and, interestingly, accelerated the decline of glyoxysomal enzymes. Kagawa and Beevers (1975) proposed that, as in the castor bean endosperm, the population of glyoxysomes is destroyed and is replaced by a new microbody population, the leaf peroxisomes which develop, as in other leaves, in the light. It was supported by evidence showing increased membrane turnover during the transition (Kagawa *et al*, 1973). This was the two population hypothesis.

The single population hypothesis, proposed initially by Trelease *et al* (1971) from similar observations on enzymes in cucumber cotyledons, and detailed ultrastructural observations, was that enzymes of the glyoxylate cycle within existing glyoxysomes were replaced by those of photorespiration when light was supplied. Thus, this is a repackaging or interconversion model. These authors found little evidence for microbody destruction or new synthesis during the transition and they estimated that the numbers of microbodies remained constant, in spite of a change in cell geometry.

It was clear that definitive evidence in favour of the single population model would be a convincing demonstration that, during the transition, the distinctive enzymes of glyoxysomes and leaf peroxisomes were present in the same organelle and such evidence was eventually forthcoming. Titus and Becker (1985) with cucumber cotyledons, Nishimura *et al* (1986) with pumpkin, and Sautter (1986) with watermelon, used refined techniques of

immunocytochemistry to show that this was indeed so. The two population hypothesis was no longer tenable: the way in which enzyme production is induced and controlled in the cytosol and targeted to peroxisomes is a continuing endeavor which is dealt with by other authors in this volume.
It should not be thought, however, that the acceptance of the re-packaging proposal solved all of the problems of the glyoxysome to leaf peroxisome transition. Two of the problems that remain are: (a) what mechanism accounts for total loss of enzymes of the glyoxylate cycle while some catalase activity remains, and (b) what is the true relationship between total microbody numbers per cell and the apparent loss of total microbody protein during greening.

An intriguing corollary to the transition of glyoxysomes to leaf peroxisomes is the more recent recognition that, during the senescence of true leaves (Tolbert *et al*, 1987; Gut and Matile, 1988; Pistelli *et al*, 1991) and cotyledons (Nishimura *et al*, 1993; de Bellis and Nishimura, 1991), enzymes of the glyoxylate cycle appear, and it has been established that this is due to the repackaging of existing leaf peroxisomes to produce functional glyoxysomes. This reverse transition has also been fully documented in flower petals by Nishimura's group (de Bellis and Nishimura, 1991). In the present volume, the synthesis and function of glyoxylate cycle enzymes in tissues other than seedlings is discussed by Cornah and Smith (Chapter 3).

In 1981, a second international conference on Peroxisomes and Glyoxysomes was held in New York and the Proceedings were published as Volume 386 of Proc. N.Y. Acad. Sci., edited by Kindl and Lazarow (1982). In that volume the progress in research with higher plants was dealt with in chapters by Newcomb, Tolbert, Beevers, Becker *et al*, Hock and Gietl, Lord and Roberts, and Kindl.

As emphasized earlier, the peroxisomes isolated from animal tissues were very limited in enzyme composition and were not known to house any major metabolic pathway. This situation changed dramatically when Lazarow and de Duve (1976) reported that rat liver peroxisomes contained precisely the same alternate β-oxidation sequence discovered by Cooper and Beevers (1969b) in glyoxysomes. This pathway, beginning with the acyl CoA oxidase, co-exists in liver with the classical mitochondrial β-oxidation sequence. The peroxisome pathway was greatly enhanced by clorfibrate, a hypolipidemic drug, which led to peroxisome proliferation (Lazarow and de Duve, 1976). These discoveries led to a renewed interest in animal peroxisomes and a major part of the 1981 conference was devoted to the alternate β-oxidation pathway in these organelles and the ability of several drugs to induce peroxisome proliferation. In the meantime, in 1978, Tanaka's group had shown that in yeast growing on alkanes there was a striking population of peroxisomes which also contained the alternate β-oxidation pathway, and as in the castor bean, no mitochondrial β-oxidation could be detected (Kawamoto *et al*, 1978).

As a personal aside, I note that in their original reports both the de Duve and Tanaka groups fully acknowledged that the peroxisome β-oxidation system was that described years earlier by Cooper and Beevers. In the subsequent rapid expansion of work in animal peroxisomes, the priority and significance of the work in plants is ignored, and the discovery of peroxisomal β-oxidation is credited to Lazarow and de Duve. *Sic transit gloria*. Or words to that effect.

Shortly after this New York conference, a second book on Peroxisomes in Plants was published in by Huang *et al* (1982). Aspects of the present brief history are covered in some detail and valuable comparative data are compiled. A somewhat tortuous argument was made in favor of abandoning the term microbody, a morphological term, not descriptive of function, which had proven so useful in the past, and used where appropriate in the present article, in favour of peroxisomes. And, if biochemical function were known, a name reflecting this would indeed be more suitable. However, when the peroxisome was named by de Duve (1965), its true function as a site of β-oxidation was unknown and a name more reflective of its metabolic importance would now, in my view, be more appropriate. Nevertheless, the name peroxisome has become accepted as the general title for this class of organelles.

In this regard, Gerhardt's (1986) work is of great interest. He showed that, what were referred to earlier as non-specialized microbodies (peroxisomes from plant tissues other than leaves and fatty seedling organs and thus not leaf peroxisomes or glyoxysomes), are also the site of the β-oxidation pathway of Cooper and Beevers and further that the same enzymes are present in leaf peroxisomes, although the levels of activity do not approach those in glyoxysomes (Gerhardt, 1986). He therefore proposed that the presence of the β-oxidation pathway is *the* distinctive general biochemical characteristic of peroxisomes from plants, as indeed it is in those from animal tissues. In my view, a suitable name reflecting this would be more appropriate than peroxisomes, which (wrongly) suggests the presence of peroxidase, a distinctive plant enzyme well known to the ancients.

In non-fatty tissues, the β-oxidation system in the peroxisomes is thought to play a role in the turnover of lipids in plant membranes (Gerhardt, 1986) and in the net breakdown during the senescence (Gerhardt, 1992). During this period, with emphasis on the participation of the peroxisomal β-oxidation in plants, attention has nevertheless been drawn to the possibility, that at least in some organs, such as the pea cotyledon, the classical mitochondrial β-oxidation pathway may play a role. The reviews of Gerhardt (1992) and of Masterson and Wood (2000) should be consulted for a detailed examination of this question.

By the time of the third U.S. Conference on Peroxisomes, held in Aspen in 1995, the investigations of the mammalian peroxisomal β-oxidation system had so expanded that they dominated the proceedings. The title of the report was appropriately, Peroxisomes: biology and role in toxicology and disease. This was volume 804 of the Ann. N.Y. Acad. Sci. edited by Reddy *et al* (1996). At one session, the problems of peroxisomal biogenesis were discussed and the emphasis was on targeting sequences and protein insertion. The two major themes were genetic diseases related to the peroxisomal β-oxidation, and the responses of this system to a variety of drugs and treatments, which lead to striking proliferation of peroxisomes. Major international attention is now being focused on findings showing that defects in the genes controlling enzymes of the pathway (e.g. acyl CoA oxidase, thiolase) lead to specific human diseases, which had been clinically recognized for many years. In addition, there are a group of five diseases which are due to defects in peroxisome assembly. Nishimura's article in this volume addresses the effects of mutations in enzymes of glyoxysomal β-oxidation in plants.

Since the initial isolation from plants in 1967, the peroxisome has emerged as an essential organelle with a variety of roles in metabolism. Of particular relevance are the functions in fatty seedlings, where peroxisomes house many of the enzymes concerned with the conversion of fat to sugar, and in leaves where they are the site of essential reactions in photorespiration. In all cells of higher plants, peroxisomes contain the enzymes of the alternate pathway of β-oxidation of fatty acids and, the organelles appear to be of importance in membrane turnover and during senescence.

Much remains to be elucidated, particularly about the biogenesis of the various peroxisomal constituents and their assembly into functional organelles. Increasingly and appropriately, the power of molecular biology is being brought to bear on these problems. A sad consequence (to me) of this new wave of progress is that the focus of work on plant peroxisomes has descended all the way from the noble castor bean to that miserable weed Arabidopsis.

REFERENCES

Baudhuin, P., Beaufay, H. and deDuve, C. (1965) Combined biochemical and morphological study of particulate fractions from rat liver. *J. Cell Biol.* **26**: 219-224.

Becker, W.M., Leaver, C.J., Weir, E.M., Titus D.E. and Reizman, H. (1978) Regulation of glyoxysomal enzymes during germination of cucumber. *Plant Physiol.* **62**:542-549.

Beevers, H. (1980) The role of the glyoxylate cycle. In *The Biochemistry of Plants.* (Stumpf, P.K. and Conn, E.E., eds.). Vol. 4: 177-130.

Beevers, H (1979) Microbodies in higher plants. *Ann. Rev. Plant Physiol.* **30**: 159-193.

Beevers, H. (1961) Metabolic production of sucrose from fat. *Nature.* **191**: 433-436.

Blobel, G. and Dobberstein, B. (1975) Transfer of proteins across membranes. *J. Cell Biol.* **67**: 835-851.

Breidenbach, R.W. and Beevers H. (1967) Association of glyoxylate cycle enzymes in a novel subcellular particle from castor bean endosperm. *Biochem. Biophys. Res. Comm.* **27**: 462-469.

Breidenbach, R.W., Kahn, A. and Beevers, H. (1968) Characterization of glyoxysomes from castor bean endosperm. *Plant Physiol.* **43**: 705-713.

Cooper, T.G. (1971) The activation of fatty acids in castor bean endosperm. *J. Biol. Chem.* **246**: 3451-3455.

Cooper, T.G. and Beevers, H. (1969a) Mitochondria and glyoxysomes from castor bean endosperm. *J. Biol. Chem.* **244**: 3507-3513.

Cooper, T.G. and Beevers, H. (1969b) β-oxidation in glyoxysomes from castor bean endosperm. *J. Biol. Chem.* **244**: 3515-3520.

Courtois-Verniquet, F. and Douce, R. (1993) Lack of aconitase in glyoxysomes and peroxisomes. *Biochem. J.* **294**: 103-107.

De Bellis, L. and Nishimura, M. (1991) Development of enzymes of the glyoxylate cycle during senescence of pumpkin cotyledons. *Plant Cell Physiol.* **32**: 551-561.

De Bellis, L., Gonzali, S. Alpi, A., Hayashi, H., Hayashi, M. and Nishimura, M. (2000) Purification and characterization of a novel pumpkin short-chain acyl-CoA oxidase which structure resembles acyl-CoA dehydrogenase. *Plant Physiol.* **123**: 327 – 334.

De Bellis, L., Tsugeki, R., Alpi, A. and Nishimura, M. (1993) Purification and characterization of aconitase isoforms from etiolated pumpkin cotyledons. *Physiol. Plant.* **88**: 485-492.

De Bellis, L., Tsugeki, R. and Nishimura, M. (1991) Glyoxylate cycle enzymes in peroxisomes isolated from petals of pumpkin during senescence. *Plant Cell Physiol.* **32**: 1227-1235.

De Duve, C. (1965) Functions of microbodies (peroxisomes). *J. Cell Biol.* **27**: 25A.

Feierabend, J. and Beevers, H. (1972a) Developmental studies on microbodies in wheat leaves. I Conditions influencing enzyme development. *Plant Physiol.* **49**: 28-32.

Feierabend, J. and Beevers, H. (1972b) Developmental studies on microbodies in wheat leaves. II Ontogeny of particulate enzyme associations. *Plant Physiol.* **49**: 33-39.

Frederick, S.E., Gruber, P.J. and Newcomb, E.H. (1975) Plant Microbodies. *Protoplasma* **84**: 1-29.

Frevert, J. and Kindl, H. (1980) A bifunctional enzyme from glyoxysome-purification of a protein possessing enoyl-CoA hydratase and 3-hydroxyacyl CoA dehydrogenase. *Eur. J. Biochem.* **107**: 79-86.

Gerhardt, B. (1992) Fatty acid degradation in plants. *Prog. Lipid Res.* **31**: 417-446.

Gerhardt, B. (1986) Basic metabolic function of the higher plant peroxisome. *Physiol. Veg.* **24**: 397-410.

Gerhardt, B. (1978) Microbodies/Peroxisomen. *pflanzlicher Zellen,* Springer Wien, New York.

Gerhardt, B. and Beevers, H. (1970) Developmental studies on glyoxysomes in Ricinus endosperm. *J. Cell Biol.* **44**: 94-102.

Gerhardt, B. and Betsche, T. (1976) The change of microbodies from glyoxysomal to peroxisomal function within fatty, greening cotyledons: hypotheses, results, problems. *Ber deutsch Bot. Ges.* **89**: 321-324.

Gut, H. and Matile, P. (1988) Apparent induction of key enzymes of the glyoxylic acid cycle in senescent barley leaves. *Planta* **176**: 548-550.

Hayashi, M. De Bellis, L., Alpi, A. and Nishimura, M. Cytosolic aconitase participates in the glyoxylate cycle in etiolated pumpkin cotyledons. (1995) *Plant Cell Physiol.* **36**: 669-680.

Hayashi, M., De Bellis, L., Hayashi, M., Yamaguchi, K., Kato, A. and Nishimura, M. (1998) Molecular characterization of a glyoxomal long-chain acyl-CoA oxidase that is synthesized as a precursor of higher molecular mass in pumpkin. *J. Biol. Chem.* **273**: 8301-8307.

Hicks, D.B. and Donaldson, R.P. (1982) Electron transport in glyoxysomal membranes. *Arch. Bioch. Biophys.* **215**: 280-288.

Hock, B., Gietl, C. and Sautter, C. (1987) Biogenesis of plant microbodies. In *Peroxisomes in Biology and Medicine.* (Fahimi, H.D. and Sies, H., eds.). pp. 417-425. Springer.

Hogg, J.F. (1969) The nature and function of peroxisomes (microbodies, glyoxysomes). *Ann. N.Y. Acad. Sci.* **168**: 208-381.

Huang, A.H.C. and Beevers, H. (1971) Isolation of microbodies from plant tissues. *Plant Physiol.* **48**: 637-641.

Huang, A.H.C., Bowman, P.D. and Beevers, H. (1974) Immunological and biochemical studies on isozymes of malate dehydrogenase and citrate synthetase in castor bean glyoxysomes. *Plant Physiol.* **54**: 364-368.

Huang, A.H.C., Trelease, R.N. and Moore, T.S. (1983) Plant Peroxisomes. Academic Press, New York.

Hutton, D. and Stumpf, P.K. (1969) Characterization of the β-oxidation system from maturing and germinating castor bean seeds. *Plant Physiol.* **44**: 508-516.

Kagawa, T. and Beevers, H. (1975) The development of microbodies (glyoxysomes and leaf peroxisomes) in cotyledons of germinating seedlings. *Plant Physiol.* **55**: 258-264.

Kagawa, T. and Gonzalez, E. (1981) Organelle-specific isozymes of citrate synthase in the endosperm of developing *Ricinus* seedlings. *Plant Physiol.* **68**: 845-850.

Kagawa, T. Lord, J.M. and Beevers, H. (1973) The origin and turnover of organelle membranes in castor bean endosperm. *Plant Physiol.* **51**: 61-65.

Kagawa, T., McGregor, D.I. and Beevers, H. (1973). Development of enzymes in the cotyledons of watermelon seedlings. *Plant Physiol.* **51**: 66-71.

Kawamoto, S., Nozaki, C., Tanaka, A. and Fukui, S. (1978) Fatty acid β-oxidation system in microbodies of alkane-grown *Candida tropicalis*. *Eur. J. Biochem.* **83**: 609-613.

Kindl, H. (1993) Fatty acid degradation in plant peroxisomes. Function and biosynthesis of the enzymes involved. *Biochimie.* **75**: 225-230.

Kindl, H. (1987) β-oxidation of fatty acids by specific organelles. In *The Biochemistry of Plants*. (Stumpf, P.K. and Conn, E.E., eds.). **9**: 31-52.

Kindl, H. (1982) Glyoxysome biogenesis via cytosolic pools in cucumber. *Ann. N.Y. Acad, Sci.* **386**: 314-328.

Kindl, H. (1982a) The biosynthesis of microbodies (peroxisomes/glyoxysomes). *Int. Rev. Cytol.* **80**: 193-229.

Kindl, H. and Lazarow, P.B. (1982b) Peroxisomes and glyoxysomes. *Ann. N.Y. Acad. Sci.* Vol 386.

Kornberg, H.L. and Beevers, H. (1957) The glyoxylate cycle as a stage in the conversion of fat to carbohydrate in castor beans. *Biochim. Biophys. Acta.* **26**: 531-537.

Lazarow, P.B. and DeDuve, C. (1976) A fatty acyl CoA oxidizing system in rat liver peroxisomes; enhancement by clofibrate, a hypolipidemic drug. *Proc. Nat. Acad. Sci.* **73**: 2043-2046.

Masterson, C. and Wood, C. (2000) Mitochondrial β-oxidation of fatty acids in higher plants. *Physiol. Plant.* **109**: 217-224.

Mettler, I. J. and Beevers, H. (1980) Oxidation of NADH in glyoxysomes by a malate-aspartate shuttle. *Plant Physiol.* **66**: 555-560.

Mollenhauer, H.H., Morre, D.J. and Kelley, A.G. (1966) The widespread occurrence of plant cytosomes resembling animal microbodies. *Protoplasma.* **62**: 44-52.

Moore, T.S., Lord, J.M., Kagawa, T. and Beevers, H. (1973) Enzymes of phospholipid metabolism in the e of castor bean endosperm. *Plant Physiol.* **52**: 50-53.

Mullin, R.T., Lisenbee, C.S., Miernyk, T.A. and Trelease, R.N. (1999) Peroxisomal membrane ascorbate peroxidase is sorted to a membranous network that resembles a subdomain of the endoplasmic reticulum. *Plant Cell.* **11**: 2167 – 2185.

Nishimura, M., Takeuchi, Y., De Bellis, L. and Hara-Nishimura, I. (1993) Leaf peroxisomes are directly transformed to glyoxysomes during senescence of pumpkin cotyledons. *Protoplasma.* **175**: 131-137.

Nishimura, M., Yamaguchi, J., Mori, H., Akazawa, T. and Yokota, S. (1986) Immunocytochemical analysis shows that glyoxysomes are directly transformed to leaf peroxisomes during greening of pumpkin cotyledons. *Plant Physiol.* **81**: 313-316.

Nito, K., Yamaguchi, K., Kondo, M., Hayashi, M. and Nishimura, M. (2001) Pumpkin peroxisomal ascorbate peroxidase is localized on peroxisomal membranes and unknown membranous structures. *Plant Cell Physiol.* **42**: 20-27.

Novikoff, A.B. and Goldfischer, S. (1969) Visualization of peroxisomes (microbodies) and mitochondria with diaminobenzidine. *J. Histochem. Cytochem.* **17**: 675-680.

Ogren, W.L. and Bowes G. (1971) Ribulose diphosphate carboxylase regulates soybean photorespiration. *Nature.* **230**: 159-161.

Olsen, L.J. and Harada, J.J. (1995) Peroxisomes and their assembly in higher plants. *Ann. Rev. Plant Phys.* **46**: 123-146.

Pistelli, L., De Bellis, L. and Alpi, A. (1991) Peroxisomal enzyme activities in attached senescing leaves. *Planta.* **184**: 151-153.

Reddy, J.K., Suga, T. Mannaerts, G. Lazarow, P. and Subramani, S. (1996) Peroxisomes: Biology and role in toxicology and disease. *Ann. N.Y. Acad. Sci.* Vol 804.

Rhodin, J. (1958) Anatomy of kidney tubules. *Int. Rev. Cytol.* **7**: 485-534.

Sautter, C. (1986) Microbody transition in greening watermelon cotyledons. Double immunocytochemical labeling of isocitrate lyase and hydroxypyruvate reductase. *Planta.* **167**: 491-503.

Somerville, C.R. and Ogren, W.L. (1982) Genetic modification of photorespiration. *Trends Bioch. Sci.* **7**: 171-174.

Titus, D.E. and Becker, W.M. (1985) Investigation of the glyoxysome-peroxisome transition in germinating cucumber cotyledons using double-label immunoelectron microscopy. *J. Cell Biol.* **101**: 1288-1299.

Tolbert, N.E. (1981) Metabolic pathways in peroxisomes and glyoxysomes. *Ann. Rev. Biochem.* **50**: 133-157.

Tolbert, N.E. (1980) Microbodies – peroxisomes and glyoxysomes. In *The Biochemistry of Plants.* (Stumpf, P.K. and Conn, E.E., eds.). **2**: 488-521.

Tolbert, N.E. (1971) Microbodies – peroxisomes and glyoxysomes. *Ann. Rev. Plant Physiol.* **21**: 45- 74.

Tolbert, N.E., Gee, R., Husic, D.W. and Dietrich, S. (1987) Peroxisomal glycolate metabolism and the C_2 oxidative photosynthetic cycle. In *Peroxisomes in Biology and Medicine.* (Fahimi, H.D. and Sies, H., eds.). pp. 213-222. Springer.

Tolbert, N.E., Oeser, A. Kisaki, T. Hageman, R.H. and Yamazaki, R.K. (1969) Peroxisomes from spinach leaves containing enzymes related to glycolate metabolism. *J. Biol. Chem.* **243**: 5179-5184.

Trelease, R.N. (1984) Biogenesis of glyoxysomes. *Ann. Rev. Plant Physiol.* **35**: 321-347.

Trelease, R.N., Becker, W.M., Gruber, P.J. and Newcomb, E.H. (1971) Microbodies (glyoxysomes and peroxisomes) in cucumber cotyledons. Correlative biochemical and ultrastructural study in light- and dark-grown seedlings. *Plant Physiol.* **48**: 461-475.

Vigil, E. (1970) Cytochemical and developmental changes in microbodies (glyoxysomes) and related organelles of castor bean endosperm. *J. Cell Biol.* **46**: 435-454.

2

MOLECULAR BIOLOGY, ENZYMOLOGY, AND PHYSIOLOGY OF β-OXIDATION

Mark A. Hooks

University of Wales, Bangor, Wales, UK

KEYWORDS

β-oxidation; fatty acids; fatty acid transport; abiotic stress; gene families; mitochondria.

INTRODUCTION

Beta-oxidation in plants is the pathway by which germinating seeds use their lipid stores to feed the newly forming plant. Once it was put into this physiological context in the 1960s, it seems to have been the pathway taken for granted. It was not even until 1981 that the pathway of β-oxidation was shown to be ubiquitous to plants cells (Gerhardt, 1981). I am fortunate to be able to contribute a review on β-oxidation at this time because, in a way, it represents a new era of research in β-oxidation. Recent interest has increased and more and more laboratories are becoming involved. This simply may reflect the availability of genetic and biological tools, such as mutants, expressible cDNA clones, and sequence information. However, an equal driving force is undoubtedly the desire to characterize plant pathways that have enormous potential for commercial exploitation. What the recent research has uncovered is a very complex system of genes and enzymes, which may have evolved for purposes beyond the basic catabolism of fatty acids. The purpose of this review is to highlight this complexity and pursue the current research of β-oxidation at the molecular level. The last

A. Baker and I.A. Graham (eds.), Plant Peroxisomes, 19–55.

comprehensive review of β-oxidation was published in 1993 and laid the enzymological framework (Kindl, 1993). This review will add the next layer to this story by recounting the recent discoveries about the enzymology and exploring the underlying genetics. The other purpose is to discuss the functions and regulation of β-oxidation. Reports that β-oxidation is involved in the metabolism of important phytohormones, such as jasmonic acid and auxin, suggests that this pathway functions in plant growth and development outside basic nutritional purposes, and may even play key roles in plant responses to stress. What will become apparent is that we still have many more questions than answers.

The Arabidopsis sequencing projects have revealed an extraordinary amount of information on β-oxidation gene families, and information from these projects will feature prominantly here. Even though the available genetic resources and technologies stemming from the sequencing projects have made the biochemical work proceed much faster, it has been – and still is – a slow process to assign enzyme function to recognised genes. The major portion of this review is devoted to describing the gene families and relating the recent advances in the enzymology. The β-oxidation pathway begins with the acyl-CoA oxidase (ACX) reaction, however, key processes priming β-oxidation are the entry of fatty acids into peroxisomes and their activation by esterification to coenzyme A. Therefore, these topics are included as well. The latter part of the review will recount what is currently known about the expression of β-oxidation genes, and what this may reveal about the possible roles of β-oxidation outside the well-known one of fatty acid mobilization. The review finishes with a brief description of other pathways through which fatty acids are modified or catabolized oxidatively. These include mitochondrial β-oxidation, α- and ω-oxidation.

FATTY ACID TRANPORT INTO PEROXISOMES

β-oxidation is the primary pathway for catabolism of fatty acids of storage lipid reserves during germination and membrane lipids during senescence. Because storage lipids in seeds are held in discreet bodies enclosed by a membrane, mechanisms must exist for the release of fatty acids from lipids and their transport to other organelles responsible for their degradation. Whether fatty acids are liberated from triglycerides by true lipases (Huang, 1992; Hoppe and Theimer, 1997) or initiated by lipoxygenase-catalysed lipid oxidation (Hause *et al*, 2000), or a combination of both, fatty acids must cross the peroxisomal membrane and become activated to the CoA ester form before entry into β-oxidation. Also, during senescence or prolonged carbohydrate starvation, membrane lipid fatty acids from various decaying organelles must be transported to peroxisomes for degradation.

There is, essentially, no information on fatty acid transport into plant peroxisomes. Some insights may be obtained from studies on other eukaryotes for which corresponding gene systems exist in plants. In animals, fatty acid binding proteins (FABPs) appear to be responsible for the intracellular movement of fatty acids (Matarese *et al*, 1989), and they can donate fatty acids to peroxisomal β-oxidation (Reubsaet *et al*, 1990). There is some evidence that increased levels of fatty acids delivered by excess FABPs may even induce acyl-CoA oxidase in the *S. cerevisiae* (Smaczska *et al*, 1994). Plants possess a family of proteins that can bind fatty acids, but noting their preference to bind phospholipids they have been termed lipid transfer proteins (LTPs, Kader, 1997). They are small (~9 kDa), basic proteins of very high abundance in plants. There are at least 15 known LTP genes in Arabidopsis (Arondel *et al*, 2000). They are differentially expressed during development, including three which have been characterized in *B. napus* that appear to be specific to germination (Soufleri *et al*, 1996). However, analysis of numerous higher plant LTP genes have revealed that each encodes a peptide with a targeting signal for extracellular secretion, and immunolocalization studies have shown the mature proteins to be present in the cell wall (Kader, 1997). Their functions remain unknown, but these findings suggest that plant LTPs do not play a role in transporting fatty acids to metabolizing organelles.

Acyl-CoA binding proteins (ACBPs) are another class of small acyl-binding proteins which are common to all eukaryotes. Mandrup *et al* (1993) have presented evidence that ACBPs may serve to regulate the intracellular pool of long-chain acyl-CoAs. This is possible in human cells where it has been shown that ACBPs can donate acyl-CoAs to acyl-CoA metabolizing organelles, although transfer to peroxisomes specifically was not demonstrated (Rasmussen *et al*, 1994). Brassicas have a family of ACBP genes (Hills *et al*, 1994), three of which have been partially characterised in Arabidopsis (Engseth *et al*, 1996; Chye *et al*, 2000). The gene isolated by Hills *et al* (1994) encodes a cytosolic ACBP, whereas the *ACBP1* and *ACBP2* isolated by Chye *et al* (2000) encode proteins with N-terminal transmembrane spanning regions. It is suggested that ACBP1 probably facilitates transfer of acyl-CoAs from the ER to the plasma membrane, whereas ACBP2 may donate acyl-CoAs to utilising enzymes. ACBP2 has the feature of ankyrin repeats, which are known to facilitate protein-protein interactions. If ACBP2 is membrane bound, it is not apparent how it may facilitate acyl-CoA transport between organelles. A cytosolic ACBP would be able to accomplish this by donating acyl-CoAs to a transporter located within the peroxisomal membrane.

There is evidence that such peroxisomal fatty acid membrane transport proteins exist. ATP-binding cassette (ABC) proteins comprise a large and diverse family of proteins with members involved in transport roles for various metabolites and proteins in all organisms (Higgins, 1992). They likely function as ATP-dependent pumps, ion channels and channel regulators (Theodoulou, 2000). Arabidopsis has sixty putative ABC proteins making it the largest protein family in plants (Davies and Coleman, 2000). Although little is known about ABC transport proteins in plants, they have been directly implicated in fatty acid transport across the peroxisomal membrane in both yeast and humans. In *S. cerevisiae*, there appear to be two mechanisms for the transport and subsequent activation of fatty acids. Medium-chain fatty acids diffuse through the membrane and are activated within the peroxisomal matrix by the acyl-CoA synthetase Faa2p, whereas long-chain fatty acids are activated externally and are transported as the CoA esters by the ABC half transport proteins Pat1p and Pat2p (Hettema *et al*, 1996; Verleur *et al*, 1997). Pat1p and Pat2p are highly similar to the peroxisomal ALD protein, an ABC half transporter (Mosser *et al*, 1993) responsible for normal peroxisomal β-oxidation of very long-chain fatty acids in humans. A similar ABC half transport protein of 70 kDa has also been found to be associated with peroxisomal membranes in human cells (Imanaka *et al*, 1999). A search of the Arabidopsis databases have identified two peroxisomal proteins with high sequence similarity to PMP70-type proteins (Davies and Coleman 2000). Furthermore, the two Arabidopsis PMP70 homologues have the SLGEQQR motifs characteristic of ABC transporters. The circumstantial evidence leads one to conclude that fatty acid transport in plants is mediated by ABC-type transporters in systems similar to those of yeast and mammals.

GENE AND ENZYME FAMILIES OF PEROXISOMAL β-OXIDATION

The majority of information presented has been obtained using Arabidopsis. This is to be expected as the Arabidopsis genome has been fully sequenced and annotated, and ESTs have been freely available for nearly a decade. Table 1 contains a list of identified Arabidopsis genes grouped into their respective families. The Table provides the chromosomal location of gene, including the BAC on which it resides, the Genbank accession number, possible expression, indication for peroxisomal targeting, and if independent evidence exists to verify function. Evidence of expression is based solely on the existence of a distinct EST in the databases.

Table 1: Gene families of lipid mobilization

Enzyme	Chromosomal Location	Accession Number	Expressed	Peroxisome identifier	Experimental	TIGR Report ID
ACS	IV (BAC T32A16)	AL078468	Yes	None	Yes	TC100870
ACS	I (BAC F15H21)	AC066689	Yes	None	No	TC93839
ACS	IV (BAC F8M12)	AF080118	No	None	No	TC101583
ACS	I (BAC F13F21)	AC007504	Yes	None	Yes	TC115944
ACS	II (BAC T8I13)	AC002337	Yes	None	No	TC90465
ACS	I (BAC T5M16)	AC010704	Yes	None	No	TC101125
ACS	II (BAC T103)	AC006951	Yes	None	No	TC91904
ACS	II (BAC T17A5)	AF024504	Yes	None	No	TC96925
FACL	III (P1 MYM9)	AP000377	Yes	None	No	TC86987
FACL	III (P1 MSL1)	AB012247	Yes	None	No	TC102546
FACL	III (F2O10)	AC013454	Yes	PTS II (RIX5HL)	No	TC95306
AMPBP	V (BAC F15A18)	AC007478	Yes	NA	No	TC88324
AMPBP	V (P1 MXA21)	AB005247	No	PTS I (SKI)	No	NP236474
AMPBP	IV (FCA-0)	Z97335	Yes	None	No	TC92494
AMPBP	I (BAC F15M4)	AC012394	Yes	None	No	TC97776
AMPBP	I (BAC T14N5)	AC004260	Yes	None	No	TC102917
AMPBP	III (BAC K14A17)	AB026636	Yes	PTS I (SRL)	No	TC88834
AMPBP	V (P1 MQK4)	AB005242	Yes	PTS I (SRM)	No	TC90220
AMPBP	I (BAC F24J8)	AC015447	Yes	None	No	TC98318
AMPBP	I (BAC F12P19)	AC009513	No	PTS I (SRL)	No	NP184114
AMPBP	I (BAC F24J8)	AC015447	No	None	No	NP281236
AMPBP	I (BAC F12P19)	AC009513	Yes	PTS I (SRL)	No	TC88319
AMPBP	I (BAC F15E12)	AC079703	No	PTS I (SRL)	No	NP281847
AMPBP	I (BAC T22E19)	AC016447	No	None	No	NP184792
ACX1	IV (FCA-6)	Z97341	Yes	PTS I (ARL)	Yes	TC100033
ACX2	V (P1 MQN23)	AB013395	Yes	PTS II (RIX5HL)	Yes	TC87515
ACX3	I (BAC F9P14)	AC025290	Yes	PTS II	Yes	TC101684
ACX4	III (AtEm1 locus)	AF049236	Yes	PTS I (SRL)	Yes	TC86810
ACX1-like	II (BAC T32F12)	AC006068	Yes	PTS I (AKL)	No	TC86273
ACX3-like	I (BAC T2D23)	AC068143	No	PTS II?	No	–
MFP1	IV (BAC F19B15)	AL078470	Yes	PTS I (SKL)	Yes	TC95126
MFP2	III (BAC F17A9)	AC016827	Yes	PTS I (SRL)	Yes	TC95129
ECH	IV (FCA-7)	Z97342	Yes	None	No	TC86837

Table 2 (continued)

Enzyme	Chromosomal Location	Accession Number	Expressed	Peroxisome identifier	Experimental	TIGR Report ID
ECH	IV (BAC F11C18)	AL049607	Yes	None	No	TC88698
ECH	V (P1 MNL12)	AB017070	Yes	PTS I (AKL)	No	TC96015
ECH	IV (FCA-5)	Z97340	Yes	None	No	TC99661
ECH	I (BAC F8A5)	AC002292	No	PTS II (RLX5HL)	No	NP043250
ECH	III (BAC T8B10)	AL138646	Yes	None	No	NP205375
3HCDH	III (BAC K7L4)	AC023839	Yes	PTS I (PRL)	No	TC91711
PKT1	V (TAC K24G6)	AB012242	Yes	None	No	TC86759
PKT2	V (TAC K24G6)	AB012242	Yes	PTS II (RQX5HL)	No	TC86760
PED1	II (BAC F25I18)	AC002334	Yes	PTS II (RQX5HL)	Yes	TC90372
PED1-like	I (BAC T1G11)	AAB80634	Yes	PTS II (RQX5HL)	No	TC96941
DCR	III (P1 MBK21)	AB024033	Yes	PTS I (SKL)	No	TC104533
DCR	III (BAC T11I18)	AC011698	Yes	None	No	TC112132
DCR	III (BAC T11I18)	AC011698	Yes	None	No	TC113566
CHY1	V (TAC K14B20)	AB018108	Yes	PTS I (AKL)	Yes	TC105629
3HIBCH	II (BAC T11J7.4)	AC002340	No	PTS I (AKL)	No	TC126136
3HIBCH	II (BAC T11J7.5)	AC002340	No	PTS I (AKL)	No	TC114313
3HIBCH	III (TAC K7M2)	AP000382	No	None	No	TC108594
3HIBCH	IV (BAC T9E8)	AL049608	Yes	None	No	TC111861
3HIBCH	I (BAC F12K11)	AC007592	Yes	None	No	TC112424
IVD	III (BAC F18N11)	AL132953	Yes	Mitochondrial	No	TC92386

Acyl-CoA synthetases

Acyl-CoA synthetases (ACS) catalyse the activation of fatty acids for degradation by coupling with a coenzyme A cofactor. ATP is used as an energy source and to form an activation intermediate. Acyl-CoA synthetase activity has been observed in preparations of membranes from various organelles. It has well defined roles in lipid biosynthesis, where fatty acids must be coupled to coenzyme A before incorporation into membrane lipids. Acyl-CoA synthetase activity has been associated with peroxisomes, where it is required to activate fatty acids for catabolism by β-oxidation. It is not surprising that a large gene family of putative acyl-CoA synthetases exists in higher plants.

ACSs belong to a superfamily of enzymes, which includes the fatty-acyl-CoA ligases and AMP binding proteins (AMPBP). The entire subfamily is referred to as the AMP binding protein superfamily, so called for a characteristic primary sequence signature referred to as the AMP-binding motif. There are twenty four distinct members of the AMPBP superfamily (Table 1). A phylogenetic analysis of twenty members of AMPBP superfamily revealed distinct groups of ACSs and AMPBPs, but with some recognizable overlap. (Shockey *et al*, 2000). In fact, there exists a high level of sequence similarity within the entire superfamily of AMPBPs, even in regions peripheral to the AMP binding motif. This suggests close functional identity in which the entire superfamily may have acyl-CoA synthetase activity. The rice superfamily of AMPBPs appears to be just as large.

All but six members of the superfamily appear to be expressed at some point during development. Shockey *et al*, (2000) conducted experiments to determine if various members of the two subgroups would complement an *S. cerevisiae* ACS double mutant, which could not grow without exogenous myristic acid in the presence of cerulenin, an inhibitor of fatty acid synthase. Four members of the ACS subgroup, 4, 5, 8 and 10 were able to complement the mutation indicating ACS activity. None of six members of the AMPBP subgroup tested was able to complement the mutations. However, the wild-type yeast ACSs in the study only activate myristate for the synthesis of structural lipids (Knoll *et al*, 1995). A separate group of ACSs activate fatty acids for catabolism. It is possible that members of the AMPBP subgroup have true ACS activity, but are involved in fatty acid activation for catabolism. One piece of evidence that supports this hypothesis is that only members of the AMPBP subgroup (except one FACL) possess clear peroxisomal targeting signals of either type PTS I or PTS II (Table 1). It will be interesting to determine if ACSs and AMPBP represent two divergent groups both possessing ACS activities. An Arabidopsis plant null in*AtACS2*, which is putatively plastidial, was analysed for alterations in the fatty acid profile. No differences were observed between the mutant and wild type plants. These results may be explained by functional redundancy inherent in such a large gene family.

A direct characterization of *B. napus* ACS genes has been done through cloning and heterologous expression in *E. coli* (Fulda *et al*, 1997). Of the five cDNA clones that were isolated, two showed ACS activity. The active clones, pMF7 (accession number X94624) and pMF45 (accession number Z72153), correspond to Arabidopsis AtACSs TC100870 and TC115944, respectively. Activity measurements were only conducted using oleic acid, thus substrate specificity profiles of the two ACS were not provided.

General studies of ACS activity have been done for a number of other plant species, such as spinach (Joyard and Stumpf, 1981), pea (Andrews and Keegstra 1983; Gerbling *et al*, 1994), elm and maize (Olsen and Huang 1988), rape (Olsen and Lusk, 1994), leek (Hlousek-Rodojcic *et al*, 1997), loblolly pine (Orchard and Anderson, 1996) and rice (Pistelli *et al*, 1996).

Acyl-CoA oxidases

Acyl-CoA oxidases (ACXs) catalyse the first oxidative reaction of the pathway with the conversion of fatty acyl-CoA into *trans*-2-enoyl-CoA (Kindl, 1987). ACXs are the best characterized family of β-oxidation genes and enzymes. It is now evident that in those plant species investigated to date, there exists a family of ACXs with overlapping substrate specificities. The first higher plant ACX was purified from cucumber and found to react primarily with long-chain fatty acids, C14 and C16 (Kirsch *et al*,1986). Although evidence from substrate specificity measurements of general peroxisomal ACX activity suggested a family of ACXs (Gerhardt, 1985), other members of the ACX family were not purified and characterised until 1996. Maize was found to have three acyl-CoA oxidases possessing substrate specificities for short- (C4-C6), medium- (C12-C14) and long-chain (C16+) fatty acyl-CoAs (Hooks *et al*, 1996). Long-chain (De Bellis *et al*, 1999) and short-chain ACXs (De Bellis *et al*, 2000) from pumpkin have been purified and characterized. The substrate specificities of these enzymes coincide with the long-chain and short-chain ACXs of maize and indicate a high degree of conserved ACX function among plant species.

The isolation and heterologous expression of ACX cDNA clones, primarily those from Arabidopsis, has provided the most insight into ACX substrate specificities. A cDNA clone labelled *AtACX1* encodes a medium- to long-chain ACX, which has substrate specificities similar to the cucumber and maize long-chain ACXs (Hooks *et al*, 1999a; Froman *et al*, 2000). A cDNA clone identified as *AtACX2* was found to encode a strict long-chain oxidase as it preferred fatty acyl-CoAs C16 and longer (Hooks *et al*, 1999a). Based on sequence comparisons, this is also the likely substrate specificity of enzymes encoded by an ACX cDNA clones isolated from *Phalaenopsis* (Do and Huang, 1997) and pumpkin (Hayashi *et al*, 1998a). The short-chain ACX (*AtACX4*) was identified by expression of an Arabidopsis cDNA clone in insect cells (Hayashi *et al*, 1999). The cDNA clone was first identified as a putative acyl-CoA dehydrogenase by Delseny and Picard (1996) during their analysis of the sequences around the *EM1* locus of chromosome III. Isolation of Arabidopsis cDNA clones encoding a medium-chain ACX

(AtACX3) (Froman *et al*, 2000; Eastmond *et al*, 2000a) revealed that overlapping substrate specificities of the family of ACXs would facilitate the complete and efficient catabolism of long-chain fatty acids. The overlapping substrate specificities of the four AtACXs would suggest that deletion of any one form may compromise, but would not prevent the complete catabolism of fatty acids during germination. This was demonstrated by the isolation of an AtACX3 mutant (Eastmond *et al*, 2000a). The degradation of medium-chain fatty acids entering β-oxidation was disrupted as indicated by the increased resistance to 2,4-dichlorophenoxybutyric acid (2,4-DB) of the *AtACX3* mutant compared to wild-type plants. However, the germination and growth behaviour of the mutant in the absence of an exogenous carbohydrate source was unaffected.

The Arabidopsis genome sequencing project has placed the chromosomal locations of the various ACX genes. *AtACX1*, *AtACX2*, *AtACX3* and *AtACX4* reside on chromosomes IV, V, I and I, respectively. The Arabidopsis genome and EST sequencing projects identified structural homologues of *AtACX1* and *AtACX3*, which are located on chromosomes II and I, respectively. These two ACXs appear to be expressed, but probably only at specific developmental stages. Sequences on chromosome III (BACs T8E24 and F3E22) show high homology to *AtACX3*. These sequences were not identified as ACXs during annotation, and subsequent analysis indicates that they would not encode full-length ACXs. Comparisons of the primary structures of the various ACXs indicated that AtACX3 is more similar to AtACX2 than to AtACX1. AtACX4 is completely unique in that it shares the highest degree of similarity with acyl-CoA dehydrogenase based on primary sequence comparison (Comella *et al*, 1999). It will prove interesting to determine those intrinsic factors dictating differences in substrate specificity among the different AtACXs and, specifically, how AtACX4 functions as an oxidase. Mining of the rice genomic sequences identified only three apparent ACX genes: one ACX1-like, one ACX2/ACX3-like and one ACX4-like. The number of identified genes corresponds to the ACX enzymology of maize where only three ACX enzyme activities were detectable and separable (Hooks *et al*, 1996).

Multifunctional proteins

Multifunctional proteins (MFPs) catalyse the conversion of *trans*-2-enoyl-CoA into 3-ketoacyl-CoA (Kindl, 1987). This conversion represents two independent enzyme reactions catalysed by the same polypeptide. The first reaction takes the *trans*-2-enoyl-CoA to 3-hydroxyacyl-CoA and the second reaction further oxidizes this to 3-ketoacyl-CoA. These are the core activities required for the oxidation of saturated acyl-CoAs, and are catalysed by independent domains enoyl-CoA hydratase (ECH) and 3-hydroxyacyl-CoA dehydrogenase (HCDH). MFP proteins are either tri- or tetrafunctional, properties that are essential for the degradation of unsaturated fatty acids, which comprise a very large proportion of plant lipids (Gühnemann-Schäfer *et al*, 1994). An epimerase activity, which catalyses the conversion of *R*-3-hydroxyacyl-CoA to *S*-3-hydroxyacyl-CoA is integral to both types of MFP. The tetrafunctional proteins also possesses a Δ^3, Δ^2-enoyl-CoA isomerase activity, which converts *trans*-3-enoyl-CoA to *trans*-2-enoyl-CoA. Two mono-functional enoyl-CoA hydro-lyases, which are involved in the β-oxidation pathways governing the degradation of unsaturated fatty acids, have been found in cucumber (Engeland and Kindl, 1991).

The multiplicity of reactivities of this enzyme class makes it an important switch among the different pathways of both saturated and unsaturated fatty acids. Long-chain unsaturated fatty acids, such as linoleic acid can be degraded by one of three mechanisms (Kindl 1993). After several rounds through the core β-oxidation activities, linoleoyl-CoA is converted to *trans*-2, 4-*cis* dienoyl-CoA. This can either be converted to *trans*-2-enoyl-CoA and re-enter core β-oxidation by the action of a dienoyl-CoA reductase and isomerase, or go through one more round of core β-oxidation to form *cis*-2-enoyl-CoA. The *cis*-2-enoyl-CoA is oxidized to *R*-3-hydroxyacyl-CoA, and subsequently converted to *S*-3-hydroxyacyl-CoA by either the 3-hydroxyacyl-CoA epimerase, or by the sequential action of enoyl-CoA hydro-lyase and enoyl-CoA hydratase. The resulting *S*-3-hydroxyacyl-CoA can enter into core β-oxidation. Using the ingenious technique of trapping β-oxidation intermediates as polyhydroxyalkanoate monomers, Allenbach and Poirier (2000) obtained evidence that substantial fatty acid flux occurs through both the epimerase and reductase/isomerase pathways in Arabidopsis. It is also possible that some flux occurs through enoyl-CoA hydro-lyase and enoyl-CoA hydratase instead of through the epimerase.

The family of MFPs has been well studied. Four distinct enzymes from cucumber have been purified and characterized as well as several of the encoding cDNAs. There are four MFP isoforms in cucumber as indicated by differences in molecular mass of the purified polypeptides. MFP I and MFP II were originally purified from cucumber cotyledons and had molecular masses of 74,000 and 76,500 Da, respectively (Behrends *et al*, 1988). A second tetrafunctional MFP of 81,000 Da was subsequently purified from cucumber cotyledons (Gühnemann-Schäfer and Kindl, 1995a). The fourth MFP was isolated from green leaves (Gühnemann-Schäfer and Kindl,

1995b). This MFP is a trifunctional enzyme of approximately 80,000 Da. Each member of the cucumber family of MFPs appears to have a broad substrate specificity (Gühnemann-Schäfer and Kindl, 1995b). Except for MFP II, each MFP exhibits activity from *trans*-2(4:1)-enoyl-CoA to *trans*-2(18:1)-enoyl-CoA. MFP II was not active with *trans*-2(18:1)-enoyl-CoA.

A search of the Arabidopsis databases finds two genes encoding MFPs. The full-length cDNAs of both have been isolated and the expression patterns determined (Richmond and Bleecker, 1999; Eastmond and Graham, 2000). One Arabidopsis MFP was identified and isolated as the result of a screen for mutants with abnormal inflorescent meristems (Richmond and Bleecker, 1999), thus the wild-type allele was named *AIM1*. The *aim1* mutant exhibited resistance to 2,4-DB characteristic of having a defect in β-oxidation. *AtMFP2* was isolated by screening an Arabidopsis cDNA library (Eastmond and Graham, 2000). Both *AIM1* and *AtMFP2* have been expressed in *E. coli* and the activity of the recombinant proteins determined (Richmond and Bleecker, 1999). Both exhibited enoyl-CoA hydratase activity, but the substrate specificities of the two enzymes were not determined. The other activities associated with the MFPs were not investigated so it is not known if they are true multifunctional proteins. Sequence comparisons with the CsMFP II indicates that the each AtMFP possesses the domains, and has sufficient sequence similarity to CsMFP II, for each to be tetrafunctional enzymes. There are also various genes identified in Arabidopsis that may encode mono-functional enoyl-CoA hydratases (Table 1). It is possible that at least two of these may be enoyl-CoA hydro-lyases. Each ECH, except TC88698 and NP205375, align within the first 385 amino acids of the cucumber MFP, and each is long enough to encompass the isomerase, epimerase, and hydratase active domain. The enzyme activities these proteins possess remain to be determined. ECHs TC88698 and NP205375 show very low sequence similarity to the cucumber MFP, although they do have characteristic ECH motifs. The putative 3-hydroxyacyl-CoA dehydrogenase (HCDH) primary sequence (Table 1) aligns between amino acids 319 and 624 of the cucumber MFP (accession number X78996), which is the NAD^+ binding domain and 3-hydroxyacyl-CoA dehydrogenase functional domain (Preisig-Müller *et al*, 1994). This suggests that the putative Arabidopsis HCDH is a mono-functional dehydrogenase. A *B. napus* MFP cDNA has been isolated also, but has not been further characterized (Geshi *et al*, 1998).

3-Ketoacyl-CoA thiolases (thiolase)

Thiolase catalyses the final step of β-oxidation. It is responsible for the cleavage of the acetyl group from the acyl-CoA thereby creating an acyl-CoA two carbons shorter that can then re-enter the β-oxidation spiral (Kindl, 1987). The thiolases are probably the least characterized gene and protein family. Thiolases have been cloned from cucumber cotyledons (Preisig-Mηller and Kindl, 1993), mango fruit (Bojorquez and Gómez-lim, 1995), rape (Olesen and Brandt, 1995), pumpkin cotyledons (Kato *et al*, 1996) and Arabidopsis (=PED1 gene) (Hayashi *et al*, 1998b; Germain *et al*, 2001). The various thiolase sequences were compared to the rape and Arabidopsis PED1 sequences, because they have have been confirmed to have thiolase activity (Olesen *et al*, 1997; Germain *et al*, 2001). The rape and Arabidopsis PED1 proteins are 96% identical, and no other thiolase showed less that 82% identical amino acids when compared to either sequence. Noting the high sequence identity of the various thiolases to the rape and PED1 enzymes, it is likely that they encode true 3-ketoacyl-CoA thiolases. Two other thiolase cDNA clones, called PKT1 and PKT2 for peroxisomal keto thiolase, have been deposited in Genbank (De Rocha and Lindsey, unpublished data, accession number AAC19122 and AAC23571). PKT2 shares 70 percent identical residues with the rape thiolase. Unlike the other families of β-oxidation enzymes, all the thiolases have PTS II-type peroxisomal targeting signals (Table 1). *PED1* and *PKTs* 1 and 2 are located on chromosomes II and V, respectively. Another thiolase, which shares 87% amino acid identity with *PED1*, is located on chromosome I. Thiolase activities with different chain-length specificities have been observed in sunflower cotyledons (Oeljeklaus and Gerhardt, 1996). Therefore, it may be possible that the substrate specificities of the different cloned thiolases vary. Because of the difficulty in synthesizing long-chain 3-ketoacyl-CoA compounds, the substrate specificities of the heterologously-expressed enzymes have not been determined. However, a mutant lacking PED1 showed dramatically decreased extractable thiolase activity using the short-chain substrate acetoacetyl-CoA and elevated levels of long-chain acyl-CoAs (Germain *et al*, 2001). These findings suggest that PED1 has a broad substrate specificity, being active with both short- and long-chain substrates.

One interesting observation about the thiolases concerns the structure and expression of the two PKT isoforms. PKT1 and PKT2 are considered to be overlapping, independent genes (De Rocha, 1999). PKT1 is structurally identical to PKT2, except that PKT1 is missing the first 43 amino acids. This contains the PTS II targeting signal peptide. PKT1 is not simply an alternatively spliced form of PKT2. An analysis of the assembled sequences for PKT1 and PKT2 present in the TIGR database clearly show EST sequences with distinct 5'-UTRs. Although they are expressed together to some degree in all principle organs, the control of their expression appears to be independent (De Rocha, 1999).

Other β-oxidation enzymes

There are enzyme families peripheral to those of core β-oxidation about which very little is known. One is the enoyl-CoA reductase family of proteins, which are involved in the degradation of unsaturated fatty acids (Kindl, 1993). Enoyl-CoA reductases act in concert with enoyl-CoA isomerases to convert *trans*-2, *cis*-4-dienoyl-CoAs to *trans*-2-enoyl-CoA as a means to eliminate a double bond at the fourth carbon, and permit re-entry of the fatty acyl-CoA into the core β-oxidation pathway. Searching the TIGR database revealed three short-chain dehydrogenase/reductase-like genes that may encode dienoyl-CoA reductases (Table 1). One isoform (TC104533) is 46% identical (53% similar) to the human peroxisomal 2,4-dienoyl-CoA reductase, and has a PTS I-type peroxisomal targeting signal. The other two putative reductases TC112132 and TC113566 share 33% and 31% amino acid identity to the same human protein.

Another family of peripheral β-oxidation enzymes comprises the acyl-CoA hydrolases, which are involved in the degradation of branched-chain amino acids (BCAA), such as valine, leucine and isoleucine. Acyl-CoA hydrolases cleave the coenzyme A cofactor from the root compound to produce the corresponding carboxylic acid, which is then converted by a chain of steps to propionyl-CoA. Structurally, they resemble the enoyl-CoA hydratase/isomerase superfamily (Zolman *et al*, 2001). The involvement of the hydrolases in BCAA β-oxidation was revealed by a genetic screen to isolate mutants disrupted in the conversion of indole-3-butyric (IBA) acid to indole-3-acetic acid (IAA, see below). These mutants were called *chy1* mutants for CoA hydrolases (Zolman *et al*, 2000). Genes for at least six isozymes of CHY1 have been identified, which are distinct from the ECHs; three are peroxisomal and three are mitochondrial (Zolman *et al*, 2001; Table 1). These findings support the evidence that BCAA catabolism occurs in peroxisomes, as evidence previously suggested (Gerbling and Gerhardt 1988; 1989), but that mitochondrial pathways may also be present (D≡schner *et al*, 1999).

GENE EXPRESSION AND FUNCTIONS OF β-OXIDATION

Development

Germination

Since the work of Beevers throughout the 1960s, the classic function of β-oxidation has always been considered to be the catabolism of fatty acids to provide carbon and energy for the newly germinated and growing seedling. The acetyl-CoA generated from β-oxidation enters the glyoxylate cycle and is converted into succinate for transport to the mitochondria. In the mitochondria, the TCA cycle shuttles organic acids into gluconeogenesis for the eventual synthesis of sucrose, the carbon supply compound for sink tissues. It is now clear that via the glyoxylate cycle acetyl-CoA is also respired to produce energy in some oilseed species where storage lipids are contained in the cotyledons, such as Arabidopsis (Eastmond *et al*, 2000b). The expression of the genes and enzymes are indicative of the nutritional role of β-oxidation during germination. The gene expression profiles of all four *AtACXs* (Hayashi *et al*, 1998; Hooks *et al*, 1999a; Eastmond *et al*, 2000a; Rylott *et al*, 2001), MFPs *AIM1* and *AtMFP2* (Richmond and Bleecker, 1999; Eastmond and Graham, 2000) and two thiolases (Preisig-Mηller and Kindl, 1993; Kato *et al*, 1996; Rylott *et al*, 2001 Germaine *et al*, 2001) have been determined in germinating oilseed species. Each gene, except *AIM1*, appears to be induced before emergence of the radical, and expression becomes maximal within two to three days of germination. The expression patterns of the β-oxidation genes parallels that of the glyoxylate cycle genes malate synthase (Graham *et al*, 1990) and isocitrate lyase (Reynolds and Smith, 1995), and the time-course of lipid disappearance in Arabidopsis seedlings (Mansfield and Briarty, 1996). This has been confirmed with a direct comparison of the expression profiles of genes involved in lipid mobilization in Arabidopsis (Rylott *et al*, 2001). Because *AtACX3* was tagged with a T-DNA harbouring a GUS reporter gene, it was possible to observe directly a high level of expression in cotyledons and hypocotyls of early post-germinative seedlings (Eastmond *et al*, 2000a). Activity and protein profiles of β-oxidation enzymes closely follow those of transcript levels indicating that gene expression is a major determinant of protein level (Germain *et al*, 2001; Rylott *et al*, 2001).

The few mutants that have been studied have already revealed some interesting information concerning the functions and requirement of certain β-oxidation enzyme activities. As mentioned previously, screening for mutants resistant to the herbicide 2,4-DB led to the identification of a mutant lacking the peroxisomal thiolase PED1 (Hayashi *et al*, 1998b). This mutant was able to germinate, but could not establish without an exogenous carbohydrate source. This showed that PED1 is critical for lipid degradation during germination. Another 2,4-DB resistant mutant, *dbr5*, has been isolated and shown to have reduced growth in the absence of exogenous sucrose indicating possible defects in lipid mobilization (Lange and Graham 2000). The *aim1* mutant also shows an interesting germination-related phenotype (Richmond and Bleecker, 1999). All seeds from a heterozygous parent germinated normally without sucrose suggesting that AIM1 is not critical for lipid mobilization. However, those seeds obtained from an *aim1* homozygous parent essentially failed to germinate with or without sucrose, which the authors conclude may be due to a maternal effect of the *aim1* mutation. In contrast to PED1 and AIM1, AtACX3 is not critical for lipid mobilization during germination (Eastmond *et al*, 2000b). Although the *acx3* mutant exhibits reduced catabolism of medium-chain acyl-CoAs, as indicated by 2,4-DB resistance, it does not require exogenous carbohydrate to germinate and grow. This mutant also shows unaltered latter stages development. It will be interesting to discover the minimal combinations of AtACX isoforms that are sufficient to support germination and early post-germinative growth. Establishment may not necessarily require a complete catabolism of fatty acids, but only catabolism to a degree sufficient to provide adequate acetyl-CoA for respiration and gluconeogenesis. The observation that the post-germinative growth of an isocitrate lyase mutant is only compromised under sub-optimal growth conditions supports the idea that anapleurotic organic acid synthesis from lipid-derived acetyl-CoA does not need to proceed efficiently under optimal conditions (Eastmond *et al*, 2000b; for review see Eastmond and Graham, 2001).

Mature tissues

As seedlings consume their lipid reserves during post-germinative growth and begin to become autotrophic high levels of lipid mobilization enzymes become unnecessary. The expression of glyoxylate cycle genes malate synthase and isocitrate lyase is rapidly repressed such that after one or two days no transcript is detectable. However, unlike the glyoxylate cycle genes, those for the β-oxidation enzymes do not disappear completely as seedlings develop into mature plants. This supports the notion that fatty acid catabolism continues in mature tissues independent of a functional glyoxylate cycle. All β-oxidation genes thus far examined show expression in mature plant tissues, and they generally appear to be coordinately expressed. For example, the *AtACX* genes show patterns of expression in that more transcript is apparent in flowers than in leaves and roots, whereas these two tissues have similar transcript levels (Hooks *et al*, 1999a; Froman *et al*, 2000; Eastmond *et al*, 2000a; Rylott *et al*, 2001). The expression patterns of the MFP genes *AIM1* (Richmond and Bleecker, 1999) and *AtMFP2* (Richmond and Bleecker, 1999; Eastmond and Graham, 2000), and the thiolase gene *PED1* (Froman *et al*, 2000; Rylott *et al*, 2001) are similar to the ACXs. There is some evidence for differential expression of β-oxidation genes, but no obvious incidences where one or more family members are not expressed in a particular tissue. However, expression levels of the different genes can vary considerably. *AtACX3* (Eastmond *et al*, 2000a), *AtACX4* (Hayashi *et al*, 1999) are expressed in siliques, whereas *AtACX1* and *AtACX2* transcript levels are too low to be detected by standard RNA gel blot hybridization. MFP *AtMFP2* is highly expressed in young seedlings, but is *AIM1* is not (Richmond and Bleecker, 1999; Eastmond *et al*, 2000b). Thiolase *PED1* also appears to be much more highly expressed in seedlings and other tissues that the PKT isoforms (Germain *et al*, 2001). Global ACX activity profiles have been determined in seedlings and leaves of Arabidopsis (Hooks *et al*, 1999a). In general, higher activities are seen with short- and medium-chain than with long-chain substrates. This coincides with gene expression studies, which show predominately higher *AtACX4* transcript levels in older tissues (Rylott *et al*, 2001), and suggests that transcription is an important determining factor of protein levels.

The diverse nature of β-oxidation is exemplified by its requirement for plant growth and development. Certain proteins do not appear to be critical after early post-germinative growth. Once the *ped1* plants become autotrophic, they are able to grow without carbohydrate supplementation and set viable seed (Hayashi *et al*, 1999). The *acx3* mutant also shows normal maturation and fertility (Eastmond *et al*, 2000a). In contrast, the MFP AIM1 protein is necessary for normal adult reproductive development (Richmond and Bleecker, 1999). Under short day growth conditions (<12 hour light) the leaves of *aim1* are smaller and twisted, but the mutant is generally indistinguishable from wild type under long days (~16 hour light). The effect of the mutation becomes much more dramatic during the transition of the inflorescence from leaf producing to flowering during long days. It results in infertility marked by the absence of any normal floral structures, and exhibits increased branching due to prematurely terminating floral meristems. Other important functions of β-oxidation may become apparent as mutants lacking the other β-oxidation enzymes are isolated and studied.

Senescence

Senescence is genetically programmed cell and tissue death during which structural molecules are degraded and the resulting nutrients transported to younger actively growing tissues or storage. Characteristic of nutrient mobilization is the degradation of lipids as a source of carbon and energy (Kowai *et al*, 1981). It is characterized by particular sequences of regulatory and functional gene expression events (Buchanan-Wollaston, 1997), including those for the degradation of membrane lipids and fatty acids. Indicative of lipid mobilization genes that are induced during senescence are those for the glyoxylate cycle and gluconeogenesis, such as malate synthase (Graham *et al*, 1992) and phosphoenolpyruvate carboxykinase (Kim and Smith, 1994), respectively. Whereas most work has focused on the glyoxylate cycle, it has been demonstrated that the β-oxidation of exogenously added palmitic acid increases in senescing leaves (Pistelli *et al*, 1992) and during artificial senescence in maize root tips induced by carbohydrate starvation (Dieuaide *et al*, 1993). Although a systematic study of β-oxidation gene expression during senescence has not been conducted, there are assorted pieces of evidence that it occurs. Those genes shown to be induced in yellowing senescent leaves are *AtACX3* and *AtACX4* (Froman *et al*, 2000; Eastmond *et al*, 2000a), *AtMFP2* (Eastmond and Graham, 2000), and *PED1* (Froman *et al*, 2000). The high level of expression in flowers observed for various genes may be the result of that tissue showing elements of rapid senescence. It will be interesting to see if senescence is altered in any of the β-oxidation mutants.

Developmental and metabolic regulation of gene expression

In contrast to animal and yeast systems, one of the least understood aspects of β-oxidation is the regulation of gene expression and those post-transcription controls dictating protein levels and enzymic activity. In mammals, peroxisomal β-oxidation genes are induced by peroxisome proliferators (PPs), which are a general class of compounds that increase both the number and size of peroxisomes in treated mammalian cells (Corton *et al*, 2000). PPs act through a family of nuclear receptors/transcription factors called PPARs for peroxisome proliferator-activated receptors. Binding of PPs to PPARs targets them to promoter elements in β-oxidation genes known as peroxisome proliferator response elements (PPREs). The first PPREs were identified in the promoter of the rat ACX gene (Osumi *et al*, 1996). Polyunsaturated fatty acids, themselves, are ligands for PPARs and can induce peroxisomal β-oxidation gene expression indicating a direct link between gene expression and the metabolic requirements for fatty acid oxidation (Forman *et al*, 1997). Yeasts also show a dietary response to long-chain fatty acids as a sole carbon source with the proliferation of peroxisomes and β-oxidation gene expression (Veenhuis *et al*, 1987). The induction of gene expression is governed by fatty acid responsive transcription factors (Gurvitz *et al*, 1999) that bind to what are collectively known as oleate responsive elements (OREs) (Einerhand *et al*, 1993). Examples of yeast genes known to contain OREs are the acyl-CoA oxidase, 3-ketoacyl-CoA thiolase and 2,4-dienoyl-CoA reductase (Karpichev and Small, 1998). The genes encoding peroxisomal β-oxidation and glyoxylate cycle enzymes are rapidly repressed when yeasts are switched to a carbohydrate-containing growth medium, and *cis*-elements mediating this repression have been identified (Wang *et al*, 1994).

The regulation of plant peroxisomal gene expression has a number of features in common with those described above. A direct effect of a PP on plant β-oxidation was observed with the treatment of Norway spruce seedlings with the commonly used herbicide 2,4,5-trichlorophenoxyacetic acid (Segura-Aguilar *et al*, 1995). This PP increased medium-chain ACX activity nearly seven-fold. Similarly to mammalian mitochondrial β-oxidation, plant peroxisomal β-oxidation is induced in response to carbohydrate starvation (Dieuaide *et al*, 1993). This is to mobilize nutrients locked in structural macromolecules. Consistent with this finding is that acyl-CoA oxidase activity with all chain length acyl-CoAs has been shown to increase in excised maize root tips during prolonged carbohydrate deprivation (Hooks *et al*, 1995). Flux through the pathway and associated enzyme activities decrease in response to carbohydrate feeding. There also appears to be a direct positive effect of fatty acids on the induction of β-oxidation enzymes. Rape plants expressing the California bay medium-chain acyl-ACP thioesterase exhibited specific induction of medium-chain ACX activity in both developing seeds and leaves (Eccleston and Ohlrogge, 1998). It was proposed that induction of β-oxidation was the result of over-production of lauric acid, which would be toxic to cells. Potential signals for induction may also be the medium-chain fatty acyl-CoAs, increased levels of which have recently been detected in extracts of developing seeds from the rape transgenic plants (Larson and Graham, 2001). In contrast, Hooks *et al*, 1999, observed no evidence for the specific induction of medium-chain ACX activity or in the expression of various genes in Arabidopsis expressing the same thioesterase. The notable difference between the rape and Arabidopsis plants was the levels of thioesterase activity, which was significantly lower in the latter. One possible explanation is that a certain threshold of fatty acid or acyl-CoA levels must be reached before induction is realized. This threshold level was attained in rape leaves, but not in those of Arabidopsis.

It is likely that an investigation of plant β-oxidation gene promoters will reveal regulatory sequences similar to those for glyoxylate cycle genes. Distinct *cis*-acting elements have been discovered within the promoters of the malate synthase (Sarah *et al*, 1996) and isocitrate lyase genes (de Bellis *et al*, 1997) that mediate the expression responses to germination and sugar signals. Nevertheless, different regulatory mechanisms must control the expression of β-oxidation genes noting their different expression patterns throughout plant development and in response to stress. The promoter of the parsley homologue of ACX2 has been isolated and shown to possess several *cis*-acting elements (see below) including a sugar-response element (Logemann *et al*, 2000). There is some information regarding protein signalling factors that may be relevant to the regulation of lipid mobilization gene expression during germination. Several genes expressed during embryo development appear responsible for repressing the expression of lipid mobilization genes (Holdsworth *et al*, 1999). The *viviparous* 1 mutant of maize (Paek *et al*, 1998) and the *lec 1* mutant of Arabidopsis (West *et al*, 1994) accumulate glyoxylate cycle transcripts in developing seed, thus demonstrating roles for these genes in suppressing germination related gene expression. Reduction in VP 1 and LEC 1 protein levels may lead to de-repression of lipid mobilization gene expression during germination. De-repression may also work through specific transcription factors. For example, AtMYB13 is a MYB-type transcription factor that is de-repressed in the *lec 1* mutant (Kirik *et al*, 1998). De-repressing the expression of such transcription factors could lead to the positive induction of lipid mobilization genes. The mutants *vp 1*, *lec 1*, and other mutants with abnormal embryo development, such as *aba* and *abi* mutants of Arabidopsis, were identified through the altered sensitivity to the hormone abscisic acid (ABA). Abscisic acid has been known for years to inhibit seed germination, and has been demonstrated to arrest Arabidopsis seed germination by directly inhibiting the mobilization of storage reserves (Garciarrubio *et al*, 1997). Although the catabolism of lipids was not specifically studied, the finding that addition of a carbohydrate source could overcome the effects of ABA suggests that this process may have been disrupted. This is similar to situations where genetic defects in β-oxidation prevent the establishment of seedlings. These results raise interesting questions about the direct effects of hormones on the expression of β-oxidation genes. Unravelling the mechanisms of hormonal regulation of lipid mobilization gene will not be easy considering that other hormones, such as the well-known germination promoting hormone gibberellic acid, are also involved. An interesting turn of the hormone story is that evidence is being slowly accumulated which reveals β-oxidation to be involved in hormone biosynthesis.

Auxin biosynthesis

Auxins are a class of phytohormones that exhibit a diverse range of biological effects (Bartel, 1997). A number of pathways of indole-3-acetic acid (IAA) biosynthesis have been charted in plants from *de novo* biosynthesis to the conversion of the amino acid tryptophan to IAA. Indole-3-butyric acid (IBA) is an auxin that is even stronger than IAA in its ability to exert physiological effects, such as root induction. IBA has been found to be a wide spread, naturally occurring auxin in plants (Epstein and Ludwig-Müller, 1993). It was first demonstrated that IBA can be converted to IAA in 1960 (Fawcett *et al*, 1960). They found that acetate was formed from indolealkenecarboxylic acid with even carbon chain lengths, whereas propionic acid was formed from those with odd carbon chain lengths. This observation implicated β-oxidation in the conversion of IBA to IAA. This conversion has been shown to occur in many plant species including Arabidopsis. A screen designed to identify mutants lacking normal physiological responses to IBA, but which remain sensitive to IAA, led to the isolation of several mutant lines that required sucrose for dark-grown seedlings to establish (Zolman *et al*, 2000). The reduced rates of fatty acid degradation in establishing seedlings further supported the hypothesis that β-oxidation was directly disrupted in these mutants. The known β-oxidation mutants, at*acx3*, *ped1,* and *aim1* are also resistant to IBA, but sensitive to IAA (Zolman *et al*, 2001). It is apparent that β-oxidation plays a significant role in the conversion of IBA to IAA, however, two aspects of IBA metabolism are not yet clear. One is the extent to which IBA conversion to IAA necessary in order to impart biological function, and the second is the regulation of the process (Normanly, 1997).

ABIOTIC STRESS RESPONSES

One of the most exciting developments in β-oxidation research within the past decade has been the discovery of the induction of certain genes in response to environmental stresses. Interestingly, the idea that β-oxidation may be involved in plant stress responses goes back to the early 1980s.

Around the time that Gerhardt (1983) had demonstrated that β-oxidation of fatty acid is ubiquitous to plant cells, Vick and Zimmerman (1984) proposed that β-oxidation was an integral part in the synthesis of the phytohormone jasmonic acid (JA). Upon wounding, for example, α-linolenic acid is released from chloroplast membranes, probably by a chloroplastic phospholipase, and subsequently oxygenated by a chloroplastic lipoxygenase (LOX2). Allene oxide synthase converts the fatty acid hydroperoxide to 12-oxophytodienoic acid (OPDA), which is then reduced to the cyclopentanone OPC-8:0 by OPDA reductase. The observation that successive two carbon units were removed from ^{18}O-labelled OPDA fed to intact plant tissues suggested the involvement of β-oxidation. An attempt to determine if β-oxidation was necessary for JA synthesis was made by studying the biological activity of unoxidizable analogues of the intermediates of the pathway (Blecert *et al*, 1995). Demonstration of the biological activity of the intermediates suggested that β-oxidation instead may be responsible for the conversion of active intermediates to JA, which can then be inactivated by conjugation to various compounds (Schaller, 2001). Important evidence supporting the hypothesized role of β-oxidation in plant stress responses came from reports showing a dramatic increase in *ACX1* homologues by drought (Grossi *et al*, 1995) and wounding (Titarenko *et al*, 1997). *ACX1* appears to be the only β-oxidation gene induced in response to wounding. An array-based expression analysis of wound-responsive Arabidopsis genes confirmed the specificity of ACX1 induction (Reymond *et al*, 2000). For β-oxidation to be involved in JA metabolism, one might expect that other members of the pathway to also be induced, unless the flux control coefficient of ACX1 is particularly large.

An additional, if not alternative, function of ACX1 may be the production of reactive oxygen species (ROS), because H_2O_2 is a direct product of the ACX reaction (Hooks *et al*, 1998). The involvement of H_2O_2 production in response to mechanical stress is unclear, but it has been documented (Low and Merida, 1996) and may serve to induce cell wall repair mechanisms or to prepare the plant for possible infection by pathogens. In Arabidopsis, maximal expression of *AtACX1* occurs approximately two hours after wounding, which falls within the time frame of the second oxidative burst. The common perception is that peroxisomal oxidases could not contribute to the oxidative burst, because of the presence of very high levels of catalase in this organelle. Several findings argue against this assumption. (i) Catalase has a high apparent K_m for H_2O_2, thus permitting diffusion of H_2O_2 from peroxisomes (Halliwell, 1974); (ii) Up to sixty percent of the hydrogen peroxide generated in rat liver peroxisomes can diffuse out of the organelle (Boveris *et al*, 1972; (iii) Peroxisomes have been visualized directly in plant tissues by staining for H_2O_2 using cerium chloride (Kausch *et al*, 1983); (iv) Peroxisomal H_2O_2 production in pear fruit has been implicated in regulating senescence, and H_2O_2 production can be increased upon addition of substrates for peroxisomal oxidase systems, or by inhibiting catalase (Brennen and Frenkel, 1977); (v) Overexpression of human peroxisomal ACX in Cos-1 cells results in H_2O_2-mediated induction of NF-κB, an important transcriptional regulator of stress responsive genes in liver (Li *et al*, 2000). It remains to be determined if induction of *ACX1* alone, or in combination with the suppression of H_2O_2 scavenging enzyme synthesis (Mittler *et al*, 1998), can lead to augmented H_2O_2 levels.

Regulation of stress-related gene expression

A considerable amount of work has been put into determining the mechanisms of wound responsive gene expression of which studies employing *AtACX1* have proven valuable. Based on pharmacological evidence, *AtACX1* is induced in Arabidopsis by the JA-independent wound-signalling pathway (Titarenko *et al*, 1997). By this pathway, wound responsive gene expression is mediated by reversible phosphorylation events (Rojo *et al*, 1998) leading to the release of Ca^{2+} from internal stores (Leon *et al*, 1998). Cytoplasmic acidification is also an early phosphorylation-dependent response to stress, and has been widely implicated in mediating resistance gene expression during plant-pathogen interactions (Lapous *et al*, 1998). Work on solanaceous species has linked ion fluxes across the plasma membrane to wound signalling mechanisms (Doherty and Bowles, 1990). Disruptions in the mechanisms that permit proton influxes have consequences detrimental to plant responses to wounding (Schaller and Oecking, 1999). *AtACX1* appears to be induced by cytosolic acidification as well as by wounding (Hooks *et al*, 1998). It will be interesting to know if ACX1 induction follows similar patterns in solanaceous species, or if it is linked to the systemin mediated signalling pathway.

It has become clear only recently that *ACX1* is not the only β-oxidation gene induced in response to abiotic stress. Transcript levels of an *AtACX2* homologue were observed to increase transiently in parsley cells upon exposure to high levels of UV irradiation (Logemann *et al*, 2000). The authors speculate that induction of this long-chain ACX may facilitate a rapid supply of acetyl-CoA for the synthesis of flavonoids. As mentioned previously, the promoter region was isolated and analysed for known *cis*-acting elements. The region proximal to the proposed TATA-box contains elements with high sequence similarity to known UV-light responsive elements of the chalcone synthase and phenylalanine ammonia lyase gene promoters, and an elicitor responsive element similar to that in the WRKY1 gene promoter.

OTHER PATHWAYS OF FATTY ACID OXIDATION

Mitochondrial β-oxidation

Plant mitochondrial β-oxidation has been reviewed recently (Masterson and Wood, 2000a). Its existence has been a controversial topic since the association of the pathway with glyoxysomes in the late 1960s (Cooper and Beevers, 1969). The review by Masterson and Wood recounts the evidence amassed thus far that covers the six criteria that must be met in order to demonstrate β-oxidation in plant mitochondria (Beevers 1961; Gerhardt 1992). Only certain points surrounding the controversy will be mentioned in this review. The conclusion that this pathway exists is based on an abundance of enzymological data, including, most importantly, that acyl-CoA dehydrogenase (ACDH) activity is present in plant protein extracts (Dieuaide *et al*, 1993, Masterson *et al*, 1998; Bode *et al*, 1999; Masterson *et al*, 2000). ACDH is the enzyme that catalyses the first step of prokaryotic β-oxidation and mitochondrial β-oxidation in animals. One further criterion that must be added is the unequivocal demonstration of a gene encoding, not only an ACDH, but any mitochondrial enzyme for long-chain β-oxidation. A gene encoding a putative ACDH with high homology to mammalian isovaleryl-CoA dehydrogenases has been isolated, and the corresponding enzyme was found to be located in mitochondria (Däschner *et al*, 1999). ESTs for rice and oat putative isovaleryl-CoA dehydrogenases have identified in the databases as well (Bode *et al*, 1999). These enzymes are involved in amino acid catabolism, which is a mitochondrial feature in animals. The substrates for these dehydrogenases are straight or branched short-chain acyl-CoAs. As mentioned above, it is likely that some partitioning of BCAA catabolism occurs between mitochondia and peroxisomes. For pea (Masterson *et al*, 1998), maize (Dieuaide *et al*, 1993; Bode *et al*, 1999) and sunflower (Bode *et al*, 1999), acyl-CoA dehydrogenase activity was obtained with medium (C8) to long (C16) straight-chain acyl-CoA substrates. Currently, no gene encoding an enzyme involved in medium- or long-chain acyl-CoA β-oxidation within higher plant mitochondria has been identified. At this time we can conclude that some aspects of amino acid catabolism take place in mitochondria. However, the existence of mitochondrial β-oxidation of longer straight-chain acyl-CoAs will remain controversial until the genes are found. It is apparent that none, at least for the acyl-CoA dehydrogenases, exist in Arabidopsis. A detailed analysis of the existing sequences for the other β-oxidation genes, such as the numerous enoyl-CoA hydratases (Table 1) may reveal mitochondrial targeting signals. It is possible that β-oxidation may occur exclusively in peroxisomes in most oilseed species, and with mitochondrial β-oxidation contributing a small portion of the degradative flux in other oilseed species, such as sunflower. There is evidence that the significant proportion of β-oxidation during the germination of the non-oilseed specie pea may occur in mitochondria (Masterson and Wood, 2000b). In order to support these claims it will be necessary for future studies to aim at measuring β-oxidation fluxes directly in highly purified mitochondria, instead of relying on differences between total and peroxisomal fatty acid degradation rates. Masterson and Wood (2000a) suggest that mitochondrial β-oxidation of long-chain acyl-CoA may be reserved to stress conditions, such as carbohydrate limitation, cold or drought, or may function to help regulate acyl-CoA pool sizes in mature plants.

α- and ω-oxidation

There is now quite clear evidence that other types of fatty acid oxidation occur in plants. The recent discovery of tobacco and Arabidopsis genes with homology to animal cyclooxygenases (Sanz *et al*, 1998; Hamberg *et al*, 1999), and corresponding activities in cucumber (Hamberg *et al*, 1999) and pea (Saffert *et al*, 2000), supports the idea of a functional α-oxidation system in higher plants. Comparative estimates of activities in crude extracts suggest that long-chain α-oxidation activity in pea is as high, if not higher, than long-chain acyl-CoA oxidase activity in either maize (Hooks *et al*, 1995), Arabidopsis (Hooks *et al*, 1999a), or pumpkin seedlings (De Bellis *et al*, 2000). This finding is of interest, noting the lack of free fatty acids in plant cells that may serve as substrates for α-oxidation. Furthermore, the lack of fatty acids with odd numbers of carbons would appear to indicate insignificant α-oxidation activity.

Omega-oxidation is another system for which recent evidence has been acquired. Omega-oxidation is the conversion of the terminal methyl groups into primary alcohols, aldehydes or carboxyl groups. Certain species, such as *Candida*, use this pathway to support growth on alkanes or fatty acids as a sole carbon source (Rehm and Reiff, 1981). In such eukaryotic microorganisms, ω-oxidation works in concert with β-oxidation to degrade long-chain alkanes and fatty acids to acetyl-CoA. Long chain alkanes are first hydroxylated at a terminal carbon, generally both, and successively oxidised to yield the fatty acid. Beta-oxidation in *Candida* species can catabolize long-chain fatty acids with ω-carbons in the various oxidised states (Casey *et al*, 1990). The alcohol dehydrogenation is catalysed by a fatty alcohol oxidase for which two genes from *C. cloacae* (FAO1 and FAO2) were recently cloned (Van hanen *et al*, 2000). Comparison with Arabidposis DNA sequences identified an expressed gene homologue resident on chromosome IV (accession number AL022580) (Van hanen *et al*, 2000). The Arabidopsis Genome Sequencing Project has identified FAO-like genes on chromosomes I (accession number AC003027) and III (accession number AB015474). This suggests that a functional ω-omega oxidation system may be present in Arabidopsis. In *Candida* species, FAO enzymes are targeted to peroxisomes by PTS I-type signals. However, the deduced Arabidopsis FAO enzymes do not possess clear peroxisomal targeting sequences. Unless protein transport is facilitated by other mechanisms (piggy backing etc.), it is not likely that interactions between α-oxidation and β-oxidation occur in higher plants as they do in *Candida*.

FUTURE DIRECTIONS

In the current literature, there are about six different functions that have been assigned to β-oxidation: (1) to provide respiratory and biosynthetic substrates during stages of lipid mobilization; (2) to maintain cellular homeostasis leading to reproductive fitness; (3) to regulate acyl-CoA pools; (4) to remove potentially harmful free fatty acids; (5) to contribute to hormone production; (6) to generate reactive oxygen species in response to abiotic stress. Apart from the first, these purported functions remain conjecture and must be proven. In addition, to deciphering its physiological roles, there are basic aspects of β-oxidation enzymology and molecular biology that require investigation. The genes for many, previously unknown enzymes have been revealed by the Arabidopsis cDNA and genome sequencing projects. It is now necessary to determine the enzymatic properties of the encoded enzymes. It is evident from Table 1 that much of this basic information is lacking. With liberal accessibility to the cDNA clones more and more expression data is being generated. Although nothing dramatic has yet been unearthed about β-oxidation gene expression in the existing publicly available microarray data, it is expected this will change as more and more data is collected. This should provide valuable information regarding gene function during development and in response to metabolic changes. Traditional expression analyses have shown that the induction and repression of gene expression during germination and senescence are specific and dramatic events, which see large changes in transcript levels. However, the underlying factors regulating these expression events are unknown. For example, what are the signalling pathways for repression by carbohydrates, and do mechanisms exist for the positive induction of gene expression by fatty acids? The generation of plant lines containing β-oxidation promoter/reporter gene fusion constructs will prove invaluable for such studies.

It is clear that some β-oxidation genes are critical for normal growth and development. What remain unknown are the enzymatic steps that primarily regulate the flux of β-oxidation under both normal and induced conditions. Results from both animal (Reubsaet *et al*, 1988; Aoyama *et al*, 1994) and plant (Kindl 1987; Holtman *et al*, 1994) studies suggest that the ACX step is primary point of flux control. Noting the number different enzymes involved, the different types of intermediates produced, the possible points of entry of fatty acids into β-oxidation, and the implication that β-oxidation is not a linear but a spiral pathway, suggests that the regulation of flux is quite complicated. It will be necessary to study flux control on a case by case basis whereby the conversion of substrate to a particular intermediate is followed.

Particularly exciting aspects of β-oxidation to emerge from future research will revolve around commercial exploitation. These aspects are discussed in more detail in separate chapters by Rylott and Larson and by Poirier in this volume. Rylott and Larson discuss the potential for manipulating plants for the production of novel fatty acids, whereas Poirier provides the state-of-the-art on channelling β-oxidation intermediates into polyhydroxyalkanoate synthesis. The success of this work will continue to rely on addressing basic questions about β-oxidation enzymology and molecular biology. It seems that β-oxidation has remained in the background for so many years. This situation is changing as the interest in the commercial exploitation of plants continues to grow.

ACKNOWLEDGEMENT

I would like to thank Elaine Murphy for helpful assistance with literature searching and documentation.

REFERENCES

Allenbach, L. and Poirier, Y. (2000) Analysis of the alternative pathways for the β-oxidation of unsaturated fatty acids using transgenic plants synthesizing polyhydroxyalkanoates in peroxisomes. *Plant Physiol.* **124**: 1159-1168.

Andrews, J. and Keegstra, K. (1983) Acyl-CoA synthetase is located in the outer membrane and acyl-CoA thioesterase in the inner membrane of pea chloroplast envelopes. *Plant Physiol.* **72**: 735-740.

Aoyama, T., Souri, M., Kamijo, T., Ushikubo, S. and Hashimoto, T. (1994) Peroxisomal acyl coenzyme-a oxidase is a rate-limiting enzyme in a very-long-chain fatty-acid β-oxidation system. *Biochem. Biophys. Res. Com.* **201**: 1541-1547.

Arondel, V., Vergnolle, C., Cantrel, C. and Kader, J.C. (2000) Lipid transfer proteins are encoded by a small multigene family in *Arabidopsis thaliana. Plant Sci.* **157**: 1-12.

Bartel, C. (1997) Auxin biosynthesis. *Ann. Rev. Plant Physiol. Plant Mol. Biol.* **48**: 51-56.

Beevers, H. (1961) Relationships to fat metabolism. In *Respiratory Metabolism in Plants.* (Brown, A.H., ed.). pp. 207-220. Row Peterson and Co. New York, NY,.

Behrends, W., Engeland, K. and Kindl, H. (1988) Characterization of two forms of the multifunctional protein acting in fatty acid oxidation. *Arch. Biochem. Biophys.* **263**: 161-169.

Blechert, S., Brodschelm, W., Holder, S., Kammerer, L., Kutchan, T.M., Mueller, M.J., Xia, Z.Q. and Zenk, M.H. (1995) The octadecanoic pathway – signal molecules for the regulation of secondary pathways. *Proc. Natl. Acad. Sci USA.* **92**: 4099-4105.

Bode, K., Hooks, M.A. and Couée, I. (1999) Identification, separation and characterization of acyl-Coenzyme A dehydrogenases involved in mitochondrial β-oxidation in higher plants. *Plant Physiol.* **119**: 1305-1314.

Bojorquez, G. and Gómez-lim, M.A. (1995) Peroxisomal thiolase mRNA is induced during mango fruit ripening. *Plant Mol. Biol.* **28**: 811-820.

Boveris, A., Oshino, N. and Chance, B. (1972) The cellular production of hydrogen peroxide. *Biochem. J.* **128**: 617-630.

Brennen, T. and Frenkel, C. (1977) Involvement of hydrogen peroxide in the regulation of senescence in pear. *Plant Physiol.* **59**: 411-416.

Buchanan-Wollaston, V. (1997) The molecular biology of leaf senescence. *J. Exp. Bot.* **48**: 181-199.

Casey, J., Dobb, R. and Mycock, G. (1990) An effective technique for enrichment and isolation of *Candida cloacae* mutants defective in alkane catabolism. *J. Gen Microbiol.* **136**: 1197-1202

Chye, M-.L., Li H-.Y. and Yung, M-.H. (2000) Single amino acid substitutions at the acyl-CoA-binding domain interrupt ^{14}[C]palmitoyl-CoA binding of ACBP2, and Arabidopsis acyl-CoA-binding protein with ankyrin repeats. *Plant Mol. Biol.* **44**: 711-721.

Comella, P., Wu, H.J., Laudie, M., Berger, C., Cooke, R., Delseny, M. and Grellet, F. (1999) Fine sequence analysis of 60 kb around the *Arabidopsis thaliana* AtEml locus on chromosome III. *Plant Mol. Biol.* **41**: 687-700.

Cooke, R. *et al.* (1996) Further progress towards a catalog of all Arabidopsis genes – analysis of a set of 5000 non-redundant EST. *Plant J.* **9**: 101-124.

Cooper, T.G. and Beevers, H. (1969) β-oxidation in glyoxysomes from castor bean endosperm. *J. Biol. Chem.* **244**: 3514-3520.

Corton, J.C., Anderson, S.P. and Stauber, A. (2000) Central role of peroxisome proliferator-activated receptors in the actions of peroxisome proliferators. *Annu. Rev. Pharmacol. Toxicol.* **40**: 491.

Däschner, K., Thalheim, C., Guha, C., Brennicke, A. and Binder, S. (1999) In plants a putative isovaleryl CoA dehydrogenase is located in mitochondria. *Plant Mol. Biol.* **39**: 1275-1282.

Davies, T.G.E. and Coleman, J.O.D. (2000) The *Arabidopsis thaliana* ATP-binding cassette proteins: an emerging superfamily. *Plant Cell Environ.* **23**: 431-443.

De Bellis, L., Ismail, I., Reynolds, S.J., Barrett, M.D. and Smith, S.M. (1997) Distinct *cis*-acting sequences are required for the germination and sugar responses of cucumber isocitrate lyase gene. *Gene.* **197**: 375-378.

De Bellis, L., Giuntini, P., Hayashi, H., Hayashi, M. and Nishimura, M. (1999) Purification and characterization of pumpkin long-chain acyl-CoA oxidase. *Physiol. Plant.* **106**: 170-176.

De Bellis, L., Gonzali, S., Alpi, A., Hayashi, H., Hayashi, M. and Nishimura, M. (2000) Purification and characterization of a novel pumpkin short-chain acyl-coenzyme A oxidase with structural similarity to acyl-coenzyme A dehydrogenases. *Plant Physiol.* **123**: 327-334.

Delseny, M. and Picard, G. (1996) Plant genome exploration. *MS-Med. Sci.* **12**: 132-136.

De Rocha, P. (1999) Promoter trapping in *Arabidopsis thaliana*: Characterization of T-DNA tagged lines. Thesis. University of Leicester, pp 263.

Dieuaide, M., Couée, I., Pradet, A. and Raymond, P. (1993) Effects of glucose starvation on the oxidation of fatty acids by maize root tip mitochondria and peroxisomes: evidence for mitochondrial fatty acid β-oxidation and acyl-CoA dehydrogenase activity in a higher plant. *Biochem. J.* **296**: 199-207.

Do, Y.Y. and Huang, P.L. (1997) Gene structure of PAC01, a petal senescence-related gene from *Phalaenopsis* encoding peroxisomal acyl-CoA oxidase homolog. *Biochem. Mol. Biol. Int.* **41**: 609-618.

Doherty, H.M. and Bowles, D.J. (1990) The role of pH and ion-transport in oligosaccharide-induced proteinase-inhibitor accumulation in tomato plants. *Plant Cell Env.* **13**: 851-855.

Eastmond, P.J., Hooks, M.A., Williams, D., Lange, P., Bechtold, N., Sarrobert, C., Nussaume, L. and Graham, I.A. (2000a) Promoter trapping of a novel medium-chain acyl-CoA oxidase, which is induced transcriptionally during Arabidopsis seed germination. *J. Biol. Chem.* **275**: 34375-34381.

Eastmond, P.J., Germain, V., Lange, P.R., Bryce, J.H., Smith, S.M. and Graham, I.A. (2000b) Post-germinative growth and lipid catabolism in oilseeds lacking the glyoxylate cycle. *Proc. Natl. Acad. Sci. USA.* **97**: 5669-5674.

Eastmond, P.J. and Graham, I.A. (2001) Re-examining the role of the glyoxylate cycle in oilseeds. *Trends Plant Sci.* **6**: 72-78.

Eastmond, P.J. and Graham, I.A. (2000) The multifunctional protein AtMFP2 is co-ordinately expressed with other genes of fatty acid β-oxidation during seed germination in *Arabidopsis thaliana* (L) Heynh. *Biochem. Soc. Trans.* **28**, 95-99.

Eccleston, V.S. and Ohlrogge, J.B. (1998) Expression of lauroyl-acyl carrier protein thioesterase in *Brassica napus* seeds induces pathways for both fatty acid oxidation and biosynthesis and implies a set point for triacylglycerol accumulation. *Plant Cell.* **10**: 613-621.

Einerhand, A.W.C., Kos, W.T., Distel, B. and Tabak, H.F. (1993) Characterization of a transcriptional control element involved in proliferation of peroxisomes in yeast in response to oleate. *Eur. J. Biochem.* **214**: 323-331.

Engeland, K, and Kindl, H. (1991) Evidence for peroxisomal fatty acid β-oxidation involving D-3-hydroxyacyl-CoAs. *Eur. J. Biochem.* **200**: 171-178.

Engeseth, N.J., Pacovsky R.S., Newman, T. and Ohlrogge, J.B. (1996) Characterization of an acyl-CoA binding protein from *Arabidopsis thaliana. Arch. Biochem. Biophys.* **331**: 55-62.

Epstein, E. and Ludwig-Müller (1993) Indole-3-butyric acid in plants: occurrence, synthesis, metabolism and transport. *Physiol. Plant.* **88**: 382-389.

Fawcett, C.H., Wain, R.L. and Wightman, F. (1960) The metabolism of 3-indolalkene-carboxyl acid and their amides, nitriles and methyl esters in plant tissues. *Proc. R. Soc. London B. Biol. Sci.* **152**: 231-254.

Forman, B.C., Chen, J. and Evans, R.M. (1997) Hypolipidaemic drugs, polyunsaturated fatty acids and eicosanoids are ligands for PPAR α and δ. *Proc. Natl Acad. Sci. USA.* **94**: 4312-4317

Froman, B.E., Edwards, P.C., Bursch, A.G. and Dehesh, K. (2000) ACX3, a novel medium-chain acyl-coenzyme A oxidase from Arabidopsis. *Plant Physiol.* **123**, 733-741.

Fulda, M., Heinz, E. and Wolter, F.P. (1997) *Brassica napus* cDNAs encoding fatty acyl-CoA synthetase. *Plant Mol. Biol.* **33**: 911-922.

Garciarrubio, A., Legaria, J.P. and Covarrubias, A.A. (1997) Abscisic acid inhibits germination of mature Arabidopsis seeds by limiting the availability of energy and nutrients. *Planta*. **203**: 182-187.

Gerbling, H. and Gerhardt, B. (1989) Peroxisomal degradation of branched-chain 2-oxo-acids. *Plant Physiol.* **91**: 1387-1392.

Gerbling, H. and Gerhardt, B. (1988) Oxidative decarboxylation of branched-chain 2-oxo fatty-acids by higher-plant peroxisomes. *Plant Physiol.* **88**: 13-15.

Gerbling, H., Axiotis, S. and Douce, R. (1994) A new acyl-CoA synthetase, located in higher plant cytosol. *J. Plant Physiol.* **143**: 561-564.

Gerhardt, B. (1992) Fatty acid degradation in plants. *Prog. Lipid Res.* **31**: 417-446.

Gerhardt, B. (1985) Substrate-specificity of peroxisomal acyl-coa oxidase. *Phytochem.* **24**: 351-352.

Gerherdt, B. (1983) Localization of β-oxidation enzymes in peroxisomes isolated from non-fatty plant tissues. *Planta*. **159**: 238-246.

Gerhardt, B. (1981) Enzyme-activities of the β-oxidation pathway in spinach leaf peroxisomes. *FEBS Lett.* **126**: 71-73.

Germain, V., Rylott, E., Larson, T.R., Sherson, S.M., Bechtold, N., Carde, J-P., Bryce, J., Graham, I. and Smith, S.M. (2001) Requirement for 3-ketoacyl-CoA thiolase-2 in peroxisome development, fatty acid β-oxidation and breakdown of triacylglycerol in lipid bodies of Arabidopsis seedlings. *Plant J.* **27**: 1-13.

Geshi, N., Rechinger, K.B. and Brandt, A. (1998) A full-length cDNA clone from *Brassica napus* encoding a multifunctional enzyme of the glyoxysomal fatty acid β-oxidation. *Plant Physiol.* **116**: 1605-1605.

Graham, I.A., Leaver, C.J. and Smith, S.M. (1992) Induction of malate synthase gene expression in senescent and and detached organs of cucumber. *Plant Cell.* **4**: 349-357.

Graham, I.A., Smith, L.M., Leaver, C.J. and Smith, S.M. (1990) Developmental regulation of expression of the malate synthase gene in transgenic plants. *Plant Mol. Biol.* **15**: 539-549.

Grossi, M., Gulli, M., Stanca, A.M. and Cattivelli, L. (1995) Characterization of two barley genes that respond rapidly to dehydration stress. *Plant Sci.* **105**: 71-80.

Gühnemann-Schäfer, K. and Kindl, H. (1995a) Fatty acid β-oxidation in glyoxysomes. Characterization of a new tetrafunctional protein (MFP III). *Biochim. Biophys. Acta.* **1256**, 181-186.

Gühnemann-Schäfer, K. and Kindl, H. (1995b) The leaf peroxisomal form of (MFP IV) of multifunctional protein functioning in β-oxidation. *Planta* **196**: 642-646.

Gühnemann-Schäfer, K., Engeland, K., Linder, D. and Kindl, H. (1994) Evidence for domain structures of the trifunctional protein and the tetrafunctional protein acting in glyoxysomal fatty acid β-oxidation. *Eur. J. Biochem.* **226**: 909-915.

Gurvitz, A., Hamilton, B., Hartig, A., Ruis, H., Dawes, I.W. and Rottensteiner, H. (1999) A novel element in the promoter of the *Saccharomyces cerevisiae* gene SPS19 enhances ORE-dependent up-regulation in oleic acid and is essential for de-repression. *Mol. Gen. Genet.* **262**: 481-492

Halliwell, B. (1974) Superoxide dismutase, catalase, and glutathione peroxidase: solutions to the problems of living with oxygen. *New Phytol.* **73**: 1075-1086.

Hamberg, M., Sanz A. and Castresana, C. (1999) α-oxidation of fatty acids in higher plants. *J. Biol. Chem.* **274**: 24503-24513.

Hause, B., Weichert, H., Hohne, M., Kindl, H. and Feussner, I. (2000) Expression of cucumber lipid-body lipoxygenase in transgenic tobacco: lipid-body lipoxygenase is correctly targeted to seed lipid bodies. *Planta.* **210**: 708-714.

Hayashi, H., de Bellis, L., Ciurli, A., Kondo, M., Hayashi, M. and Nishimura, M. (1999) A novel acyl-CoA oxidase that can oxidize short-chain acyl-CoA in plant peroxisomes. *J. Biol. Chem.* **274**: 12715-12721.

Hayashi, H., de Bellis, L., Yamaguchi, K., Kato, A., Hayashi, M. and Nishimura, M. (1998a) Molecular characterisation of a glyoxysomal long-chain acyl-CoA oxidase that is synthesized as a precursor of higher molecular mass in pumpkin. *J. Biol. Chem.* **273**: 8301-8307.

Hayashi, M., Toriyama, K., Kondo, M. and Nishimura, M. (1998b) 2,4-dichlorophenoxy butyric acid-resistant mutants of Arabidopsis have defects on glyoxysomal fatty acid β-oxidation. *Plant Cell.* **10**: 183-195.

Hettema, E.H., van Roermund, C.W.T., Distel, B., van den Berg, M., Vilela, C., Rodrigues-Pousada, C. Wanders, R.J.A. and Tabak, H.F. (1996) The ABC transporter proteins Pat1 and Pat2 are required for import of long-chain fatty acids into peroxisomes of *Saccharomyces cerevisiae*. *EMBO J.* **15**: 3813-3822.

Higgins, C.F. (1992) ABC transporters: from microorganisms to man. *Ann. Rev. Cell Biol.* **8**: 67-113.

Hills, M.J., Dann, R., Lydiate, D. and Sharpe, A. (1994) Molecular cloning of a cDNA from *Brassica napus* L. for a homologue of acyl-CoA-binding protein. *Plant Mol. Biol.* **25**: 917-920.

Holdsworth, M., Kurup, S. and McKibbin, R. (1999) Molecular and genetic mechanisms regulating the transition from embryo development to germination. *Trends Plant Sci.* **4**: 275-279

Holtman, W.L., Heistek, J.C., Mattern, K.A., Bakhuizen, R. and Douma, A.C. (1994) β-oxidation of fatty-acids is linked to the glyoxylate cycle in the aleurone, but not in the embryo of germinating barley. *Plant Sci.* **99**: 43-53.

Hooks, M.A., Bode, K. and Couée, I. (1996) Higher-plant medium-chain and short-chain acyl-CoA oxidases – identification, purification and characterization of two novel enzymes of eukaryotic peroxisomal β-oxidation. *Biochem. J.* **320**: 607-614.

Hooks, M.A., Bode, K. and Couée, I. (1995) Regulation of acyl-CoA oxidases in maize seedlings. *Phytochem.* **40**: 657-660.

Hooks, M.A., Fleming, Y. and Graham, I.A. (1999b) No induction of β-oxidation in leaves of transgenic Arabidopsis over-producing lauric acid. *Planta.* **207**: 385-392.

Hooks, M.A., Gallagher, L. and Graham, I.A. (1998) Specific induction of an H_2O_2-generating acyl-CoA oxidase by cytosolic acidification. In *Advances in Plant Lipid Research* (Sanchez, J., ed.). Secretariado de Publicaciones, Universidad de Sevilla, Seville, Spain.

Hooks, M.A., Kellas, F. and Graham, I.A. (1999a) Long-chain acyl-CoA oxidases of Arabidopsis. *Plant J.* **20**: 1-13.

Hoppe, A. and Theimer, R.R. (1997) Enzymes for lipolysis and fatty acid metabolism in different organelle fractions from rape seed cotyledons. *Planta.* **202**: 227-234.

Huang, A.H.C. (1992) Oil bodies and oleosins in seeds. *Ann. Rev. Plant Physiol. Plant Mol. Biol.* **43**: 177-200.

Imanaka, T., Aihara, K., Takano, T., Yamashita, A., Sato, R., Suzuki, Y., Yokota, S. and Osumi, T. (1999) Characterization of the 70-kDa peroxisomal membrane protein, an ATP-binding cassette transporter. *J. Biol. Chem.* **274**: 11968-11976.

Joyard, J. and Stumpf, P.K. (1981) Synthesis of long-chain acyl-CoA in chloroplast envelope membranes. *Plant Physiol.* **67**: 250-256.

Kader J-.C. (1997) Lipid-transfer proteins: a puzzling family of plant proteins. *Trends Plant Sci.* **2**: 66-70.

Karpichev, I.V. and Small, G.M. (1998) Global regulatory functions of Oaf1p and Pip2p (Oaf2p), transcription factors that regulate genes encoding peroxisomal proteins in *Saccharomyces cerevisiae. Mol. Biol. Cell.* **18**: 6560-6570.

Kato, A., Hayashi, M., Takeuchi, Y. and Nishimura, M. (1996) cDNA cloning and expression of a gene for 3-ketoacyl-CoA thiolase in pumpkin cotyledons. *Plant Mol. Biol.* **31**: 843-852.

Kausch, A.P., Wagner, B.L. and Horner, H.T. (1983) Use of the cerium chloride technique and energy-dispersive X-ray microanalysis in plant peroxisome identification. *Protoplasma.* **118**: 1-9.

Kim, D.J. and Smith, S.M. (1994) Molecular cloning of cucumber phosphoenolpyruvate carboxykinase and developmental regulation of gene expression. *Plant Mol. Biol.* **26**: 423-434.

Kindl, H. (1993) Fatty acid degradation in plant peroxisomes: Function and biosynthesis of the enzymes involved. *Biochimie.* **75**: 225-230.

Kindl, H. (1987) β-oxidation of fatty acids by specific organelles. In: *The Biochemistry of Plants. A Comprehensive Treatise.* (Stumpf P.K. and Conn E.E., eds.). Academic Press, London. **9**: 31-52.

Kirik, V., Kolle, K., Wohlfarth, T., Misera, S. and Baumlein, H. (1998) Ectopic expression of a novel MYB gene modifies the architecture of the Arabidopsis inflorescence. *Plant J.* **13**: 729-742.

Kirsch, T., Löffler, H-G. and Kindl, H. (1986) Plant acyl-CoA oxidase. *J. Biol. Chem.* **261**: 8570-8575.

Knoll, L.J., Johnson, D.R. and Gordon, J.I. (1995) Complementation of Saccharomyces strains containing fatty acid activation gene (FAA) deletions with a mammalian acyl-CoA synthetase. *J. Biol. Chem.* **270**: 10861-10867.

Kowai, A., Matsuzaki, T., Suzuki, F. and Kawashima, N. (1981) Changes in total and polar lipids and their fatty acid composition in tobacco leaves during growth and senescence. *Plant Cell Physiol.* **22**: 1059-1065.

Lange, P.R. and Graham, I.A. (2000) *Arabidopsis thaliana* mutants disrupted in lipid mobilization. *Biochem. Soc. Trans.* 762-765.

Lapous, D., Mathieu, Y., Guern, J. and Lauriere, C. (1998) Increase of defence gene transcripts by cytoplasmic acidification in tobacco cell suspensions. *Planta.* **205**: 452-458.

Larson, T. and Graham, I.A. (2001) A novel technique for the sensitive quantification of acyl-CoA esters from plant tissues. *Plant J.* **25**: 115-125.

Leon, J., Rojo, E., Titarenko, E. and Sanchez-Serrano, J.J. (1998) Jasmonic acid-dependent and -independent wound signal transduction pathways are differentially regulated by Ca^{2+}/calmodulin in *Arabidopsis thaliana. Mol. Gen. Gen.* **258**: 412-419.

Li, Y., Tharappel, J.C, Cooper, S., Glenn, M., Glauert, H.P. and Spear, B.T. (2000) Expression of the hydrogen peroxide-generating enzyme fatty acyl-CoA oxidase activates NF-κB. *DNA Cell Biol.* **19**: 113-120.

Logemann, E., Tavernaro, A., Schulz, W., Somssich, I.E. and Hahlbrock, K. (2000) UV light selectively co-induces supply pathways from primary metabolism and flavonoid secondary product formation in parsley. *Proc. Natl. Acad. Sci. USA.* **97**: 1903-1907.

Low, P.S. and Merida, J.R. (1996) The oxidative burst in plant defence: Function and signal transduction. *Physiol. Plant.* **96**: 533-542.

Mandrup, S., Jepsen, R., Skott, H., Rosendal, J., Hojrup, P., Kristiansen, K. and Knudsen, J. (1993) Effect of heterologous expression of acyl-CoA binding protein on acyl-CoA level and composition in yeast. *Biochem. J.* **290**: 369-374.

Mansfield, S.G. and Briarty, L.G. (1996) The dynamics of seedling and cotyledon cell development in *Arabidopsis thaliana* during reserve mobilization. *Int. J. Plant Sci.* **157**: 280-295

Masterson, C., Mulholland, I., Artuso, A. and Wood, C. (1998) Acyl-CoA dehydrogenase activity in pea cotyledon mitochondria. In *Plant Mitochondria: From Gene to Function*. (Møller, I.M., Gardström, P., Glimelius, K. and Glaser, E., eds.). pp. 71-75. Backhuys Publishers, Leiden.

Masterson, C. and Wood, C. (2000a) Mitochondrial β-oxidation of fatty acids in higher plants. *Physiol. Plant.* **109**: 217-224.

Masterson, C. and Wood, C. (2000b) Contribution of mitochondria and peroxisomes to palmitate oxidation in pea tissues. *Biochem. Soc. Trans.* **28**: 757-760.

Masterson, C., Blackburn, A. and Wood, C. (2000) Acyl-CoA dehydrogenase activity in pea cotyledon tissue during germination and initial growth. *Biochem. Soc. Trans.* **28**: 760-762.

Matarese, V., Stone, R.L., Waggoner D.W. and Bernlohr, D.A. (1989) Intracellular fatty acid trafficking and the role of cytosolic lipid binding proteins. *Prog. Lipid Res.* **28**: 245-272.

Mittler, R., Xuqiao, F. and Cohen, M. (1998) Post-transcriptional suppression of cytosolic ascorbate peroxidase expression during pathogen-induced programmed cell death in tobacco. *Plant Cell.* **10**: 461-473.

Mosser, J., Douar, A-.M., Sarde C-.O., Kioschis, P., Feil, R., Moser, H., Poustka, A-.M. *et al.* (1993) Putative X-linked adrenoleukodystrophy gene shares unexpected homology with ABC transporters. *Nature.* **361**: 726-730.

Normanly, J. (1997) Auxin metabolism. *Physiol.Plant.* **100**: 431-442.

Oeljeklaus, S. and Gerhardt, B. (1996) Occurrence of 3-ketoacyl-CoA thiolases with different chain length specificity in glyoxysomes of sunflower cotyledons. *Mol. Biol. Cell.* **7**: 2884-2884

Olesen, C. and Brandt, A. (1995) A full-length cDNA encoding 3-ketoacyl-CoA thiolase from *Brassica napus*. *Plant Physiol.* **110**: 714.

Olesen, C., Thomsen, K.K., Svendsen, I. and Brandt, A. (1997) The glyoxysomal 3-ketoacyl-CoA thiolase precursor from *Brassica napus* has enzymatic activity when synthesized in *Escherichia coli*. *FEBS Lett.* **412**: 138-140.

Olsen, J.A. and Huang, A.C. (1988) Glyoxysomal acyl-CoA synthetase and oxidase form germinating elm, rape and maize seed. *Phytochem.* **27**: 1601-1603.

Olsen, J.A. and Lusk, K.R. (1994) Acyl-CoA synthetase activity associated with rape seed lipid body membranes. *Phytochem.* **36**: 7-9.

Orchard, S.G. and Anderson, J.W. (1996) Substrate specificity of the short chain fatty acyl-CoA synthetase of *Pinus radiata*. *Phytochem.* **41**: 1465-1472.

Osumi, T., Osada, S. and Tsukamoto, T. (1996) Analysis of peroxisome proliferator-responsive enhancer of the rate acyl-CoA oxidase gene. *Ann. NY. Acad. Sci.* **804**: 202-213.

Paek, N.C., Lee, B.M., Bai, D.G. and Smith, J.D. (1998) Inhibition of germination gene expression by *Viviparous*-1 and ABA during maize kernel development. *Mol. Cells.* **8**: 336-342.

Piestelli, L., Gerhardt, B. and Alspi, A. (1996) β-oxidation of fatty acids by the unspecialized peroxisomes from rice coleoptile. *Plant Sci.* **118**: 25-30.

Pistelli, L., Perata, P. and Alpi, A. (1992) Effect of leaf senescence on glyoxylate cycle enzyme activities. *Aust. J. Plant Physiol.* **19**: 723-729.

Preisig-Müller, R. and Kindl, H. (1993) Thiolase mRNA translated *in vitro* yields a peptide with a putative N-terminal presequence. *Plant Mol. Biol.* **22**, 59-66.

Preisig-Müller, R., Gühnemann-Schäfer, K. and Kindl, H. (1994) Domains of the tetra-functional protein acting in glyoxysomal fatty acid oxidation. *J. Biol. Chem.* **269**: 20475-20481.

Rasmussen, J.T., Færgeman, N.J., Kristiansen, K. and Knudsen, J. (1994) Acyl-CoA binding protein (ACBP) can mediate intermembrane acyl-CoA transport and donate acyl-CoA for β-oxidation and glycerolipid synthesis. *Biochem. J.* **299**: 165-170.

Rehm, H. and Reiff, J. (1981) Mechanisms and occurrence of microbial degradation of long-chain alkanes. *Adv. Biochem. Eng.* **19**, 175-215.

Reubsaet, F.A.G., Veerkamp, J.H., Bruckwilder, M.L.P., Trijbels, J.M.F. and Monnens, L.A.H. (1990) The involvement of fatty-acid binding-protein in peroxisomal fatty-acid oxidation. *FEBS Lett.* **267**: 229-230.

Reubsaet, F.A.G., Veerkamp, J.H., Bukkens, S.G.F., Trijbels, J.M.F. and Monnens, L.A.H. (1988) Acyl-CoA oxidase activity and peroxisomal fatty-acid oxidation in rat tissues. *Biochim. Biophys. Acta.* **958**: 434-442.

Reymond, P., Weber, H., Damond, M. and Farmer, E.E. (2000) Differential gene expression in response to mechanical wounding and insect feeding in Arabidopsis. *Plant Cell.* **12**: 707-719.

Reynolds, S.J. and Smith, S.M. (1995) Regulation of expression of the cucumber isocitrate lyase gene in cotyledons upon seed germination and by sucrose. *Plant Mol. Biol.* **29**: 885-896.

Richmond, T.A. and Bleecker, A.B. (1999) A defect in β-oxidation causes abnormal inflorescence development in Arabidopsis. *Plant Cell.* **11**: 1911-1924.

Rojo, E., Titarenko, E., Leon, J., Berger, S., Vancanneyt, G. and Sanchez-Serrano, J.J. (1998) Reversible protein phosphorylation regulates jasmonic acid-dependent and -independent wound signal transduction pathways in *Arabidopsis thaliana*. *Plant J.* **13**: 153-165.

Rylott, E.L., Hooks, M.A. and Graham, I.A. (2001) Co-ordinate regulation of genes involved in storage lipid mobilization in *Arabidopsis thaliana*. *Biochem. Soc. Trans.* **29**: 283-287.

Saffert, A., Hartmann-Schreier, J., Schön, A. and Schreier, P. (2000) A dual function α-dioxygenase-peroxidase and NAD^+ oxidoreductase active enzyme from germinating pea rationalizing α-oxidation of fatty acids in plants. *Plant Physiol.* **123**: 1545-1551.

Sanz, A., Moreno, J.I. and Castresana, C. (1998) PIOX, a new pathogen-induced oxygenase with homology to animal cyclo-oxygenase. *Plant Cell.* **10**: 1523-1537.

Sarah, C.J., Graham, I.A., Reynolds, S.J., Leaver, C.J. and Smith, S.M. (1996) Distinct *cis*-acting elements direct the germination and sugar responses of the cucumber malate synthase gene. *Mol. Gen. Gen.* **250**: 153-161.

Schaller, A. and Oecking, C. (1999) Modulation of plasma membrane H^+-ATPase activity differentially activates wound and pathogen defence responses in tomato plants. *Plant Cell.* **11**: 263-272.

Schaller, F. (2001) Enzymes of the biosynthesis of octadecanoid-derived signalling molecules. *J. Exp. Bot.* **52**: 11-23.

Segura-Aguilar, J., Hakman, I. and Rydström, J. (1995) Studies on the mode of action of the herbicidal effect of 2,4,5-trichlorophenoxyacetic acid on germinating Norway spruce. *Env. Exp. Bot.* **35**: 309-319.

Shockey, J., Schnurr, J. and Browse J. (2000) Characterisation of the AMP-binding protein gene family in *Arabidopsis thaliana*: will the real acyl-CoA synthetases please stand up? *Biochem. Soc. Trans.* **28**: 955-957.

Smaczynska, I., Skoneczny, M. and Kurlandzka, A. (1994) Studies on the effect of heterologous fatty acid-binding protein on acyl-CoA oxidase induction in *Saccharomyces cerevisiae*. *Biochem. J.* **301**: 615-620.

Soufleri, I.A., Vergnolle, C., Miginiac, E. and Kader J-.C. (1996) Germination-specific lipid transfer protein cDNAs in *Brassica napus* L. *Planta.* **199**: 229-237.

Theodoulou, F. (2000) Plant ABC transporters. *Biochim. Biophys. Acta.* **1465**: 79-103.

Titarenko, E., Rojo, E., Leon, J. and Sanchez-Serrano, J.J. (1997) Jasmonic acid-dependent and -independent signalling pathways control wound-induced gene activation in *Arabidopsis thaliana*. *Plant Physiol.* **115**: 817-826.

Van hanen, S., West, M., Kroon, J.T.M., Lindner, N., Casey, J., Cheng, Q., Elborough, K.M. and Slabas, A.R. (2000) A consensus sequence for long-chain fatty acid alcohol oxidases from *Candida* identifies a family of genes involved in lipid ω-oxidation in yeast with homologues in plants and bacteria. *J. Biol. Chem.* **275**: 4445-4452.

Veenhuis, M., Mateblowski, M., Kunau, W-H. and Harder, W. (1987) Proliferation of microbodies in *Saccharomyces cerevisiae*. *Yeast.* **3**, 77-84.

Verleur, N., Hettema, E.H., van Roermund, C.W.T., Tabak, H.F. and Wanders, R.J.A. (1997) Transport of activated fatty acids by the peroxisomal ATP-binding cassette transporter Pxa2 in a semi-intact yeast cell system. *Eur. J. Biochem.* **249**: 657-661.

Vick, B.A. and Zimmerman, D.C. (1984) Biosynthesis of jasmonic acid by several plant-species. *Plant Physiol.* **75**: 458-461.

Wang, T.W., Luo, Y. and Small, G.M. (1994) The *POX1* gene encoding peroxisomal acyl-CoA oxidase in *Saccharomyces cerevisiae* is under the control of multiple regulatory elements. *J. Biol. Chem.* **269**: 24480-24485.

West, M.A.L., Yee, K.M., Danao, J., Zimmerman, J.L., Fischer, R.L., Goldberg, R.B. and Harada, J.J. (1994) Leafy cotyledon 1 is an essential regulator of late embryogenesis and cotyledon identity in Arabidopsis. *Plant Cell.* **6**: 1731-1745.

Zolman, B.K., Monroe-Augustus, M., Thompson, B., Hawes, J.W., Krukenburg, K.A., Matsuda, S.P.T. and Bartel, B. (2001) *chy1*, an Arabidopsis mutant with impaired β-oxidation is defective in a peroxisomal β-hydroxyisobutyryl-CoA hydrolase. *J. Biol. Chem.* (in press).

Zolman, B.K., Yoder, A. and Bartel, B. (2000) Genetic analysis of indole-3-butyric acid responses in Arabidopsis reveals four mutant classes. *Genetics.* **156**, 1323-1337.

3

SYNTHESIS AND FUNCTION OF GLYOXYLATE CYCLE ENZYMES

Johanna E. Cornah and Steven M. Smith

University of Edinburgh, Edinburgh, UK

KEYWORDS

Germination; starvation; senescence; mutants; isocitrate lyase; malate synthase.

INTRODUCTION

The glyoxylate cycle in plants has been the subject of much research for several decades, and many hundreds of primary research papers have been published. There have been numerous reviews covering different aspects of the synthesis and function of glyoxylate cycle enzymes (Beevers, 1979; Trelease, 1984; Escher and Widmer 1997; Eastmond and Graham, 2001). Here we choose firstly to summarize the main features of the glyoxylate cycle as understood from the many studies of its role in lipid metabolism in seedlings of oilseed species. Then we focus on aspects that have been discovered since the previous treatise on plant peroxisomes (Huang *et al*, 1983) and highlight those aspects which are still to be understood. Among these topics are the discovery that aconitase is a cytosolic enzyme while other key enzymes of the cycle are peroxisomal, and that the metabolic pathway of the cycle is still unknown, principally because we do not know how reducing equivalents are exported from the peroxisome. We review the role of the glyoxylate cycle in seed and pollen development, senescence and starvation, highlighting different potential functions. The importance of genetic aproaches for future research is illustrated by studies of knock-out

A. Baker and I.A. Graham (eds.), Plant Peroxisomes, 57–101.

mutants of both isocitrate lyase and malate synthase in*Arabidopsis thaliana*. Finally, the progress in understanding the complex regulation of gene expression and enzyme synthesis through the application of molecular technologies is reviewed.

ROLE OF THE GLYOXYLATE CYCLE IN LIPID METABOLISM DURING SEEDLING GROWTH

Background

The glyoxylate bypass was discovered in bacteria during studies of growth on acetate and other 2-carbon (C2) compounds as the sole carbon source (Kornberg and Krebs, 1957). It was found that isocitrate lyase (ICL) and malate synthase (MS) (Table 1) together provide a bypass of the decarboxylation steps of the citric acid cycle, allowing the net conversion of two moles of acetate into one of succinate (Figure 1). Oxaloacetate (OAA) can subsequently serve as a gluconeogenic substrate through the activities of phospho*enol*pyruvate carboxykinase (PEPCK, EC 4.1.1.49), glycolytic enzymes and fructose-1,6-bisphosphatase (FBPase, EC 3.1.3.11). When *Escherichia coli* is growing on acetate as sole carbon source, isocitrate dehydrogenase of the citric acid cycle becomes phosphorylated by a specific kinase, so reducing its activity and diverting carbon through the glyoxylate bypass (Figure 1). At the same time that the pathway of metabolism during growth of bacteria on C2 substrates was being elucidated, the means by which triacylglyceride (TAG) fuels seedling growth in oilseed species was also being studied in Oxford. It was recognised that the glyoxylate cycle provides the common solution (Kornberg and Beevers 1957). The identification of all key enzyme activities in the endosperm of germinating castor bean (*Ricinus communis*) seeds and the conversion of radiolabelled actetate into sugars established that the glyoxylate cycle functions here to convert acetyl coenzyme A from fatty acid β-oxidation into C4 acids for subsequent gluconeogenesis (Beevers, 1961). Subsequently, the glyoxylate cycle was found to provide the means for fungi to grow on C2 carbon sources (Flavell and Fincham 1968; Armit *et al*, 1976). Thus, the glyoxylate cycle serves the same fundamental role in bacteria, fungi and higher plants. However, since eukaryotic metabolism is compartmentalized in different organelles, the organization of the glyoxylate cycle and the control of carbon flux through respiratory and gluconeogenic pathways is significantly different in fungi and plants to that of *E. coli* (see below).

Table 1: The five key enzymes commonly considered to constitute the glyoxylate cycle

Enzyme name(s) [and abbreviation]	Reaction	EC number	Polypeptide M_r (kDa)	Quaternary structure	Peroxi-somal targeting
Citrate synthase [CS]	Oxaloactate + acetyl CoA + $H_2O \rightarrow$ [citroyl CoA] $\rightarrow$ citrate + CoA	4.1.3.7	Precursor 55 Mature 46-49	Dimer	PTS2
Aconitase (or citrate hydrolyase, or aconitate hydratase) [ACO]	Citrate $\leftrightarrow$ [*cis*-aconitate + H_2O] $\leftrightarrow$ isocitrate	4.2.1.3	98-100	Monomer 4Fe-4S cluster	None
Isocitrate lyase [ICL]	Isocitrate $\leftrightarrow$ glyoxylate + succinate	4.1.3.1	64-66	Tetramer	PTS1
Malate synthase [MS]	Glyoxylate + acety CoA + $H_2O \rightarrow$ malate + CoA	4.1.3.2	63-65	Octamer	PTS1
NAD^+-Malate dehydrogenase [MDH]	Malate + $NAD^+ \leftrightarrow$ oxaloacetate + NADH + H^+	1.1.1.37	Precursor 38 Mature 33	Dimer	PTS2

Figure 1: The glyoxylate cycle in Escherichia coli during growth on acetate

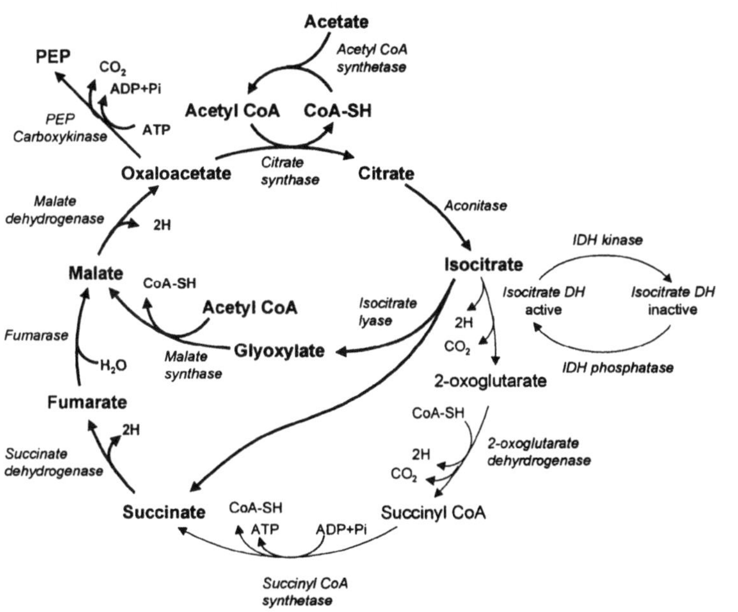

Figure 1. The metabolic scheme shows regulation by phosphorylation and dephosphorylation of isocitrate dehydrogenase (IDH). Thickness of reaction arrows indicates relatively reduced flux through the decarboxylating enzymes of the TCA cycle.

Role of the glyoxylate cycle in triacylglyceride metabolism in oilseeds

Many seeds contain a major store of TAG which provides carbon and energy for germination and seedling growth (Bewley and Black, 1994). The TAG is contained in lipid bodies within the cytosol of storage tissue cells. In dicotyledonous species the storage tissue is typically either the cotyledons (e.g. cucurbits, oilseed rape, *Arabidopsis thaliana*, sunflower, cotton, soybean) or endosperm (e.g. castor bean, oil palm). A further important distinction is that in those species exhibiting epigeal germination, the cotyledons grow, undergoing a transition from heterotrophic storage organ to photoautotrophic organ (Figures 2 and 4). In hypogeal dicots or endospermic species, the storage organ withers and dies. In cereals, the major form of stored carbon is starch (in the endosperm), but an appreciable amount of TAG is stored in the scutellum and in the aleurone layer. The significance of these differences in the site of storage of TAG and the growth of the seedling, is that the way in which the TAG is utilized and its relative importance is not the same in different species. Consequently the role of the glyoxylate cycle and its importance may be different in different species. In the case of castor bean the glyoxylate cycle is apparently required to quantitatively convert acetyl CoA into sucrose, whereas in other species TAG can be converted to sucrose or respired in different proportions (Table 2).

Table 2: Fate of TAG during seed germination and seedling growth and the involvement of the glyoxylate cycle

Species	Site of TAG storage	ICL and MS during seedling growth	Fate of TAG	References
Castor bean *Ricinus communis*	Endosperm	Yes	Quantitatively converted to sucrose for export to embryo	Beevers, 1961
Marrow *Cucurbita pepo*	Cotyledons (epigeal)	Yes	Mostly converted to sucrose for export to embryonic axis. Some respired	ap Rees, 1990
Arabidopsis thaliana	Cotyledons (epigeal)	Yes	Some converted to sucrose for export to embryonic axis. Some respired.	Eastmond *et al*, 2000
Arabidopsis thaliana icl mutant	Cotyledons (epigeal)	Only MS	Appreciable amounts respired and converted to organic acids	Eastmond *et al*, 2000
Sunflower *Helianthus annus*	Cotyledons (epigeal) a. Early germination b. Seedling	a. No b. Yes	a. Mostly respired b. Some converted to sucrose for export to embryonic axis, some respired	Raymond *et al*, 1992
Lettuce *Lactuca sativa*	Cotyledons (epigeal) a. Early germination b. Seedling	a. No b. Yes	a. Mostly respired b. Some converted to sucrose for export to embryonic axis, some respired	Raymond *et al*, 1985 Salon *et al*, 1988
Barley *Hordeum vulgare*	Embryo	No	Apparently respired	Holtman *et al*, 1994
Barley *Hordeum vulgare*	Aleurone a. Early germination b. Seedling growth	a. Little b. Yes	a. Mostly respired b. Converted to sucrose for export, and some respired	Newman and Briggs, 1976 Holtman *et al*, 1994
Maize *Zea mays*	Scutellum	Yes	Presumably some converted to sucrose, and some respired	Longo *et al*, 1975

Typically, in dicotyledonous oilseed species, there is a co-ordinate increase in the activities of enzymes of lipolysis, fatty acid β-oxidation, glyoxylate cycle and gluconeogenesis in the TAG-storing tissue immediately after seed germination (Figure 2). Subsequently these activities decline as the TAG decreases in amount and the seedling establishes photosynthetic competence (represented by GO and HPR activities in Figure 2). These changes in enzyme activity appear to be the result of changes in enzyme synthesis brought about by changes in mRNA amounts. Thus within a period of a few days there is a co-ordinated regulation of gene expression which leads to the mobilization of TAG, gluconeogenesis and subsequently establishment of green seedlings. Factors regulating gene expression are discussed later.

Figure 2: Schematic representation of developmental changes in TAG and peroxisomal enzymes in cotyleons of an epigeal plant during germination and seedling growth in the light

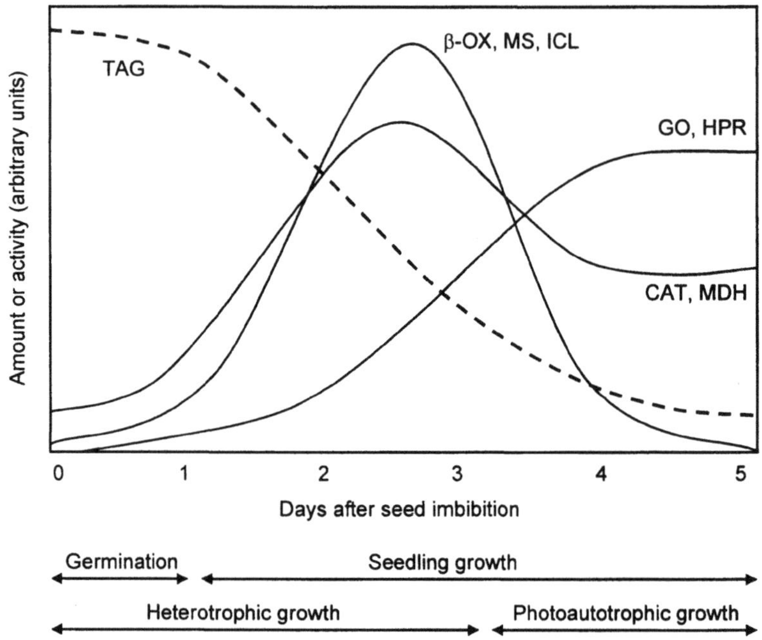

Figure 2. This representation is a compilation of data from many different experiments in plant species such as Arabidopsis thaliana, cucumber, sunflower and cotton. The changes are indicative of both the amounts and the activities of the enzymes mentioned. TAG, triacylglycerol; β-ox, enzymes of fatty acid β-oxidation; MS, malate synthase; ICL, isocitrate lyase; GO, glycolate oxidase; HPR, hydroxypyruvate reductase; CAT, catalse; MDH, malate dehydrogenase.

Sub-cellular location of glyoxylate cycle enzymes

Fractionation of plant tissue and separation by density gradient centrifugation established that glyoxylate cycle enzymes CS, ICL, MS and MDH are localized in peroxisomes (microbodies) together with catalase and enzymes of fatty acid β-oxidation (see Huang *et al*, 1983, also Chapter 2). The presence of CS, ICL and MS in peroxisomes has subsequently been confirmed by immunocytochemical methods (Nishimura *et al*, 1986; Sautter, 1986; Nishimura *et al*, 1993). Since such peroxisomes were considered to be specialized for glyoxylate cycle metabolism, they were named 'glyoxysomes' (Breidenberg and Beevers 1967). Vicentini and Matile (1993) have also proposed that peroxisomes present in senescing tissues, also containing glyoxylate cycle enzymes, should be called 'gerontosomes'. However, it is now recognized that there is no clear distinction between different forms of peroxisome, that their functions simply reflect the complement of enzymes which they contain, and that this can readily change (Titus and Becker, 1985; Nishimura *et al*, 1986; Sautter, 1986; Nishimura *et al*, 1993; Onyeocha *et al*, 1993; Olsen *et al*, 1993; Hayashi *et al*, 1996). Therefore, we prefer to use the generic term, peroxisome.

While cell fractionation studies clearly located key glyoxylate cycle enzymes in peroxisomes, ACO was difficult to detect (Huang *et al*, 1983). Despite this, it was assumed that lack of ACO was artefactual, resulting either from an inherent instability of the peroxisomal enzyme, or from its solubilization during peroxisome preparation. However, re-evaluation in the 1990s led to the clear conclusion that peroxisomes lack ACO, and that operation of the glyoxylate cycle necessitates involvement of a cytosolic ACO (Courtois-Verniquet and Douce, 1993; DeBellis *et al*, 1994; Hayashi *et al*, 1995) and shuttling of citrate and isocitrate across the peroxisomal membrane (Figure 3). Pumpkin and soybean have several ACO isozymes, including cytosolic and mitochondrial forms (De Bellis *et al*, 1995; Cots and Widmer, 1999). In both cases one isoform changes in activity in parallel with that of ICL and MS and is assumed to be a participant in the glyoxylate cycle (Hayashi *et al*, 1995; Cots and Widmer, 1999). While only a single *ACO* gene was originally reported for *Arabidopsis thaliana* (Peyret *et al*, 1995), the complete genome sequence predicts at least three genes (The Arabidopsis Genome Initiative, 2000).

Figure 3: The malate-asparate shuttle for export of reducing equivalents from the peroxisome

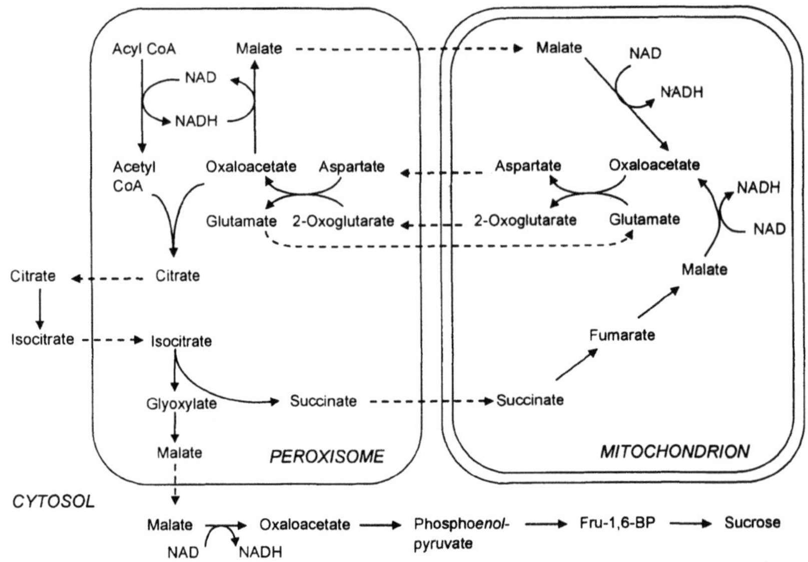

Figure 3. This scheme is based on that proposed by Mettler and Beevers (1980). The stoichiometry of reaction intermediates is not shown. Solid arrows represent metabolic reactions, dashed arrows indicate metabolite transport. This scheme involves participation of peroxisomal and mitochondrial forms of apsartate:2-oxoglutarate aminotransferase. Other enzymes are as shown in Figure 1.

The cytosolic location for ACO is assumed to be because this Fe-S-containing enzyme is highly sensitive to inactivation by reactive oxygen species (Verniquet *et al*, 1991) and so is partitioned away from the oxidizing environment of the peroxisome (Chapter 7). ACO is also implicated in Fe homeostasis, which may necessitate a cytosolic localization. However, the realization that ACO is cytosolic raises some important issues that need to be addressed. The cytosol contains appreciable $NADP^+$-dependent isocitrate dehydrogenase (IDH) activity (Galvez and Gadal, 1995), which could potentially divert isocitrate from the glyoxylate cycle to organic acid or amino acid metabolism, or isocitrate could potentially enter the mitochondrion and be respired. The potential advantages or disadvantages of these putative alternative fates for such isocitrate are unknown. NADP-IDH activity in pumpkin cotyledons changes in parallel with glyoxylate cycle activity (Nieri *et al*, 1995), suggesting a possible link. Diversion of isocitrate from the glyoxylate cycle might provide a means of generating cytosolic NADPH, but would require that the cycle is replenished by glyoxylate or oxaloacetate. The means by which isocitrate destined for the glyoxylate cycle is channelled back into the peroxisome is unknown. There is good evidence for the spatial organization of enzymes in leaf peroxisomes and for metabolite channelling in photorespiration (Heupel and Heldt, 1994; Reumann *et al*, 1994), so this could also be the case for glyoxylate cycle function.

Although it is generally accepted that ICL is a peroxisomal enzyme in higher plants, a cytosolic location has been clearly established in the unicellular alga *Chlorella fusca* (Pacy and Thurston, 1987) and in yeast, *Saccharomyces cerevisiae* (Taylor *et al*, 1996; Chaves *et al*, 1997). In such cases glyoxylate, rather than isocitrate, must enter the peroxisome from the cytosol. In a further variant, ICL has been reported to be in the matrix of mitochondria isolated from cotyledons of germinating *Vigna cylindrica* seeds (Kim *et al*, 1988). While the mitochondria also contained some MS, the majority was found in the peroxisome. It is proposed that the glyoxylate cycle operates in the mitochondrion in this plant (Kim *et al*, 1988). This would be consistent with the situation in some nematode worms, where the glyoxylate cycle is mitochondrial (McKinley *et al*, 1979). ICL has also been detected in the mitochondrial fraction of pea and tobacco leaves (Hunt and Fletcher, 1977; Zelitch, 1988) and cultured rose cells (Hunt *et al*, 1978). These striking observations are all based on enzymological assays of fractionated organelles. It will be necessary to investigate them with independent methods involving molecular characterisation of the proteins and their subcellular localization using reporter constructs or immunocytochemical methods, to assess their validity. In cucumber (*Cucumis sativus*) (Reynolds and Smith 1995a) and *Arabidopsis thaliana* (Eastmond *et al*, 2000), only one *ICL* gene is apparent, with no indication of a mitochondrial targeting sequence, but other plants may be different. The evolution of the subcellular distribution of glyoxylate cycle enzymes is a potentially fascinating and important research area.

Establishing how the glyoxylate cycle works requires knowledge of NADH oxidation

Peroxisomal fatty acid β-oxidation produces one mole of NADH for each oxidation cycle. If the glyoxylate cycle operates with the same sequence of reactions as proposed for bacteria (Figure 1), MDH will also catalyse the reduction of NAD^+ to NADH. This raises the question of how NADH becomes re-oxidised to maintain β-oxidation and glyoxylate cycle activities (Lord and Beevers 1972). The possibility that NADH is exported for oxidation outside the peroxisome is now considered unlikely because the peroxisomal membrane appears to be impermeable to NAD^+ and NADH (Donaldson, 1982; see Chapter 8) and evidence has been obtained to show that the yeast peroxisomal membrane is impermeable to NADH *in vivo* (Van Roermund *et al*, 1995). Two alternative mechanisms have been proposed to explain NADH oxidation. These have been reviewed in detail by Escher and Widmer (1997) and so will simply be outlined here.

The first mechanism involves the export of reducing equivalents in the form of malate, which can then be oxidized in the cytosol or mitchondrion, and OAA returned to the peroxisome (Mettler and Beevers, 1980). Essential features of this model are that peroxisomal MDH operates in the direction of OAA reduction rather than malate oxidation, that OAA is transported back to the peroxisome in the form of aspartate, and that a glutamate-oxoglutarate shuttle maintains the nitrogen balance (Figure 3). Active aspartate aminotransferases are present in both peroxisome and mitochondrion to facilitate this scheme. This metabolic scheme indicates an important role for cytosolic MDH, consistent with studies in yeast (Minard and McAllister-Henn, 1991).

The second mechanism has been proposed following discovery of an NADH dehydrogenase associated with the peroxisomal membrane, which is able to transfer electrons to ferricyanide and cytochrome c *in vitro* (Hicks and Donaldson, 1982; Donaldson and Fang, 1987; Luster and Donaldson, 1987). The membrane contains cytochromes b_5 and P450 which may be components of a short electron transport chain transferring electrons from the peroxisomal matrix to an acceptor in the cytosol. The cytosolic electron acceptor is not known, but dehydroascorbate is one candidate (Bowditch and Donaldson, 1990). Without a full understanding of this potential electron transfer system, it is difficult to evaluate whether it is capable of catalyzing the oxidation of NADH during the periods of most active TAG utilization during seedling growth.

Of all the unanswered questions concerning glyoxylate cycle function in plants, this is arguably the most fundamental, since we still do not know the path of carbon metabolism through the cycle. In future this will be resolved through transgenic manipulation and the analysis of mutants, particularly focusing on the roles of MDH, aspartate aminotransferases, and the proteins that constitute the membrane electron transfer system.

GLYOXYLATE CYCLE ENZYMES AT OTHER STAGES OF DEVELOPMENT AND IN STARVATION

Glyoxylate cycle enzymes in seed development

The glyoxylate cycle enzymes ICL and MS have been detected in developing seeds of several species (Table 3). In cotton, MS activity was reported in the absence of ICL in maturing embryos (Choinski and Trelease 1978; Miernyk and Trelease, 1981). Consistent with this observation, MS but not ICL was detected in mature seeds of many different oilseed species (Miernyk *et al*, 1979). Based on these observations, together with studies of organic acid metabolism in maturing embryos, Miernyk and Trelease (1981) proposed that MS is involved in citrate synthesis, with the required acetate derived from fatty acid β-oxidation and the glyoxylate derived principally from glycine. Subsequently Turley and Trelease (1990) reported that mRNAs, proteins and enzyme activities could be detected for *both* ICL and MS in developing cotton seeds. However, ICL appeared before MS and reached an appreciably lower activity. The earlier failure to detect ICL was attributed to the presence of ICL inhibitor(s) in the embryo extracts. Such inhibitors have also been reported in sunflower (*Helianthus annus* L.) (Fusseder and Theimer, 1984).

Table 3: Examples of glyoxylate cycle enzymes ICL and MS in developing seeds

Species	Tissue	Enzyme	Measurement[1]	References
Cotton (*Gossypium hirsutum*)	embryo	MS (ICL[2])	activity	Choinsky and Telease 1978; Miernyk and Trelease 1981
Cotton (*Gossypium hirsutum*)	embryo	ICL, MS	mRNA, protein, activity	Turley and Trelease, 1990
Sunflower (*Helianthus annus*)	embryo	ICL, MS	activity	Fusseder and Theimer, 1984
Sunflower (*Helianthus annus*)	embryo	ICL	mRNA	Allen *et al*, 1988
Cucumber (*Cucumis sativus*)	embryo	ICL, MS	activity	Frevert *et al*, 1980
Oilseed rape (*Brassica napus*)	embryo	ICL, MS	mRNA, protein, activity	Comai *et al*, 1989b; Ettinger and Harada, 1990
Oilseed rape (*Brassica napus*)	embryo	ICL, MS	activity	Eccleston and Ohlrogge, 1998
Oilseed rape (*Brassica napus*)	embryo	ICL, MS	protein	Chia and Rawsthome, 2000
Maize (*Zea mays*) *vp-1* and *aba* mutants	kernel	ICL, MS	mRNA	Paek *et al*, 1998

[1]*Activity was measured enzymically in crude cell free extracts or isolated peroxisomes, protein by Western blot and mRNA by Northern blot.* [2]*ICL was assayed but not detected.*

It is apparent that ICL and MS are synthesized in low amounts in developing seeds in several oilseed species, but that their relative amounts or timing of synthesis can vary. In some cases protein or mRNA levels do not correlate with measurements of enzyme activities, which can be explained partly by the presence of ICL inhibitors, but may also imply some forms of post-transcriptional control (Ettinger and Harada, 1990). Much inconsistency in reports of the amounts of these enzymes in developing seeds could be the result of the different methods employed, the species of plant, or factors which alter carbon metabolism in the developing seed, such as the plant growth conditions (also see below for an effect of transgenic modification). It is reported that *ICL* gene expression is repressed by abscisic acid (ABA) (Ihle and Dure 1972) while that of MS is not (Choinski *et al*, 1981). ICL and MS mRNAs were not detected in developing wild type maize kernels, but both were detected in ABA-deficient and ABA-insensitive mutant embryos (Paek *et al*, 1998). Thus, maternal or environmental factors that change ABA content may influence the timing and relative amounts of ICL and MS in developing seeds. Turley and Trelease (1990) and Skadsen and Scandalios (1989) further suggest that two or more different programmes regulate glyoxylate cycle enzyme synthesis in developing embryos, one controlling tissue-specific induction and the other controlling the temporal levels of expression of the induced genes.

Function of glyoxylate cycle enzymes in seed development

The proposal that MS is involved in citrate synthesis in cotton seeds (Miernyk and Trelease, 1981) would appear to be plausible, but does not explain the presence of ICL. In yeast (*Saccharomyces cerevisiae*) there are at least two *MS* genes, one of which (*DAL7*) functions specifically to metabolize glyoxylate produced from urea during purine catabolism (Rai *et al*, 1999). Miernyk and Trelease (1981) could detect only very low rates of conversion of allantoic acid or glyoxylurea to glyoxylate in maturing cotton seeds, and so concluded that MS probably does not function in purine catabolism. While none of the studies referred to here investigated other key glyoxylate cycle enzymes (e.g. CS, ACO and MDH), operation of the glyoxylate cycle in seed development would appear to be the most obvious function for ICL and MS.

It is apparent that in crucifers, TAG decreases in amount during seed maturation (Eastmond and Rawsthorne, 2000), and that β-oxidation is active (Richmond and Bleecker, 1999; Poirier *et al*, 1999). High level expression of a medium-chain thioesterase in transgenic oilseed rape leads to enhanced synthesis of lauric acid in developing seeds, but at the same time medium-chain acyl-CoA oxidase activity is increased several-fold, and ICL and MS activities increase six- and thirty-fold (Eccleston and Ohlrogge, 1998). It is apparent that fatty acid β-oxidation and the glyoxylate cycle are increased in activity when an excess of a particular fatty acid is produced. To determine if the glyoxylate cycle and gluconeogenesis are active in developing oilseed rape seeds, Chia and Rawsthorne (2000) fed [^{14}C]-acetate to isolated wild type embryos of different ages and followed the fate of the radiolabel. At early stages when TAG synthesis is maximal, radiolabel was preferentially found in the organic fraction, whereas at later stages when ICL, MS and PEPCK were prevalent, a much greater proportion of ^{14}C was recovered in the aqueous fraction including glucose. This implies a gluconeogenic role for the glyoxylate cycle. However, the amount of carbon from the glyoxylate cycle entering other pathways or compounds is yet to be determined, so other functions for the glyoxlate cycle may be established. A requirement for gluconeogenesis may be explained by the inability of the maturing seed to import sucrose from the maternal tissues, and the need for the seed to contain some sucrose and raffinose-family oligosaccharides (Bentsink *et al*, 2000) to aid in the dessication process, or for use in germination.

Glyoxylate cycle enzymes in senescence and starvation

The appearance of ICL and MS activities in starved cells was first reported by Kudielka and Theimer (1983a, b), following withdrawal of sucrose from the incubation medium of cultured *Pimpinella anisum* cells. Subsequently Gut and Matile (1988) reported that ICL and MS activities could be detected in detached ('senescing') barley leaves after incubation in the dark for up to eight days. Since then, there have been many reports of ICL and MS enzyme activity or gene expression in starved or senescing tissues of many plant species (Table 4). In many cases the literature refers to starvation of detached organs as a form of senescence, whereas here we restrict use of the term 'senescence' to those situations in which mature organs undergo a developmentally-determined programme of degenerative events while still attached to the plant. In such cases, the physiology of the organ is different to that of a detached, starved organ, since there is still the opportunity for the transport of water, minerals, metabolites and hormones between the senescing organ and the remainder of the plant. This relationship is believed to be important for the recovery of resources from the senescing organ for use elsewhere in the plant (Gan and Amasino, 1997). Differences in the pattern of gene expression in detached and naturally-senescing barley leaves indicate that detachment leads to expression of wound- and osmotic stress-induced genes (Becker and Apel, 1993). While the depletion of carbohydrate content is a generally common feature of senescent and starved organs, an increase in sucrose content has been reported in some senescing leaves (Crafts-Brandner *et al*, 1984; Nooden *et al*, 1997; Wingler *et al*, 1998). Starvation is reversible by addition of sugar, or in the case of photosynthetic tissues, by exposure to light. Senescence is also reversible to an extent but exogenous sugars are only partially effective, whereas cytokinin treatment or changes in source-sink relationships within the plant (such as by removal of sink organs) can be effective. Despite the importance of this distinction between senescence and starvation, it is likely that many of the physiological changes that take place in starved and senescent organs are similar, including breakdown of chlorophyll, lipids and protein, each of which can potentially fuel the glyoxylate cycle.

In those cases in which starvation of cultured cells is reported to increase ICL and MS activity or synthesis (Table 4), repression by metabolisable sugars was also demonstrated. In two cases, an increase in enzyme activity in response to acetate is also reported (see below). Many of the experiments with detached organs (Table 4) also report repression of ICL and MS synthesis by sugars. These observations demonstrate the metabolic responsiveness of these genes (see later for full discussion).

Few of the studies referred to in Table 4 report on glyoxylate cycle enzymes other than ICL and MS, or on the key gluconeogenic enzyme PEPCK. Kudielka and Theimer (1983a) showed that CS, MDH and catalase activities were readily detectable in cultured cells before starvation, but did not increase in amount upon starvation. However, addition of acetate did increase their activities, but the peroxisomal isozymes were not specifically assayed. Pistelli *et al* (1992) found that peroxisomal CS activity increased slightly, but peroxisomal MDH activity changed little during starvation of beet leaves. Peroxisomal MDH mRNA did not increase in amount in starved cucumber cotyledons (Kim and Smith 1994a) or callus cultures (Graham *et al*, 1994a). Vicentini and Matile (1993) detected CS in peroxisomes from starved rape cotyledons. The general picture which emerges is that peroxisomal CS and MDH are present in appreciable amounts in non-starved tissues, and change little in activity during starvation. A similar situation seems to exist in senescing tissues (Pistelli *et al*, 1991; De Bellis *et al*, 1991; Kim and Smith, 1994a). Little information on ACO activity in starvation and senescence is available, but one isoform increases in amount in wounded soybean tissue (Cots and Widmer, 1999).

Table 4: Synthesis of ICL and MS in response to starvation and senescence

Species	Tissue	Enzyme	Measurement[1]	References
Starvation of cells cultured *in vitro*				
Anise (*Pimpinella anisum*)	Suspension culture	ICL, MS	Activity	Kudielka and Theimer 1983a,b
Rice (*Oryza sativa*)	Suspension culture	ICL	Activity	Lee and Lee, 1996
Carrot (*Daucus carota*)	Suspension culture			Lee *et al*, 1998
Cucumber (*Cucumis sativus*)	Callus culture	ICL, MS	mRNA	Graham *et al*, 1994a
Cucumber (*Cucumis sativus*)	Protoplasts	ICL, MS	GUS reporter	Graham *et al*, 1994b; Reynolds and Smith 1995b
Cucumber (*Cucumis sativus*)	Protoplasts	ICL, MS	Protein	McLaughlin and Smith, 1994;
Cucumber (*Cucumis sativus*)	Pairy roots	ICL, MS	mRNA, GUS reporter	Ismail *et al*, 1997 De Bellis *et al*, 1997
Starvation of detached organs				
Barley (*Hordeum vulgare*)	Leaves	ICL, MS	Activity	Gut and Matile 1988; De Bellis *et al*, 1990
Barley (*Hordeum vulgare*)	Leaves	ICL	Protein	Chen *et al*, 2000b
Cucumber (*Cucumis sativus*)	Leaves	MS	mRNA, protein	Graham *et al*, 1992
Nicotiana plumbaginifolia	Leaves (transgenic)	Cucumber MS	GUS reporter	Graham *et al*, 1992
Spinach (*Spinacea oleracea*)	Leaves	ICL, MS	Activity	Landolt and Matile, 1990
Rice (*Oryza sativa*)	Leaves	ICL, MS	Activity	De Bellis *et al*, 1990
Leaf beet (*Beta vulgaris*)	Leaves	ICL, MS	Activity, MS protein	De Bellis *et al*, 1990; Pistelli *et al*, 1992
Pea (*Pisum sativum*)	Leaves	ICL, MS	Activity	Pastori and Del Rio, 1994
Cucumber (*Cucumis sativus*)	Cotyledons	ICL, MS	mRNA, protein	McLaughlin and Smith, 1994; Kim and Smith, 1994a,b
Rape (*Brassica napus*)	Cotyledons	ICL	Activity	Vicentini and Matile 1993
Pumpkin (*Cucurbita pepo*)	Cotyledons	ICL, MS	Activity, MS protein	De Bellis *et al*, 1990; De Bellis and Nishimura 1991
Cucumber (*Cucumis sativus*)	Roots	ICL, MS	mRNA, GUS reporter	Graham *et al*, 1992; Ismail *et al*, 1997
Maize (*Zea mays*)	Root tips	MS (ICL[2])	Activity	Dieuaide *et al*, 1992
Pumpkin (*Cucurbita pepo*)	Fruits	ICL, MS	Activity, protein	Pistelli *et al*, 1996
Dark treatment or defoliation of whole plants				
Rice (*Oryza sativa*)	Leaves	ICL, MS	Activity	Pistelli *et al*, 1991
Wheat (*Triticum aestivum*)	Leaves	ICL, MS	Activity	Pistelli *et al*, 1991
Cucumber (*Cucumis sativus*)	Leaves	MS	GUS reporter	Graham *et al*, 1992
Barley (*Hordeum vulgare*)	Leaves	ICL	Protein	Chen *et al*, 2000b
Cucumber (*Cucumis sativus*)	Cotyledons	ICL, MS	mRNA	Birkhan and Kindl, 1990

Table 4 (continued)

Nicotiana plumbaginifolia (transgenic)	Whole seedlings	Cucumber ICL, MS	GUS reporter	Reynolds and Smith 1995b; Sarah *et al*, 1996
Cucumber (*Cucumis sativus*)	Roots (defoliated)	ICL, MS	mRNA	Ismail *et al*, 1997
Natural senescence				
Rice (*Oryza sativa*)	Leaves	ICL, MS	Activity	Pistelli *et al*, 1991
Wheat (*Triticum aestivum*)	Leaves	ICL, MS	Activity	Pistelli *et al*, 1991
Tomato (*Lycopersicon esculentum*)	Leaves	ICL, MS	Protein	Nieri *et al*, 1997
Sweet potato (*Ipomoea batatas*)	Leaves	ICL	mRNA	Chen *et al*, 2000a
Cucumber (*Cucumis sativus*)	Leaves, Cotyledons	MS	mRNA	Graham *et al*, 1992
Pumpkin (*Cucurbita pepo*)	Cotyledons	ICL, MS	Activity, MS protein	De Bellis and Nishimura 1991
Cucumber (*Cucumis sativus*)	Cotyledons	ICL, MS	mRNA, protein	Kim and Smith 1994a,b; McLaughlin and Smith, 1995; Chen *et al*, 2000b
Cucumber (*Cucumis sativus*)	Petals	MS	mRNA	Graham *et al*, 1990, 1992
Pumpkin (*Cucurbita pepo*)	Petals	ICL, MS	Activity, MS protein and mRNA	De Bellis et al., 1991

[1]Activity was measured enzymically in crude cell free extracts or isolated peroxisomes, protein by Western blot, mRNA by Northern blot and GUS reporter by fluorimetry. [2]ICL was assayed but not detected.

Function of the glyoxylate cycle in starvation and senescence

The consensus is that glyoxylate cycle enzymes are present in starved plant tissues, although at a level appreciably lower than found in seedlings of oilseeds, and often requiring isolation of peroxisomes before their activity can be measured effectively. By feeding with [^{14}C]-acetate or citrate, Pistelli *et al* (1995) showed that peroxisomes isolated from starved pumpkin cotyledons could catalyze all the steps of the glyoxylate cycle, when supplemented with ACO. Thus, the glyoxylate cycle appears to be active during starvation. This raises, firstly, the question of the origin of the substrates which feed into the cycle, and secondly, the fate of the products.

The metabolic changes that take place in starved plant cells are well documented (Journet *et al*, 1986; Roby *et al*, 1987; Brouquisse *et al*, 1991; Dieuaide *et al*, 1992; Aubert *et al*, 1996; Yu, 1999). Initially in the adaptation phase, the cellular carbohydrate levels and respiration rate fall. In the subsequent survival phase (which is reversible by addition of sugar), lipid and protein are consumed, and Pi, phosphorylcholine and free amino acids accumulate. Finally, starved cells enter a terminal phase of ammonia release, disorganization and death. Throughout starvation there is a defined programme of autophagy in which organelle sacrifice is prioritized: plastids, ribosomes and endoplasmic reticulum are consumed relatively early, whereas the plasma membrane, mitochondria and peroxisomes persist. The survival phase is characterised by an activation of β-oxidation (Dieuaide *et al*, 1992) and proteolysis leading to accumulation of amino acids, particularly Asn (Genix *et al*, 1990; Brouquisse *et al*, 1992). Starvation of detached green tissues follows a similar pattern, characterised most obviously by chlorophyll and galactolipid breakdown (Gut and Matile 1988; 1989).

Not only is peroxisomal β-oxidation activated, but mitochondrial β-oxidation is also reported to be active in starved plant cells (Dieuaide*et al*, 1993; Masterson and Wood 2000). Peroxisomal β-oxidation is likely to be responsible for metabolism of fatty acids from membrane lipids, and to supply acetyl CoA to the glyoxylate cycle. Mitochondrial oxidative pathways are responsible for metabolism of skeletons from amino acids such as leucine and valine and to feed acetyl CoA into the Krebs cycle (Daeschner *et al*, 2001; Masterson and Wood, 2000). The glyoxylate cycle could also be supplied with glyoxylate derived directly from glycine or glycolate, or potentially from breakdown of purines, hydroxyproline, pentoses and other compounds during starvation.

The next question concerns the fate of the 4-carbon acids that are products of the glyoxylate cycle. Studies of detached barley leaves fed with [^{14}C]-oleic acid showed that some radioactivity was recovered in sugars, suggesting a gluconeogenic role for the glyoxylate cycle (Wanner *et al*, 1991). In support of this proposal, Kim and Smith (1994b) demonstrated the presence of PEPCK in starved and senescing cucumber cotyledons, and Ismail *et al* (1997) argued that gluconeogenesis may serve an important role, since glycolytic and pentose phosphate pathway intermediates may be required for biosynthesis, even in starvation. However, further consideration casts doubt on the conclusion that gluconeogenesis is important in starvation. The apparent gluconeogenesis detected in barley leaves (Wanner *et al*, 1991) also occurred in leaves before ICL and MS were detected, and most of the ^{14}C was found in CO_2 and other fractions, suggesting that the glyoxylate cycle was fueling respiration and other pathways. Graham *et al* (1994a), studying glyoxylate cycle enzymes in starved cucumber callus cells, found that addition of malate or succinate did not replenish sugar levels, suggesting that gluconeogenesis was not active. More recently, Chen *et al* (2000b) have provided evidence that the glyoxylate cycle does not serve a gluconeogenic role in starved barley leaves or cucumber cotyledons. They found that no PEPCK could be detected in barley leaves either before or after starvation, and that while starved cucumber cotyledons contained PEPCK, it was not located in the mesophyll cells with ICL, but was located separately in the vasculature and trichomes. Pyruvate Pi dikinase (PPDK) which would provide an alternative gluconeogenic step in conjunction with NAD(P)-malic enzyme, was detected in starved cucumber cotyledons and barley leaves, but it decreased in amount during this time. These observations argue against the glyoxylate cycle providing substrates for gluconeogenesis in starved cells.

Instead it is proposed that the glyoxylate cycle serves an anaplerotic ('topping up') role to maintain Krebs cycle intermediates, when there is no supply of carbohydrate which would otherwise supply oxaloacetate through the activity of PEP carboxylase (Graham *et al*, 1994a; Eastmond and Graham, 2001). Chen *et al* (2000b) also argue that respiration plays a key role in both starvation and senescence, and that the glyoxylate cycle may serve an anaplerotic role in the absence of substrate for PEP carboxylase. It has also been suggested that an important function could be the provision of carbon skeletons for amino acid metabolism (Ismail *et al*, 1997; Chen *et al*, 2000b). These different possible functions of the glyoxylate cycle are discussed in a wider context later.

In senescence, there is much less evidence upon which to formulate a role for glyoxylate cycle enzymes. De Bellis *et al* (1991) showed that CS and MDH increase slightly during petal senescence, so we might assume that the complete glyoxylate cycle operates. The cycle is probably fuelled principally by acetyl CoA from peroxisomal β-oxidation of membrane-derived fatty acids. There is a more obvious function for gluconeogenesis in senescing tissue relative to starved tissue, since the conversion of lipid into sucrose would allow the carbon to be transported out of the senescent organ for use elsewhere in the plant. PEPCK is present in senescent cucumber cotyledons (Kim and Smith, 1994b). However, it is located in the vasculature and trichomes, whereas ICL is in the mesophyll, arguing against a gluconeogenic role for the glyoxylate cycle (Chen *et al*, 2000b). PPDK is also present in cucumber cotyledons, but in low amount (Chen *et al*, 2000b). Perhaps the amount of carbon available is only sufficient to fuel the necessary respiration in the senescent organ. Clearly, more work is required to understand the role of the glyoxylate cycle in senescence.

Glyoxylate cycle enzymes at other stages of growth and development

Expression of *ICL* and *MS* genes has been clearly demonstrated in developing *Brassica napus* pollen (Zhang *et al*, 1994). Gene expression correlates with a period of development when pollen lipid content declines, but is not apparently activated during pollen germination when lipid reserves are utilised. While the presence of a functional glyoxylate cycle has yet to be confirmed, it is possible that it has a similar function in pollen development to that in seed development. The apparent absence of *ICL* and *MS* gene expression in germinating pollen might reflect the fact that sugar can be imported from the stigmatic tissues.

ICL activity has been detected in apparently healthy, non-senescing leaves of several species including wheat, maize (Godavari *et al*, 1973), pea (Hunt and Fletcher, 1977) and tobacco (Zelitch, 1988). Godavari *et al* (1973) showed that ICL activity was increased by dark treatment of leaves, raising the possibility that the low levels in apparently healthy leaves could be due to inadvertant ageing or wounding. Expression of the gene for ICL has been detected in developing tomato leaves using a cDNA probe (Janssen, 1995), and both ICL and MS were detected in developing and senescent leaves, but not in mature leaves of tomato (Nieri *et al*, 1997). The presence of ICL and MS in young tomato leaves should not be due to wounding or ageing. The function of ICL and MS in young expanding tomato leaves is unknown and intriguing. It would be particularly valuable to determine in which cells ICL and MS are located, and to determine if PEPCK or PPDK are also present. It might be that the glyoxylate cycle functions in a particular aspect of organic or amino acid metabolism in such leaves as has been suggested for senescing cucumber and barley leaves (Chen *et al*, 2000b).

IMPORTANT INFORMATION FROM KNOCK-OUT MUTANTS OF *ARABIDOPSIS THALIANA*

Arabidopsis thaliana stores appreciable amounts of triglyceride in its cotyledons (up to 40% of seed dry weight), but no starch, making it ideally suited to studies of glyoxylate cycle function. ICL and MS are encoded by single genes, and knock-out mutants of each have been isolated (Eastmond *et al*, 2000; Germain *et al*, 2000). Seed germination (radical emergence) in the mutants appears normal, but seedling growth is compromised by lack of a functioning glyoxylate cycle. It is likely that seed germination is fuelled by small amounts of soluble carbohydrates (e.g. sucrose and raffinose) and organic acids, and that this supply is rapidly exhausted. In the dark, mutant seedling growth (hypocotyl extension in particular) is much reduced compared to wild type (Figure 4), and after extended periods maintained in the dark, mutant seedlings lose the capacity to turn green and grow when transferred to the light, more quickly than do wild types (Eastmond *et al*, 2000). When mutant seedlings in the dark are provided with an exogenous supply of sucrose, they grow as rapidly and extensively as wild type, indicating that they are otherwise simply limited by a lack of carbohydrate, which the glyoxylate cycle would normally help to provide. Mutant seedlings in the light grow as effectively as wild type, provided that the available light is of sufficient intensity and duration each day (Eastmond *et al*, 2000). Apparently the role of the glyoxylate cycle becomes progressively less important as more light becomes available for seedling photosynthesis. All of these observations are consistent with the long held understanding that the glyoxylate cycle functions to provide carbon for gluconeogenesis.

Studies of the fate of [^{14}C]-acetate supplied to seedlings over a four hour period show that, whereas about 25% is recovered in sugars in the wild type, the amount recovered in sugars in the *icl* mutant is negligible. This confirms the gluconeogenic role of the glyoxylate cycle. Analysis of the *icl* mutant shows that TAG is used up, particularly when seedling growth is promoted by the addition of exogenous sucrose. The acetate feeding studies further show that whereas in wild type about 35% is released as CO_2 and 30% enters organic and amino acids, the corresponding figures for the *icl* mutant are 40% and 47%, showing that respiration of acetate and organic acid synthesis are appreciable. These results indicate that TAG is respired in the absence of the glyoxylate cycle. Such a result is not necessarily surprizing, since there are naturally-occuring situations in which TAG is utilized in the absence of the glyoxylate cycle, such as germinating sunflower and lettuce seeds, and barley embryos (Table 2). Since β-oxidation of fatty acids from TAG can only take place in the peroxisome (Germain *et al*, 2001), carbon must be transferred from the peroxisome to the mitochondrion even in the absence of a functional glyoxylate cycle.

Figure 4: Growth of icl and ms mutant seedlings

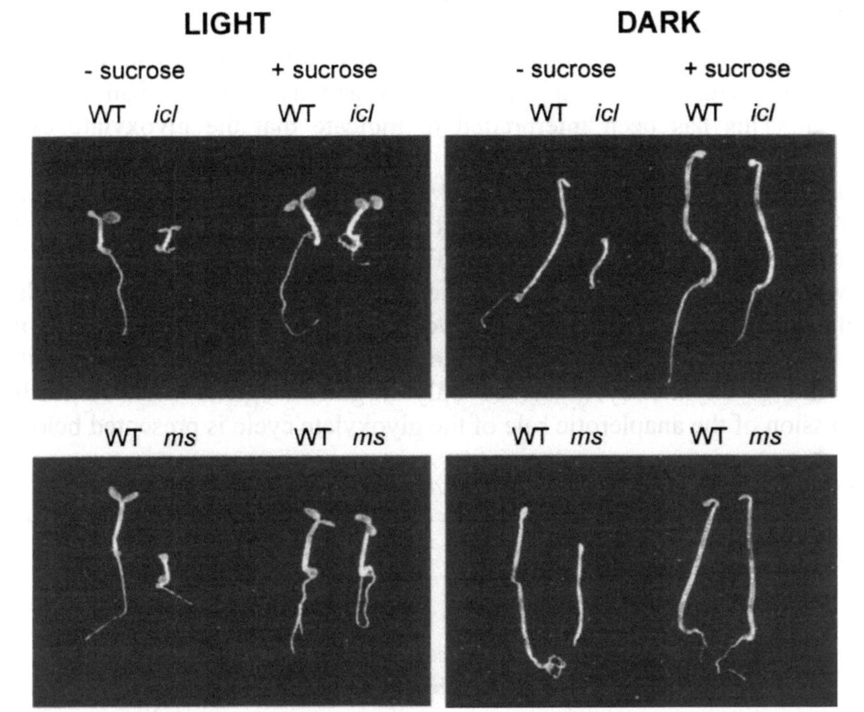

Figure 4. Wild type and mutant seedlings were germinated and grown for 5 days either in low light (50 µmol m^{-2} s^{-1}) or in the dark, in the presence or absence of 1% (w/v) sucrose. Representative seedlings are shown in each case. The icl mutant is Columbia ecotype while ms is Wassilewskija.

These observations collectively suggest that in many circumstances, a modified TCA cycle can operate between three sub-cellular compartments to achieve the respiration of acetate from peroxisomal fatty acid β-oxidation (Figure 5). Potentially, acetate (acetyl units) or citrate could leave the peroxisome for the mitochondrion. While in yeast evidence for acetylcarnitine export has been obtained (van Roermund *et al*, 1999), no such evidence exists for plants. In plants, citrate apparently leaves the peroxisome under normal circumstances because ACO is cytosolic, so this is a likely route for net export from the peroxisome. A peroxisomal CS mutant would allow us to distinguish these two possibilities (Eastmond and Graham, 2001; Thorneycroft *et al*, 2001). It is likely that in some plants (e.g. *Arabidopsis thaliana*) carbon is exported from the peroxisome for respiration in the mitochondrion, even in the presence of a functional glyoxylate cycle. However, in germinating castor bean, in which TAG is quantitatively converted into sucrose for export from the endosperm, it is apparent that very little of the TAG is respired (Table 2), and it is likely that lack of a glyoxylate cycle would be seedling-lethal.

Arabidopsis thaliana icl mutant seedlings grow better and utilise their TAG more completely when supplied with exogenous sucrose, than they do without. This has been interpretted to indicate that the glyoxylate cycle serves an anaplerotic role (Eastmond *et al*, 2000). It was argued that the respiration of TAG is restricted by lack of an anaplerotic reaction to maintain TCA cycle activity in the *icl* mutant (Eastmond and Graham, 2001). An alternative explanation is simply that TAG utilization depends upon seedling growth which in turn depends upon the gluconeogenic role of the glyoxylate cycle. Since gluconeogenesis and glycolysis operate simultaneously in the cotyledons of marrow seedlings (Thomas and ap Rees, 1971), it is possible that the glyoxylate cycle serves only a gluconeogenic role. A further discussion of the anaplerotic role of the glyoxylate cycle is presented below.

Figure 5: Operation of the TCA cycle between peroxisome, cytosol and mitochondrion

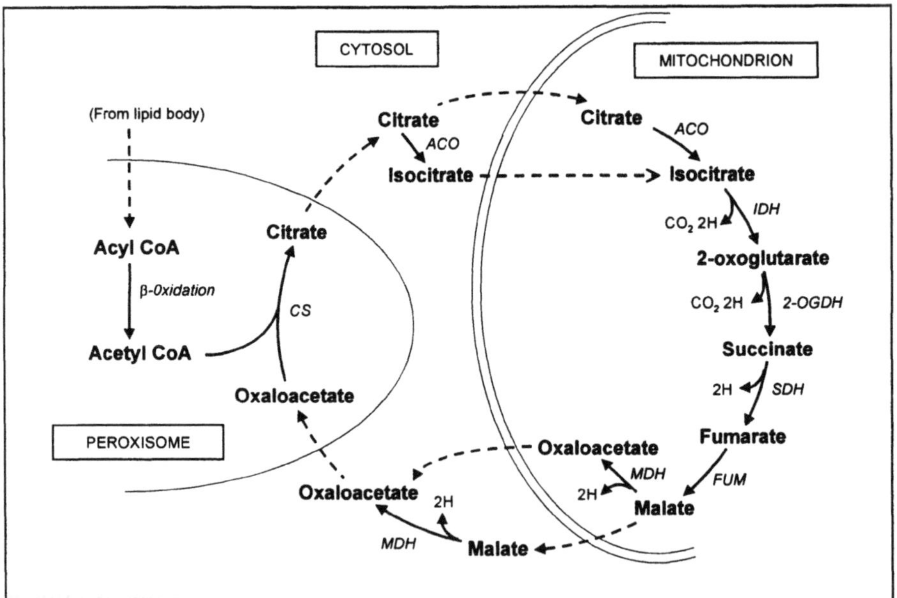

Figure 5. This scheme indicates how peroxisomal fatty acid β-oxidation and respiration may be achieved. It is not known which of the alternative sites for aconitase (ACO) and malate dehydrogenase (MDH) activities would function in such a scheme. CS, citrate synthase; IDH, isocitrate dehydrogenase; 2-OGDH, 2-oxoglutarate dehydrogenase; SDH, combined succinyl CoA synthetase and succinate dehydrogenase; FUM, fumarase. Solid arrows indicate metabolic reactions, dashed arrows indicate metabolite transport.

Knock-out mutants can tell us about the importance of the glyoxylate cycle throughout growth and development and in starvation. In *Arabidopsis thaliana icl* and *ms* mutants, pollen viability and seed production are not obviously impaired, so that the glyoxylate cycle does not serve an essential role. The *icl* mutant is more sensitive than wild type to germination and seedling growth in prolonged darkness (Eastmond *et al*, 2000). This does not necessarily imply a greater sensitivity to starvation, but could reflect the failure of the mutant to make effective use of TAG in seedling growth. Consistent with this, we find little evidence that light-grown*icl* or *ms* mutant seedlings are more sensitive to dark starvation than wild types (V.Germain, S. Footitt, J.H.Bryce and S.M.Smith, unpublished observations). However, *ICL* and *MS* gene expression is not readily activated by starvation in *Arabidopsis thaliana*, so the glyoxylate cycle may be relatively less important in starvation in this plant than in others. Nevertheless, the mutants available now provide the opportunity to assess the importance of the glyoxylate cycle in growth and development under a range of environmental conditions.

OVERVIEW OF GLYOXYLATE CYCLE FUNCTIONS IN PLANTS

It is seen that glyoxylate cycle enzymes are synthesized in many different situations in plants throughout growth and development and in starvation. The role of the glyoxylate cycle in the conversion of TAG into sucrose in seed germination and seedling growth has long been established, but it is now also apparent that the glyoxylate cycle serves other functions, both in terms of the substrates feeding into the cycle and the fate of the 4C acids produced. There are situations in which gluconeogenesis does not apparently occur, or is a minor activity (starved cells, senescence). Instead, C4 acids may act as precursors for amino acids, amides or organic acids, or are respired in the Krebs cycle (Figure 6). It is argued that in starved cells, entry of C4 acids from the glyoxylate cycle into the Krebs cycle constitutes an anaplerotic ('topping up') reaction, replacing the role fulfilled by PEPC when carbohydrate is available (Graham *et al*, 1994a; Chen *et al*, 2000b; Eastmond and Graham, 2001). This concept is consistent with the anaplerotic role played by the glyoxylate cycle in micro-organisms growing on acetate. However, there is an important difference between micro-organisms and plants, in that plant mitochondria are not restricted to the oxidation of acetate: they can respire TCA cycle intermediates, because NAD^+-malic enzyme provides the means to convert such intermediates into acetate (ap Rees, 1990). Thus, succinate produced by the glyoxylate cycle can be completely oxidised in the mitochondrion, in a pathway that is not necessarily an anaplerotic reaction, and does not require one (Figure 7). By comparison, yeast growing on fatty acids is obliged to respire acetyl units directly, which will require an anaplerotic reaction (the glyoxylate cycle) to replenish TCA cycle intermediates withdrawn for biosynthesis. Therefore, in plants, the extent to which the glyoxylate cycle fulfills an anaplerotic role relative to a direct respiratory role, depends upon the activity of NAD^+-malic enzyme, the extent to which acetate enters the Krebs cycle directly, and the extent to which intermediates are withdrawn. The relative magnitude or importance of these alternative roles is unknown, and is expected to be different under different circumstances. We should see the role of the glyoxylate cycle as much more varied and flexible than is portrayed in current plant biology textbooks.

Figure 6: General scheme for glyoxylate cycle functions in plants

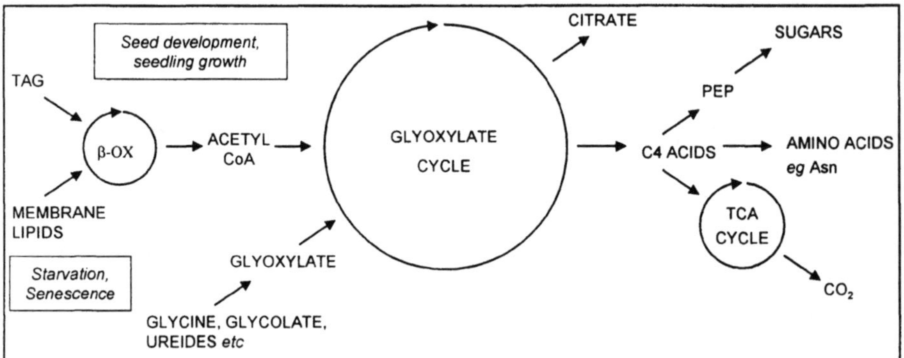

Figure 6. Substrates from different sources can feed carbon into the cycle and the products of the cycle can have different fates, depending upon growth stage or metabolic status. The association of the indicated growth stages with particular metabolic activities are intended to be illustrative.

Figure 7: Pathway for the complete respiration of succinate

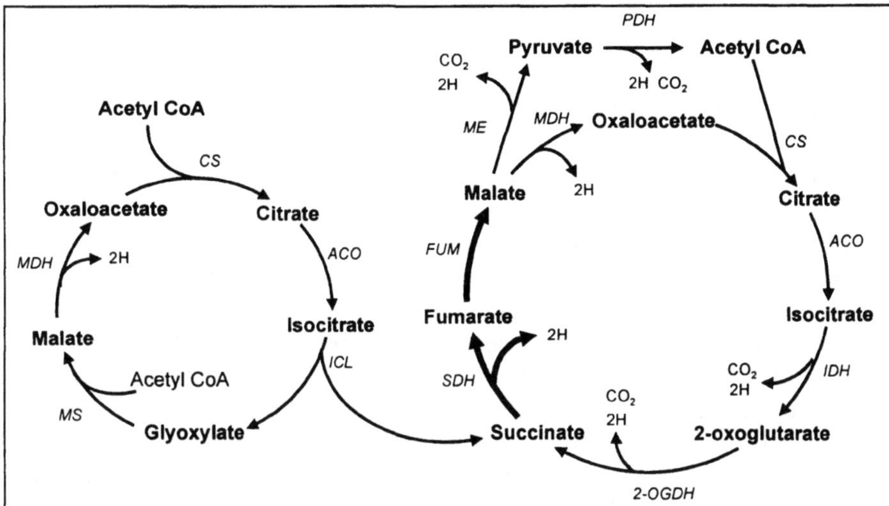

Figure 7. For convenience the glyoxylate cycle is shown in simplified form. Abbreviations as for Figure 5, plus: ME, NAD^+-malic enzyme; PDH, pyruvate dehydrogenase complex. Thickness of arrows indicates relative molar proportions of intermediates.

ENZYME SYNTHESIS, ASSEMBLY AND REGULATION IN HIGHER PLANTS

Import into peroxisomes

Glyoxylate cycle enzymes are synthesised on free cytosolic ribosomes and imported post-translationally into the peroxisome (Chapter 11). ACO is an exception since it functions in the cytosol (see above). CS and MDH are imported by means of cleavable amino-terminal (type 2) Peroxisomal Targeting Sequences (PTS2), whereas ICL and MS employ C-terminal PTS1 tripeptides (see Chapter 11). Different isoforms of CS and MDH are imported into mitochondria, using cleavable N-terminal targetting sequences which are different to PTS2 sequences (Gietl, 1992; Koyama *et al*, 2000). The mechanism of import of glyoxylate cycle enzymes has not been extensively studied, but experiments with ICL indicate that it is transported in oligomeric form, since polypeptides from which the PTS1 sequence has been removed can be piggy-backed into the organelle by intact ICL polypeptides (Lee *et al*, 1997). Whether other enzymes are imported in oligomeric form remains to be determined. The quaternary structures of glyoxylate cycle enzymes are shown in Table 1.

Assembly and organization of enzymes

Numerous cell fractionation studies have indicated the association of glyoxyxlate cycle enzymes with the peroxisomal membrane, including CS, MS and MDH (Köller and Kindl, 1977). However, MS in particular appears to have a propensity to aggregate and to have amphipathic properties that may cause it to associate with membranes *in vitro* (Kruse and Kindl, 1983). Subsequently MS, like ICL, has been shown to be distributed throughout the matrix (Chapman *et al*, 1989). Nevertheless, the possibility that the glyoxylate cycle enzymes are organised spatially in the peroxisome deserves examination. The spatial organization of enzymes facilitating metabolite channelling is well established in microbial and animal systems (Hrazdina and Jensen, 1992; Welch and Easterby, 1994). Direct evidence for spatial organization of enzymes in leaf peroxisomes has also been obtained (Heupel and Heldt, 1994; Reumann *et al*, 1994), but evidence for such organization of glyoxlate cycle enzymes is indirect. The utilization of a cytosolic ACO raises the question of whether isocitrate is channelled back into the peroxisome, but there is no evidence for this. In *E.coli*, isocitrate dehydrogenase is inhibited by phosphorylation when the glyoxylate by-pass is active, but it is not known if the same is true of plant cytosolic isocitrate dehydrogenase. ICL and catalase co-purify from maize scutella. It was suggested that catalase might be required to protect the glyoxylate produced by ICL from oxidation to formate and carbon dioxide by hydrogen peroxide (Khan *et al*, 1992). Similarly, MS co-purifies with MDH from soybean cotyledons (Guex *et al*, 1995), suggesting channelling of malate from MS to MDH, although this would be inconsistent with the Mettler and Beevers (1980) model for the role of MDH. Clearly there would be much merit in further investigating protein-protein interactions and metabolite channelling in order to understand how the glyoxylate cycle functions.

Regulation of enzyme activity

ICL is activated by phosphorylation in *Escherichia coli* (Robertson *et al*, 1988). ICL is phosphorylated in *Saccharomyces cerevisiae* in response to glucose treatment, but this reduces activity (Lopez-Boado *et al*, 1988). It is reported that castor and cucumber ICL enzymes are phosphorylated by a peroxisomal protein kinase, but the function is unknown (Finnessy *et al*,

1994). MS of castor bean also appears to be phosphorylated by a peroxisomal protein kinase which appears to be stimulated by calcium, but again the function of such phosphorylation is unknown (Yang *et al*, 1988). One possibility is that phosphorylation targets these proteins for degradation (Lopez Boado *et al*, 1988). MS is specifically degraded in peroxisomes of greening pumpkin cotyledons but the mechanism is unknown (Mori and Nishimura, 1989).

There are several reports of inhibitors of ICL activity in plant extracts (Godavari *et al*, 1973; Frevert *et al*, 1980; Fusseder and Theimer, 1984; Theimer, 1976) although they have not been identified and it is not known if they function *in vivo*. Often it is possible to detect ICL activity in isolated peroxisomes when activity is not detected in crude extracts (Godavari *et al*, 1973; Hunt and Fletcher, 1977; Nieri *et al*, 1997). ICL activity in the megagametophyte of loblolly pine declines during seedling growth even though the ICL protein remains (Mullen and Gifford, 1997). Such observations may explain some cases in which MS activity, but not ICL activity is detected (Dieuaide *et al*, 1992; Choinski and Trelease 1978; Miernyk and Trelease, 1981).

There are no consistent reports of metabolites that may regulate ICL or MS activity at physiological concentrations, so fine control of glyoxylate cycle activity may not be an important feature under normal circumstances. Instead, the control of enzyme synthesis and breakdown may be the key to the regulation of glyoxylate cycle activity. ICL is reported to have a high metabolic control coefficient (~0.66) during the period of rapid conversion of TAG into sugar in castor been endosperm (Runquist and Kruger,1999).

The special case of dual functions for aconitase

In animals, fungi and bacteria it has been shown that ACO has two mutually-exclusive functions. The same is likely to be true for plants (Navarre *et al*, 2000). The 4Fe-4S cluster is required for ACO activity, but in the presence of reactive oxygen species (ROS) including nitric oxide, one labile Fe is readily lost, and potentially the whole Fe-S cluster (Peyret *et al*, 1995; Navarre *et al*, 2000). This not only eliminates ACO activity, but converts the protein into one which binds to mRNAs encoding proteins of Fe acquisition or utilization (e.g. ferritin and transferrin receptor). These mRNAs have Iron Response Elements (IRE) in either 5' or 3' untranslated region, through which the binding protein (BP) regulates translation. Thus, in the presence of ROS, ACO is converted into IRE-BP which leads to regulation of cellular Fe levels, which may in extreme cases participate in oxidation reactions associated with cell death. The effect of inactivation of mitochondrial ACO in animal cells is to slow respiration. Since the cytosolic ACO is also implicated in the response to ROS, it would be predicted that in the event of treatments that elevate ROS in plant cells (e.g. pathogen attack), the glyoxylate cycle would be slowed or stopped.

CONTROL OF GENE EXPRESSION

Metabolic and developmental control

In *E. coli*, the *ace* operon encoding ICL, MS and IDH kinase/phosphatase is subject to glucose repression and activation by acetate (Maloy and Nunn, 1982). The same is true of *ICL* and *MS* genes in filamentous fungi (Armit *et al*, 1976; King and Casselton, 1977).

Such observations of metabolic regulation of glyoxylate cycle gene expression in other organisms led to the consideration of similar mechanisms operating in plants. Kudielka and Theimer (1983a, b) showed that MS and ICL activities increased in cultured cells of *Pimpinella* when sucrose was withdrawn from the medium, and increased further when acetate was subsequently added. In a study of diauxic growth in cultured rice cells, Lee and Lee (1996) also clearly showed induction of glyoxylate cycle enzyme activities during growth on acetate and repression by glucose. Remarkably though, in the presence of both acetate and glucose, the cells first used the acetate for growth, before adapting to growth on glucose. It was suggested that acetate inhibited glucose uptake into the cells (Lee and Lee, 1996).

The appearance of MS and ICL in detached barley leaves (Gut and Matile 1988) and expression of *MS* and *ICL* genes in senescing and starved cucumber organs (Graham *et al*, 1992), are also consistent with the metabolic regulation of these genes in plants. A model in which expression is repressed by sugars and activated by acetate from lipid catabolism was proposed (Graham *et al*, 1992). Subsequently a role for sugars in repressing expression was clearly established using detached cucumber organs, cultured cells and protoplasts (McLaughlin and Smith, 1994; Graham *et al*, 1994a, b; Reynolds and Smith, 1995b). However, no activating effect of acetate could be demonstrated (Graham *et al*, 1994a), possibly due to the toxic nature of this compound to some plant cells.

The repression of ICL and MS synthesis by sugars has been established in a range of cucumber vegetative tissues and cultured cells (Graham *et al*, 1992, 1994a; McLaughlin and Smith, 1994; De Bellis *et al*, 1997; Ismail *et al*, 1997), and in barley leaves (Chen *et al*, 2000b). However, it is not clear if this is generally the case in other plant species. Not all species seem to respond to starvation by activation of *MS* and *ICL* gene expression. For example, under the conditions in which dark-incubated detached cucumber or barley leaves accumulate *MS* and *ICL* mRNAs, these mRNAs are barely detectable in *Brassica, Arabidopsis thaliana* and tobacco leaves (Zhang *et al*, 1996; S.M. Smith, unpublished results). Either metabolism itself, or the metabolic responsiveness of *MS* and *ICL* genes, varies appreciably between plant species.

In germinating seeds, there are differences in the reported responsiveness of ICL and MS synthesis to repression by sugar. In castor bean seeds, glucose partially represses appearance of ICL in the endosperm, but very strongly represses ICL appearance in excised cotyledons (Lado *et al*, 1968). Longo and Longo (1970) showed that germination of peanut seeds in the presence of glucose appreciably reduced the activity of ICL and MS in cotyledons. However, application of exogenous sucrose to germinating seeds of *Arabidopsis thaliana* (Eastmond *et al*, 2000; Germain *et al*, 2001) or cucumber (S.M. Smith, unpublished observations) does not appear to appreciably repress *MS* and *ICL* gene expression or enzyme activities. In addition, the cucumber PEPCK and peroxisomal MDH genes are co-ordinately expressed with *MS* and *ICL* genes during seedling growth, but are not repressed by sugars even in starved cells (Kim and Smith, 1994a, b). These observations point to a developmental regulation of glyoxylate cycle gene expression, independent of sugar regulation.

The possibility that products of TAG breakdown (e.g. acetate or acetyl CoA) may activate expression of *MS* and *ICL* genes (Graham *et al*, 1992) does not find support from studies of a 3-ketoacyl-CoA thiolase mutant in which TAG breakdown in germinating seeds is blocked, since MS and ICL appear as normal in the seedlings (Germain *et al*, 2001). It is likely, therefore, that both metabolic and developmental signals act upon *MS* and *ICL* genes to regulate expression throughout growth and development. This conclusion is clearly established for cucumber *MS* and *ICL* genes in which distinct *cis*-acting elements have been identified (see below).

Regulation of expression by sugars

The mechanism by which sugars regulate gene expression in plants is currently the subject of much research (Jang and Sheen, 1997; Sheen *et al*, 1999; Gibson, 2000; Smeekens, 2000). Graham *et al* (1994a) studied *MS* and *ICL* gene expression in cultured cucumber callus tissue and showed that mRNA levels increase when sucrose is removed from the medium. Activation of gene expression correlated with a threshold sugar content of the cells. Only sugars which are phosphorylated by hexokinase were found to be effective in repressing gene expression, including 2-deoxyglucose and mannose which are metabolized slowly or inappropriately after phosphorylation. These observations, led to the proposal that hexokinase plays a key role in the sugar signalling pathway, as has been proposed in yeast (Carlson, 1998; Johnston, 1999). A similar conclusion was reached from observations of photosynthetic gene expression in maize protoplasts (Jang and Sheen, 1994). It is clear that hexokinase is required for sugar sensing in plants, but the proposal that it is a sugar sensor (Jang and Sheen, 1997) is not proven (Halford *et al*, 1999). Other potential components of a signal transduction pathway for metabolic control of *ICL* and *MS* gene expression are unknown, but the SNF1-type protein kinase and PRL1 interacting protein are candidiates (Smeekens, 2000).

Control by abscisic acid and gibberellic acid

There are numerous examples which demonstrate the apparent repression of ICL and MS synthesis by abscisic acid (ABA) and activation by gibberellic acid (GA). ABA inhibits synthesis of ICL in cotyledons of developing cotton seeds (Ihle and Dure, 1972). In developing maize kernels, while ICL and MS mRNAs were not detected in wild type, both were detected in developing embryos of ABA-insensitive (vp1) and ABA-defficient (vp7 and vp10) mutants (Paek *et al*, 1998). ICL and MS appear in the aleurone cells of germinating barley (Newman and Briggs, 1976) and wheat (Doig *et al*, 1975). In both cases this requires either the presence of the embryo or exogenous GA. In castor bean, ABA inhibits the appearance of ICL and MS mRNAs and enzyme activities, while GA promotes their appearance (Marriott and Northcote, 1975, 1977; Martin and Northcote, 1982; Martin *et al*, 1984; Dommes and Northcote, 1985; Rodriguez *et al*, 1987). Since ABA has a general repressive effect on germination and GA has a promotive effect, it is not clear whether these hormones have direct effects on expression of *ICL* and *MS* genes. There are no reports of their effects on *ICL* and *MS* gene expression in starvation or senescence.

Transcriptional and post-transcriptional control

There have been numerous studies demonstrating that changes in ICL and MS mRNA levels correlate with changes in the amounts of the respective proteins or enzyme activities. In general, it is apparent that these changes are brought about primarily by changes in transcription rate. Such a conclusion is based on studies employing the GUS reporter gene linked to *ICL* or *MS* gene promoter fragments from *Cucumis sativus* and *Brassica napus*, and by a relatively small number of nuclear 'run-on' experiments in *Brassica napus*.

Cucumber contains single genes encoding ICL and MS (Graham *et al*, 1989; Reynolds and Smith, 1995a). The cucumber *MS* and *ICL* gene promoters linked to GUS are expressed in the appropriate temporal and spatial pattern in seedlings of transgenic *Nicotiana plumbaginifolia* (Graham *et al*, 1990; Reynolds and Smith, 1995b; Sarah *et al*, 1996; De Bellis *et al*, 1997). Both promoters also show activation by starvation and repression by exogenous sugar in transgenic *Nicotiana* seedlings (Reynolds and Smith, 1995b; Sarah *et al*, 1996; De Bellis *et al*, 1997), electroporated cucumber protoplasts (Graham *et al*, 1994b; Reynolds and Smith, 1995b) and cucumber transgenic hairy roots (Ismail *et al*, 1997; De Bellis *et al*, 1997). Thus post-germinative expression and sugar responses of the cucumber genes are primarily regulated by transcription. Expression was not investigated in developing seeds and pollen, or in senescence, partly because the level of cucumber gene expression in transgenic Nicotiana is rather low. *Brassica napus* has several *MS* genes (Comai *et al*, 1989a) and at least six *ICL* genes (Zhang *et al*, 1993). Zhang *et al* (1993) showed that different *ICL* genes are expressed at several stages in development, implying that they are each regulated in a similar manner. Individual *Brassica napus ICL* and *MS* gene promoters linked to GUS were shown to be expressed appropriately in developing seeds, seedlings and pollen of transgenic *Brassica napus,* and in some cases *Arabidopsis thaliana* (Zhang *et al*, 1994, 1996). Expression during starvation or senescence were not investigated largely because *ICL* and *MS* gene expression is very low in *Brassica napus* and *Arabidopsis thaliana* in such situations. Collectively, these results indicate transcriptional regulation of *ICL* and *MS* expression throughout growth and development. This conclusion is confirmed for developing *Brassica napus* seeds and seedlings by 'run-on' experiments with isolated nuclei (Comai *et al*, 1989b, Comai and Harada, 1990). However, the correlation between reporter gene expression or nuclear 'run-on', and endogenous mRNA levels are not always absolute (Comai *et al*, 1989b, 1990; Reynolds and Smith, 1995b; Sarah *et al*, 1996), suggesting some post-transcriptional regulation. There is some evidence that cucumber ICL and MS mRNAs are more stable in starved cells than in sugar-replete cells (Ismail, 1997). Ettinger and Harada (1990) further show that in *Brassica napus* embryos and seedlings, MS mRNA produces relativley much more protein than does the ICL mRNA, suggesting translational or post-translation effects.

Deletion analysis and mutagenesis of promoter fragments have been employed to determine if *ICL* and *MS* gene expression throughout development is mediated by common *cis*-acting elements, and hence transcription factors. This is a powerful way of determining if there are different mechanisms of controlling expression of these genes. The clear conclusion is that there are multiple mechanisms. Distinct *cis*-acting elements in the *Brassica napus ICL* promoter have important roles in directing expression in developing embryos compared to seedlings (Zhang *et al*, 1996). Similarly, in cucumber *ICL* and *MS* promoters, the *cis*-acting elements that control expression in seedlings are distinct from those mediating the response to sugars (Graham *et al*, 1994b; Reynolds and Smith, 1995b, Sarah *et al*, 1996; De Bellis *et al*, 1997). Since *ICL* and *MS* genes are co-ordinately regulated, comparison of their promoter sequences is expected to identify conserved DNA elements which control expression. A 191 bp fragment of the *MS* promoter that is sufficient for the metabolic response in cucumber protoplasts, contains a region of about 20 bp that is conserved in the *ICL* gene, and binds to a cucumber protein *in vitro* (Graham *et al*, 1994b). While this DNA sequence may be required for the sugar response of the *MS* gene, it is not sufficient (Sarah *et al*, 1996). Further comparison of *ICL* and *MS* promoters reveals several other conserved sequences, one of which appears to be required for the sugar response of both genes (Sarah *et al*, 1996; De Bellis *et al*, 1997). A protein which binds to this sequence has been identified (M. Barrett, W. Lindsay C.J. Leaver and S.M. Smith, unpublished), opening up a way to identify components of a sugar-signalling pathway.

FUTURE RESEARCH

Although the glyoxylate cycle in plants has been under active investigation for nearly fifty years, there is still much that we do not understand about its operation, physiological role and evolution. The manipulation of all of the genes encoding glyxoylate cycle enzymes is now a realistic goal for the near future. Indeed the study of knock-out mtutants in glyoxylate cycle enzymes in *Arabidopsis thaliana* has already begun to provide some of the answers to some of the fundamental, unanswered questions. Ecophysiological and evolutionary studies are expected to follow. Futher work can also provide knowledge of key aspects of peroxisome structure and function, in particular the mechanisms of protein import, spatial arrangement of glyoxylate cycle enzymes and channelling of intermediates. Futhermore, studies of gene expression will provide a route to understanding metabolite sensing and signalling in plants. This, in turn, may help us to elucidate the true physiological role(s) of the glyoxylate cycle throughout plant development and enviromental challenges. We confidently expect another fifty years of exciting research.

ACKNOWLEDGEMENTS

We thank Luigi De Bellis for checking the accuracy of some of the information in this chapter. Not all relevant work could be cited in the space available, so we apologise to those whose research is not mentioned.

REFERENCES

Allen, R.D., Trelease, R.N. and Thomas, T.L. (1988) Regulation of isocitrate lyase gene expression in sunflower. *Plant Physiol.* **86**: 527-532.

ap Rees, T. (1990) Carbon metabolism in mitochondria. In: *Plant Metabolism.* (Dennis and Turpin, eds.). pp. 106-123. Longman, Harlow, Essex.

Armit, S. McCulloch, W. and Roberts, C.F. (1976) Analysis of acetate non-utilizing (*acu*) mutants in *Aspergillus nidulans. J. Gen. Microbiol.* **92**: 263-282.

Aubert, S., Gout, E., Bligny, R., Marty Mazars, D., Barrieu, F., Alabouvette, J., Marty, F. and Douce, R. (1996) Ultra-structural and biochemical characterization of autophagy in higher plant cells subjected to carbon deprivation: Control by the supply of mitochondria with respiratory substrates. *J. Cell Biol.* **133**: 1251-1263.

Becker, W. and Apel, K. (1993) Differences in gene expression between natural and artificially induced leaf senescence. *Planta.* **189**: 74-79.

Beevers, H. (1979) Microbodies in higher plants. *Ann. Rev. Plant Physiol.* **30**: 159-193.

Beevers, H. (1961) The metabolic production of sucrose from fat. *Nature.* **191**: 433-436.

Bentsink, L., Alonso-Blanco, C., Vreugdenhil, D., Tesnier, K., Groot, S.P. and Koorneef, M. (2000) Genetic analysis of seed-soluble oligosaccharides in relation to seed storability in Arabidopsis. *Plant Physiol.* **124**: 1595-1604.

Bewley, J.D. and Black, M. (1994) *Seeds: Physiology of develoment and germination* (2nd ed.). Plenum Press, New York.

Birkhan, R. and Kindl, H. (1990) Re-activation of the expression of glyoxysomal genes in green plant tissue. *Z. Naturforsch.* **45**: 107-111.

Bowditch, M.I. and Donaldson, R.P. (1990) Ascorbate free-radical reduction by glyoxysomal membranes. *Plant Physiol.* **94**: 531-537.

Breidenberg, R.W. and Beevers, H. (1967) Association of the glyoxylate cycle enzymes in a novel subcellular particle from castor bean endosperm. *Biochem. Biophys. Res. Comm.* **27**: 462-469.

Brouquisse R., James, F., Pradet, A. and Raymond, P. (1992) Asparagine metabolism and nitrogen distribution during protein degradation in sugar-starved maize root tips. *Planta.* **188**: 384-395.

Brouquisse, R., James, F., Raymond, P. and Pradet A. (1991) Study of glucose starvation in excised maize root tips: *Plant Physiol.* **96**: 619-626.

Carlson, M. (1998) Regulation of glucose utilization in yeast. *Curr. Opin. Genet. Dev.* **8**: 560-564.

Chapman, K.D., Turley, R.B. and Trelease, R.N. (1989) Relationship between cotton seed malate synthase aggregation behaviour and suborganellar location in glyoxysomes and endoplasmic reticulum. *Plant Physiol.* **89**: 352-359.

Chaves, R.S., Herrero, P., Ordiz, I., DelBrio, M.A. and Moreno, F. (1997) Isocitrate lyase localisation in *Saccharomyces cerevisiae* cells. *Gene.* **198**: 165-169.

Chen, H-J., Hou, W-C., Jane, W-N. and Lin, Y-H. (2000a) Isolation and characterisation of a isocitrate lyase gene from senescent leaves of sweet potato (*Ipomoea batatas* cv. Taintong 57). *J. Plant Physiol.* **157**: 669-676.

Chen, Z-H., Walker, R.P., Acheson, R.M., Tesci, L.I., Wingler, A., Lea, P.J. and Leegood, R.C. (2000b) Are isocitrate lyase and phosph*enol*pyruvate carboxykinase involved in gluconeogenesis during senescence of barley leaves and cucumber cotyledons? *Plant Cell Physiol.* **41**: 960-967.

Chia, T. and Rawsthorne, S. (2000) Fatty acid breakdown in developing embryos of *Brassica napus* L. *Biochem. Soc. Trans.* **28**: 753-754.

Choinski, J.S., Trelease, R.N. and Doman, D.C. (1981) Control of enzyme-activities in cotton cotyledons during maturation and germination. 3. *In vitro* embryo development in the presence of abscisic acid. *Planta.* **152** (5): 428-435.

Choinski, J.S. and Trelease, R.N. (1978) Control of enzyme activities in cotton cotyledons during maturation and germination. II. Glyoxysomal enzyme development in embryos. *Plant Physiol.* **62**: 141-145.

Comai, L. and Harada, J.J. (1990) Transcriptional activities in dry seed nuclei indicate the timing of the transition from embryogeny to germination. *Proc. Natl. Acad. Sci. USA.* **87**: 2671-2674.

Comai, L., Baden, C.S. and Harada, J.J. (1989a) Deduced sequence of a malate synthase polypeptide encoded by a subclass of the gene family. *J. Biol. Chem.* **264**: 2778-2782.

Comai, L., Dietrich, R.A., Maslyar, D.J., Baden, C.S. and Harada, J.J. (1989b) Co-ordinate expression of transcriptionally-regulated isocitrate lyase and malate synthase genes in *Brassica napus* L. *Plant Cell.* **1**: 293-300.

Cots, J. and Widmer, F. (1999) Germination, senescence and pathogenic attack in soybean (*Glycine max* L.): identification of the cytosolic aconitase participating in the glyoxylate cycle. *Plant Sci.* **149**: 95-104.

Courtois-Verniquet, F. and Douce, R. (1993) Lack of aconitase in glyoxysomes and peroxisomes. *Biochem. J.* **294**: 103-107.

Crafts-Brandner, S.J., Below, F.E., Harper, J.E. and Hageman, R.H. (1984) Differential senescence of maize hybrids following ear removal. 1. Whole plant. *Plant Physiol.* **74**: 360-367.

Daeschner, K., Couee, I. and Binder, S. (2001) The mitochondrial isovaleryl-coenzyme A dehyrodgenase of Arabidopsis oxidizes intermediates of leucine and valine catabolism. *Plant Physiol.* **126**: 601-612.

De Bellis L. and Nishimura, M. (1991) Development of enzymes of the glyoxylate cycle during senescence of pumpkin cotyledons. *Plant Cell Physiol.* **32**: 555-561.

De Bellis, L., Hayashi, M., Nishimura, M. and Alpi, A. (1995) Subcellular and developmental changes in distribution of aconitase isoforms in pumpkin cotyledons. *Planta.* **195**: 464-468.

De Bellis L., Hayashi, M., Biagi, P.P., Haranishimura, I., Alpi, A. and Nishimura, M. (1994) Immunological analysis of aconitase in pumpkin cotyledons – the absence of aconitase in glyoxysomes. *Physiol. Plantarum.* **90**: 757-762.

De Bellis, L., Ismail, I., Reynolds, S. J., Barrett, M.D. and Smith, S. M. (1997) Distinct *cis*-acting sequences are required for the germination and sugar responses of the cucumber isocitrate lyase gene. *Gene.* **197**: 375-378.

De Bellis L., Picciarelli, P., Pistelli, L. and Alpi, A. (1990) Localisation of glyoxylate-cycle marker enzymes in peroxisomes of senescent leaves and green cotyledons. *Planta.* **180**: 435-439.

De Bellis, L., Tsugeki, R., Alpi, A and Nishimura, M. (1993) Purification and characterization of aconitase isoforms from etiolated pumpkin cotyledons. *Physiol. Plantarum.* **88**: 485-492.

De Bellis, L., Tsugeki, R., Nishimura, M. (1991) Glyoxylate cycle enzymes in peroxisomes isolated from petals of pumpkin (*Cucurbita* sp) during senescence. *Plant Cell Physiol.* **32**: 1227-1235.

Dieuaide, M., Brouquisse, R., Pradet, A. and Raymond, P. (1992) Increased fatty acid β-oxidation after glucose starvation in maize root tips. *Plant Physiol.* **99**: 595-600.

Dieuaide, M., Couée, I., Pradet, A. and Raymond P. (1993) Effects of glucose starvation on the oxidation of fatty acids by maize root tip mitochondria and peroxisomes: evidence for mitochondrial fatty acid β-oxidation and acyl-CoA dehydrogenase activity in a higher plant. *Biochem J.* **296**: 199-207.

Doig, R.I., Colborne, A.J., Morris, G. and Laidman, D.L. (1975) The induction of glyoxysomal enzyme activities in the aleurone cells of germinating wheat. *J. Exp. Bot.* **26**: 387-398.

Dommes, J. and Northcote, D.H. (1985) The action of exogenous abscisic and gibberellic acids on gene-expression in germinating castor beans. *Planta.* **165**: 513-521.

Donaldson, R.P. (1982) Nicotinamide cofactors (NAD and NADP) in glyxoxsomes, mitochondria and plastids isolated from castor bean endosperm. *Arch. Biochem. Biophys.* **215**: 274-279.

Donaldson, R.P. and Fang, T.K. (1987) β-oxidation and glyoxylate cycle coupled to NADH – cytochrome c and ferricyanide reductases in glyoxysomes. *Plant Physiol.* **85**: 792-795.

Eastmond, P.J. and Graham, I.A. (2001) Re-examining the role of the glyoxylate cycle in oilseeds. *Trends Plant Sci.* **6**: 72-77.

Eastmond, P., Germain, V., Lange, P., Bryce, J.H., Smith, S.M. and Graham, I.A. (2000) Post-germinative growth and lipid catabolism in oilseeds lacking the glyoxylate cycle. *Proc. Natl. Acad. Sci. USA.* **97**: 5669-5674.

Eastmond, P.J. and Rawsthorne, S. (2000) Co-ordinate changes in carbon partitioning and plastidial metabolism during the development of oilseed rape embryo. *Plant Physiol.* **122**: 767-774.

Eccleston, V.S. and Ohlrogge, J.B. (1998) Expression of lauroyl-acyl carrier protein thioesterase in *Brassica napus* seeds induces pathways for both fatty acid oxidation and biosynthesis and implies a set point for triacylglycerol accumulation. *Plant Cell.* **10**: 613-621.

Escher, C.L. and Widmer, F. (1997) Lipid mobilization and gluconeogenesis in plants: Do glyoxylate cycle enzyme activities constitute a real cycle? A hypothesis. *Biol. Chem.* **378**: 803-813.

Ettinger W.F and Harada, J.J. (1990) Translational or post-translational processes affect differentially the accumulation of isocitrate lyase and malate synthase proteins and enzyme-activities in embryos and seedlings of *Brassica napus. Arch. Biochem. Biophys.* **281**: 139-143.

Finnessy, J.J., Trelease, R.N. and Randall, D.D. (1994) Glyoxysomal isocitrate lyase is a phosphoprotein. *Plant Physiol.* **105**: supp. p92.

Flavell, R.B. and Fincham, J.R.S. (1968) Acetate non-utilizing mutants of *Neurospora crassa* L. Mutant isolation, complementation studeis and linakes relationships. *J. Bacteriol.* **95**: 1056-1062.

Frevert. J., Koller, W. and Kindl, H. (1980) Occurrence and biosynthesis of glyoxysomal enzymes in ripening cucumber seeds. *Z. Physiol. Chem.* **361**: 1557-1565.

Fusseder, A. and Theimer, R.R. (1984) Lipolytic and glyoxysomal enzyme-activities in cotyledons of ripening and germinating sunflower (*Helianthus Annuus* L) seeds. *Z. Pflanzenphysiol.* **114**: 403-411.

Galvez, S. and Gadal, P. (1995) On the function of the NADP-dependent isocitrate dehydrogenase isozymes in living organisms. *Plant Sci.* **105**: 1-14.

Gan, S. and Amasino, R.M. (1997) Making sense of senscence. *Plant Physiol.* **113**: 313-319.

Genix, P., Bligny, R., Martin, J-B. and Douce, R. (1990) Transient accumulation of asparagine in sycamore cells after a long period of sucrose starvation. *Plant Physiol.* **94**: 717-722.

Germain, V., Footitt, S., Dieuaide-Noubhani M., Raymond, P., Renaudin J-P., Bryce, J.H. and Smith, S.M. (2000) Role of malate synthase and the glyoxylate cycle in oilseed plants. *Plant Mol. Biol. Rep.* **18**: S20-26.

Germain, V., Rylott, E.L., Larson, T.R., Sherson, S.M., Bechtold, N., Carde, J-P., Bryce, J.H., Graham, I.A. and Smith, S.M. (2001) Requirement for 3-ketoacyl-CoA thiolase-2 in peroxisome development, fatty acid β-oxidation and breakdown of triacylglycerol in lipid bodies of Arabidopsis seedlings. *Plant J.* **27**: in press.

Gibson, S.I. (2000) Plant sugar-response pathways. Part of a complex regulatory web. *Plant Physiol.* **124**: 1532-539.

Gietl, C. (1992) Partitioning of malate-dehydrogenase isoenzymes into glyoxysomes, mitochondria and chloroplasts. *Plant Physiol.* **100**: 557-559.

Godavari, H.R., Badour, S.S. and Waygood, E.R. (1973) Isocitrate lyase in green leaves. *Plant Physiol.* **51**: 863-867.

Graham I. A., Smith, L. M., Brown, J. W. S., Leaver, C. J. and Smith, S. M. (1989) The malate synthase gene of cucumber. *Plant Mol. Biol.* **13**: 673-684.

Graham, I. A., Smith, L. M., Leaver, C. J. and Smith, S. M. (1990) Developmentally-regulated expression of the malate synthase gene in transgenic plants. *Plant Mol. Biol.* **15**: 539-549.

Graham, I. A., Leaver, C. J. and Smith, S. M. (1992) Induction of malate synthase gene expression in senescent and detached organs of cucumber. *Plant Cell.* **4**: 349-357.

Graham, I.A., Denby, K.J. and Leaver, C.J. (1994a) Carbon catabolite repression regulates glyoxylate cycle gene expression in cucumber. *Plant Cell.* **6**: 761-772.

Graham, I.A., Baker, C.J. and Leaver, C.J. (1994b) Analysis of the cucumber malate synthase gene promoter by transient expression and gel retardation assays. *Plant J.* **6**: 893-902.

Guex, N., Henry, H., Flach, J., Richter, H. and Widmer, F. (1995) Glyoxysomal malate-dehydrogenase and malate synthase from soybean cotyledons (*Glycine max* L.): enzyme association, antibody production and cDNA cloning. *Planta.* **197**: 369-375.

Gut, H. and Matile, P. (1988) Apparent induction of key enzymes of the glyoxylic-acid cycle in senescent barley leaves. *Planta.* **176**: 548-550.

Gut, H. and Matile, P. (1989) Breakdown of galactolipids in senescent barley leaves. *Bot. Acta.* **102**: 31-36.

Halford, N.G., Purcell, P.C. and Hardie, D.G. (1999) Is hexokinase really a sugar sensor in plants? *Trends Plant Sci.* **4**: 117-120.

Hayashi, M., Aoki, M., Kato, A., Kondo, M. and Nishimura, M. (1996) Transport of chimaeric proteins that contain a carboxy-terminal targeting signal into plant microbodies. *Plant J.* **10**: 225-234.

Hayashi, M., DeBellis, L., Alpi, A. and Nishimura, M. (1995) Cytosolic aconitase participates in the glyoxylate cycle in etiolated pumpkin cotyledons. *Plant Cell Physiol.* **36**: 669-680.

Heupel, R. and Heldt, H.W. (1994) Protein organization in the matrix of leaf peroxisomes – A multi-enzyme complex involved in photorespiratory metabolism. *Eur. J. Biochem.* **220**: 165-172.

Hicks, D.B. and Donalson, R.P. (1982) Electron transport in glyoxysomal membranes. *Arch. Biochem. Biophys.* **215**: 280-288.

Holtman, W.L., Heistek, J.C., Mattern, K.A., Bakhuizen, R. and Douma, A.C. (1994) β-oxidation of fatty acids is linked to the glyoxylate cycle in the aleurone, but not in the embryo of germinating barley. *Plant Sci.* **99**: 43-53.

Hrazdina, G. and Jensen, R.A. (1992) Spatial-organization of enzymes in plant metabolic pathways. *Ann. Rev. Plant. Phys.* **43**: 241-267.

Huang, A.H.C., Trelease, R.N. and Moore, T.S. (eds.) (1983) *Plant peroxisomes.* Academic Press, New York.

Hunt, L. and Fletcher, J.S. (1977) Intracellular location of isocitrate lyase in leaf tissue. *Plant Sci. Lett.* **10**: 243-247.

Hunt, L., Skvarla, J.J. and Fletcher, J.S. (1978) Subcellular localisation of isocitrate lyase in non-green tissue culture cells. *Plant Physiol.* **61**: 1010-1013.

Ihle, J.N. and Dure, L.S. (1972) The developmental biochemistry of cottonseed embryogenesis and germination. III Regulation of the biosynthesis of enzymes utilised in germination. *J. Biol. Chem.* **247**: 5048-5055.

Ismail, I. (1997) Sugar regulation of malate synthase and isocitrate lyase gene expression in cucumber. PhD Thesis, University of Edinburgh, UK.

Ismail, I., De Bellis, L., Alpi, A. and Smith, S.M. (1997) Expression of glyoxylate cycle genes in roots responds to carbohydrate supply and can be activated by shading or defoliation of the shoot. *Plant Mol. Biol.* **35**: 633-640.

Jang J.C. and Sheen, J. (1997) Sugar sensing in higher plants. *Trends Plant Sci.* **2**: 208-214.

Jang, J.C. and Sheen, J. (1994) Sugar sensing in higher plants. *Plant Cell.* **6**: 1665-1679.

Janssen, B-J. (1995) A cDNA clone for isocitrate lyase from tomato. *Plant Physiol.* **108**: 1339.

Johnston, M. (1999) Feasting, fasting and fermenting – glucose sensing in yeast and other cells. *Trends Genet.* **15**: 29-33.

Journet, E.P., Bligny, R. and Douce, R. (1986) Biochemical changes during sucrose deprivation in higher plant cells. *J. Biol. Chem.* **261**: 3193-3199.

Khan, A.S., Van Driessche, E., Kanarek, L. and Beeckmans, S. (1992) The purification and physicochemical characteristics of maize (*Zea mays* L.) isocitrate lyase. *Arch. Biochem. Biophys.* **297**: 9-18.

Kim, D-J. and Smith, S. M. (1994a) Expression of a single gene encoding microbody NAD-malate dehydrogenase during glyoxysome and peroxisome development in cucumber cotyledons. *Plant Mol. Biol.* **26**: 1833-1841.

Kim, D-J. and Smith, S. M. (1994b) Molecular cloning of cucumber phospho*enol*pyruvate carboxykinase and developmental regulation of gene expression. *Plant Mol. Biol.* **26**: 423-434.

Kim. K-W., Nagai, K., Mukaida, H., Tezuka, T. and Yamamoto, Y. (1988) Organic-acid metabolism in cellular organelles of *Vigna cylindrica* (mitorisasage) cotyledons. *Plant Cell Physiol.* **29**: 51-59.

King, H.B.and Casselton, L.A. (1977) Genetics and function of isocitrate lyase in *Coprinus*. *Mol. Gen. Genet.* **157**: 319-325.

Köller, W. and Kindl, H. (1977) Glyoxylate cycle enzymes of the glyoxysomal membrane from cucumber cotyledons. *Arch. Biochem. Biophys.* **181**: 236-248.

Kornberg, H.L. and Beevers, H. (1957) The glyoxylate cycle as a stage in the conversion of fat to carbohydrate in castor beans. *Biochim. Biocphys. Acta.* **26**: 531-537.

Kornberg, H.L. and Krebs, H.A. (1957) Synthesis of cell constituents from C2-units by a modified tricarboxylic acid cycle. *Nature.* **179**: 988-991.

Koyama, H., Kawamura, A., Kihara, T., Hara, T., Takita, E. and Shibata, D. (2000) Overexpression of mitochondrial citrate synthase in *Arabidopsis thaliana* improved growth on a phosphorous-limited soil. *Plant Cell Physiol.* **41**, 1030-1037.

Kruse and Kindl, H. (1983) Malate synthase – aggregation, de-aggregation, and binding of phospholipids. *Arch. Biochem. Biophys.* **233**: 618-628.

Kudielka, R.A. and Theimer, R.R. (1983a) De-repression of glyoxylate cycle enzyme activities in anise suspension culture cells. *Plant Sci. Lett.* **31**: 237-244.

Kudielka, R.A. and Theimer, R.R. (1983b) Repression of glyoxysomal enzyme-activities in anise (*Pimpinella anisum* L.) suspension-cultures. *Plant Sci. Lett.* **31**: 245-252.

Lado, P., Schwendimann, M. and Marré, E. (1968) Repression of isocitrate lyase synthesis in seeds germinated in the presence of glucose. *Biochim. Biophys. Acta.* **157**: 140-148.

Landolt, R. and Matile, P. (1990) Glyoxysome-like microbodies in senescent spinach leaves. *Plant Sci.* **72**: 159-163.

Lee, M.S., Mullen, R.T. and Trelease, R.N. (1997) Oilseed isocitrate lyase lacking their essential type I peroxisomal targeting signal are piggybacked to glyoxysomes. *Plant Cell.* **9**: 185-197

Lee, S.H., Chae, H.S., Lee, T.K., Kim, S.H., Shin, S.H., Cho, B.H., Cho, S.H., Kang, B.G. and Lee, W.S. (1998) Ethylene mediated phospholipid catabolic pathway in glucose-starved carrot suspension cells. *Plant Physiol.* **116**: 223-229.

Lee, T.K. and Lee, W.S. (1996) Diauxic growth in rice suspension cells grown on mixed carbon sources of acetate and glucose. *Plant Physiol.* **110**: 465-470.

Longo, C.P. and Longo G.P. (1970) The development of glyoxysomes in peanut cotyledons and maize scutella. *Plant Physiol.* **45**: 249-254.

Longo, C.P., Bernasconi, E. and Longo, G.P. (1975) Solubilization of enzymes from glyoxysomes of maize scutellum. *Plant Physiol.* **55**: 1115-1119.

Lopez-Boado,Y.S., Herrero, P., Fernandez, T., Fernandez, R. and Moreno, F. (1988) Glucose-stimulated phosphorylation of yeast isocitrate lyase i*n vivo. J. Gen. Micobiol.* **134**: 2499-2505.

Lord and Beevers. H. (1972) The problem of reduced nicotinamide adenine dinucleotide oxidation in glyoxysomes. *Plant Physiol.* **49**: 249-251.

Luster, D.G. and Donaldson, R.P. (1987) Glyoxysomal membrane electron transport proteins. In: *Peroxisomes in biology and medicine.* (Fahimi, D.D. and Sies, H. eds.). pp. 189-193. Spinger-Verlag, Heidelberg.

Maloy, S.R. and Nunn, W.D. (1982) Genetic regulation of the glyoxylate shunt in *Escherichia coli* K-12. *J. Bacteriol.* **149**: 173-180.

Marriott, K.M. and Northcote, D.H. (1977) The influence of abscisic acid, adensosine 3', 5' cyclic phosphate and giberellic acid on the induction of isocitrate lyase activity in the endosperm of germinating castor bean seeds. *J. Exp. Bot.* **28**: 219-224.

Marriott, K.M. and Northcote, D.H. (1975) The induuction of enzyme activity n the endosperm of germinating castor bean seeds. *Biochem. J.* **152**: 65-70.

Martin, C. and Northcote, D.H. (1982) The action of exogenous gibberellic-acid on isocitrate lyase-messenger RNA in germinating castor bean-seeds. *Planta.* **154**: 174-183.

Martin, C., Beeching, J.R. and Northcote, D.H. (1984) Changes in levels of transcripts in endosperms of castor beans treated with exogenous gibberellic acid. *Planta.* **162**: 68-76.

Masterson, C. and Wood, C. (2000) Mitochondrial β-oxidation of fatty acids in higher plants. *Physiol. Plantarum.* **109**: 217-224.

McKinley, M.P., Field, L.A. and Trelease, R.N. (1979) Multiple forms of isocitrate lyasae in the matirx of *Turbatrix aceti* mitochondria. *Arch. Biochem. Biophys.* **197**: 253-263.

McLaughlin J.C. and Smith, S.M. (1995) Glyoxylate cycle enzyme synthesis during the irreversible phase of senescence of cucumber cotyledons. *J. Plant Physiol.* **146**: 133-138.

McLaughlin, J.C. and Smith, S.M. (1994) Metabolic regulation of glyoxylate cycle enzyme synthesis in detached cucumber cotyledons and protoplasts. *Planta.* **195**: 22-28.

Mettler and Beevers, H. (1980) Oxidation of NADH in glyoxysomes by a malate-aspartate shuttle. *Plant Physiol.* **66**: 555-560.

Miernyk, J.A. and Trelease, R.N. (1981) Role of malate synthase in citric acid synthesis by maturing cotton embryos: a proposal. *Plant Physiol.* **67**: 875-881.

Miernyk, J.A., Trelease, R.N. and Choinski, J.S. (1979) Malate synthase activity in cotton and other ungerminated oilseeds. *Plant Physiol.* **63**: 1060-1071.

Minard, K.I. and McAlister-Henn L. (1991) Isolation, nucleotide-sequence analysis, and disruption of the MDH2 gene from *Saccharomyces cerevisiae* – evidence for three isozymes of yeast malate-dehydrogenase. *Mol. Cell Biol.* **11**: 370-380.

Mori, H. and Nishimura, M. (1989) Glyoxysomal malate synthase is specifically degraded in microbodies during greening of pumpkin cotyledons. *FEBS Lett.* **244**: 163-166.

Mullen, R.T. and Gifford, D.J. (1997) Regulation of two loblolly pine (*Pinus taeda* L) isocitrate lyase genes in megagametophytes of mature and stratified seeds and during post-germinative growth. *Plant Mol. Biol.* **33**: 593-604.

Navarre, D.A., Wendehenne, D., Durner, J., Noad, R. and Klessig, D.F. (2000) Nitric oxide modulates the activity of tobacco aconitase. *Plant Physiol.* **122**: 573-582.

Newman, J.C. and Briggs, D.E. (1976) Glyceride metabolism and gluconeogenesis in barley endosperm. *Phytochem.* **15**: 1453-1458.

Nieri, B, Ciurli, A., Pistelli, L., Smith, S.M., Alpi, A and De Bellis, L. (1997) Glyoxylate cycle enzymes in seedlings and mature plants of tomato. *Plant Sci.* **129**: 39-47.

Nieri, B., DeBellis, L., Biagi P.P. and Alpi, A. (1995) NADP(+)-isocitrate dehydrogenase in germinating and senescing pumpkin cotyledons. *Physiol. Plant.* **94**: 351-355.

Nishimura, M., Takeuchi, Y., DeBellis, L. and Haranishimura, I. (1993) Leaf peroxisomes are directly transformed to glyoxysomes during senescence of pumpkin cotyledons. *Protoplasma.***175**: 131-137.

Nishimura, M., Yamaguchi, J., Mori, H., Akazawa, T. and Yokota, S. (1986) Immunocytochemical analysis shows that glyoxysomes are directly transformed to leaf peroxisomes during greening of pumpkin cotyledons. *Plant Physiol.* **80**: 313-316.

Noodén, L.D., Guiamet, J.J. and John, I. (1997) Senescence mechanisms. *Physiol. Plant.* **101**: 746-753.

Olsen, L.J., Ettinger, W.F., Damsz, B., Matsudaira, K., Webb, M.A. and Harada, J.J. (1993) Targeting of glyoxysomal proteins to peroxisomes in leaves and roots of a higher plant. *Plant Cell.* **5**: 941-952.

Onyeocha, I., Behari, R., Hill, D. and Baker, A. (1993) Targeting of castor bean glyoxysomal isocitrate lyase to tobacco leaf peroxisomes. *Plant Mol. Biol.* **22**: 385-396.

Pacy, J. and Thurston, C.F. (1987) Distribution of isocitrate lyase in induced and de-adapting cells of the green alga *Chlorella fusca*. *J. Gen. Microbiol.* **133**: 3341-3346.

Paek, N.C., Lee, B.M., Bai, D.G and Smith, D.G. (1998) Inhibition of germination gene expression by viviparous-1 and ABA during maize kernel development. *Mol. Cells.* **8**: 336-342.

Pastori, G.M., del Rio, L.A. (1994) An activated-oxygen-mediated role for peroxisomes in the mechanism of senescence of *Pisum sativum* L. leaves. *Planta.* **193**: 385-391.

Peyret, P., Perez, P. and Alric, M. (1995) Structure, genomic organization, and expression of the *Arabidopsis thaliana* aconitase gene – plant aconitase show significant homology with mammalian iron-responsive element-binding protein. *J. Biol. Chem.* **270**: 8131-8137.

Pistelli, L., De Bellis, L. and Alpi, A. (1995) Evidences of glyoxylate cycle in peroxisomes of senescent cotyledons. *Plant Sci.* **109**: 13-21.

Pistelli, L., De Bellis, L. and Alpi, A. (1991) Peroxisomal enzyme activities in attached senescing leaves. *Planta.* **184**: 151-153.

Pistelli, L., Nieri, B., Smith, S. M., Alpi, A. and De Bellis, L. 1996. Glyoxylate cycle enzyme activities are induced in senescent pumpkin fruits. *Plant Sci.* **119**: 23-29.

Pistelli, L., Perata, P. and Alpi, A. (1992) Effect of leaf senescence on glyoxylate cycle enzyme activities. *Aust. J. Plant Physiol.* **19**: 723-729.

Poirier, Y., Ventre, G. and Caldelari, D. (1999) Increased flow of fatty acids toward β-oxidation in developing seeds of Arabidopsis deficient in diacylglycerol acyltransferase activity or synthesizing medium-chain-length fatty acids. *Plant Physiol.* **121**: 1359-1366.

Rai, R., Daugherty, J.R., Cunningham, T.S. and Cooper, T.G. (1999) Overlapping positive and negative GATA factor binding sites mediate inducible DAL7 gene expression in *Saccharomyces cerevisiae*. *J. Biol. Chem.* **274**: 28026-28034.

Raymond, P., Carre-Nemesio, A.M. and Pradet, A. (1985) Metabolism of [^{14}C]-glucose and [^{14}C]-acetate by lettuce embryos during early germination. *Physiol. Plant.* **64**: 529-534.

Raymond, P., Spiteri, A., Dieuaide, M., Gerhardt, B. and Pradet, A. (1992) Peroxisomal β-oxidation of fatty acids and citrate formation by a particulate fraction from early germinating sunflower seeds. *Plant Physiol. Biochem.* **30**: 153-161.

Reumann, S. Heupel, R. and Heldt, H.W. (1994) Compartmentation studies on spinach leaf peroxisomes. 2. Evidence for the transfer of reductant from the cytosol to the peroxisomal compartment via a malate shuttle. *Planta.* **193**: 167-173.

Reynolds, S.J. and Smith S.M. (1995a) The isocitrate lyase gene of cucumber: Isolation, characterisation and expression in cotyledons following seed germination. *Plant Mol. Biol.* **27**: 487-497.

Reynolds S.J. and Smith, S.M. (1995b) Regulation of expression of the cucumber isocitrate lyase gene in cotyledons upon seed germination and by sucrose. *Plant Mol. Biol.* **29**: 885-896.

Richmond, T.A. and Bleecker, A.B. (1999) A defect in β-oxidation causes abnormal inflorescence development in Arabidopsis. *Plant Cell.* **11**: 1911-1923.

Robertson, E.F., Hoyt, J.C. and Reeves, H.C. (1988) Evidence of histidine phosphorylation in isocitrate lyase from *Escherichia coli. J. Biol. Chem.* **263**: 2477-2482.

Roby, C., Martin, J.B., Bligny, R. and Douce, R. (1987) Biochemical changes during sucrose deprivation in higher plant cells – p-31 nuclear magnetic resonance studies. *J. Biol. Chem.* **262**: 5000-5007.

Rodriguez, D., Dommes, J. and Northcote, D.H. (1987) Effect of abscisic and gibberellic acids on malate synthase transcripts in germinating castor bean seeds. *Plant Mol. Biol.* **9**: 227-235.

Runquist, M. and Kruger, N.J. (1999) Control of gluconeogenesis by isocitrate lyase in endosperm of germinating castor bean seedlings. *Plant J.* **19**: 423-431.

Salon, C., Raymond, P. and Pradet, A. (1988) Quantification of carbon fluxes through the tricarboxylic acid cycle in early germinating lettuce embryos. *J. Biol. Chem.* **263**: 12278-12287.

Sarah, C.J., Graham, I.A., Reynolds, S.J., Leaver, C.J. and Smith, S.M. (1996) Distinct *cis*-acting elements direct the germination and sugar responses of the cucumber malate synthase gene. *Mol. Gen. Genet.* **250**: 153-161.

Sautter, C. (1986) Microbody transition in greening watermelon cotyledons double immunocytochemical labelling of isocitrate lyase and hydroxypyruvate reductase. *Planta.* **167**: 491-503.

Sheen, J., Zhou, L. and Jang, J.C. (1999) Sugars as signalling molecules. *Curr. Opin. Plant Biol.* **2**: 410-418.

Skadsen, R.W. and Scandalios, J.G. (1989) Pre-translational control of the levels of glyoxysomal protein gene expression by the embryonic axis in maize. *Dev. Genet.* **10**: 1-10.

Smeekens, S. (2000) Sugar-induced signal transduction in plants. *Ann. Rev. Plant Phys. Plant Mol. Biol.* **51**: 49-81.

Taylor, K.M., Kaplan, C.P., Gao, X.P. and Baker, A. (1996) Localization and targeting of isocitrate lyases in *Saccharomyces cerevisiae. Biochem. J.* **319**: 255-262.

The Arabidopsis Genome Initiative (2000) Analysis of the genome sequence of the flowering plant *Arabidopsis thaliana. Nature.* **408**: 796-815.

Theimer, R.R. (1976) A specific inactivator of glyoxysomal isocitrate lyase from sunflower (*Helianthus annus* L.) cotyledons. *FEBS Lett.* **62**: 292-300.

Thomas, S.M. and ap Rees, T. (1971) Glycolysis during gluconeogenesis in cotyledons of *Cucurbita pepo. Phytochem.* **11**: 2187-2194.

Thorneycroft, D., Sherson, S.M. and Smith, S. M. (2001) Using gene knock-outs to investigate plant metabolism. *J. Exp. Bot.* **52**: 1593-1601.

Titus, D.E. and Becker, W.M. (1985) Investigation of the glyoxysome peroxisome transition in germinating cucumber cotyledons using double-label immunoelectron microscopy. *J. Cell Biol.* **101**: 1288-1299.

Trelease, R.N. (1984) Biogenesis of glyoxysomes. *Ann. Rev. Plant Phys.* **35**: 321-347.

Turley, R.B. and Trelease, R.N. (1990) Development and regulation of three glyoxylate cycle enzymes during cotton seed maturation and growth. *Plant Mol. Biol.* **14**: 137-146.

Van Roermund, C.W.T., Hettema, E.H., van den Berg, M., Tabak, H.F. and Wanders, R.J.A. (1999) Molecular characterisation of carnitine-dependent transport of acetyl-CoA from peroxisomes to mitochondria in *Saccharamyces cerevisiae* and identificaiton of a plasma membrane carnitine transporter, Agp2p. *EMBO J.* **18**: 5843-5852.

Van Roermund, C.W.T., Elgersma, Y., Singh, N., Wanders, R.J.A. and Tabak, H.F. (1995) The membrane of peroxisomes in *Saccharomyces cerevisiae* is impermeable to NAD(H) and acetyl-CoA under *in vivo* conditions. *EMBO J.* **14**: 3480-3486.

Verniquet, F., Gaillard, J., Neuburger, M. and Douce, R. (1991) Rapid inactivation of plant aconitase by hydrogen-peroxide. *Biochem. J.* **276**: 643-648.

Vicentini, F. and Matile, P. (1993) Gerontosomes, a multifunctional type of peroxisome in senescent leaves. *J. Plant Physiol.* **142**: 50-56.

Wanner, L., Keller, F. and Matile, P. (1991) Metabolism of radiolabelled galactolipids in senescent barley leaves. *Plant Sci.* **78**: 199-206.

Welch, G.R. and Easterby, J.S. (1994) Metabolic channelling versus free diffusion – transition-time analysis. *Trends Biochem. Sci.* **19**: 193-197.

Wingler, A., von Schaewen, A., Leegood, R.C., Lea, P.J. and Quick, W.P. (1998) Regulation of leaf senescence by cytokinin, sugars, and light – Effects on NADH-dependent hydroxypyruvate reductase. *Plant Physiol.* **116**: 329-335.

Yang, Y-P., Randall, D.D. and Trelease, R.N. (1988) Phosphorylation of glyoxysomal malate synthase from castor oil seeds *Ricinus communis* L. *FEBS Lett.* **234**: 275-279.

Yu, S-M. (1999). Cellular and genetic responses of plants to sugar starvation. *Plant Physiol.* **121**: 687-693.

Zelitch, I. (1988) Synthesis of glycolate from pyruvate via isocitrate lyase by tobacco leaves in light. *Plant Physiol.* **86**: 463-468.

Zhang, J.Z., Gomezpedrozo, M., Baden, C.S. and Harada, J.J. (1993) Two classes of isocitrate lyase genes are expressed during late embryogeny and post-germination in *Brassica napus* L. *Mol. Gen Genet.* **238**: 177-184.

Zhang, J.Z., Laudencia-Chingcuanco, D.L., Comai, L., Li, M. and Harada, J.J. (1994) Isocitrate lyase and malate synthase genes from *Brassica napus* L. are active in pollen. *Plant Physiol.* **104**: 857-864.

Zhang, J.Z., Santes, C.M., Engel, M.L., Gasser, C.S. and Harada, J.J. (1996) DNA sequences that activate isocitrate lyase gene expression during late embryogenesis and during post-germinative growth. *Plant Physiol.* **110**: 1069-1079.

4

PLANT CATALASES

Michael Heinze and Bernt Gerhardt

Westfälische Wilhelms-Universität, Münster, Germany

KEYWORDS

Catalase, core, isoforms, subunits, peroxisome, import, photoinactivation, protein purification, protein biogenesis, expression, structure.

INTRODUCTION

The main objectives of this article are to survey current advances about selected aspects of biochemistry, molecular biology and physiological role of catalases in higher plants. The reader is also referred to other recent reviews dealing with specific aspects of catalases (Willekens *et al*, 1995; Scandalios *et al*, 1997; Zámocký and Koller, 1999; Nicholls *et al*, 2001). Catalase is a very effective enzyme with a high turnover number but a weak affinity for its substrate hydrogen peroxide. Catalase works as a single enzyme in plants at high hydrogen peroxide concentrations or in co-operation with other antioxidative enzymes to prevent the generation of reactive oxygen species. Therefore, the biological function of catalases becomes more and more complex. Catalases work in a bifunctional mode (see later). It is also necessary to distinguish between monofunctional catalases (typical catalases) and catalase-peroxidases like ascorbate peroxidases in plants which participate in a network system of hydrogen peroxide metabolizing enzymes. The substrate hydrogen peroxide is proposed to act as a signal molecule in cellular signal transduction and also influences processes like plant pathogen response. In this chapter, we focus on catalase acting in non-

A. Baker and I.A. Graham (eds.), Plant Peroxisomes, 103–140.

stressed plant cells. Catalases of algae will not be included into this review and data resulting from studies on catalases of heterotrophic organisms are only integrated when needed in context. The catalases (EC 1.11.1.6) covered in this chapter belong to the typical, monofunctional tetrameric haem-containing catalases (Zámocký and Koller, 1999; Nicholls *et al*, 2001). One exception will be the peroxisomal ascorbate peroxidases, treated at the end of the chapter. The frontier between monofunctional catalases and catalase-peroxidases is not quite obvious in literature (Zámocký and Koller, 1999).

MONOFUNCTIONAL HAEM-CONTAINING CATALASES

Monofunctional haem-containing catalases, also denoted typical catalases, are the largest group of plant catalases. They are tetrameric enzymes with one haem group per subunit, containing iron in the centre of the haem. They differ from catalase-peroxidases, which are mainly dimeric.

Subcellular localization

When examined, catalase activity has been demonstrated in plant tissues. Within the plant cell, catalase is localized in peroxisomes and the enzyme is widely used as marker for this cell organelle (Huang *et al*, 1983). Catalase comprises as much as 10-25% of the total peroxisomal protein (Tolbert, 1980). Plant peroxisomes occur in three function types; glyoxysomes, leaf peroxisomes and non-specialized peroxisomes. Catalase proteins in higher plants have been demonstrated for all three peroxisomal function types. Peroxisomal catalases have been found in various tissues, e.g. in castor bean endosperm, spinach leaves, soybean nodules, sunflower cotyledons and potato tubers (Huang *et al*, 1983; Tenberge *et al*, 1997). The peroxisomal localization of catalase has been demonstrated biochemically following subcellular fractionation, as well as by electron microscopical detection using staining techniques with 3,3'-diaminobenzidine or immunogold labelling. The staining of catalase with 3,3'-diaminobenzidine depends on the peroxidatic activity of the enzyme and is sensitive to the catalase inhibitor 3-amino-1,2,4-triazole (Novikoff and Goldfischer, 1969). This activity may not be displayed by all catalase isoforms. Regarding the localization of catalase within the peroxisome, the enzyme is a component of the organelle matrix as well as of the electron-dense inclusion(s) of the organelle matrix if the peroxisome contains such a sub-compartment. Enrichment of catalase in the crystalline inclusions (cores) in comparison to the organelle matrix has been demonstrated by immunogold labelling for the peroxisomes of sunflower cotyledons (Tenberge and Eising, 1995) and has been suggested based on catalase staining with 3,3'-diaminobenzidine for the peroxisomes of other plant tissues. On the other hand, catalase enrichment in cores could not be demonstrated biochemically (Huang and Beevers, 1973). This apparently contradictory result can be explained today by the distinct difference in enzymatic activity between the catalases localized in cores and matrix (see below).

There are a few reports on (additionally) extraperoxisomal localization of catalase in plant tissues. The enzyme has been purified from PS II membranes prepared from spinach leaves, and biochemically characterized (Sheptovitsky and Brudvig, 1996). The chloroplast catalase is a dimer with a molecular weight of 63 kD for the subunits and a native molecular weight of about 130 kD. The pH optimum is at pH 8.2.

Catalase has been detected also in the apoplast, e.g. in xylem walls of sunflower cotyledons by immunoelectron and immunofluorescence microscopy (Eising *et al*, 1998), in isolated cell wall fractions from horseradish by activity measurements (O_2 evolution from H_2O_2; Elstner and Heupel, 1976), and in the apoplast of maize roots on the basis of H_2O_2 decomposition detected by a staining method (Salguero and Böttger, 1995). Plant cell wall catalases may be involved in the destruction of hydrogen peroxide arising during the lignification process (Eising *et al*, 1998, Bestwick *et al*, 1997). Based on cell fractionation studies, extra-peroxisomal localization of catalase has been reported to occur in mitochondria and the cytosol of germinating maize seedlings (Scandalios *et al*, 1997, Scandalios, 1974, Scandalios *et al*, 1980). However, cell fractionation studies are a rather unsuitable method to draw secure conclusions about cytosolic localization of an enzyme, particularly when it is housed in a very fragile organelle like the peroxisome.

Biochemistry

Basic reaction mechanism

The knowledge of the reaction mechanism of catalase results preferentially from studies on mammalian and bacterial catalases (Nicholls *et al*, 2001; Schonbaum and Chance, 1976; Dounce, 1983; Jouve *et al*, 1997). Catalase is a tetrameric, H_2O_2-degrading enzyme containing one haem (ferric protoporphyrine IX in most cases) per subunit. The enzyme catalyzes the dismutation of hydrogen peroxide to water and molecular oxygen (catalytic reaction) as well as the reduction of hydrogen peroxide to water using a variety of two- and one-electron donors (peroxidatic activity, Sichak and Dounce, 1986). The peroxidatic activity of the typical catalases is generally minor in comparison to the catalytic activity and the activity of actual peroxidases.

Both overall reactions catalyzed by catalase involve firstly a two electron oxidation of the resting enzyme (ferric catalase) with hydrogen peroxide as electron acceptor, resulting in the formation of the so-called compound I and water. Compound I contains an oxoferryl group and a π-cationic porphyrin radical. If compound I is subject to a fast two-electron reduction by a second molecule of hydrogen peroxide, the ferric-catalase will be regenerated, oxygen evolved and water formed, i.e. the catalytic reaction occurs by the combined two steps. In the peroxidatic reaction, which occurs at low hydrogen peroxide concentrations, compound I is reduced either directly by a two-electron donor other than hydrogen peroxide (e.g. short-chain alcohols) to the ferric-catalase or by an one-electron donor (e.g. artificial phenolic compounds or endogenous donors) to compound II which has to be reduced subsequently by a second one-electron transfer to the ferric catalase. Compound II can also be converted to a hyperoxidized catalase form (Compound III) at high hydrogen peroxide concentrations.

Compound II and III are inactive forms of the enzyme and their formation results in a (reversible) inhibition of the catalytic activity. The formation of compound II is prevented by oxidation of the NADPH tightly bound to the surface of certain catalases (four NADPH per catalase tetramer; Zámocký and Koller, 1999; Nicholls *et al*, 2001; Kirkman *et al*, 1999; Kirkman and Gaetani, 1984; Fita and Rossmann, 1995). How the oxidation of NADPH, which is not in close contact with the catalytic site, prevents the formation of coumpound II is not yet fully understood. The NADP formed is replaced by another molecule of NADPH.

Properties

Catalase was one of the first enzymes to be isolated in a pure state. Animal catalases and bacterial catalases were purified from many different sources (Schonbaum and Chance, 1976; Hochman and Goldberg, 1991). In contrast, catalases have been isolated only from a few plant sources, e.g. spinach leaves (*Spinacia oleracea,* Galston *et al*, 1951; Galston, 1955), maize (*Zea maize* L., Chandlee *et al*, 1983), papaya fruit (*Carica papaya* L., Chan *et al*,

1978), lentil leaves (*Lens culinaris* L., Schiefer *et al*, 1976), cucumber cotyledons (*Cucumis sativus* L., Lamb *et al*, 1978; Kindl, 1982), sweet potato roots (*Ipomoea batatas* L., Esaka and Asahi, 1982), wheat germ (*Triticum aestivum* L., Garcia *et al*, 2000), tobacco leaves (*Nicotiana* spec., Havir *et al*, 1996), *Zanthedeschia aethiopica* (Trindade *et al*, 1988), pumpkin cotyledons (*Cucurbita pepo* L., Yamaguchi *et al*, 1984), potato tubers (*Solanum tuberosum*, Beaumont *et al*, 1990), cotton seeds (*Gossypium hirsutum*, Kunce *et al*, 1988), loblolly pine megagametophytes (*Pinus taeda* L., Mullen and Gifford, 1993) pea leaves (*Pisum sativum* L., Corpas *et al*, 1999) and sunflower cotyledons (*Helianthus annuus* L., Eising and Gerhardt, 1986; Kleff *et al*, 1997). Properties of isolated catalases from selected plant tissues are listed in Table 1. The number and different properties of isoforms for plant species are described later.

Table 1: Properties of purified plant catalases

Species	Molecular weight Subunit [kD]	Molecular weight Native Protein [kD]	pH range	Specific activity [µkat/mg]	Absorption spectra λ_{max}	A_{280}/A_{405}
Sweet Potato roots	60	240	6.0 to 8.5	470	280, 405, 510, 545, 620	1,49
Potato tubers	56 ± 2	224 ± 8	6.0 to 8.0	50	280, 403, 500, 535, 620	1,13
Lentil leaves	54	225	6.5 to 8.5	492	NE	1.5
Wheat germ	NE	240	6.0 to 8.0	CAT-1: 51 CAT2: 147	NE	NE
Spinach leaves	72	300	5.3 to 8.9	473	280, 405, 510, 545, 620	1.5
Zantedeschia ethiopica leaves	54	220	6.0 to 8.0	2.44	280, 405	0.92
Pea leaves	57	NE	NE	15883	252, 400	NE
Maize	60	240	7.0 to 9.0	CAT-1 4668 CAT-2 266 CAT-3 72	NE	NE
Cucumber cotyledons	54,5	225	NE	87	NE	NE
Pumpkin cotyledons	55, 59	230, 215	6.5 to 10	59 kD: 200 55 kD: 3333	280, 405, 510, 530, 625	1.0
Castor bean endosperm	54, 56	about 200	NE	1500 (endosperm) 14,8 (hypocotyl)	280, 406, 620	1.9 (endosperm)1.4 (hypocotyl)
Loblolly pine megagameto-phytes	59	235	NE	36916	NE	1.5
Sunflower cotyledons	55 59	265	6.0 to 8.0	1252 (day 2) 334 (day 7.5)	280, 405, 505, 539, 624	1,45

NE: Not examined

Plant catalases are usually composed of four identical subunits, and their molecular weights are found to be from 220 to 240 kD (Kunce *et al*, 1988). Absorption spectra for pumpkin catalases display two major peaks at 280 and 405 nm (Soret band). The ratio of absorption peaks at 280 and 405 nm was calculated to be 1.0 for pumpkin catalase (Yamaguchi and Nishimura, 1984) being lower than the values 1.49 for sweet potato catalase and 1.5 for spinach leaf catalase. The value indicated a high haem content for pumpkin catalase of about four haem groups per catalase molecule. Galston *et al* (1951) expected two haem prosthetic groups per spinach catalase molecule, but according to Gregory (1968), the enzyme contained four haematin groups per subunit. However, Yamaguchi *et al* (1986) and Eising and Gerhardt (1986) also proved that catalases isolated from pumpkin and sunflower cotyledons contained four haem groups per catalase molecule, even if the absorbance ratio A_{280}/A_{405} for these catalases was higher than 1.0. One exception among plant catalases is another catalase isolated from spinach leaves (Hirasawa *et al*, 1987) with subunits of 55 kD and a native molecular weight of 125 kD containing two moles of iron per one mole of catalase. The two irons were attributed to protohaem and novel haem-bearing spectral properties differing from the spinach catalase described by Galston *et al* (1951).

The specific catalytic activities of purified plant catalases are mainly in the range from 470 μkat/mg (green lentil leaves; Esaka and Asahi, 1982) and 1500 μkat/mg (etiolated pumpkin cotyledons; Yamaguchi and Nishimura, 1984). In sunflower cotyledons, the specific activity depends on the developmental stage of the cotyledons. It changes from 1252 μkat/mg at day 1.5 to 334 μkat/mg at day 7.5 of development (Eising and Gerhardt, 1986). The specific activity of sunflower catalase also depends on the type of purified catalase. Specific activities of core catalase isoforms from sunflower were in the range of 10 μkat/nmol haem being significantly lower than activities of matrix catalase isoforms (100 μkat/nmol haem).

Peroxidatic activity of plant catalases is in general lower than catalytic activity. Total peroxidatic activity in extracts of twenty one day old *Nicotiana sylvestris* seedlings grown in air was about 1 μmol ethanol to acetaldehyde per min and g fresh weight and considerably lower than the catalytic activity (about 3000 μmol H_2O_2 min^{-1} g^{-1} fresh weight, Havir and McHale, 1987). The ratio of peroxidatic to catalytic activity was determined to be in the range from 0.25 to 0.42 for typical catalases from maize, tobacco and barley (Havir and McHale, 1989). In contrast, for maize CAT-3, tobacco CAT-3 and barley EP-CAT with enhanced peroxidatic activity, the ratios of peroxidatic activity to catalytic activity were 17.6, 9.2 and 11.8.

Purified catalases of wheatgerm exhibit low stability, while other catalases from plant sources are described to be rather stable (e.g. pumpkin catalase, Yamaguchi and Nishimura, 1984). Experiments where the effect of pH on the activity of plant catalases was investigated suggest a pH optimum between 6 and 8 for most of the catalases. In comparison with the basic part of the pH curve, where a slight decrease in activity was found, activity decreased rapidly in the acidic part of the pH curve (Chan *et al*, 1978; Esaka and Asahi, 1982; Chandlee *et al*, 1983; Garcia *et al*, 2000; Galston, 1955). At high temperatures, most catalases are rather unstable. Spinach catalase was completely inactivated after incubation for 10 minutes at 60°C (Galston *et al*, 1955).

Catalases show a weak affinity to their substrate hydrogen peroxide, so catalases only work efficiently at high hydrogen peroxide concentrations with high V_{max}. For purified potato catalase, the specific molecular rate constant at 25 °C was determined to be $10^{-6} M^{-1} s^{-1}$, and the turnover number to be 11200 s^{-1} (Beaumont *et al*, 1990). These data are in agreement with most other catalases (Scandalios *et al*, 1997; Nicholls *et al*, 2001).

In some plant systems, distinct catalase proteins with different properties and spatial and temporal distinct expression patterns occur (see later). It was proven for different catalases e.g. from wheat (Garcia *et al*, 2000) and pumpkin (Yamaguchi *et al*, 1986) that they have different properties (Table 2). The K_m values for plant catalases are in accordance with a K_m of 50 mM towards H_2O_2 determined for catalase from the bacterium *Klebsiella pneumoniae* (Hochman and Goldberg, 1991).

Table 2: K_m and V_{max} values (H_2O_2) for different catalase species in wheat and pumpkin

source	catalase	K_m (mM) towards H_2O_2	V_{max} (mMol H_2O_2/ min/mg protein)	IEP
wheat	CAT-1	38.9	3501	6.3
wheat	CAT-2	7.69	NE	5.8
pumpkin	59-kD species	60	1140	6.1
pumpkin	55-kD species	124	114	6.6

NE: not examined IEP: isoelectric point

Isoforms

Catalases from many plant species bear multiple isoforms. More than thirty years ago, Scandalios (1965) first demonstrated that multiple isoforms of catalase were the products of distinct, unlinked genes. Subsequently, multiple catalase isoforms were found among various species examined. Prominent examples are maize (Scandalios *et al*, 1984), pumpkin (Yamaguchi *et al*, 1986), *Nicotiana tabacum* (Havir and McHale, 1987), sunflower (Eising *et al*, 1990, 1996), cotton (Ni *et al*, 1990), *Nicotiana plumbaginifolia* (Willekens *et al*, 1994), *Arabidopsis thaliana* (Zhong *et al*, 1994; Frugoli *et al*, 1996), *Pinus taeda* (Mullen and Gifford, 1993), tomato (Gianinetti *et al*, 1993) castor bean (Ota *et al*, 1992), pea (Corpas *et al*, 1999) and potato tuber (Beaumont *et al*, 1990). The five catalase isoforms described for pea leaves (Corpas *et al*, 1999) probably arise from one single catalase gene (Isin and Allen, 1991). In many plant species investigated so far, catalase isoforms arise by tetramerization of subunits with different properties, e.g. different charges or different molecular masses. In many plant species, the different types of subunits are expressed and regulated by different genes (for regulation, see later). In cotton, two catalase genes encode two subunit types with identical size (57 kD) but different charges (Ni and Trelease, 1991a) leading to a system of five different isoforms by homo- and hetero-tetramerization. (Ni *et al*, 1990). Most isoforms from plant catalases appear in a pI-range from 5.5 to 6.5, e.g. pea isoforms, denoted as CAT1-5 displaying isoelectric points of 6.41, 6.36, 6.16, 6.13 and 6.09, respectively.

Isoform systems with more than five isoforms were also described. Greening sunflower cotyledons exhibit up to ten isoforms (Eising *et al*, 1996) and mustard displays at least twelve isoforms (Drumm and Schopfer, 1974), but biosynthesis and regulation of these isoforms was not investigated. Whereas the number of catalase isoforms was collected, mainly by DAB-staining, the aim of recent analyses is to find isoforms with specific patterns of regulation and specific biological functions, and to find out the correlation between expression of catalase coding genes and isoforms. In plants, complex isoform systems were mainly characterized for Arabidopsis, maize, *Nicotiana tabacum*, castor bean, and sunflower. (For accession numbers of the catalase genes for plant species, see Scandalios *et al*, 1997). In Arabidopsis, three catalase genes and at least six isoforms were found (McClung, 1997). In maize, three unlinked structural genes *CAT1* to *CAT3* are coding for three biochemically different isoforms CAT-1 to CAT-3 with different spatial and temporal regulation manners (Chandlee *et al*, 1983). For tobacco catalase, at least ten isoforms with isoelectric points between 6.0 and

7.6 were found. Catalase isolated from tobacco leaves mainly contained subunits with masses of 57 kD and 55 kD. 57-kD-homotetramers built basic isoforms with high specific activities, whereas 55-kD-isoforms built acidic isoforms with low specific activities, but a high degree of peroxidatic activity (Havir *et al*, 1996; Durner and Klessig, 1996). Recently, distinct catalase isoforms with high peroxidatic activity were described for tobacco, barley and maize (Havir and McHale, 1989; Havir *et al*, 1996) giving hints that distinct catalase isoforms in plants may have specific physiological functions. Two catalase genes were described for castor bean (Suzuki *et al*, 1994). However, three types of subunits were found (54 to 56 kD) leading to tissue-specific isoforms with differing specific activities (Ota *et al*, 1992). For sunflower cotyledons, the biogenesis of ten isoforms from four different catalase-mRNAs is described later. Different properties and groups of isoforms were found in this species, especially different specific activities and different photoinactivation kinetics.

Photoinactivation

Light not only acts as a regulator or stimulus of catalase synthesis in plants, but also impairs the activity of the enzyme. Besides the photosynthetic D1 protein (Aro *et al*, 1993), catalase is a well investigated plant protein inactivated by light. The haem-containing catalase enzyme from human and different animal and plant species is inactivated by ultraviolet B, ultraviolet A and near visible light *in vitro* (Giordani *et al*, 1997; Cheng *et al*, 1981; Grotjohann *et al*, 1997) Photoinactivation of catalase is presumably due to irreversible photo-oxidation. The state of photo-oxidation depends on the presence of reactive oxygen species and leads to an inactivation of catalase (Kono and Fridovich, 1982). In the literature, there exist different point of views about the mechanisms involved in photoinactivation. While Orr (1967) reported a cleavage of catalase *in vitro* when the enzyme was exposed to oxygen radicals, Feierabend and Engel (1986) did not find a cleavage of the protein moiety during photoinactivation, but a major portion of enzyme-bound haem was dissociated. In a photoinactivated sample with 16% residual activity, about 65% of the catalase-bound haem was lost (Feierabend and Engel, 1986). Feierabend and Dehne (1996) observed only a minor haem destruction during photoinactivation of bovine liver catalase, whereas destruction of rye catalase haem was considerably higher. Similar photoinactivation as described for bovine liver and rye catalases was observed for peroxisomal catalases isolated from pea, cucumber, maize and sunflower. Photoinactivation seems to be a general feature of plant catalases. Photoinactivation of catalase is mediated by blue light absorption of the haem group (Cheng *et al*, 1981; Grotjohann *et al*, 1997; Shang and Feierabend, 1999). Experiments with rye leaves demonstrated that catalase is not only inactivated *in vitro*, but also in intact leaves. Additionally, at high photon flux densities (1000 μmol $m^{-2}s^{-1}$ photosynthetic active radiation) a significant inactivation also occurred in red light (Shang and Feierabend, 1999). In leaves, catalase photoinactivation may be mediated indirectly by photo-oxidative events initiated in the chloroplasts. Photoinactivation of catalase *in vivo* can be detected when new catalase synthesis is prevented by translation inhibitors (Feierabend and Engel, 1986) or by stress conditions suppressing protein synthesis (Hertwig *et al*, 1992; Streb and Feierabend, 1996). When catalase protein synthesis is suppressed, its enzymatic activity can be largely depleted after one day of light exposure (Schäfer and Feierabend, 2000).

Photo-oxidation of catalase *in vivo* does not occur exclusively under stress conditions, but was also observed under normal conditions being described for needles of Norway spruce (Schittenhelm *et al*, 1994). A physiological function of catalase photoinactivation may be the distinct regulation of the amount of physiologically active catalase protein. Catalase from rye seedlings displays a rapid turnover in light, therefore the steady-state level of active catalase protein depends on continuous new synthesis. At a photosynthetic photon flux density of 520 μmol m^{-2} s^{-1}, the half-life of catalase in rye leaves was three to four hours (Hertwig *et al*, 1992). Furthermore, a photoprotective function of catalase for glycolate oxidase was proposed. Both enzymes, catalase and glycolate oxidase are photosensitive. They are associated in a multi-enzyme complex in the peroxisomal matrix (Heupel and Heldt, 1994). Photo-destruction of catalase localized around glycolate oxidase in this complex may function as a protective for glycolate oxidase guaranteeing the enzymatic activity of this enzyme also under light stress (Schäfer and Feierabend, 2000).

Recently, highly photoresistant catalase proteins were described for the alpine high mountain plants *Homogyne alpina, Ranunculus glacialis* and *Soldanella alpina* (Streb *et al*, 1997). These plants have to survive under stress conditions like exposure to excessive light during the day, high temperature amplitudes and sometimes water stress. In this respect, substrates of the peroxidatic mode of catalase, e.g. alcohols, are known to protect the enzyme very efficiently from photoinactivation (Cheng *et al*, 1981; Giordani *et al*, 1997). The synthesis of low light-sensitive catalases bears a strategy for the plants to avoid photo-oxidation under conditions when photoinactivation of typical catalases cannot be compensated by rapid protein synthesis. Catalases from high mountain plant show a ten-fold higher stability than typical catalases. Stable catalase forms with a high resistance to photoinactivation may guarantee a stable ground level of hydrogen peroxide decomposition even under stress conditions.

In sunflower cotyledons, groups of isoforms with different properties have been observed. The photosensitivity of group A isoforms (CAT 2 to CAT 5) is tenfold reduced in comparison with group B isoforms (CAT 6 through CAT 8). Group A isoforms differ from group B isoforms in subunit composition and further biochemical properties. Group A isoforms contain both 59-kD and 55-kD subunits, whereas group B isoforms are composed exclusively of 55-kD catalase subunits. 59-kD subunits of group A are much more resistant to photodegradation than the 55-kD subunits of group B isoforms. Irradiated group A isoforms reach a stable plateau of residual activity, whereas group B isoforms are inactivated completely. Grotjohann *et al* (1997) suggested for sunflower catalase that photoinactivation precedes protein degradation. Photoinactivated catalase from sunflower still contains four haem groups per molecule. Specific physiological functions of the sunflower catalase isofoms are suggested. Group A isoforms with low photosensitivity and turnover guarantee a minimum level of catalase activity. Their function is in accordance with the stable catalase forms of high mountain plants (Streb *et al*, 1997). In contrast, group B isoforms with a high photosensitivity and turnover may achieve rapid changes in total catalase activity during cotyledon growth (Eising and Gerhardt, 1987). However, selective subcompartmentation of the two catalase isoform groups between matrix and core in sunflower peroxisomes and identification of different catalase-coding mRNA species led to new insights into the biogenesis of sunflower catalase isoforms (see later).

Catalase-enriched peroxisomal inclusions (cores)

In the peroxisomal matrix of many plant species, inclusions showing a high electron density occur depending on the developmental stage of the plant organ or tissue. Different types of peroxisomal inclusions may appear, presumably consisting of protein, which are a characteristic feature of distinct plant species (for reviews see: Huang *et al*, 1983; Frederick *et al*, 1968, 1975). Major groups of inclusions are amorphous types on the one hand and crystalline types on the other hand (Huang *et al*, 1983). Inclusions may appear as narrow threads with repeating substructure, linear aggregates or mainly dominating as crystalline cores. Crystalline cores were described before plant microbodies were named peroxisomes for pea leaves and *Avena sativa* coleoptile (Sitte, 1958; Thornton and Thimann, 1964). Whereas cores in animal peroxisomes contain urate oxidase forming the crystalline core ultrastructure itself (Alvares *et al.*, 1992), cores from plant tissues have not been found to contain this enzyme (van den Bosch and Newcomb, 1986) except in the amorphous nucleoids in vascular parenchyma cells of soybean nodules (Vaughn and Stegink, 1987). In plant cores, catalase is by far the predominant protein (Frederick *et al*, 1975; Kleff *et al*, 1997). Yet, the physiological function of cores is still obscure. In older literature, a storage function of inactivated protein was proposed for the cores (Cronshaw, 1964; Marinos, 1965; Fagerberg, 1984). Vigil (1970) postulated a crystallization of cores by condensation of matrix catalase.

In a wide range of plant species, cores appear in a development-dependent manner. Today, cores have been described for many plant organs and tissue types. The inclusions have been observed in leaves, stems and hypocotyls of plants, but also appear in roots and storage tissues, e.g. potato tubers. Describing size and properties of plant cores may be a first step to elucidate the function. Visualization of cores in plant tissues was traditionally performed by transmission electron microscopy (TEM), e.g. for castor bean, tobacco and sunflower (Frederick *et al*, 1975; Tenberge and Eising, 1995). Tenberge *et al* (1997) developed an isolation procedure for cores from plant tissues allowing the properties of these inclusions to be investigated in more detail. Recently, size and ultrastructure of peroxisomal cores from five day old sunflower cotyledons were demonstrated by high resolution scanning electron microscopy (HRSEM, Heinze *et al*, 2000). Figure 1 shows a HRSEM-micrograph of such a core. Sunflower cores are quadrangular blocks, mainly with square sizes between 400 and 600 nm for two edge lengths, whereas the third edge length is reduced up to 100 nm. After measuring 15 cores, the corresponding core volumes were calculated being in the range from 0.01 to 0.1 μm^3. For the first time, the core surface was visualized as containing cubical units with edge lengths of about 20 nm. Between the units, interstices of about 4 nm appeared, possibly creating a diffusion space for smaller molecules, e.g. isocitrate lyase (Gao *et al*, 1996). The regular ultrastructure for sunflower cores was also shown by light optical diffraction (Heinze *et al*, 2000) justifying reference to cores as protein crystals.

Figure 1: High resolution scanning electon micrographs of peroxisomal cores from sunflower cotyledons

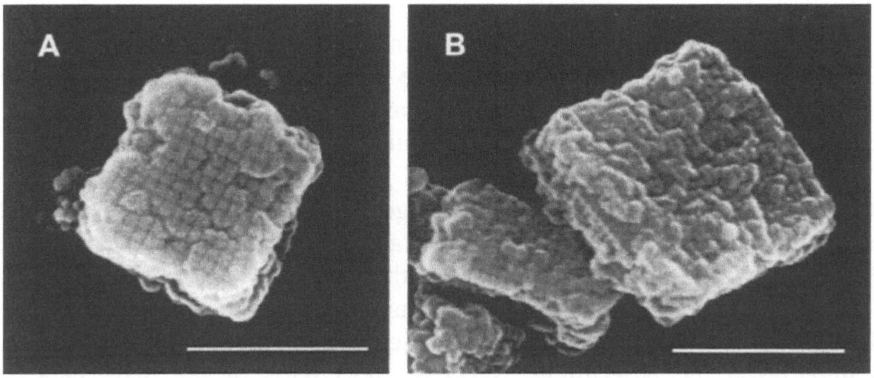

Figure 1. A. Cores are composed of regular arranged square units with edge lengths of about 20 nm. B. Cores are quadrangular blocks with two edge length of about 500 nm and a third edge length being greatly reduced. Micrographs recorded with an Hitachi S-5000 microscope, acceleration voltage 5 kV. Bars correspond to 400 nm.

Biochemical analyses of core fractions from sunflower cotyledons and potato tubers presented evidence that catalases eluted from core fractions are special, biologically active catalase forms differing in properties from typical matrix catalases. For core catalases, specific activities are reduced up to ten-fold and photosensitivity to blue light *in vitro* was strongly reduced. Whereas matrix catalase was completely inactivated with a half life of about thirty minutes, core catalase reached a plateau of about 30% residual activity (Eising *et al*, 1998). A broader pH and temperature stability provided further evidence that core catalase is the more stable catalase form in sunflower peroxisomes.

Molecular biology

Amino acid sequencing and homologies

No amino acid sequence information could be obtained many plant catalases, indicating that they are N-terminally blocked. Animal catalases and some yeast catalases also have a blocked N-terminus (van Eyk *et al*, 1992). For some plant species like pea, exclusively one type of catalase cDNA was cloned and sequenced (Isin and Allen, 1991). Most plant species contain multiple catalase proteins. Catalase proteins display a high degree of homology which is highest in the N-terminal portion, especially from residue 60 to residue 85, and declines towards the C-terminal (von Ossowsky *et al*, 1993; Schultes *et al*, 1994). In general, amino acid sequence identities between catalase proteins of different plant species are in the range 85% to 90%. Identities between different catalase proteins in the same plant species reach up to 99%. Most plant catalases have a length of 489 to 496 amino acids, e.g. *Nicotiana tabaccum* (489 residues, Chen *et al*, 1993), maize CAT-2 (491 residues, Skadsen and Scandalios, 1986), Arabidopsis leaf and most other plant catalases (492 residues, Frugoli *et al*, 1996), cottonseed (492 residues, Ni *et al*, 1990), sunflower catalases (492 residues, Kleff *et al*, 1994). An exception is maize CAT-2 with 529 residues (Redinbaugh *et al*, 1988). Important sites for haem-binding and catalyses are conserved for all plant catalases. Most plant catalase sequences start with the residue methionine at the N-terminus. Whereas most parts of the plant catalase sequences are rather conserved, some regions appear dividing plant catalases into two groups. The most characteristic example is the region from residue 325 to 329 (sunflower). The sequence of this region in one group of catalases is NH_2-CPAII-COOH, the sequence of other catalases including species containing peroxisomal cores is NH_2-NPGLV-COOH. Probably the sequence in this region is diagnostic of core catalases.

Targeting signals

For many peroxisomal matrix proteins it was clearly proven that they are imported into peroxisomes via the typical PTS1 or PTS2 pathway (Chapter 11). In contrast, different views were published about the import mechanism of catalases in plants and other organisms, because a typical carboxyterminal SKL-tripeptide for the PTS1-pathway or an aminoterminal PTS2-signal were lacking. Ni and Trelease (1991a) postulated that some conserved serines located near the amino terminus might function as a targeting signal for peroxisomal catalases. In contrast, González (1991) postulated an internal SKL-related motif required for catalase import. The C-terminal sequences from seven plant and three animal catalases were analyzed leading to the result that the chemical properties of amino acids in some positions of the catalase sequences from different species were identical. In conclusion, an internal SKL-related motif with a distance of 9 to 14 amino acids to the carboxy terminus was proposed. Suzuki *et al* (1994) confirm the conserved nature of the internal SKL-motif for catalases from different species, but the relevance of this sequence motif for protein import was not clearly proven. Whereas little effort was made for some time to investigate catalase import into plant peroxisomes, the tripeptide ANL-COOH was identified as a targeting signal for rat liver catalase and it was clearly shown that this tripeptide and not the internal SKL-motif was required for targeting catalase to peroxisomes (Trelease *et al*, 1996b). Also, the tetrapeptide KANL-COOH was found to be required for import of human catalase (Purdue and Lazarow, 1996). From this and further work (Mullent *et al*, 1997a) it has become apparent that catalases are imported via a PTS1 pathway in which the penultimate residue is non-basic and the residue at the position can have a large side chain. The basic position can have a large side chain. The basic amino acid at position -4 is important for targeting the last three amino acids are necessary, but not sufficient (see Chapter 11). The transfer of peroxisomal proteins through the peroxisomal membrane is assisted by chaperones, mainly from the HSP70 and HSP40 type. The involvement of chaperones in peroxisomal protein import was detected for bovine liver catalase (Hook and Harding, 1997) and for plant isocitrate lyase (Crookes and Olsen, 1998). The relevance of chaperones for import of plant catalases is still obscure. A DnaJ protein is anchored on the cytosolic side of the glyoxysomal membrane of cucumber cotyledons binding a cytosolic HSP70 isoform but no other HSP70 species. The resulting multiprotein complex is proposed to be relevant for peroxisomal import (Diefenbach and Kindl, 2000), but the involvement of chaperones in catalase import has not been clearly proven. A classical view of the catalase import mechanism described the separate import of catalase haem and catalase apoprotein monomers synthesized on free polysomes in the cytosol (Middelkoop *et al*, 1993). In the matrix, oligomerization and incorporation of haem groups (one haem group per monomer) take place (Lazarow and de Duve, 1973; Eising *et al*, 1990). Catalase haem, imported into peroxisomes by a mechanism which is not completely understood, is synthesized in mitochondria (Surpin and Chory, 1997).

Besides the classical import model, the import of native oligomeric proteins into peroxisomes was described (Rachubinski and Subramani, 1995), but the mechanism by which this occurs remains speculative (Pool *et al*, 1998). It is still not understood how an oligomeric or folded protein is able to pass through the peroxisomal membrane. Peroxisomes in plant cells are able to incorporate oligomeric and folded proteins into their matrix and both the PTS1 and the PTS2 pathway can mediate this process (Kato *et al*, 1999). Import of tetrameric catalase was also described for peroxisomes from animal tissues (McNew and Goodman, 1994, 1996; Subramani, 1996). Studies on human catalase biosynthesis demonstrated an oligomerization of catalase and association with the peroxisomal membrane in a few minutes, whereas incorporation into the organelles needed up to some hours. An interaction of PTS1 and PTS2 receptors was proposed for human peroxisomes (Rachubinski and Subramani, 1995) suggesting links between these two pathways, probably being in agreement with the oligomeric protein import model.

Heterologous expression

All four catalase mRNAs from sunflower (HNNCATA1 to HNNCATA4) were expressed in *E. coli*. Expression of HNNCATA1 and HNNCATA2 led to an overexpression of a catalase with a molecular weight of 55 kD for the subunits. This result corresponded to the size of matrix catalase from sunflower cotyledons (Kleff *et al*, 1997). As a result of expression of HNNCATA3 and HNNCATA4, a 59-kD-product appeared as detected for catalase isolated from core fractions of sunflower (Kleff *et al*, 1997). Plant catalases were classified based on expression properties (Willekens *et al*, 1995). Class I catalases displayed a high expression rate in photosynthetic tissues. Examples are cotton subunit 2 catalase, *Nicotiana plumbaginifolia* Cat1, maize CAT-2 and *Arabidopsis thaliana* CAT2. Class II catalases show highest expression rates in vascular tissues, e.g. *N. plumbaginifolia* Cat2, Castor bean CAT2, maize CAT-3, tomato TOMCAT1 and potato Cat2St. Class III catalases are expressed mainly in seeds and young seedlings, e.g. cotton SU1 catalase, *N. plumbaginifolia* Cat3, castor bean CAT1 and maize CAT-1.

Post-translational modifications

There has been extensive discussion about post-translational modifications and processings of plant catalases. For some other peroxisomal proteins such as cucumber malate dehydrogenase (Gietl, 1990) and 3-ketoacyl–CoA thiolase (Osumi *et al*, 1991), cleavable transit or signal sequences have been described. Catalases from some plant species are also processed, while catalases from a variety of other species undergo no processing steps. Evidence for processing *in vivo* of plant catalases is presented for maize catalase CAT-2. The primary translation product had a molecular weight of 56 kD, compared to that of 54 kD for purified CAT-2 protein (Skadsen and Scandalios, 1986). In germinating pumpkin seeds, catalase is synthesized as a precursor with a molecular mass of 59 kD, whereas the mature catalase contains 55 kD-subunits (Yamaguchi *et al*, 1986). Both the precursor and mature forms of catalase from pumpkin are localized in the microbodies, i.e. glyoxysomes and leaf peroxisomes (Yamaguchi *et al*,

1984). Homotetramerization of the two subunit types lead to two separate isoforms. It was shown that 55 kD- and 59 kD-catalases differ in tertiary structure. It is very unlikely that structural modification of the two molecules results from carbohydrate conjugation, because catalase was found to be a non-glycosylated protein (Yamaguchi *et al*, 1986). Two catalase translation products resulting from two genes were described for castor bean. One of these products will be proteolytically processed and form catalase homotetramers distinguishable from homotetramers of the non-processed form (Ota *et al*, 1992). Kleff *et al* (1994) also sugggested proteolytic processing for sunflower matrix catalase (accession number L28740). Recent investigations presented evidence that this sunflower catalase is not processed. Cotton catalases are synthesized as monomers with molecular weights of 57 kD and 64 kD, but experimental data strongly indicate a translation of catalase monomers on free ribosomes in cytosol without any cleavable transit or signal sequences (Kunce *et al*, 1988).

Molecular structure

Molecular structures of plant catalases have not been published as yet. Molecular structures of catalases have been determined for seven animal, yeast and bacterial catalases (Nicholls *et al*, 2001) including bovine liver catalase (Murthy *et al*, 1981), catalase A from *Saccharomyces cerevisiae* (Berthet *et al*, 1997), typical bacterial catalases from *Proteus mirabilis* (Gouet *et al*, 1995) and *Micrococcus lysodeikticus* (Murshudov *et al*, 1992) and 'large subunit' catalases from *Penicillium vitale* (Vainshtein *et al*, 1986) and *E. coli* HPII (Bravo *et al*, 1995) with sizes for the subunits between 75 and 84 kD. Even though the published structures are from different organisms, the amino acid sequences have different lengths and the subunits are differing in size, common structural features being characteristic to catalases appear and it is expected that these features are also present in plant catalases. All catalases investigated were homotetramers, the haem group of each subunit was deeply buried in an interior region near to a β-sheet domain. Channels to allow the substrate(s) to penetrate almost 3 nm into the protein to reach the active centre and to direct the reaction products to the surface have been described (Nicholls *et al*, 2001). The molecular structure of the subunits is also rather conserved just displaying moderate differences between different catalases (Zámocký and Koller, 1999). By alignment of sequences with published structures and plant catalases, positions relevant for haem binding or enzymatic catalysis can be assigned to the definite amino acids. Residues important for haem binding or enzymatic catalysis are strictly conserved among all catalases. For haem binding, these are the highly conserved positions V64, R102, T105, F141, P326, R344 and Y348 (numbers correspond to the sunflower matrix catalase-1 sequence, accession number L28740). For enzymatic catalysis, mainly the residues H65, S104, and N138 are important (Zámocký and Koller, 1999) and also H65, S104, V106, N138, F143, R355 (Vainshtein *et al*, 1986). All eukaryotic catalases display a rather conserved region of about 380 to 390 amino acids (Klotz *et al*, 1997; Nicholls *et al*, 2001) mainly localized between residues 53 and 434 for sunflower catalases.

The molecular structure of all catalases analyzed so far displays at least four different regions. At the N-terminus, the sequence bears an arm with a loop structure standing far outside from the centre of the molecule (Reid *et al*, 1981). This N-terminal arm is important for the association of subunits. After the N-terminal arm has finished at position 65 (sunflower), an antiparallel β-barrel domain is connected, containing residues 66 to 311 in sunflower. The third domain in the C-terminal half is a wrapping domain connecting the β-barrel region with the α-helical section localized from position 312 to 424. The α-helical section is localized at the C-terminus from position 425 to 492. It was suggested that the C-terminal end of the connection domain (residues 398 to 424) forms a loop important for tetramerization of catalases (Nicholls *et al*, 2001).

The catalase purified from tobacco leaves has been reported to contain NADPH as well as to bind NADPH (Durner and Klessig, 1996). In contrast, the catalase purified from potato tuber peroxisomes which is probably a core catalase appears to be devoid of NADPH (Beaumont *et al*, 1990). There was also no indication of Compound II accumulation when assaying the potato tuber catalase at low hydrogen peroxide concentration. The amino acid residue of catalase corresponding to His 304 of bovine liver catalase (corresponding to residue 295 of sunflower catalases) is important for the binding of NADPH to a particular catalase (Zámocký and Koller, 1999). Catalases containing histidine or glutamine at this position bind NADPH tightly and weakly, respectively. Alignment of forty five sequences of plant catalases, including eight tobacco catalases, results without exception, in glutamic acid at the position in question, suggesting that plant catalases do not contain/bind NADPH. Deduced from molecular structure, position 155 is relevant for binding of salicylic acid. According to Rüffer *et al* (1995), isoleucine at position 155 interferes with the binding of salicylic acid to the haem iron, whereas valine deters the sterical conflict. According to the alignment of plant catalases, one half of the sequences including sunflower matrix catalases displays valine, while the other half including sunflower core catalases displays isoleucine and should not bind salicylic acid.

Model of catalase biogenesis in sunflower cotyledons

Catalase from sunflower cotyledons is mainly localized in peroxisomes. Peroxisomes from sunflower cotyledons contain crystalline inclusions, also denoted as cores. These cores appear in a time-dependent manner. They have also been observed in dark-grown cotyledons, but prominent cores have been described for peroxisomes of five day old, greening cotyledons. These cores are mainly composed of crystalline catalase protein differing from matrix catalase protein (Kleff *et al*, 1997). Recent analysis proved that the peroxisome core and matrix subcompartment each contain two different catalase gene products. The catalase encoding cDNA sequences in sunflower (HNNCATAs) were enumerated HNNCATA1 to HNNCATA4. HNNCATA1 corresponds to the sequence HNNCATA (GenBank accession number L28740). Table 3 displays some features of the four catalase encoding cDNAs. The deduced amino acid sequences exhibit lengths of 492 amino acid residues. Core and matrix catalase proteins are distinguishable by the molecular weights of the subunits, determined after SDS-PAGE. Matrix catalase proteins display subunits with a molecular mass of 55 kD differing from core catalase proteins with a molecular mass for the subunits of 59 kD. The masses for the sunflower catalases were confirmed by heterologous expression of the catalases in *E. coli*. Concluding, the proteins encoded by HNNCATA1 and HNNCATA2 were named matrix catalase-1 and matrix catalase-2, the proteins encoded by HNNCATA3 and HNNCATA4 were named core catalase-1 and core catalase-2.

A special feature for all four sunflower catalase proteins described so far is the amino acid tyrosine at the carboxy terminus (Table 3). This is an exception as no other plant catalase amino acid sequence described so far ends with tyrosine at this site. In this respect, tyrosine may be relevant for import. Mullen *et al* (1997b) described a relevance of C-terminal tyrosine for the import of chloramphenicol-acyltransferase into plant glyoxysomes. Interestingly, the -4 amino acid is different for matrix and core catalases in sunflower. Matrix catalases have lysine at this position, while for core catalases, this residue is arginine. Mullen *et al* (1997a) presented evidence for import relevance of the -4 residue in plant catalases. They found import was strongly reduced for plant catalases if arginine was replaced by lysine. These findings may provide the first evidence for different import mechanisms of core and matrix catalases. Walton *et al* (1995) found that stable folded proteins and also gold particles with a diameter of 9 nm were imported into peroxisomes. Possibly, core catalase is also imported by such mechanisms differing from the classical matrix catalase import, i.e. core catalases may be imported as oligomers but matrix catalases as monomers.

Table 3: Properties of peroxisomal catalase cDNA species from sunflower cotyledons

Properties	HNNCATA1	HNNCATA2	HNNCATA3	HNNCATA4
Total length [bp]	1735	1736	1860	1871
GenBank accesion number	L28740	AF243517	AF243518	AF243519
Total length Non-coding regions	256 bp	257 bp	381 bp	393 bp
Length 5'-Non-coding region	16 bp	51 bp	37 bp	37 bp
Length 3'-Non-coding region	240 bp	206 bp	344 bp	355 bp
Polyadenylation signal	yes	no	no	yes
Deduced mass for subunits [kD]	56.755	56.931	56.968	56.915
Mass for subunits [kD] Heterologous expression	55	55	59	59
Deduced IEP	6.59	6.53	6.59	6.62
C-terminal sequence	KPNY–COOH	KPSY–COOH	RPNY–COOH	RPNY–COOH

In sunflower cotyledons, four different mRNA species encode for up to ten separatable catalase isoforms (Eising *et al*, 1990, Eising *et al*, 1996). Core catalase isoforms (group A, CAT 1 to CAT 5) are more acidic than matrix catalase isoforms (group B, CAT 6 to CAT 10, Eising *et al*, 1996). Isoforms CAT 9 and CAT 10 are only visible under specific conditions. Figure 2 displays a first working model for the biosynthesis of isoforms in sunflower cotyledons. Two subunit-types with an equivalent mass of 59 kD, each combine to form five core catalase isoforms by homo- and heterotetramerization. In the same way, five matrix catalase isoforms are generated from two mRNA species encoding for two different 55-kD subunit types. The interaction between core and matrix catalase subunits has been excluded. Evidence for the specific tetramerization of core and matrix catalases came from studies on the mechanisms of catalase tetramerization in general. Catalase subunits perform 'arm-exchange', connecting the N-terminal arm of one subunit with a rather proline-rich region at the C-terminal of the wrapping domain of a second subunit, forming an interwoven dimer structure (Bergdoll *et al*, 1997). This process continues, resulting in a tetramer. It is suggested that the amino acid composition of the N-terminal arm and of the loop region leads to a specific secondary structure permitting the reliable and selective interaction of catalase subunits. Frequently, residue exchanges detected for both core catalases in comparison to the matrix catalases in the denoted regions may modify the structure of core catalase subunits. The modified structure seems to permit selective tetramerization of core catalase subunits excluding heterotetramerization of core- and matrix catalase subunits, because a stable dimer- or tetramer association of core-/matrix-heterotetramers is lacking.

Core catalase isoforms have a slow turnover, low specific activities and are less liable to photoinaction in comparison with matrix catalases. However, it was proven that sunflower core catalases are enzymatically active proteins which are encoded by specific mRNA species being biogenetically independent (Kleff *et al*, 1997; Eising *et al*, 1998).

Figure 2: Model for biogenesis of catalase isoforms in sunflower cotyledons

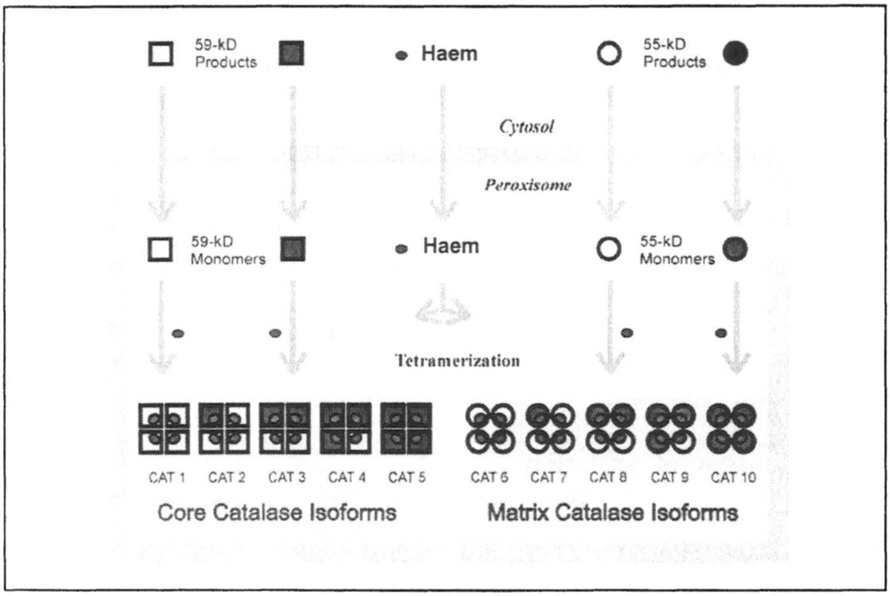

Physiology

Function

A main function of plant catalases is to detoxify hydrogen peroxide arising from reactions of peroxisomal oxidases (Gerhardt, 1986). The first step of peroxisomal β-oxidation is catalyzed by an acyl-CoA-oxidase (EC 1.3.3.6) producing hydrogen peroxide which has to be detoxified by catalase

(Gerhardt, 1987). In green leaves, hydrogen peroxide arising during photorespiration is detoxified (Schäfer and Feierabend, 2000). Hydrogen peroxide as a substrate for catalase is also produced by superoxide dismutase (EC 1.15.1.1, del Río *et al*, 1996). The substrates of superoxide dismutase are $O_2^{\bullet -}$ radicals produced in the peroxisomal matrix by xanthine oxidase (EC 1.1.3.22) or in the peroxisomal membrane (del Río *et al*, 1998).

Plant catalases play roles in resistance to oxidative stress (McClung, 1997) and in signal transduction involving H_2O_2 as a second messenger (Low and Merida, 1996). H_2O_2 and catalase are involved in the plant anti-oxidative defense system (del Río *et al*, 1998) where different cellular components like peroxisomes, mitochondria and chloroplasts are involved (see Chapters 7 and 8). Hydrogen peroxide can easily diffuse through membranes of cell organelles and cause oxidative stress far away from its origin. In this respect, an important function for plant catalases is to detoxify H_2O_2 directly at its origin, mainly localized in peroxisomes.

A significant role of catalase with enhanced peroxidatic activity is still obscure (Havir and McHale, 1990). In sunflower cotyledons, one function of stable core catalase is probably guaranteeing a stable minimum of catalase activity under conditions when matrix catalases are photoinactivated (Eising *et al*, 1998).

Turnover and regulation

For the regulation of catalase protein turnover, distinction has to be made between regulation of gene expression and protein synthesis and degradation. Regulation of active catalase proteins in plants is a complex system, because total catalase activity in many plant organs is the sum of the activity of different catalase isoforms. It has been demonstrated for the peroxisomal catalase of sunflower cotyledons that turnover of catalase apoprotein and haem is, essentially, co-ordinate (Eising and Süselbeck, 1991).

It has yet to be established if different catalase isoforms in a plant species are the product of one or more than one catalase gene. In castor bean, one single catalase form was initially reported (González *et al*, 1993), but in further studies, multiple catalase forms and two catalase genes, *cat1* and *cat2* were identified (Suzuki *et al*, 1994). Plant systems with multiple catalase genes and/or multiple catalase isoforms are listed in former reviews (Scandalios *et al*, 1997). Most systems are regulated by synthesis of catalase isoforms in a tissue-specific or time-dependent manner. Hypocotyls and roots from *Ricinus communis* L. contain only catalase subunits of 56 kD, while the endosperm and cotyledons contain two different subunits of 54 and 56 kD. Further examples of plant systems with complex regulation are Arabidopsis and maize. In Arabidopsis, three catalase genes (*CAT1 to CAT3*) and at least six isoforms occur (Frugoli *et al*, 1996). Expression is also controlled in a tissue-specific and time-dependent manner (Boldt and Scandalios, 1997). *CAT1* and *CAT3* mRNAs are mainly localized in bolts and leaves, and *CAT2* mRNA exclusively in leaves. Two isoforms were detected in roots and seven in blossoms. Furthermore, the three Arabidopsis genes respond in a different way to light; *CAT3* shows no reaction, *CAT1* mRNA is slightly and *CAT2* mRNA strongly induced. Except *CAT1*, the genes are also influenced by a circadian rhythm; CAT2 mRNA expression is maximal at dawn and *CAT3* mRNA at dusk (Zhong and McClung, 1996).

Expression of cottonseed catalases is mainly regulated at the post-transcriptional level (Ni and Trelease, 1991a; Ni and Trelease, 1991b). The accumulation of cotton subunit 1 mRNA is not light dependent, whereas the accumulation of subunit 2 mRNA occurs only after exposure of seedlings to light. In maize, three unlinked structural genes, *CAT1* to *CAT3*, code for three biochemically different isoforms, CAT1 to CAT3. This system was elucidated in former articles (Chandlee *et al*, 1983, Redinbaugh *et al*, 1988, Guan and Scandalios, 1996, Scandalios *et al*, 1997) and will not be discussed in detail here. Additionally, Guan *et al* (2000) presented evidence that H_2O_2 and abscisic acid are involved in CAT1 expression. Expression of CAT1 increased when concentrations of these substances were enhanced (Guan and Scandalios, 1998). Abscisic acid probably influences directly the level of H_2O_2 and activates expression of *cat1*.

Evolution

Evidence was provided that typical catalases arose from catalase-peroxidases by a gene duplication event (Welinder, 1991). Sequences of typical catalases are rather conserved (Okada *et al*, 1987, von Ossowski *et al*, 1991), so they are suitable for phylogenetic reconstructions. Von Ossowski *et al* (1993) analyzed amino acid sequences of twenty prokaryotic, fungal, animal and plant catalases. The homology was highest in the N-terminal part and declined to the C-terminal part. Ossowski *et al* (1993) proposed that plant catalases arose independently from animal and fungal catalases. Guan and Scandalios (1996) compared the sequences of three maize catalases with sixteen other plant catalases and other eukaryotic and prokaryotic sequences.

Phylogenies showed that all plant catalases are derived from a common ancestral gene and can be divided into three distinct groups. The first major group includes most of the dicot catalases. The second group also contains dicot catalases, whereas the third group is a monocot-specific catalase class. Klotz *et al* (1997) enlarged this phylogenetic comparison with up to seventy sequences from different organisms. They grouped sunflower matrix catalase-1 to group I plant catalases. Both sunflower matrix catalases show high homology to group I plant catalases. However, sunflower core catalases show the highest homology to group II plant catalases, but founded evolutionary analyses for these catalase species were not performed, yet. Iwamoto *et al* (1998) proposed an ancestral catalase gene with seven introns and an intron loss and isozyme divergence after separation of monocots and dicots. Frugoli *et al* (1998) described evolution of the Arabidopsis catalase gene family as a series of gene duplications before the divergence of monocots and dicots leading to intron loss.

PEROXISOMAL ASCORBATE PEROXIDASES

Ascorbate peroxidase activity has been demonstrated in various cell compartments of plant cells, including the peroxisomes, as has the existence of compartment-specific isoforms of the enzyme. Plant ascorbate peroxidases have been classified as catalase-peroxidases which form one of the three evolutionary-based groups of the catalase family (Zámocký and Koller, 1999). The bifunctional catalytic behaviour of the members of this group has not been demonstrated for peroxisomal ascorbate peroxidases, i.e. catalytic activity has not been reported for the peroxisomal ascorbate peroxidases.

Ascorbate peroxidase uses two molecules of ascorbate as electron donor for the reduction of one molecule of H_2O_2 to water. The concomitantly formed two molecules of the monodehydroascorbate radical can be reduced to ascorbate by the monodehydroascorbate reductase using different electron donors (e.g. NAD(P)H, reduced ferredoxin) or can disproportionate non-enzymatically to ascorbate and dehydroascorbate. The latter can be reduced to ascorbate by the ascorbate-glutathione cycle. Thus, H_2O_2 detoxification by the ascorbate peroxidase consumes reducing power once generated by energy expence in the plant. Monodehydroascorbate reductase has been demonstrated in peroxisomes (membrane-bound enzyme; Bowditch and Donaldson, 1990, Bunkelmann and Trelease, 1996, Ishikawa *et al*, 1998) as have been the components of the ascorbate-glutathione cycle (Jiménez *et al*, 1997, see also Chapters 7 and 8).

So far, peroxisomal ascorbate peroxidase has been partially purified only as recombinant protein (Ishikawa *et al*, 1998). The specific activity of the enzyme preparation amounted to 0.7 nkat (mg protein)$^{-1}$ with the substrate H_2O_2. Ascorbate peroxidase activities based on organelle protein and H_2O_2 decomposition ranged from 0.1 to 1.3 nkat (mg peroxisomal protein)$^{-1}$ (Ishikawa *et al*, 1998; Jiménez *et al*, 1997; Yamaguchi *et al*, 1995). Where determined in parallel, the ascorbate peroxidase activity was two to three orders of magnitude lower than that of the catalase. Apparent K_m values of 74 μM and 1.9 mM for H_2O_2 and ascorbate, respectively, have been reported for the purified recombinant ascorbate peroxidase of etiolated spinach cotyledons (Ishikawa *et al*, 1998). Since catalase has a far lower affinity for H_2O_2 (measured K_m values range from 47x10^3 to 110x10^4 μM, Halliwell, 1974), ascorbate peroxidase can scavange H_2O_2 more efficiently than catalase at low H_2O_2 concentrations. The ascorbate/dehydroascorbate content of pea leaf peroxisomes isolated in polar media has been determined to be 1.6 μg (mg organelle protein)$^{-1}$ (Jiménez *et al*, 1997). This corresponds to a peroxisomal ascorbate concentration of about 2 mM estimated on the basis of the liver peroxisome dry weight of 2.4x10^{-11} mg (de Duve and Baudhuin, 1966) and an average volume of 1 μm^3 of the cotton cotyledon glyoxysome (Kunce *et al*, 1984).

The peroxisomal ascorbate peroxidase does not accept NAD(P)H, glutathione or cytochrome c as electron donor. The enzyme is inhibited by azide and cyanide, indicating that the peroxisomal ascorbate peroxidase is a typical haem protein (Ishikawa *et al*, 1998; Yamaguchi *et al*, 1995).

Ascorbate peroxidase is a protein of the peroxisome membrane, which has been identified using antibodies recognizing a 30 to 32 kD prominent peroxisome membrane polypeptide (Bunkelmann and Trelease, 1996; Ishikawa *et al*, 1998; Yamaguchi *et al*, 1995) on SDS gels, rather than by its enzymic activity (Ishikawa *et al*, 1998; Jiménez *et al*, 1997). The native enzyme is probably a homodimer (Bunkelmann and Trelease, 1996). The monomer contains a hydrophobic, putative membrane-spanning region near the C-terminus (Bunkelmann and Trelease, 1996; Ishikawa *et al*, 1998; Zhang *et al*, 1997). The isolated cDNAs encoding peroxisomal ascorbate peroxidases possess an open reading frame of 858 to 864 bp coding for proteins containing 286 to 288 amino acid residues and exhibiting predicted molecular masses of 31 to 32 kD (Bunkelmann and Trelease, 1996; Ishikawa *et al*, 1998; Zhang *et al*, 1997, GenBank accession number U69138). A putative full-length cDNA possessed 1258 bp (Bunkelmann and Trelease, 1996). Recently, a peroxisomal ascorbate peroxidase from barley with a length of 291 amino acids was cloned (Shi *et al*, 2001, accession number

BAB62533). Peroxisomal ascorbate peroxidases from different plant species showed high identity (>80%) on the deduced amino acid sequences (Bunkelmann and Trelease, 1996; Ishikawa *et al*, 1998). Homology to cytosolic ascorbate peroxidases is also high, but sequences from cytosolic proteins lack the putative membrane-spanning region at the C-terminus (Bunkelmann and Trelease, 1996; Ishikawa *et al*, 1998). On the other hand, a putative membrane-spanning region at the C-terminus is also present in the thylakoid-bound ascorbate peroxidase of spinach chloroplasts (Ishikawa *et al*, 1998). The peroxisomal ascorbate peroxidase synthesized on polysomes in the cytosol is not directly targeted to its functional site. The enzyme appears to be inserted post-translationally into a distinct ER subdomain and sorted to the peroxisomes by way of the ER (Mullen *et al*, 1999).

Regarding the orientation of the active site of the ascorbate peroxidase in the peroxisome membrane, exposure of the active site to the peroxisome matrix (Bunkelmann and Trelease, 1996) as well as exposure to the cytosolic side of the organelle (Ishikawa *et al*, 1998; Jiménez *et al*, 1997; Yamaguchi *et al*, 1995) has been proposed based on the results of protease treatment of the organelles and latency studies. Accordingly, the function of the ascorbate peroxidase is seen either in the scavenging of intraperoxisomal H_2O_2 at low H_2O_2 concentrations (when catalase is rather inefficient; see above) (Bunkelmann and Trelease, 1996) or in the detoxification of H_2O_2 leaking through the peroxisome membrane (Ishikawa *et al*, 1998, Yamaguchi *et al*, 1995; Kunce *et al*, 1984). In the former case, a mechanism for the intraperoxisomal reoxidation of the intraperoxisomally formed NADH, i.e. NADH oxidation by the monodehydroascorbate reductase or by the ascorbate-glutathione cycle (see above) could concomitantly result (for further mechanisms postulated for the reoxidation of intraperoxisomal NADH see Mullen and Trelease, 1996, and Chapters 5, 7 and 8 in this volume).

Due to their catalytic properties, ascorbate peroxidases are considered in the broader sense to be important defense enzymes protecting plant cells from oxidative stress damage. There are results suggesting that the peroxisomal ascorbate peroxidase specifically protects the peroxisome against oxidative stress resulting from peroxisome metabolism, as other ascorbate peroxidase isoforms appear to do with respect to their housing compartment (Wang *et al*, 1999; Yoshimura *et al*, 2000). Recently, a co-operation of catalases and ascorbate peroxidases for the detoxification of hydrogen peroxide was suggested (Mizuno *et al*, 1998).

CONCLUSION

Catalases are present in all plants where they have been looked for. In the 1970s, scientists were mainly concerned to describe catalase properties and reaction mechanisms in general, mainly based on animal or yeast catalases. In the 1980s, the investigations focussed on the description of multiple catalase genes and catalase isoforms for many plant species. In the 1990s until the present, investigations focussed on regulation of expression of catalases in plant species. Hydrogen peroxide was detected to be a second messenger in plants and influences of substances like abscisic acid or salicylic acid on activity of catalases were discussed. Additionally, detoxification of hydrogen peroxide via ascorbate glutathione cycle was proven (Jiménez *et al*, 1997). In this respect, the enzyme catalase is embedded into the complex system of antioxidant defense with different manners of regulation. Also, the function of hydrogen peroxide and catalase in stressed plants and pathogen response was discussed, getting more and more complex (del Río *et al*, 1998). Further studies will probably be focused on specific functions of separate catalase isoforms in this respect. On the other hand, displaying the molecular structure of plant catalases may help to understand the complex process of tetramerization and regulation of catalase expression in distinct plant tissues or organs.

REFERENCES

Alvares, K., Widrow, R.J., Abu-Jadweh, G.M., Schmidt, J.V., Yelandi, A.V., Rao, M.S. and Reddy, J.K. (1992) Rat urate oxidase produced by recombinant baculovirus expression: formation of peroxisome crystalloid core-like structures. *Proc. Natl. Acad. Sci. USA.* **89**: 4908-4912.

Aro, E.M., Virgin, I. and Andersson, B. (1993) Photoinhibition of photosystem II. Inactivation, protein damage and turnover. *Biochim. Biophys. Acta.* **1143**: 113-134.

Beaumont, F., Jouve, H.M., Gagnon, J., Gaillard, J. and Pelmont, J. (1990) Purification and properties of a catalase from potato tubers (*Solanum tuberosum*). *Plant Sci.* **72**: 19-26.

Bergdoll, M., Remy, M.H., Cagnon, C., Masson, J.M. and Dumas, P. (1997) Proline-dependent oligomerization with arm exchange. *Structure.* **5**(3): 391-401.

Berthet, S., Nykyri, L.M., Bravo, J., Maté, M.J., Berthet-Colominas, C., Alzari, P.M., Koller, F. and Fita, I. (1997) Crystallization and preliminary structural analysis of catalase A from *Saccharomyces cerevisiae. Prot. Sci.* **6**: 481-483.

Bestwick, C.S., Brown, I.R., Bennett, M.H.R. and Mansfield, J.W. (1997) Localization of hydrogen peroxide accumulation during the hypersensitive reaction of lettuce cells to *Pseudomonas syringae* pv *phaseolica*. *Plant Cell.* **9**: 209-221.

Boldt, R. and Scandalios, J.G. (1997) Influence of UV-light on the expression of the *cat2* and *cat3* catalase genes in maize. *Free Rad. Biol. Med.* **23**(3): 505-514.

Bowditch, M.I. and Donaldson, R.P. (1990) Ascorbate free-radical reduction by glyoxysomal membranes. *Plant Physiol.* **94**: 531-537.

Bravo, J., Verdaguer, N., Tormo, J., Betzel, C., Switala, J., Loewen, P.C. and Fita, I. (1995) Crystal structure of catalase HPII from *Escherichia coli*. *Structure.* **3**(5): 491-502.

Bunkelmann, J. and Trelease, R.N. (1996) Ascorbate peroxidase. A prominent membrane protein in oilseed glyoxysomes. *Plant Physiol.* **110**: 589-598.

Chan, H.T., Tam, S.Y. and Koide, R.T. (1978) Isolation and characterization of catalase from papaya. *J. Food Sci.* **43**: 989-991.

Chandlee, J.M., Tsaftaris, A.S. and Scandalios, J.G. (1983) Purification and partly characterization of three genetically defined catalases of maize. *Plant Sci. Lett.* **29**: 117-131.

Chen, Z., Silva, H. and Klessig, D.F. (1993) Active oxygen species in the induction of plant systemic acquired resistance by salicylic acid. *Science.* **262**: 1883-1886.

Cheng, L., Kellogg, E.W. III and Packer, L. (1981) Photoinactivation of catalase. *Photochem. Photobiol.* **34**: 125-129.

Corpas, F.J., Palma, J.M., Sandalio, L.M., López-Huertas, E., Romero-Puertas, M.C., Barroso, J.B. and del Río, L.A. (1999) Purification of catalase from pea leaf peroxisomes: identification of five different isoforms. *Free Rad. Res.* **31**: S235-S241.

Cronshaw, J. (1964) Crystal bodies of plant cells. *Protoplasma.* **59**: 318-325.

Crookes, W.J. and Olsen, L.J. (1998) The effect of chaperones and the influence of protein assembly on peroxisomal protein import. *J. Biol. Chem.* **273**: 17236-17242.

de Duve, C. and Baudhuin, P. (1966) Peroxisomes (microbodies and related particles). *Physiol. Rev.* **46**: 323-357.

del Río, L.A., Palma, J.M., Sandalio, L.M., Corpas, F.J., Pastori, G.M., Bueno, P. and López-Huertas, E. (1996) Peroxisomes as a source of superoxide and hydrogen peroxide in stressed plants. *Biochem. Soc. Trans.* **24**: 434-442.

del Río, L.A., Pastori, G.M., Palma, J.M., Sandalio, L.M., Sevilla, F., Corpas, F.J., Jiménez, A., López-Huertas, E. and Hernández, J.A. (1998) The activated oxygen role of peroxisomes in senescence. *Plant Physiol.* **116**: 1195-1200.

Diefenbach, J. and Kindl, H. (2000) The membrane-bound DnaJ protein located at the cytosolic site of glyoxysomes specifically binds the cytosolic isoform 1 of Hsp70, but not other Hsp70 species. *Eur. J. Biochem.* **267**: 746-754.

Dounce, A.L. (1983) A proposed mechanism for the catalytic action of catalase. *J. Theor. Biol.* **105**: 553-567.

Drumm, H. and Schopfer, P. (1974) Effect of phytochrome on development of catalase activity and isoenzyme pattern in mustard (*Sinapis alba* L.) seedlings. A re-investigation. *Planta* **120**: 13-30.

Durner, J. and Klessig, D.F. (1996) Salicylic acid is a modulator of tobacco and mammalian catalases. *J. Biol. Chem.* **271**(45): 28492-28501.

Eising, R. and Gerhardt, B. (1987) Catalase degradation in sunflower cotyledons during peroxisome transition from glyoxysomal to leaf peroxisomal function. *Plant Physiol.* **84**: 225-232.

Eising, R. and Gerhardt, B. (1986) Activity and haematin content of catalase from greening sunflower cotyledons. *Phytochem.* **25**(1): 27-31.

Eising, R. and Süselbeck, B. (1991) Turnover of catalase haem and apoprotein moieties in cotyledons of sunflower seedlings. *Plant Physiol.* **97**: 1422-1429.

Eising, R., Heinze, M., Kleff, S. and Tenberge, K.B. (1998) Subcellular distribution and photooxidation of catalase in sunflower. In: *Antioxidants in Higher Plants: Biosynthesis, characteristics, actions and specific functions in stress defense.* (Noga, G. and Schmitz, M. eds.). pp. 53-63. Shaker-Verlag, Aachen.

Eising, R., Kleff, S., Ruholl, C. and Tenberge, K.B. (1996) Turnover of catalase in sunflower cotyledons. In: *Proceeedings of the International Compositae Conference, Kew, 1994*, Vol. 2. Biology and Utilization. (Caligary, P.D.S. and Hind, D.J.N. vol. eds.). Royal Botanic Gardens, Kew, pp. 45-60.

Eising, R., Trelease, R.N. and Ni, W. (1990) Biogenesis of catalase in glyoxysomes and leaf-type peroxisomes of sunflower cotyledons. *Arch. Biochem. Biophys.* **278**: 258-264.

Elstner, E.F. and Heupel, A. (1976) Formation of hydrogen peroxide by isolated cell wall from horseradish (*Armoracia lapathifolia* Gilib). *Planta.* **130**: 175-180.

Esaka, M. and Asahi, T. (1982) Purification and properties of catalase from sweet potato root microbodies. *Plant Cell Physiol.* **23**(2): 315-322.

Fagerberg, W.R. (1984) Cytochemical changes in palisade cells of developing sunflower leaves. *Protoplasma.* **119**: 21-30.

Feierabend, J. and Dehne, S. (1996) Fate of the porphyrin cofactors during the light-dependent turnover of catalase and the photosystem II reaction-centre protein D1 in mature rye leaves. *Planta.* **198**: 413-422.

Feierabend, J. and Engel, S. (1986) Photoinactivation of catalase *in vitro* and in leaves. *Arch. Biochem. Biophys.* **251**: 567-576.

Fita, I. and Rossmann, M.G. (1995) The NADPH binding site on beef liver catalase. *Proc. Natl. Acad. Sci. USA.* **82**: 1604-1608.

Frederick, S.E., Gruber, P.J. and Newcomb, E.H. (1975) Plant microbodies. *Protoplasma*. **84**: 1-29.

Frederick, S.E., Newcomb, E.H., Vigil, E.L. and Wergin, W.P. (1968) Fine structural characterization of plant microbodies. *Planta*. **8**: 229-252.

Frugoli, J.A., McPeek, M.A., Thomas, T.L. and McClung, R. (1998) Intron loss and gain during evolution of the catalase gene family in angiosperms. *Genetics*. **149**: 355-365.

Frugoli, J.A., Zhong, H.H., Nuccio, M.L., McCourt, P., Thomas, T.L. and McClung, C.R. (1996) Catalase is encoded by a multi-gene family in *Arabidopsis thaliana* Heynh. *Plant Physiol*. **112**(1): 327-336.

Galston, A.W. (1955) Plant catalase. *Meth. Enzymol*. **2**: 789-791.

Galston, A.W., Bonnichsen, R.K. and Arnon, D.I. (1951) The preparation of highly purified spinach leaf catalase. *Act. Chem. Scand*. **5**: 781-790.

Gao, X., Marrison, J.L., Pool, M.R., Leech, R.M. and Baker, A. (1996) Castor bean isocitrate lyase lacking the putative peroxisomal targeting signal 1 ARM is imported into plant peroxisomes both *in vitro* and *in vivo*. *Plant Physiol*. **112**: 1457-1464.

Garcia, R., Kaid, N., Vignaud, C. and Nicolas, J. (2000) Purification and some properties of catalase from wheatgerm (*Triticum aestivum* L.). *J. Agric. Food Chem*. **48**: 1050-1057.

Gerhardt, B. (1987) Fatty acid β-oxidation in higher plants. In: *The Metabolism, Structure, and Function of Plant Lipids*. (Stumpf, P.K., Mudd, J.B. and Nes, W.D. eds.). pp. 399-404, Plenum, New York.

Gerhardt, B. (1986) Basic metabolic function of the higher plant peroxisome. *Physiol. Vég*. **24**(3): 397-410.

Gianinetti, A., Cantoni, M., Lorenzoni, C., Salamini, F. and Marocco, A. (1993) Altered levels of antioxidant enzymes associated with two mutations in tomato. *Physiol. Plant*. **89**: 157-164.

Gietl, C. (1990) Glyoxysomal malate dehydrogenase from watermelon is synthesized with an amino terminal transit peptide. *Procl. Natl. Acad. Sci. USA*. **87**: 5773-5777.

Giordani A, Morlière P, Aubailly M and Santus R (1997) Photoinactivation of cellular catalase by ultraviolet radiation. *Redox Report*. **3**(1): 49-55.

González, E. (1991) The C-terminal domain of plant catalases: implications for a glyoxysomal targeting sequence. *Eur . J. Biochem*. **199**: 211-215.

González, E., Brush, M., Lee, V. and Wainwright, I. (1993) Evidence for a single catalase gene in castor bean. *Plant Physiol. Biochem*. **31**(3): 379-386.

Gouet, P., Jouve, H.M. and Dideberg, O. (1995) Crystal structure of *Proteus mirabilis* PR catalase with and without bound NADPH. *J. Mol. Biol*. **249**: 933-954.

Gould, S.J., Keller, G.A. and Subramani, S. (1987) Identification of a peroxisomal targeting signal at the carboxy terminus of firefly luciferase. *J. Cell Biol.* **105**: 2923-2931.

Gregory, R.P.F. (1968) An improved preparative method for spinach catalase and evaluation of some of its properties. *Biochim. Biophys. Acta.* **159**: 429-439.

Grotjohann, N., Janning, A. and Eising, R. (1997) *In vitro* photoinactivation of catalase isoforms from cotyledons of sunflower (*Helianthus annuus* L.). *Arch. Biochem. Biophys.* **346**(2): 208-218.

Guan, L. and Scandalios, J.G. (1998) Effects of the plant growth regulator abscisic acid and high osmoticum on the developmental expression of the maize catalase genes. *Physiol. Plant.* **104**: 413-422.

Guan, L. and Scandalios, J.G. (1996) Molecular evolution of maize catalases and their relationship to other eukaryotic and prokaryotic catalases. *J. Mol. Evol.* **42**: 570-579.

Guan, L.M., Zhao, J. and Scandalios, J.G. (2000) *Cis*-elements and *trans*-factors that regulate expression of the maize *cat1* antioxidant gene in response to ABA and osmotic stress: H_2O_2 is the likely intermediary signalling molecule of response. *Plant J.* **22**(2): 87-95.

Halliwell, B. (1974) Oxidation of formate by peroxisomes and mitochondria from spinach leaves. *Biochem. J.* **138**(1) : 77-85.

Havir, E.A. and McHale, N.A. (1990) Purification and characterization of an isozyme of catalase with enhanced-peroxidatic activity from leaves of *Nicotiana sylvestris. Arch. Biochem. Biophys.* **283**(2): 491-495.

Havir, E.A. and McHale, N.A. (1989) Enhanced peroxidatic activity in specific catalase isozymes of tobacco, barley, and maize. *Plant Physiol.* **91**: 812-815.

Havir, E.A. and McHale, N.A. (1987) Biochemical and developmental characterization of multiple forms of catalase in tobacco leaves. *Plant Physiol.* **84**: 450-455.

Havir, E.A., Brisson, L.F. and Zelitch, I. (1996) Distribution of catalase isoforms in *Nicotiana tabacum. Phytochem.* **41**(3): 699-702.

Heinze, M., Reichelt, R., Kleff, S. and Eising, R. (2000) High resolution scanning electron microscopy of protein inclusions (cores) purified from peroxisomes of sunflower (*Helianthus annuus* L.) cotyledons. *Cryst. Res. Technol.* **35**(6-7): 877-886.

Hertwig, B., Streb, P. and Feierabend, J. (1992) Light-dependence of catalase synthesis and degradation in leaves and the influence of interfering stress conditions. *Plant Physiol.* **100**: 1547-1553.

Heupel, R. and Heldt, H.W. (1994) Protein organization in the matrix of of leaf peroxisomes. A multi-enzyme complex involved in photorespiratory metabolism. *Eur. J. Biochem.* **220**: 165-172.

Hirasawa, M., Gray, K.A., Shaw, R.W. and Knaff, D.B. (1987) Spectroscopic properties of spinach catalase. *Biochim. Biophys. Acta.* **911**: 37-44.

Hochman, A. and Goldberg, I. (1991) Purification and characterization of a catalase-peroxidase and a typical catalase from the bacterium *Klebsiella pneumoniae*. *Biochim. Biophys. Acta.* **1077**(3): 299-307.

Hook, D.W.A. and Harding, J.J. (1997) Molecular chaperones protect catalase against thermal stress. *Eur. J. Biochem.* **247**: 380-385.

Huang, A.H.C. and Beevers, H. (1973) Localization of enzymes within microbodies. *J. Cell Biol.* **58**: 379-389.

Huang, A.H.C., Trelease, R.N. and Moore, T.S. (1983) *Plant peroxisomes*. Academic Press, New York.

Ishikawa, T., Yoshimura, K., Sakai, K., Tamoi, M., Takeda, T. and Shigeoka, S. (1998) Molecular characterization and physiological role of a glyoxysome-bound ascorbate peroxidase from spinach. *Plant Cell Physiol.* **39**(1): 23-34.

Isin, S. and Allen, R. (1991) Isolation and characterization of a pea catalase cDNA. *Plant Mol. Biol.* **17**: 1263-1265.

Iwamoto, M., Maekawa, M., Saito, A., Higo, H. and Higo, K. (1998) Evolutionary relationship of plant catalase genes inferred from exon-intron structures: Isozyme divergence after the separation of monocots and dicots. *Theor. Appl. Genet.* **97**: 9-19.

Jiménez, A., Hernández, J.A., del Río, L.A. and Sevilla, F. (1997) Evidence for the presence of the ascorbate-glutathione cycle in mitochondria and peroxisomes of pea leaves. *Plant Physiol.* **114**: 275-284.

Jouve, H.M., Andreoletti, P., Gouet, P., Hajdu, J. and Gagnon, J. (1997) Structural analysis of compound I in haemoproteins: Study on *Proteus mirabilis* catalase. *Biochimie.* **79**: 667-671.

Kato, A., Hayashi, M. and Nishimura, M. (1999) Oligomeric proteins containing N-terminal targeting signals are imported into peroxisomes in transgenic Arabidopsis. *Plant Cell Physiol.* **40**(6): 586-591.

Kindl, H. (1982) Glyoxysome biogenesis via cytosolic pools in cucumber. In: *Peroxisomes and glyoxysomes.* (Kindl, H. and Lazarow, P.B. eds.). *Ann. N.Y. Acad. Sci.* **386**: 314-328.

Kirkman, H.N. and Gaetani, G.F. (1984) Catalase: a tetrameric enzyme with four tightly bound molecules of NADPH. *Proc. Natl. Acad. Sci. USA.* **81**(14): 4343-4347.

Kirkman, H.N., Rolfo, M., Ferraris, A.M. and Gaetani, G.F. (1999) Mechanisms of protection of catalase by NADPH. *J. Biol. Chem.* **274**(20): 13908-13914.

Kleff, S., Trelease, R.N. and Eising, R. (1994) Nucleotide and deduced amino acid sequence of a putative higher molecular weight precursor for catalase in sunflower cotyledons. *Biochim. Biophys. Acta.* **1224**: 463-466.

Kleff, S., Sander, S., Mielke, G. and Eising, R. (1997) The predominant protein in peroxisomal cores of sunflower cotyledons is a catalase that differs in primary structure from the catalase in the peroxisomal matrix. *Eur. J. Biochem.* **245**: 402-410.

Klotz, M.G., Klassen, G.R. and Loewen, P.C. (1997) Phylogenetic relationships among prokaryotic and eukaryotic catalases. *Mol. Biol. Evol.* **14**(9): 951-958.

Kono, Y. and Fridovich, I. (1982) Superoxide radical inhibits catalase. *J. Biol. Chem.* **257**(10): 5751-5754.

Kunce, C.M., Trelease, R.N. and Doman, D.C. (1984) Ontogeny of glyoxysomes in maturing and germinated cotton seeds – a morphometric analysis. *Planta*. **161**: 156-164.

Kunce, C.M., Trelease, R.N. and Turley, R.B. (1988) Purification and biosynthesis of cottonseed (*Gossypium hirsutum* L.) catalase. *Biochem. J.* **251**: 147-155.

Lamb, J.E., Riezman, H., Becker, W.M. and Leaver, C.J. (1978) Regulation of glyoxysomal enzymes during germination of cucumber. *Plant Physiol.* **62**: 754-760.

Lazarow, P.B. and de Duve, C. (1973) The synthesis and turnover of rat liver peroxisomes. V. Intracellular pathway of catalase synthesis. *J. Cell Biol.* **59**: 507-524.

Low, P.S. and Merida, J.R. (1996) The oxidative burst in plant defense: function and signal transduction. *Physiol. Plant.* **96**: 533-542.

Marinos, N.G. (1965) Comments on the nature of a crystal-containing body in plant cells. *Protoplasma*. **60**: 31-33.

McClung, C.R. (1997) Regulation of catalases in Arabidopsis. *Free Rad. Biol. Med.* **23**(3): 489-496.

McNew, J.A. and Goodman, J.M. (1994) An oligomeric protein is imported into peroxisomes *in vivo*. *J. Cell Biol.* **127**(5): 1245-1257.

Middelkoop, E., Wiemer, E.A.C., Tycho Schonmaker, D.E., Strijland, A. and Tager, J.M. (1993) Topology of catalase assembly in human skin fibroblasts. *Biochim. Biophys. Acta.* **1220**: 15-20.

Mizuno, M., Kamei, M. and Tsuchida, H. (1998) Ascorbate peroxidase and catalase co-operate for protection against hydrogen peroxide generated in potato tubers during low temperature storage. *Biochem. Mol. Biol. Intern.* **44**(4): 717-726.

Mullen, R.T. and Gifford, D.J. (1993) Purification and characterization of catalase from loblolly pine (*Pinus taeda* L.). *Plant Physiol.* **103**: 477-483.

Mullen, R.T. and Trelease, R.N. (1996) Biogenesis and membrane properties of peroxisomes: Does the boundary membrane serve and protect? *Trends Plant Sci.* **1**(11): 389-394.

Mullen, R.T., Lee, M.S. and Trelease, R.N. (1997A) Identification of the peroxisomal targeting signal for cottonseed catalase. *Plant J.* **12**: 313-322.

Mullen, R.T., Lee, M.S., Flynn, R. and Trelease, R.N. (1997b) Implications for the role of accessory residues upstream of the type I peroxisomal targeting signal. *Plant Physiol.* **115**: 881-889.

Mullen, R.T., Lisenbee, C.S., Miernyk, J.A. and Trelease, R.N. (1999) Peroxisomal membrane ascorbate peroxidase is sorted to a membranous network that resembles a subdomain of the endoplasmic reticulum. *Plant Cell.* **11**: 2167-2185.

Murshudov, G.N., Melik-Adamyan, W.R., Grebenko, A.L., Barynin, V.V., Vagin, A.A., Vainshtein, B.K., Dauter, Z. and Wilson, K.S. (1992) Three-dimensional structure of catalase from *Micrococcus lysodeikticus* at 1.5 Å resolution. *FEBS Lett.* **312**(2-3), 127-131.

Murthy, M.R., Reid, T.J., Sicignano, A., Tanaka, N. and Rossmann, M.G. (1981) Structure of beef liver catalase. *J. Mol. Biol.* **152**: 465-499.

Ni, W. and Trelease, R.N. (1991a) Two genes encode the two subunits of cottonseed catalase. *Arch. Biochem. Biophys.* **289**(2): 237-243.

Ni, W. and Trelease, R.N. (1991b) Post-transcriptional regulation of catalase isozyme expression in cotton seeds. *Plant Cell.* **3**: 737-744.

Ni, W., Trelease, R.N. and Eising, R. (1990) Two temporally synthesized charge subunits interact to form the five isoforms of cottonseed catalase. *Biochem. J.* **269**: 233-238.

Nicholls, P., Fita, I. and Loewen, P.C. (2001) Enzymology and structure of catalases. *Adv. Inorg. Chem.* **51**: 51-106.

Novikoff, A.B. and Goldfischer, S. (1969) Visualization of peroxisomes (microbodies) and mitochondria with diaminobenzidine. *J. Histochem. Cytochem.* **17**: 675-680.

Okada, H., Ueda, M., Sugaya, T., Atomi, H., Mozaffar, S., Hishida, T., Teranishi, Y., Okazaki, K., Takechi, T., Kamiryo, T. and Tanaka, A. (1987) Catalase gene of the yeast *Candida tropicalis. Eur. J. Biochem.* **170**: 105-110.

Orr, C.W.M. (1967) Studies on ascorbic acid. I. Factors influencing the ascorbate-mediated inhibition of catalase. *Biochemistry.* **6**(10): 2995-3000.

Osumi, T., Tsukamoto, T., Hata, S., Yokota, S., Miura, S., Fujiki, Y., Hijakata, M., Miyazawa, S. and Hashimoto, T. (1991) Amino-terminal pre-sequence of the precursor of peroxisomal 3-ketoacyl-CoA thiolase is a cleavable signal peptide for peroxisomal targeting. *Biochem. Biophys. Res. Comm.* **181**(3): 947-954.

Ota, Y., Ario, T., Hayashi, K., Nakagawa, T., Hattori, T., Maeshima, M. and Asahi, T. (1992) Tissue-specific isoforms of catalase subunits in castor bean seedlings. *Plant Cell Physiol.* **33**(3): 225-232.

Pool, M.R., López-Huertas, E. and Baker, A. (1998) Characterization of intermediates in the process of plant peroxisomal protein import. *EMBO J.* **17**(23): 6854-6862.

Purdue, P.E. and Lazarow, P.B. (1996) Targeting of human catalase to peroxisomes is dependent upon a novel COOH-terminal peroxisomal targeting sequence. *J. Cell Biol.* **134**: 849-862.

Rachubinski, R.A. and Subramani, S. (1995) How proteins penetrate peroxisomes. *Cell.* **83**: 525-528.

Redinbaugh, M.G., Wadsworth, G.J. and Scandalios, J.G. (1988) Characterization of catalase transcripts and their differential expression in maize. *Biochim. Biophys. Acta.* **951**: 104-116.

Reid ,T.J. III, Murthy, M.R., Sicignano, A., Tanaka, N., Musick, W.D. and Rossmann, M.G. (1981) Structure and haem environment of beef liver catalase at 2.5 Å resolution. *Proc. Natl. Acad. Sci. USA.* **78**(8): 4767-4771.

Rüffer, M., Steipe, B. and Zenk, M.H. (1995) Evidence against specific binding of salicylic acid to plant catalase. *FEBS Lett.* **377**: 175-180.

Salguero, J. and Böttger, M. (1995) Secreted catalase activity from roots of developing maize (*Zea mays* L.) seedlings. *Protoplasma.* **184**: 72-78.

Scandalios, J.G. (1974) Subcellular localization of catalase variants coded by two genetic loci during maize development. *J. Hered.* **65**: 28-32.

Scandalios, J.G. (1965) Subunit dissociation and recombination of catalase isozymes. *Proc. Natl. Acad. Sci. USA.* **53**: 1035-1040.

Scandalios, J.G., Guan, L.M. and Polidoros, A. (1997) Catalases in plants: gene structure, properties, regulation, and expression. In: *Oxidative stress and the molecular biology of antioxidant defences* (Scandalios, J.G., ed.). pp. 343-406. Plainview, New York, Cold Spring Harbor Laboratory Press.

Scandalios, J.G., Tsaftaris, A.S., Chandlee, J.M. and Skadsen, R.W. (1984) Expression of the developmentally regulated catalase (*Cat*) genes in maize. *Dev. Genet.* **4**: 281-293.

Scandalios, J.G., Tong, W.F. and Roupakias, D.G. (1980) *Cat3*, a third gene locus coding for a tissue-specific catalase in maize: genetics, intracellular location and some biochemical properties. *Mol. Gen. Genet.* **179**: 33-41.

Schäfer, L. and Feierabend, J. (2000) Photoinactivation and protection of glycolate oxidase *in vitro* and in leaves. *Z. Naturforsch.* **55**c: 361-372.

Schiefer, S., Teifel, W. and Kindl, H. (1976) Plant microboby proteins. I. Purification and characterization of catalase from leaves of *Lens culinaris. Physiol. Chem.* Hoppe-Seyler **357**: 163-175.

Schittenhelm, J., Toder, S., Fath, S., Westphal, S. and Wagner, E. (1994) Photoinactivation of catalase in needles of Norway spruce. *Physiol. Plant.* **90**: 600-606.

Schonbaum, G.R. and Chance, B. (1976) Catalase. In: *The Enzymes*. Boyer, P.D. (ed.). Vol. XIII, S. pp. 363-408, Third edition. Academic Press, New York, San Francisco, London.

Schultes, N.P., Zelitch, I., McGonigle, B. and Nelson, T. (1994) The primary leaf catalase gene from *Nicotiana tabacum* and *Nicotiana sylvestris. Plant Physiol.* **106**: 399-400.

Shang, W. and Feierabend, J. (1999) Dependence of catalase photoinactivation in rye leaves on light intensity and quality and characterization of a chloroplast-mediated inactivation in red light. *Photosynth. Res.* **59**: 201-213.

Sheptovitsky, Y.G. and Brudvig, G.W. (1996) Isolation and characterization of spinach photosystem II membrane-associated catalase and polyphenol oxidase. *Biochemistry.* **35**: 16255-16263.

Shi, W.M., Muramoto, Y., Ueda, A. and Tabake, T. (2001) Cloning of peroxisomal ascorbate peroxidase gene from barley and enhanced thermotolerance by overexpression in *Arabidopsis thaliana*. *Gene.* **273**(1): 23-27.

Sichak, S.P. and Dounce, A.L. (1986) Analysis of the peroxisomal mode of catalase. *Arch. Biochem. Biophys.* **249**(2): 286-295.

Sitte, P. (1958) Die Ultrastruktur von Wurzelmeristemzellen der Erbse (*Pisum sativum* L.). Eine elektronenmikroskopische Studie. *Protoplasma.* **49**: 447-522.

Skadsen, R.W. and Scandalios, J.G. (1986) Evidence for processing of maize catalase 2 and purification of its messenger RNA aided by translation of antibody-bound polysomes. *Biochemistry.* **25**: 2027-2032.

Streb, P. and Feierabend, J. (1996) Oxidative stress responses accompanying photoinactivation of catalase in NaCl-treated rye leaves. *Bot. Acta.* **109**: 125-132.

Streb, P., Feierabend, J. and Bligny, R. (1997) Resistance to photoinhibition of photosystem II and catalase and antioxidative protection in high mountain plants. *Plant Cell Environ.* **20**: 1030-1040.

Subramani, S.(1996) Protein translocation into peroxisomes. *J. Biol. Chem.* **271**: 32483-32486.

Subramani, S. (1993) Protein import into peroxisomes and biogenesis of the organelle. *Ann.Rev. Cell Biol.* **9**: 445-478.

Surpin, M. and Chory, J. (1997) The co-ordination of nuclear and organellar genome expression in eukaryotic cells. *Essays Biochem.* **32**: 113-125.

Suzuki, M., Ario, T., Hattori, T., Nakamura, K. and Asahi, T. (1994) Isolation and characterization of two tightly linked catalase genes from castor bean that are differentially regulated. *Plant Mol. Biol.* **25**: 507-516.

Tenberge, K.B. and Eising, R. (1995) Immunogold labelling indicates high catalase concentration in amorphous and crystalline inclusions of sunflower (*Helianthus annuus* L.) peroxisomes. *Hist. J.* **27**: 184-195.

Tenberge, K.B., Ruholl, C., Heinze, M. and Eising, R. (1997) Purification and immuno-electron microscopical characterization of crystalline inclusions from plant peroxisomes. *Protoplasma.* **196**: 142-154.

Thornton, R.M. and Thimann, K.V. (1964) On a crystal-containing body in cells of the oat coleoptile. *J. Cell Biol.* **20**: 345-350.

Tolbert, N.E. (1980) Microbodies – peroxisomes and glyoxysomes. In: *The Biochemistry of Plants.* (Stumpf, P.K. and Conn, E.E. eds.). pp. 359-388. Academic Press, New York.

Trelease, R.N., Lee, M.S., Banjoko, A. and Bunkelmann, J. (1996a) C-terminal polypeptides are necessary and sufficient for *in vivo* targeting of transiently-expressed proteins to peroxisomes in suspension-cultured plant cells. *Protoplasma*. **195**: 156-167.

Trelease, R.N., Xie, W., Lee, M.S. and Mullen, R.T. (1996b) Rat liver catalase is sorted to peroxisomes by its C-terminal tripeptide Ala-Asn-Leu, not by the internal Ser-Lys-Leu motif. *Eur. J. Cell Biol.* **71**: 248-258.

Trindade, H., Karmali, A. and Pais, M.S. (1988) One step purification of catalase from leaves of *Zantedeschia aethiopica*. *Biochimie*. **70**: 1769-1763.

Vainshtein, B.K., Melik-Adamyan, W.R., Barynin, V.V., Vagin, A.A., Grebenko, A.I., Borisov, V.V., Bartels, K.S., Fita, I. and Rossmann, M.G. (1986) Three-dimensional structure of catalase from *Penicillium vitale* at 2.0 Å resolution. *J. Mol. Biol.* **188**: 49-61.

van den Bosch, K.A. and Newcomb, E.H. (1986) Immunogold localization of nodule-specific uricase in developing soybean root nodules. *Planta*. **167**: 425-436.

van Eyk, A.D., Litthauer, D. and Oelofsen, W. (1992) The isolation and partial characterization of catalase and a peroxidase active fraction from human white adipose tissue. *Eur. J. Biochem.* **24**(7): 1101-1109.

Vaughn, K.C. and Stegink, S.J. (1987) Peroxisomes of soybean (*Glycine max*) root nodule vascular parenchyma cells contain a 'nodule-specific' urate oxidase. *Physiol. Plant.* **71**: 251-256.

Vigil, E.L. (1970) Cytochemical and developmental changes in microbodies (glyoxysomes) and related organelles of castor bean endosperm. *J. Cell Biol.* **46**: 435-454.

von Ossowski, I., Hausner, G. and Loewen, P.C. (1993) Molecular evolutionary analysis based on the amino acid sequence of catalase. *J. Mol. Evol.* **37**: 71-76.

Von Ossowski, I., Mulvey, M.R., Leco, P.A., Borys, A. and Loewen, P.C. (1991) Nucleotide sequence of *Escherichia coli katE*, which encodes catalase HPII. *J. Bacteriol.* **173**: 514-520.

Walton, P.A., Hill, P.E. and Subramani, S. (1995) Import of stably folded proteins into peroxisomes. *Mol. Biol. Cell.* **6**: 675-683.

Wang, J., Zhang, H. and Allen, R.D. (1999) Overexpression of an *Arabidopsis* peroxisomal ascorbate peroxidase gene in tobacco increases protection against oxidative stress. *Plant Cell Physiol.* **40**(7): 725-732.

Welinder, K.G. (1991) Bacterial catalase-peroxidases are gene duplicated members of the plant peroxidase superfamily. *Biochim. Biophys. Acta*. **1080**(3): 215-220.

Willekens, H., Inzé, D., van Montagu, M. and van Camp, W. (1995) Catalase in plants. *Mol. Breed.* **1**: 207-228.

Willekens, H., Langebartels, C., Tiré, C., van Montagu, M., Inzé, D. and van Camp, W. (1994) Differential expression of catalase genes in *Nicotiana plumbaginifolia*. *Proc. Natl. Acad. Sci. USA*. **91**: 10450-10454.

Yamaguchi, J. and Nishimura, M. (1984) Purification of glyoxysomal catalase and immunochemical comparison of glyoxysomal and leaf peroxisomal catalase in germinating pumpkin cotyledons. *Plant Physiol.* **74**: 261-267.

Yamaguchi, K., Mori, H. and Nishimura, M. (1995) A novel isoenzyme of ascorbate peroxidase localized on glyoxysomal and leaf peroxisomal membranes in pumpkin. *Plant Cell Physiol.* **36**: 1157-1162.

Yamaguchi, J., Nishimura, M. and Akazawa, T. (1986) Purification and characterization of haem-containing low-activity form of catalase from greening pumpkin cotyledons. *Eur. J. Biochem.* **159**: 315-322.

Yamaguchi, J., Nishimura, M. and Akazawa, T. (1984) Maturation of catalase precursor proceeds to a different extent in glyoxysomes and leaf peroxisomes of pumpkin cotyledons. *Proc. Natl. Acad. Sci. USA.* **81**: 4809-4813.

Yoshimura, K., Yabuta, Y., Ishikawa, T. and Shigeoka, S. (2000) Expression of spinach ascorbate peroxidase isoenzymes in response to oxidative stress. *Plant Physiol.* **123**: 223-233.

Zámocký, M. and Koller, F. (1999) Understanding the structure and function of catalases: clues from molecular evolution and *in vitro* mutagenesis. *Prog. Biophys. Mol. Biol.* **72**: 19-66.

Zhang, H., Wang, J., Nickel, U., Allen, R.D. and Goodman, H.M. (1997) Cloning and expression of an Arabidopsis gene encoding a putative peroxisomal ascorbate peroxidase. *Plant Mol. Biol.* **34**: 967-971.

Zhong, H.H. and McClung, C.R. (1996) The circadian clock gates expression of two Arabidopsis catalase genes to distinct and opposite circadian phases. *Mol. Gen. Genet.* **251**(2): 196-203.

Zhong, H.H., Young, J.C., Pease, E.A., Hangarter, R.P. and McClung, R.C. (1994) Interactions between light and the circadian clock in the regulation of CAT2 expression in Arabidopsis. *Plant Physiol.* **104**: 889-898.

5

THE PHOTORESPIRATORY PATHWAY OF LEAF PEROXISOMES

Sigrun Reumann

Universität Göttingen, Göttingen, Germany

KEYWORDS

Aspartate aminotransferase, bioinformatics, catalase, glutamate:glyoxylate aminotransferase, glycolate oxidase, hydroxypyruvate reductase, malate dehydrogenase, metabolite transport, photorespiration, porin-like channel, serine:glyoxylate aminotransferase, targeting.

INTRODUCTION

Photorespiration is a light-dependent process that results in the uptake of O_2 and the release of CO_2. Photorespiration is linked to photosynthesis by the dual function of ribulose-1,5-bisphosphate carboxylase/oxygenase (RubisCO) which uses the mutually competitive substrates CO_2 and O_2 for reaction with ribulose-1.5-bisphosphate (Ru1.5BP). Oxygenation of ribulose-1.5-bisphosphate leads to the production of 2-P-glycolate, which is salvaged in the photorespiratory pathway (C_2 cycle). Several enzymes of the photorespiratory pathway are localized in leaf peroxisomes and the recycling of 2-P-glycolate is the main function of leaf peroxisomes known so far. Apart from the photorespiratory C_2 cycle, leaf peroxisomes have an important physiological role in the metabolism of active oxygen species (see Chapter 7), mediate fatty acid β-oxidation (Chapter 1 this volume; Gerhardt,

A. Baker and I.A. Graham (eds.), Plant Peroxisomes, 141–189.

1981, 1983; Gühnemann-Schäfer and Kindl, 1995), and are involved in the biosynthesis of glycine betaine.

A number of review articles have been written on photorespiration and leaf peroxisomal metabolism (Gerhardt, 1987; Tolbert, 1980b; Lorimer and Andrews, 1981; Huang *et al*, 1983, Ogren, 1984; Husic *et al*, 1987; Douce and Heldt, 2000). Recently, considerable progress has been made in several areas of photorespiratory metabolism. It has been found, for instance, that most matrix proteins of peroxisomes are targeted to their destination by one of two peroxisomal targeting signals (PTS): the C-terminal peroxisomal targeting signal 1 (PTS1), the so-called "SKL" motif ([SAC][KRH]L; Gould *et al*, 1987, 1988, 1989, 1990; Swinkels *et al*, 1992; Hayashi *et al*, 1997), or the N-terminal nonapeptide PTS2 ([RK][LVI]x5[HQ][LA]; Osumi *et al*, 1991; Swinkels *et al*, 1991; Glover *et al*, 1994; Gietl *et al*, 1994; Kato *et al*, 1998; Flynn *et al*, 1998; see also Chapters 10 and 11). The targeting signals of several photorespiratory enzymes, such as those of glycolate oxidase, catalase, and malate dehydrogenase, have been studied in detail and allow more insights into the degree of sequence conservation of the targeting patterns.

The recently completed genome sequence of *Arabidopsis thaliana* L. enables the identification of novel homologues and isoforms of peroxisomal proteins in this higher plant. The substrate specificity of some novel isoforms of photorespiratory enzymes can be predicted by multiple sequence alignment and by analysis of the conservation of residues forming the active site. The expression rate of novel homologues can often be estimated by analysis of the number of corresponding *e*xpressed *s*equence *t*ags (ESTs) in various non-normalized cDNA libraries ("digital northern", Mekhedov *et al*, 2000). Thanks to algorithms that have been deduced from targeting sequences from proteins of plastids (von Heijne *et al*, 1989), mitochondria (Isaya *et al*, 1991, Baker and Schatz; 1987, Roise *et al*, 1988), peroxisomes (Gould *et al*, 1988, 1990; Osumi *et al*, 1991; Swinkels *et al*, 1992; Glover *et al*, 1994; Hayashi *et al*, 1997; Kato *et al*, 1998, Flynn *et al*, 1998) and of secretory proteins (Emanuelsson *et al*, 2000), the subcellular localization of many unknown Arabidopsis homologues can often be predicted with high accuracy (TargetP, Emanuelsson *et al*, 2000; PSORT: Nakai and Kanehisa, 1992).

Compartmentalization studies revealed that photorespiratory metabolism in leaf peroxisomes is largely compartmentalized by an organised arrangement of matrix enzymes in a multi-enzyme complex, rather than by the surrounding membrane. Instead of forming a strict permeability barrier, the membrane of plant peroxisomes is equipped with a porin-like channel that is well-suited to facilitate diffusion of a broad range of small negatively charged metabolites across the membrane.

This article focuses on four topics: (i) new biochemical and molecular data obtained in the past ten years dealing with the cloning of new genes and the regulation of gene expression of photorespiratory enzymes; (ii) studies of specific targeting signals of photorespiratory enzymes; (iii) analysis of the recently completed genome sequence of *Arabidopsis thaliana* L. for homologues of photorespiratory enzymes; and (iv) transfer of reducing equivalents, compartmentalization and metabolite transport of plant peroxisomes.

GENERAL ASPECTS OF PHOTORESPIRATION

Photorespiration results in the light-dependent uptake of O_2 and release of CO_2. The metabolic pathway of photorespiration, in which sugars are oxidized to CO_2 in the light, is known as the "oxidative photosynthetic carbon cycle" or the photorespiratory cycle (C_2 cycle, Tolbert, 1980b, 1981, 1997). The C_2 cycle is linked to the reductive photosynthetic carbon cycle (or C_3 cycle) of photosynthesis because RubisCO catalyzes the first step in both O_2 uptake via the C_2 cycle and CO_2 fixation via the C_3 cycle. Carboxylation of Ru1.5BP produces two molecules of 3-P-glycerate, whereas oxygenation of Ru1.5BP produces one molecule of 3-P-glycerate and one of 2-P-glycolate. By the enzymes of the photorespiratory pathway, two molecules of 2-P-glycolate are metabolized to CO_2 and 3-P-glycerate, and the latter is fed into the C_3 cycle. Photorespiration is distinct from mitochondrial respiration in its light dependence and the fact that O_2 consumption is accompanied with a consumption of ATP (Husic *et al*, 1987).

The recycling of 2-P-glycolate occurs in a complex pathway that consists of sixteen enzymes and at least six translocators that are distributed over the three cell organelles, the chloroplast, the leaf peroxisome and the mitochondrion. The photorespiratory pathway involves several shuttles to support transamination, ammonia re-fixation and the transfer of reductants from the site of generation (chloroplast and mitochondrion) to the leaf peroxisome.

In summary, photorespiration is counter-productive to CO_2 fixation in releasing CO_2 in the light and thus results in a decline in plant productivity. The relationship between photosynthesis and photorespiration is determined by the concentrations of oxygen and CO_2. It can be estimated that under physiological conditions the ratio of carboxylation to oxygenation is 1:0.4, leading to a release of photorespiratory CO_2 of 25% of the rate of net CO_2 assimilation (Sharkey, 1988). The question of why the control of photorespiration has not been achieved by modification of RubisCO has, to date, not been answered satisfactorily. Photorespiration is thought to protect the plant from photoinhibition in a low CO_2 environment by consuming ATP and NADPH and limiting the amount of toxic oxygen species (Heber and Krause, 1980; Ogren, 1984). Alternatively, the oxygenase activity of RubisCO has been suggested to be an unavoidable consequence of the RuBP carboxylase reaction mechanism, so that photorespiration is inevitable (Andrews and Lorimer, 1978). However, the ratio of carboxylase to oxygenase activity increased during the natural evolution of photosynthesis from photosynthetic bacteria to angiosperms although only to low extent (Somerville and Ogren, 1982a; Ogren, 1984), indicating a certain variability in nature and the possibility that additional advantageous changes can be introduced in RubisCO by genetic engineering. Yet, research targeted at modifying RubisCO has not been reported to be agronomically successful (Tolbert, 1997).

THE PEROXISOMAL ENZYMES OF THE PHOTORESPIRATORY PATHWAY

The oxidative photosynthetic carbon cycle begins in the chloroplast with the bifunctional enzyme RubisCO (EC 4.1.1.39) catalyzing the reaction of Ru1.5BP with molecular oxygen and yielding the products 3-P-glycerate and 2-P-glycolate (Figure 1). Hydrolysis of the phosphate group of 2-P-glycolate to glycolate is catalyzed by the stromal enzyme phosphoglycolate phosphatase (EC 3.1.3.18; Richardson and Tolbert, 1961; Hardy and Baldy, 1986; Belanger and Ogren, 1987). The enzymes from tobacco and maize are oligomers with subunits of 21 kDa (homotetramer, Christellar and Tolbert, 1978) and 32 kDa (homodimer, Hardy and Baldy, 1986), respectively. The isolation of an Arabidopsis mutant lacking P-glycolate phosphatase

(Somerville and Ogren, 1979) provided unequivocal evidence that 2-P-glycolate is the precursor of glycolate and photorespiratory CO_2. Somerville and Ogren mutagenized Arabidopsis with EMS (*e*thyl *m*ethane *s*ulphonate) and screened for conditionally lethal mutants which are phenotypically indistinguishable from the wild-type when grown in air enriched to 1% CO_2 to suppress oxygenation of Ru1.5BP, but which are inviable in normal atmospheric conditions of 0.03% CO_2 (for review, see Somerville and Ogren, 1982a; Somerville, 2001). By using this approach, the authors also identified mutants deficient in serine-glyoxylate aminotransferase (see later), glycine decarboxylase, and serine hydroxymethyl transferase (Somerville and Ogren, 1980, 1981, 1982b). A phosphoglycolate phosphatase gene has yet to be cloned from higher plants.

Glycolate is exported from the chloroplast in exchange for glycerate by a translocator that has similar affinities for glycolate, glyoxylate, D-glycerate and D-lactate (Howitz and McCarty, 1985 a, b; 1986). Glycolate and glycerate can be transported together, by counter exchange for each other, or individually by H^+ symport (or OH^- antiport), as demonstrated for the transporter from pea and spinach chloroplasts (Howitz and McCarty, 1988, 1991). This flexible transport mechanism allows export of two moles of glycolate in exchange of one mole of glycerate as well as a non-stoichiometric transport of the two intermediates, if considerable amounts of the amino acids serine and glycine are taken from the photorespiratory C_2 cycle for biosynthetic reactions. The corresponding translocator gene remains to be cloned.

Figure 1: Reactions and transport of intermediates of the photorespiratory C_2 cycle

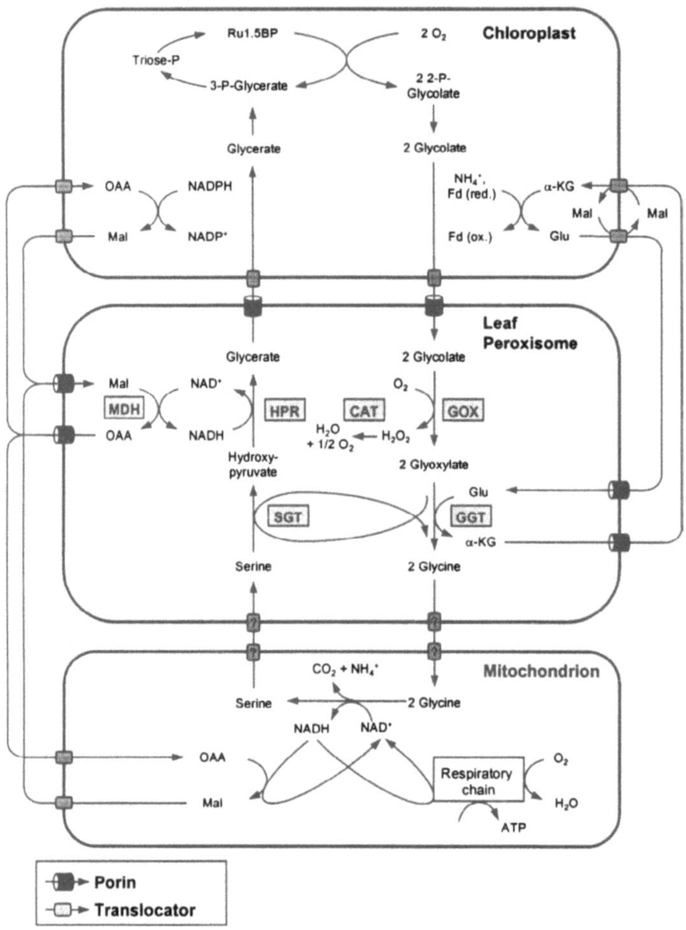

Figure 1: α-KG: α-ketoglutarate; OAA: oxaloacetate. CAT: catalase; GGT: glutamate:glyoxylate aminotransferase; Glu: glutamate; GOX: glycolate oxidase; HPR: hydroxypyruvate reductase; Mal: Malate; MDH: malate dehydrogenase; Ru1.5BP: ribulose-1.5-bisphosphate; SGT: serine:glyoxylate aminotransferase.

Glycolate oxidase

Glycolate diffuses to leaf peroxisomes and probably passes through the membrane via the electrophysiologically characterized porin-like channel (Reumann *et al*, 1995, 1998; see later). Glycolate is oxidized internally by the matrix enzyme glycolate oxidase (GOX, EC 1.1.3.1), which is an α-hydroxyacid oxidase and uses flavin mononucleotide, FMN, as cofactor. The enzyme catalyzes the transfer of electrons to O_2 and generates the products glyoxylate and H_2O_2. Glycolate oxidase (M_r = 40-43 kDa) was first purified and cloned from *Spinacia oleracea* L. by Volokita and Somerville (1987), followed by cloning of the genes from *Lens culinaris* L. (Ludt and Kindl, 1990) and *Cucurbita* sp. (Tsugeki *et al*, 1993). Similar to hydroxypyruvate reductase, GOX activity and transcript abundance are also regulated by development and light (Ludt and Kindl, 1990; Tsugeki *et al*, 1993).

The crystal structure of the spinach enzyme has been resolved to 2Å (Lindqvist and Bränden, 1985, 1989; Lindqvist, 1989), and it has been shown that the polypeptide chain is folded into an eight-fold α/β-barrel structural motif corresponding to the FMN domain which was first described for triosephosphate isomerase. Four subunits interact strongly with each other forming a tight tetramer, whereas the two tetramers are held together only by weak interactions (Lindqvist, 1989), consistent with the former biochemical data on the native molecular mass (Hall *et al*, 1985).

All known GOX enzymes contain a C-terminal tripeptide (ARL> or PRL>) that closely resembles a conserved PTS1 with the typical characteristics of the amino acids "small-basic-hydrophobic sidechains", e.g. ([SAC] [KRH]L>, Gould *et al*, 1987, 1988; 1989; [SACP][KR][ILM]>, Hayashi *et al*, 1997). A proline residue at position -3 of the PTS1, however, was shown to yield only a poor signal for targeting of the passenger protein chloramphenicol acetyltransferase to glyoxysomes of tobacco BY-2 cells (Mullen *et al*, 1997). Indeed, GOX is an example for a PTS1-containing protein of which the targeting is not only determined by the three C-terminal amino acids. The amino acids preceding the PTS1 can provide additional information, and other sequences can substitute the known targeting signals at least *in vitro*. The last six amino acids of spinach GOX (RAVARL>) were found to be sufficient for targeting of the passenger protein β-glucuronidase to peroxisomes in transgenic tobacco (Volokita, 1991). To analyze the kinetics and the efficiency of protein import into plant peroxisomes, an *in vitro* assay for protein import into sunflower and pumpkin glyoxysomes has been set up by Horng *et al* (1995) and Brickner *et al* (1997), respectively. Horng *et al* (1995) demonstrated that a mutant version of spinach GOX lacking the C-terminal 53 amino acids was also imported into peroxisomes *in vitro* in a temperature- and ATP-dependent manner, whereas a fusion protein containing the last twenty amino acids of GOX attached to dihydrofolate reductase did not. Brickner and Olsen (1998) used GOX as a standard protein to investigate the energy-dependence of protein import and showed that import of GOX is dependent on ATP or GTP hydrolysis and that a protonmotive force is not absolutely required for peroxisomal protein import.

The complete genome sequence of *Arabidopsis thaliana* L. (Lin *et al*, 1999; Mayer *et al*, 1999; Arabidopsis Genome Initiative, 2000; Tabata *et al*, 2000; Salanoubat *et al*, 2000; Theologis *et al*, 2000) reveals five genes that are homologous to spinach GOX. Surprisingly, some of the genes are predicted to be expressed in non-green tissue and to catalyze oxidative reactions in metabolic pathways other than the photorespiratory C_2 cycle (Reumann *et al*, in preparation). Two Arabidopsis proteins (At_GOX1, At_GOX2) can be identified as close homologues of GOX from *Spinacia olearacea* L. and *Cucurbita sp.* by their maximum sequence similarity (about 90% identity) and by high rates of expression of their genes in leaves, as indicated by a large number of EST clones in the EST collection derived from above-ground organs. A third GOX homologue (At_GOX3; AKL>, 85% sequence identity with At_GOX1/2) is also expected to be localized in peroxisomes due to the presence of a PTS1, but seems to be expressed in roots. Because all active site residues of this enzyme are conserved as compared to spinach GOX, the predicted substrate of this enzyme is also glycolate.

Two additional Arabidopsis genes encode for proteins that are more distantly related to the photorespiratory enzymes but can be identified as homologues by conserved patterns and their conserved exon-intron structure (At_HAOX1/2, 2-hydroxy acid oxidase, SML>, about 60% identity with At_GOX1-3; Reumann *et al*, in preparation). Both genes seem to be expressed. However, the substrate specificity of the enzymes appears to be altered since two amino acids of the active site of spinach GOX (Lindqvist and Braenden, 1989; Lindqvist *et al*, 1991; Stenberg *et al*, 1995) have been substituted by point mutations. The amino acid W-108 is substituted for methionine and Y-24 for phenylalanine in both Arabidopsis GOX homologues (Figure 2). A substitution of Y-24 to F-24 in spinach GOX resulted in a ten-fold increase in the K_m for glycolate (Stenberg *et al*, 1995). Similar substitutions have also been reported in mammalian GOX homologues that have recently been cloned. These GOX homologues, which share about 50% sequence identity with plant GOX, are also peroxisomal enzymes of the same size with a PTS1, but represent 2-hydroxy acid oxidases of different substrate specificity. Only the isoform HAOX1 is specific for glycolate, whereas HAOX2 and HAOX3 use long-chain and medium-chain 2-hydroxy acids, respectively (Kohler *et al*, 1999; Jones *et al*, 2000; Williams *et al*, 2000). It is, therefore, reasonable to speculate that the latter two GOX homologues from Arabidopsis are not specific for glycolate but have a function closer to those of mammalian HAOX2 and HAOX3 and use a substrate of longer chain length. For this reason, the acronym At_HAOX1/2 is recommended instead of At_GOX4/5. Targeting of At_HAOX1/2 to peroxisomes by a variant of the PTS1 SML> may be possible despite the methionine residue at postion -2 ([CASP][KR][ILM]>, Hayashi *et al*, 1987) and needs to be verified experimentally. A possible function of At_HAOX1/2 in α-oxidation of fatty acids, similar to that suggested for the mammalian homologues (Jones *et al*, 2000), may be worthwhile reinvestigating. In line with the result that the mammalian enzymes are involved in lipid metabolism, an EST clone corresponding to At_HAOX1 has been identified in an EST collection from developing seeds.

Figure 2: Sequence conservation of active site residues in Arabidopsis and mammalian homologues of spinach glycolate oxidase (GOX)

	Active site residues (incl. substrate binding)						FMN binding								
	24	108	129	157	254	257	78	106	127	155	230	252	285	289	309
So_GOX	Y	W	Y	D	H	R	T	S	Q	T	K	S	D	R	R
Csp_GOX	Y	W	Y	D	H	R	T	S	Q	T	K	S	D	R	R
At_GOX1	Y	W	Y	D	H	R	T	S	Q	T	K	S	D	R	R
At_GOX2	Y	W	Y	D	H	R	T	S	Q	T	K	S	D	R	R
At_GOX3	Y	W	Y	D	H	R	T	S	Q	T	K	S	D	R	R
At_HAOX1	F	M	Y	D	H	R	T	S	Q	T	K	S	D	R	R
At_HAOX2	F	M	Y	D	H	R	T	S	Q	T	K	S	D	R	R
Hs_HAOX1	Y	W	Y	D	H	R	T	S	Q	T	K	S	D	R	R
Hs_HAOX2	F	F	Y	D	H	R	T	S	Q	T	K	S	D	R	R
Hs_HAOX3	F	Y	Y	D	H	R	T	S	Q	T	K	S	D	R	R

Figure 2: The given numbers of the amino acids of the active sites and the FMN binding site refer to GOX from Spinacia oleracea L. (Lindqvist, 1989; Lindqvist and Braenden, 1989; Stenberg et al, 1995; Kohler et al., 1999). Sp_GOX: GOX of Spinacia oleracea, Swissprot acc. number P05414; Csp_GOX: GOX of Cucurbita sp., PIR acc. number T10242; At_GOX1: GOX of Arabidopsis thaliana, DDBJ acc. number BAB01334; At_GOX2: GOX of A. thaliana, DDBJ acc. number BAB01333; At_GOX3: GOX of A. thaliana, EMBL acc. number CAA16716; At_HAOX1: GOX homologue of A. thaliana, DDBJ acc. number BAB02977; At_HAOX2: GOX homologue of A. thaliana, DDBJ acc. number BAB02979; Hs_HAOX1: short-chain HAOX of Homo sapiens, Swissprot acc. number Q9UJM8; Hs_HAOX2: long-chain HAOX of H. sapiens, Swissprot acc. number Q9NYQ3; Hs_HAOX3: medium-chain HAOX of H. sapiens, Swissprot acc. number Q9NYQ2.

Catalase

The highly reactive by-product of glycolate oxidation, H_2O_2, is immediately detoxified by disproportionation to water and oxygen as catalyzed by catalase (CAT, EC 1.11.1.6). Catalase is the principal H_2O_2-scavenging enzyme in plants and is characterized by a very high reaction rate, but a relatively low affinity for H_2O_2. The high content of CAT of up to 10-25% of the total amount of leaf peroxisomal protein (Tolbert, 1980a) may guarantee effective detoxification of H_2O_2, but auxiliary enzymes, such as ascorbate peroxidase, are also operating (Bunkelmann and Trelease, 1996; Yamaguchi *et al*, 1995, see Chapter 7). Plant CAT is a tetrameric haem-containing enzyme with an apparent subunit molecular mass of about 55 kDa. The gene was first cloned from maize (Bethards *et al*, 1987), followed by a large number of other plant species. By using an *in vivo* protein import system with glyoxysomes of tobacco BY-2 cells (Banjoko and Trelease, 1995), it was demonstrated that CAT from cottonseed is sorted to plant peroxisomes by a degenerate PTS1 (PSI>) (Mullen *et al*, 1997). Targeting of the tripeptide strictly depends on the context of adjacent C-terminal amino acid residues, especially the fourth last amino acid, arginine.

Because CAT plays an essential role not only in the photorespiratory C_2 cycle but also in the metabolism of glyoxysomes, gerontosomes (a peroxisome type found in scenescing tissue that is related to glyoxysomes), and nodule-specific peroxisomes, it is not surprising that the number of CAT isoforms exceeds that of other peroxisomal enzymes by far. For instance, five isoforms have been separated in maize (Williamson *et al*, 1993), six in Arabidopsis (Frugoli *et al*, 1996), eight in sunflower (Grotjohann *et al*, 1997), and twelve in mustard (Drumm and Schopfer, 1974). Three genes have been cloned from tobacco (Willekens *et al*, 1994) and Arabidopsis (Frugoli *et al*, 1996).

For several higher plant species, the multiple CAT genes are reported to be differentially expressed. Of the three CAT isoforms from pumpkin, for instance, only one (Cat3) has been described to be a constitutive peroxisomal enzyme, whereas the two other isoenzymes (Cat1 and Cat2, 90% sequence identity to each other, 70% identity to Cat3) seem to be specific for glyoxysomes and leaf peroxisomes, respectively (Esaka *et al*, 1997). The five tetrameric CAT isoenzymes of cotton seedlings are hetero-oligomers that consist of two different subunits in different ratios. The two subunits are encoded by two distinct genes and are temporally synthesized, with subunit 1 and 2 being predomoinant in CAT of glyoxysomes and leaf peroxisomes, respectively (Ni *et al*, 1990; Ni and Trelease, 1991). In pumpkin and other species, an additional isoform of higher apparent molecular mass (M_r = 59 kDa) has been distinguished that forms heterooligomers together with the 55-kDa form (Yamaguchi *et al*, 1984; Eising *et al*, 1990; Grotjohann *et al*, 1997). Catalase is concentrated in the so-called peroxisomal cores, i.e. electron-dense crystalline or pleomorph structures in the matrix of peroxisomes. The recent finding that this core CAT of predominantly 59 kDa from sunflower peroxisomes has a different primary structure from other CAT isoforms, as deduced from peptide mapping, and is larger than the matrix form (55 kDa), contradicts the previous view that the formation of the cores occurs via simple condensation of matrix CAT (Kleff *et al*, 1997).

In addition to the regulation by post-translational processing, CAT is light sensitive and undergoes photoinactivation with subsequent degradation similar to the D1 reaction centre protein of photosystem II (Feierabend and Engel, 1986; Hertwig *et al*, 1992). Different CAT isoforms from sunflower cotyledons reveal a different sensitivity to photoinactivation (Eising and Gerhardt, 1987, 1989; Grotjohann *et al*, 1997). In summary and in contrast to most other enzymes of the photorespiratory C_2 cycle, post-translational regulation of CAT activity seems to be more pronounced than regulation at the transcriptional level. Nevertheless, expression of CAT genes is also regulated by other specific stimuli. A circadian oscillation of the mRNA abundance was demonstrated for CAT2 and CAT3 from Arabidopsis (Zhong *et al*, 1994; Zhong and McClung, 1996). Furthermore, expression of *CAT* genes is stimulated by environmental stress conditions, including wounding, ozone, ultraviolet B and chilling-induced oxidative stress. The relationship between environmental stress and *CAT* expression advocate a new concept of the importance of CAT in plant defense responses. The protein that binds the defense-mediating signal salicylic acid was demonstrated to be identical to CAT (Chen *et al*, 1993 a, b). The function of CAT may not be restricted to removal of H_2O_2 originating from peroxisomal reactions, but may also involve detoxification of H_2O_2 from other subcellular compartments. Thus, CAT may be essential for the antioxidant defense during biotic and abiotic stresses that generate AOS in various compartments (Willekens *et al*, 1997). More research is required to fully understand the postulated role of CAT in the signal transduction of the defense response. For further details on regulation and post-translational modification of CAT and its physiological role see Chapter 4.

Serine:glyoxylate aminotransferase

Glyoxylate deriving from GOX activity is subsequently transaminated to glycine. This transamination is achieved by at least two alternative ways, as catalyzed by serine:glyoxylate aminotransferase (SGT; EC. 2.6.1.45) and glutamate:glyoxylate aminotransferase (GGT, EC 2.6.1.2), which normally operate in a 1:1 ratio (Rehfeld and Tolbert, 1972). Serine-glyoxylate aminotransferase couples the amination of glyoxylate to the deamination of serine, which is the product of mitochondrial glycine oxidation two steps down the oxidative photosynthetic carbon cycle. Both SGT and GGT catalyze physiologically irreversible reactions (Rehfeld and Tolbert, 1972; Nakamura and Tolbert, 1983). Serine:glyoxylate aminotransferase has been characterized and partially purified from leaves of several species (Yu *et al*, 1984). Its native molecular mass was found to range from about 60 to 180 kDa with a subunit molecular mass of about 48 kDa (Ireland and Joy, 1983; Noguchi and Hayashi, 1980; Hondred *et al*, 1985). The enzyme from cucumber cotyledons appears to consist of two different subunits of 45 and 47 kDa, which share only little immunological cross-reactivity (Hondred *et al*, 1985). The activity of SGT was shown to co-purify with that of serine-pyruvate aminotransferase, indicating that both activities are located on one protein (Rehfeld and Tolbert, 1972; Smith, 1973). Similar to other peroxisomal enzymes, the activity of SGT is regulated developmentally and by light (Schnarrenberger *et al*, 1971; Noguchi and Fujiwara, 1982; Hondred *et al*, 1987) Upon induction, the patterns of HPR and SGAT protein and mRNA were found to be very similar (Hondred *et al*, 1987), suggesting that expression of the two genes is co-ordinately regulated.

The first higher plant gene that encodes for a peroxisomal aminotransferase with physiological K_m values for serine and glyoxylate has only recently been identified by Liepman and Olsen (2001). The gene was cloned from Arabidopsis and has been termed alanine-glyoxylate aminotransferase 1 (At_AGT1, AF063901, 401 aa, 44 kDa, SRI>) based on 30% sequence identity shared with mammalian AGT1. The C-terminal tripeptide SRI> is consistent with the plant peroxisomal PTS1 [SACP][KR][ILM]>, which has been found to be efficient in targeting the passenger protein β-glucuronidase into microbodies of transgenic Arabidopsis (Hayashi *et al*, 1997). With respect to its physiological role, it needs to be emphasized that At_AGT1 is thought to act *in vivo* as a serine-glyoxylate aminotransferase. In line with previous results (Rehfeld and Tolbert, 1972; Smith, 1973; Nakamura and Tolbert, 1983), the recombinant protein expressed heterologously in *E. coli* had an unphysiologically high K_m for the substrate alanine whereas only the amino donor serine and the amino acceptors glyoxylate and pyruvate had K_m values in the physiological range (Liepman and Olsen, 2001). The substrate specificity is also reflected by sequence homology as the closest homologues of At_AGT1 are SGT enzymes of methylotrophic bacteria (about 50% sequence identity), namely *Methylobacterium extorquens* AM1 (Chistoserdova and Lidstrom, 1994) and *Hyphomicrobium methylovorum* (Hagishita *et al*, 1996). These methylotrophs are able to grow on C_1 compounds, such as methanol and methylamine, and contain the serine pathway, including the genes for SGT, hydroxypyruvate reductase and serine hydroxymethyl transferase, for generation of the acceptor molecule glycine and assimilation of C_1 compounds via tetrahydrofolate. Interestingly, plant SGT does not belong to the large subgroup I of aminotransferases (Prosite PDOC00098; Metha *et al*, 1993) that comprises most prokaryotic and eukaryotic aspartate aminotransferases, including aspartate aminotransferase from plant peroxisomes (see later), but belongs to subgroup V of aminotransferases (Prosite: PDOC00514; corresponds to subgroup IV defined by Metha *et al*, 1993), which includes specific prokaryotic aspartate aminotransferases (e.g. *Methanobacterium thermoformicicum*, Tanaka *et al*, 1994). Thus, the different aminotransferases of plant peroxisomes, SGT and aspartate aminotransferase, have a different evolutionary origin.

Consistent with the expected important function of the enzyme in the photorespiratory pathway, *At_AGT1* shows highest expression in photosynthetic tissue. The enzyme was shown to be a dimer and was localized to peroxisomes by cofractionation of the endogenous activity with the fraction of leaf peroxisomes from Arabidopsis (Liepman and Olsen, 2001). Analysis of the photorespiratory air-sensitive *sat* mutant, which is deficient in SGT activity (Somerville and Ogren, 1980), revealed a point mutation (Pro to Leu) in the single *At_AGT1* gene (Liepman and Olsen, 2001). It is interesting to note that the mammalian isoform AGT1 is targeted simultaneously to peroxisomes and mitochondria by the variable use of two alternative transcription and translation initiation sites which determine whether or not the region encoding the N-terminal mitochondrial targeting sequence is contained within the open reading frame (Figure 3) (Oda *et al*, 1990; Takada *et al*, 1990; Holbrook *et al*, 2000).

Figure 3: Different targeting mechanisms of matrix proteins to peroxisomes

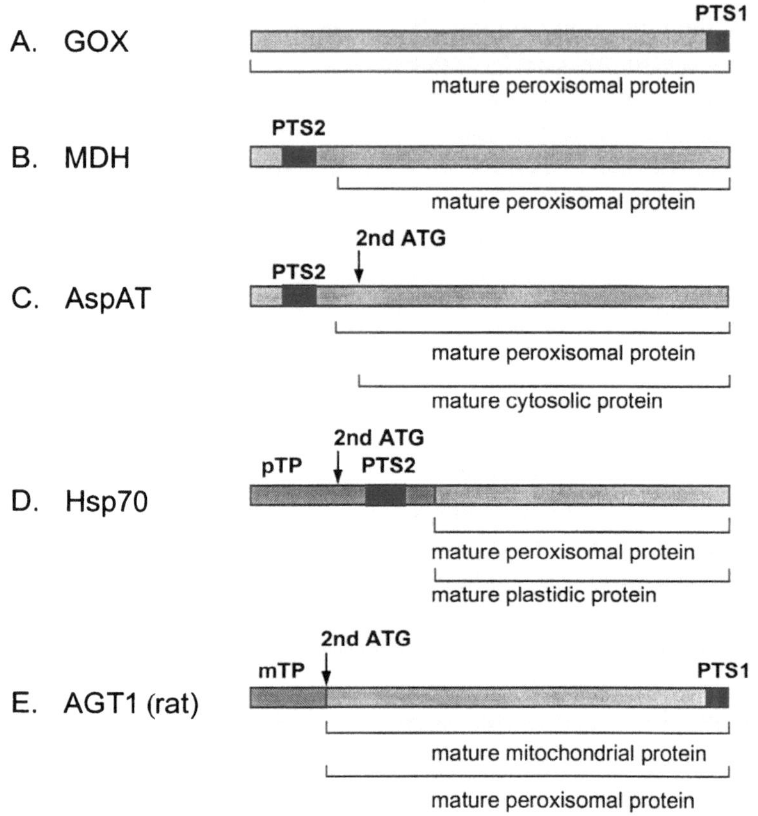

Most matrix proteins, for instance GOX, contain a C-terminal PTS1 (A), whereas some matrix proteins, such as MDH, contain an N-terminal PTS2 (B) for targeting to peroxisomes. Aspartate aminotransferase (AspAT) can be targeted to peroxisomes as well as to the cytosol (C). Hsp70 from watermelon is targeted to peroxisomes and plastids (Wimmer et al, 1997)(D). The exact location of the two processing sites is not yet known. Alanine-glyoxylate aminotransferase (AGT1) of omnivorous vertebrates (e.g. rats) is targeted to peroxisomes and mitochondria (E). For the latter three proteins, alternative targeting is determined by the variable use of two start codons.

Glutamate:glyoxylate aminotransferase

The second aminotransferase, glutamate:glyoxylate aminotransferase (GGT; EC 2.6.1.2), is reported to be less specific with respect to its amino donor and uses either glutamate or alanine (Nakamura and Tolbert, 1983; Noguchi and Hayashi, 1981; Noguchi and Fujiwara, 1982). Moreover, the aminotransferase was shown to be identical to alanine:2-oxoglutarate aminotransferase because both enzyme activities co-purified and were not additive (Noguchi and Hayashi, 1981). Glutamate:glyoxylate aminotransferase purified from spinach peroxisomes has a molecular mass of about 100 kDa determined for the native protein and a subunit molecular mass of 48 kDa (Noguchi and Fujiwara, 1982). The corresponding gene has not yet been cloned. However, analysis of the Arabidopsis genome allows some predictions. In mammals, AGT2 represents a second although mitochondrial isoform that also catalyzes a transaminase reaction with the substrates alanine and glyoxylate. The enzyme AGT2 belongs to the subgroup III of aminotransferases (Prosite: PDOC00519; corresponds to subgroup II defined by Metha *et al*, 1993) and is only distantly related to AGT1. Plant homologues of AGT2 with peroxisomal targeting signals are therefore good candidates for the photorespiratory enzyme GGT. Several Arabidopsis genes encode for enzymes that are related to mammalian AGT2, (about 40% sequence identity) and these comprise a small gene family. Similar to mammalian AGT2, the motif of the aminotransferase III family is conserved in all proteins. These putative transaminases contain not only predicted N-terminal mitochondrial targeting signals (PSORT, TargetP, Mitoprot), but also a C-terminal peptide that represents a PTS1 according to Hayashi *et al* (1997) (SRL>, SKM>). Experimental evidence is required to ascertain whether alternative forms of transcriptional and translational initiation target one of these isoforms to both mitochondria and peroxisomes and to demonstrate that one of these homologues of mammalian AGT2 has a low K_m for glyoxylate and glutamate and possibly alanine.

Having reached the mitochondrion, one glycine molecule is oxidized to yield CO_2, an ammonium ion (NH_4^+), and a C_1 unit bound to tetrahydrofolate (THF). This C_1 unit is, in turn, added to a second glycine molecule to form serine. The interconversion of glycine and serine involves two enzyme systems: the glycine decarboxylase complex and serine hydroxymethyltransferase. For a detailed description of these reactions and enzymes the reader is referred to recent reviews (Douce and Neuburger, 1999; Douce and Heldt, 2000; Douce *et al*, 2001).

Hydroxypyruvate reductase

Subsequent to the mitochondrial oxidation of glycine to serine, the latter enters the peroxisomal matrix and is deaminated to hydroxypyruvate by the above mentioned aminotransferase, SGT (At_AGT1). Hydroxypyruvate is reduced to D-glycerate by the NADH-dependent hydroxypyruvate reductase (HPR, EC 1.1.1.29, synonyms are glycerate dehydrogenase and NADH-glyoxylate reductase), which is a commonly used marker enzyme for leaf peroxisomes. The enzyme is also active with glyoxylate but the high K_m of about 10 mM as compared to the low K_m for hydroxypyruvate (120 μM, Tolbert *et al*, 1970), indicates that hydroxypyruvate is the only physiologically relevant substrate. Purified HPR from spinach and cucumber were shown to be homodimers of about 100 kDa with a subunit mol mass of 39 to 47 kDa (Kohn *et al*, 1970; Titus *et al*, 1983; Greenler *et al*, 1989). Genes encoding HPR were first cloned from cucumber (Greenler *et al*, 1989; Schwartz *et al*, 1991) and later from pumpkin (Hayashi *et al*, 1996) and Arabidopsis (Mano *et al*, 1997). The genes encode proteins of about 42 kDa that are probably targeted to peroxisomes by a PTS1 (see below). Mammalian glyoxylate/hydroxypyruvate reductase (Rumsby and Cregeen, 1999) is only distantly related to plant HPR with about 30% sequence identity over the entire length of the protein.

In many plant species, photostimulation of the activities of peroxisomal enzymes has been observed, of which that of HPR has been studied best. The activity of HPR was early shown to be developmentally regulated and enhanced by white light (Schnarrenberger *et al*, 1971; Tchang *et al*, 1984 a, b; Hondred *et al*, 1987) and to be a result of enhanced gene expression (Hondred *et al*, 1987; Greenler *et al*, 1989; Greenler and Becker, 1990). Expression of *HPR* is undetectable early during seedling development, but begins several days after germination and increases to high levels in the light, coinciding with metabolic transition in the cotyledons from heterotrophy to autotrophy. In the dark, transcripts and activity also appear after germination, but do not accumulate to high levels (Hondred *et al*, 1987). The white light-stimulated expression of cucumber *HPR* was also found to be down-regulated by elevated CO_2 (Bertoni and Becker, 1996).

The response of peroxisomal enzymes to light is mediated at least in part by phytochrome (Schopfer *et al*, 1976) although other photoreceptors have also been implicated (Tchang *et al*, 1984a, b; Feierabend, 1975). Bertoni and Becker (1993) investigated the effects of light fluence and wavelength on expression of *HPR* in detail and found that a phytochrome-dependent component is involved in *HPR* regulation in dark-adapted green cotyledons, as indicated by red light induction and partial far red light reversibility. By the use of promotor deletion constructs fused to the reporter gene GUS, distinctive promotor elements necessary for light-regulated and leaf-specific expression were identified (Sloan *et al*, 1993; Daniel and Becker, 1995).

Cytokinins regulate many aspects of growth and development via differential gene expression. The transcription of *HPR* from pumpkin is responsive to cytokinin (Andersen *et al*, 1996). Two cytokinin-responsive and protein-binding elements, including an as-1 TGACG motif, which is related to a known element of auxin- and salicylic acid-regulated plant promotors, were identified in the promotor region of *Hpr-A* (Jin *et al*, 1998). Furthermore, evidence could be provided that cytokinin and light do not share identical signal transduction pathways in regulating *Hpr-A* expression. Mature chloroplasts are not only required for the proper expression of many photogenes, but also for that of some nuclear-encoded extraplastidic enzymes involved in biochemical pathways closely associated with plastid metabolism, including GOX and HPR (Bajracharya *et al*, 1987; Feierabend and Schubert, 1978). Photo-oxidative destruction of chloroplasts by exposure to white light of norflurazon-treated cucumber (*Cucumis sativus* L.) seedlings with inhibited synthesis of carotenoid pigments led to a reduction in the levels of transcript and protein of HPR (Schwartz *et al*, 1992). It was concluded from these and other results that intact chloroplasts are required for maximal expression of *HPR* in the light and that plastids affect *HPR* expression at a pre-translational level.

In addition to peroxisomal HPR, leaves also contain a cytosolic HPR that prefers NADPH over NADH as a co-substrate and is slightly smaller (subunit: 38 kDa, dimer: about 70 kDa) and immunologically only weakly related to the peroxisomal isoform (Kleczkowski and Randall, 1988; Kleczkowski *et al*, 1988; for review see Givan and Kleczkowski, 1992). Even though the complete oxidative pentose phosphate pathway is only located in plastids (Schnarrenberger *et al*, 1995), the enzymes of the oxidative part are present in both the plastid and the cytosol and can generate NADPH in the cytosol. (Schnarrenberger *et al*, 1973; von Schaewen *et al*, 1995). Peroxisomal and cytosolic HPR can be distinguished by their sensitivity to inhibitors, such as oxalate (Klezkowski *et al*, 1991). Even though cytosolic HPR is not believed to be involved in the main route of carbon flow through the glycolate pathway, it can serve in an auxiliary function to reduce hydroxypyruvate and/or glyoxylate which had leaked from the peroxisomes to the cytosol. Supporting evidence for this hypothesis was provided by barley mutants lacking peroxisomal NADH-dependent HPR. The mutants, which had only 5% of wild-type HPR activity in leaf tissue, were fully viable and able to fix carbon at a rate of 75% of that of wild-type plants. In these mutants the reduction of hydroxypyruvate is most likely catalyzed by cytosolic HPR (Murray *et al*, 1989; Kleczkowski *et al*, 1990).

The results of cloning the cytosolic isoform of HPR are still unclear. The HPR gene from cucumber, which terminates with the amino acids GNA> (Greenler *et al*, 1989), lacks a C-terminal PTS1. In pumpkin, two kinds of HPR cDNA clones (*HPR1* and *HRP2*) were isolated from screening a cDNA library from green cotyledons and their nucleotide sequences were almost identical except for the 3' end, which encodes a PTS1 in *HPR1*. Only HPR1 was shown to be transported to peroxisomes, whereas HPR2 remained in the cytosol (Hayashi *et al*, 1996; Mano *et al*, 1999). Light-regulated alternative splicing has been suggested to give rise to the two kinds of HPR mRNA (Mano *et al*, 1999). The physiological function of HPR2 still needs to be defined. This type of transcriptional processing, however, does not seem to be a general mechanism of HPR regulation in all higher plants because the single HPR gene from Arabidopsis (SKL>) is not prone to alternative splicing and the encoded enzyme is exclusively localized in leaf peroxisomes (Mano *et al*, 1997).

Malate dehydrogenase

The membrane of leaf peroxisomes appears to be impermeable to NADH, the cofactor of hydroxypyruvate reduction (Yu and Huang, 1986; Reumann *et al*, 1994, 1996). Reducing equivalents need to be imported from the cytosolic pool via malate and provided internally by the activity of peroxisomal malate dehydrogenase (MDH, EC 1.1.1.37). Peroxisomal MDH is a dimeric enzyme with a subunit molecular mass of about 33 kDa (Liu and Huang, 1976; Walk and Hock, 1977; Walk *et al*, 1977; for review see Gietl, 1992 a, b). The first peroxisomal isoform of MDH was cloned from glyoxysomes of *Citrullus lanatus* (watermelon) (Gietl, 1990), followed by the homologues from *Cucumis sativus*, *Glycine max.*, and *Cucurbita* sp. (Kim and Smith, 1994; Guex *et al*, 1995; Kato *et al*, 1998). The peroxisomal isoforms from dicots share about 95% to 75% sequence identity with each other, but lower sequence identity with peroxisomal MDH from fungi and Trypanosoma (45%) as compared to isoforms from other plant cell organelles (about 60%).

The peroxisomal isoforms from plants were shown to contain an N-terminal cleavable transit peptide of 37 amino acids (Gietl, 1990; Gietl *et al*, 1994) and to harbour a functional PTS2 motif (RI-x5-HL) similar to the targeting motifs of 3-ketoacyl-CoA thiolase and citrate synthase (R[ILQ]-x5-[HQ]L), Gietl, 1996; Kato *et al*, 1998). Interestingly, peroxisomal enzymes are sometimes targeted to their destination by different targeting signals in different organisms. Even though all plant peroxisomal isoforms of MDH identified so far contain an N-terminal PTS2, those from *S. cerevisiae* and *Trypanosoma* are targeted to peroxisomes and glycosomes, respectively, by a C-terminal PTS1.

Evidence that glyoxysomal MDH is identical to the isoform of leaf peroxisomes was first provided by Kim and Smith (1994). Cucumber glyoxysomal MDH was shown to be encoded by a single gene by Southern blot analysis. Moreover, the expression of *MDH* concurred with that of isocitrate lyase and malate synthase during seed germination, and later that of HPR during cotyledon development (Kim and Smith, 1994). In Arabidopsis, one isoform of peroxisomal MDH has been analyzed experimentally (Berkemeyer *et al*, 1998) and a second (pMDH2) can been identified by the genome sequence, as indicated by the presence of the conserved PTS2 at the N-terminus and the absence of mitochondrial and plastidic targeting sequences. The expression of *pMDH1* seems to exceed that of *pMDH2* by far as indicated by the number of EST clones (Reumann *et al*, in preparation). The substrate specificity of the second pMDH isoform needs to be determined.

Aspartate aminotransferase

Aspartate aminotransferase (AspAT, EC 2.6.1.1), also known as glutamate-oxaloacetate aminotransferase, is a pyridoxal-phosphate-dependent aminotransferase, characterized by the consensus pattern of the aminotransferase class-I pyridoxal-phosphate attachment site (Prosite: PDOC00098, Metha *et al*, 1993), with the cofactor being tightly linked to a lysine residue. The enzyme catalyzes the reversible transfer of the amino group from L-aspartate to 2-oxoglutarate to form oxaloacetate and

L-glutamate. The aminotransferase plays a key role in both nitrogen and carbon metabolism, and distinct peroxisomal isoenzymes are thought to be involved in shuttling reducing equivalents between subcellular compartments (Rehfeld and Tolbert, 1972; Givan, 1980). Isoforms of AspAT are located in the cytosol, the mitochondria, and plastids, and the genes have been cloned (for review, see Wadsworth, 1997). The three-dimensional crystalline structures of the cytosolic and mitochondrial isoforms from vertebrates have revealed that the enzymes are dimers of identical subunits with approximate molecular masses of 45 kDa (McPhalen *et al*, 1992).

An AspAT isoenzyme has also been detected in peroxisomes of some plant species (Huang *et al*, 1976; Liu and Huang, 1977; Rehfeld and Tolbert, 1972). In glyoxysomes, the enzyme has been postulated to be involved in the export of reducing equivalents out of glyoxysomes via a malate-aspartate shuttle (Mettler and Beevers, 1980). Whether also AspAT of leaf peroxisomes (Huang *et al*, 1976; Yamazaki and Tolbert, 1970; Rehfeld and Tolbert, 1972) participates in the import of reducing equivalents into leaf peroxisomes, is still a matter of debate. From the results of an *in vitro* assay of the photorespiratory C_2 cycle under physiological conditions, in which the activity of leaf peroxisomal AspAT was not high enough to convert OAA internally to aspartate, it was concluded that a malate-aspartate shuttle does not operate in leaf peroxisomes to a major extent (Reumann *et al*, 1994, see also Reumann, 2000).

Identification and cloning of the peroxisomal enzyme turned out to be very difficult because of a dual targeting mechanism. It could first be demonstrated for *Saccharomyces cerevisiae* that cytosolic AspAT can also be targeted to peroxisomes (Verleur *et al*, 1997). Similarly, Gebhardt *et al* (1998) characterized a single soybean cDNA encoding the cytosolic as well as the glyoxysomal AspAT. The gene contains two in-frame start codons and a putative PTS2 succeeding the first ATG. The second start codon aligns with previously reported start codons of plant cytosolic cDNAs. This suggests that the glyoxysomal isoform is produced by translational initiation from the first ATG codon and directed to peroxisomes by an N-terminal modified PTS2 (RL-x5-HF). It should, however, be noted that the precursor lacks a conserved cysteine residue located at or adjacent to the processing site, which is found in most PTS2-targeted proteins (Kato *et al*, 1998). The cytosolic isozyme is produced by translational initiation from the second ATG codon (Figure 3). Heterologous expression of the open reading frames initiated by each of the putative start codons produced proteins with AspAT activity. However, it has yet to be determined whether alternative forms of transcriptional initiation and/or translational initiation result in the variable expression of N-terminal targeting sequence and are responsible for the dual targeting of AspAT isozymes in soybean. A putative PTS2 on the protein initiated from the first start codon as well as the two start codons were indeed found to be conserved in cytosolic AspATs from several plant species (Gebhardt *et al*, 1998), suggesting that several species may utilize a single gene to generate both cytosolic and glyoxysomal or peroxisomal forms of AspAT.

In Arabidopsis, five AspAT isoforms have been characterized (Schultz *et al*, 1998; Wilkie and Warren, 1998), of which one isoform (ASP3) is a candidate for the peroxisomal enzyme and contains a consensus similar to a PTS2 (RI-x5-HL)(Schultz and Coruzzi, 1995; Gebhardt *et al*, 1998; Schultz *et al*, 1998).The corresponding gene is expressed in relatively high amounts in many tissues, including leaves of light-grown and dark-adapted plants (Schultz and Coruzzi, 1995). Because expression of cytosolic and peroxisomal AspAT probably cannot be distinguished at the mRNA level, a role of AspAT in photorespiratory metabolism cannot directly be deduced from Northern blot analysis and more research at the protein level is required.

The end product of the peroxisomal reactions, D-glycerate, leaves the peroxisome probably by diffusion through the porin-like channel and is transported to the chloroplast stroma by the glycolate/glycerate translocator, where its phosphorylation is catalyzed by glycerate kinase. The resulting P-glycerate can re-enter the reductive photosynthetic carbon cycle to generate ribulose-P_2 or, after reduction to triose-P in the chloroplast, can be used for synthesis of storage carbohydrates.

TRANSFER OF REDUCING EQUIVALENTS FROM THE MITOCHONDRIA AND CHLOROPLASTS TO LEAF PEROXISOMES

Import of redox equivalents into leaf peroxisomes for hydroxypyruvate reduction

The reduction of hydroxypyruvate to glycerate by peroxisomal HPR in the course of the photorespiratory C_2 cycle is dependent on NADH. The enzymes of fatty acid β-oxidation are active in leaf peroxisomes and able to generate NADH in the matrix (Gerhardt, 1981). However, these activities are not thought to be able to support the high rates of hydroxypyruvate reduction required during photosynthesis (Gerhardt, 1981, 1987). Three alternative mechanisms of redox transfer have been discussed: direct transfer of NADH across the membrane of leaf peroxisomes (Schmitt and Edwards, 1983) and indirect import of reducing equivalents via a malate-aspartate shuttle (Huang *et al*, 1976; Yamazaki and Tolbert, 1970; Rehfeld and Tolbert, 1972), as suggested for export of redox equivalents out of glyoxysomes (Mettler and Beevers, 1980), or via a malate-oxaloacetate shuttle (Yu and Huang, 1986).

Even though high rates of glycerate formation from serine were observed when NADH was added as reductant to isolated leaf peroxisomes (Schmitt and Edwards, 1983), it turned out that these results were mainly due to artificial leakage of hydroxypyruvate out of the organelle in the absence of internal reductant (Reumann *et al*, 1994). Under these conditions, hydroxypyruvate is directly reduced to glycerate by soluble HPR deriving from broken organelles using external NADH (Reumann *et al*, 1994). After correction for this external reduction, the maximal velocity of NADH-dependent glycerate production was found to be one order of magnitude lower than with malate as reductant. Taking the physiological substrate concentrations into account, it has been concluded that the demands of peroxisomal hydroxypyruvate reduction in a photosynthesizing leaf are only met by oxidation of malate (Reumann *et al*, 1994, 1996).

As mentioned above, AspAT is present in leaf peroxisomes (Yamazaki and Tolbert, 1970; Gebhardt *et al*, 1998). When glycerate formation from serine was investigated under conditions that allowed peroxisomal AspAT to convert oxaloacetate to aspartate, only a relatively low rate of glycerate formation and significant levels of oxaloacetate were observed, indicating that the matrix acitvity of AspAT is not high enough to play a significant role in the import of redox equivalents (Reumann *et al*, 1994). From these results, it has been concluded that peroxisomes are supplied with reducing equivalents indirectly via a malate-oxaloacetate shuttle. However, under certain physiological conditions, such as during the degradation of fatty acids in the dark period, AspAT may play a role in the export of redox equivalents (Mettler and Beevers, 1980).

Provision of redox equivalents by mitochondria and chloroplasts

NADH produced in the mitochondria by the glycine decarboxylase complex can be re-oxidized by the respiratory chain to synthesize ATP. Indeed, mitochondrial ATP synthesis is required during photosynthesis to supply energy to the cytosol of mesophyll cells. Alternatively, NADH can be re-oxidized by oxaloacetate and the high activity of mitochondrial MDH and exported as malate. In this way, reducing equivalents can be delivered from the mitochondria to the peroxisomes via a malate-oxaloacetate shuttle. The NADH to NAD ratio in the mitochondria is more than fifty times higher than in the cytosol (Heineke *et al*, 1991) and in the peroxisomes (Raghavendra *et al*, 1998) and, therefore, strongly favors diffusion of malate from mitochondria to peroxisomes via this shuttle.

To allow passage of malate across the inner membrane, mitochondria possess a highly active malate-oxaloacetate transporter with a K_m in the lower micromolar range (Ebbighausen *et al*, 1985; Hanning and Heldt, 1993). Hanning *et al* (1999) determined in inhibitor studies that oxaloacetate is transported in an obligatory counter-exchange with malate, 2-oxoglutarate, succinate, citrate, or aspartate. A second function of this translocator is probably the export of citrate via the citrate-OA shuttle. NADH can also be exported by a malate-aspartate shuttle found in plant mitochondria. At physiological oxaloacetate concentrations, however, the malate-aspartate shuttle accounts for only about 20% of the rate of NADH export of the malate-oxaloacetate shuttle (Krömer, 1995).

Alternatively, redox equivalents for hydroxypyruvate reduction can be provided by the chloroplasts. Chloroplasts export reducing equivalents at high rate by the malate-oxaloacetate shuttle (Hatch *et al*, 1984). In this case, malate is generated in the stroma by oxidation of NADPH, which is generated by non-cyclic electron transport, catalyzed by NADP-dependent MDH. Malate and oxaloacetate are transported across the inner envelope membrane via a specific translocator, probably in a counter-exchange. Activation of chloroplastic MDH by thioredoxin makes it possible to export excessive reducing equivalents for the reduction of hydroxypyruvate in the peroxisome and other reactions.

In the photorespiratory cycle, the amount of NADH generated by the oxidation of glycine in the mitochondria is equimolar to the NADH required for hydroxypyruvate reduction in the peroxisome. It has been demonstrated that mitochondrial ATP synthesis is required during photosynthesis to supply energy to the cytosol of mesophyll cells (Krömer and Heldt, 1991b). Mitochondria use about half of the generated NADH for ATP synthesis and the remainder to deliever about half of the reducing equivalents required for peroxisomal hydroxypyruvate reduction. Chloroplasts provide the remaining portion (Krömer and Heldt, 1991a, b; Hanning and Heldt, 1993).

COMPARTMENTALIZATION OF PHOTO-RESPIRATORY METABOLISM BY A SPECIFIC MATRIX STRUCTURE

Metabolic compartmentalization of other cell organelles

Eukaryotic cell organelles were evolved to create specialized compartments that are equipped with specific enzymes and in which definite concentrations of metabolites and inorganic ions can be maintained. Metabolic compartmentalization of cell organelles is mostly achieved by the organelle membrane forming a permeability barrier. The inner membrane of mitochondria and plastids represents the main permeability barrier, where passage of metabolites is controlled by metabolite translocators of high specificity (for reviews of plastid translocators see Flügge and Heldt, 1991; Flügge, 1999; for such of mitochondrial translocators see Palmieri *et al*, 1996). The outer membranes of mitochondria and plastids contain large diffusion channels, such as the voltage-dependent anion-selective channel in mitochondria (VDAC; Colombini, 1979; Zalman *et al*, 1980; for review see

Benz, 1994a), a VDAC homologue in non-green plastids (Fischer *et al*, 1994), and a porin-like general diffusion channel in chloroplasts (Flügge and Benz 1984; Pohlmeyer *et al*, 1998; Röhl *et al*, 1999). Recent investigations have demonstrated, however, that the permeability of these general diffusion channels seems to be more dynamic than previously thought and that specific channels have complementary function (Pohlmeyer *et al*, 1997; Bölter *et al*, 1999; for review see Flügge, 2000; Bölter and Soll, 2001).

Moreover, microcompartmentation of certain metabolites can be achieved by an organized arrangement of sequential enzymes of certain metabolic pathways in functional complexes, so-called multi-enzyme complexes or metabolons. Multi-enzyme complexes allow substrate channelling, i.e. the direct transfer of an intermediate between the active sites of two sequential enzymes (for review see Srere, 1987; Ovadi, 1991; Miles *et al*, 1999). Presumably, several reaction sequences of primary metabolism are regarded as mediated by multi-enzyme complexes, e.g. the citric acid cycle (Robinson *et al*, 1985; 1987; Srere 1987; Sumegi *et al*, 1990), the Calvin cycle (Gontero *et al*, 1988; Rault *et al*, 1993), glycine decarboxylation (Neuburger *et al*, 1986; Oliver, 1994), and fatty acid biosynthesis (Roughan and Ohlrogge, 1996).

Metabolic compartmentalization of leaf peroxisomes by a multienzyme complex

Metabolic compartmentalization can be investigated by analyzing the enzymes for their latency. Latency is generally defined as the percentage of total enzyme activity assayed in the presence of detergent that is not measurable in a suspension of intact organelles. Studies with leaf peroxisomes from spinach (*Spinacia oleracea* L.) revealed that the latency of most enzymes of the photorespiratory C_2 cycle was not markedly reduced by the application of an osmotic shock to the organelles, even though this treatment led to rupture of the membrane (Heupel *et al*, 1991). The matrix enzymes of osmotically-shocked peroxisomes remained associated in the matrix structure, visible in electron micrographs as peroxisomal particles of pleomorphic morphology that are only slightly smaller in size than those of intact leaf peroxisomes. These observations indicate that the latency of photorespiratory enzymes cannot be explained by a restricted access of one substrate to the active centres of the enzymes by the membrane.

With metabolically highly competent leaf peroxisomes from spinach, the entire sequence of peroxisomal reactions of the C_2 cycle can be simulated *in vitro* by supplementing isolated peroxisomes with those intermediates at physiological concentrations (Heineke *et al*, 1991; Winter *et al*, 1994) that are generated *in vivo* by chloroplasts and mitochondria, namely glycolate, serine, glutamate and malate (Figure 1), and analyzing the kinetics of glycerate formation. In such an experiment, a lag phase in the formation of glycerate due to the time-dependent accumulation of intermediates might have been expected (glyoxylate, NADH, and hydroxypyruvate). In contrast, glycerate formation started at a constant rate from the very beginning when using not only intact but also osmotically shocked leaf peroxisomes (Heupel *et al*, 1991; Heupel and Heldt, 1994). Since glycerate formation requires the activities of at least four peroxisomal enzymes, this linear kinetic can only be explained by a direct 'metabolite channelling' of products from one enzyme to the next in the reaction sequence without the release of intermediates into the bulk phase. In line with this hypothesis, the intermediates of peroxisomal metabolism, i.e. glyoxylate, H_2O_2 and hydroxypyruvate, are not released during glycolate oxidation from osmotically shocked organelles. These results, together with the membrane-independent latency of leaf peroxisomal enzymes, demonstrated that the compartmentalization of the photorespiratory C_2 cycle of leaf peroxisomes is not a function of the membrane, but is due to enzymes being arranged in a multi-enzyme complex (Heupel *et al*, 1991; Heupel and Heldt, 1994).

Even though channelling of intermediates of photorespiratory metabolism to this extent was an unexpected result, this mode of compartmentalization appears advantageous for several reasons. Two harmful intermediates of peroxisomal metabolism, the strong oxidant H_2O_2, and the weak acid glyoxylate, which are strong inhibitors of thioredoxin-activated enzymes, such as RubisCO and stromal fructose bisphosphatase and sedoheptulose bisphosphatase (Flügge *et al*, 1980; Cook *et al*, 1985; Campbell and Ogren, 1990), can pass the phospholipid bilayer membrane by unspecific diffusion without the mediation of special proteins and, therefore, require an alternative mechanism for compartmentalization. In the absence of auxiliary H_2O_2 scavenging systems, the high apparent K_m of CAT over 1 M (Huang *et al*, 1983) may result in high steady state levels of H_2O_2 in the peroxisomal matrix promoting diffusion into the cytosol. By micro-compartmentation, however, high concentrations of H_2O_2 can be restricted to the micro-environment at the active site of CAT. By metabolite channelling, the flux rate of metabolites through pathways can increase to a maximum, which is especially advantageous if several enzymes compete for the same substrates (Reumann, 2000). Finally, micro-compartmentalization allows the single peroxisomal membrane to be equipped with a porin-like channel for transfer of a broad range of negatively charged metabolites across the membrane.

TRANSPORT OF METABOLITES ACROSS THE MEMBRANE OF LEAF PEROXISOMES

Localization and electrophysiological characterization of a porin-like channel in the membrane of leaf peroxisomes

The high rate of glycolate production during photosynthesis (Sharkey, 1988) leads to high flux rates of photorespiratory intermediates, such as glycolate, glutamate, serine, malate, and glycerate, across the membrane of leaf peroxisomes (Figure 1), and requires an efficient transport system. Analogous transport pathways are necessary in non-plant peroxisomes. An unspecific permeability of the peroxisomal membrane from mammals has been reported (van Veldhoven *et al*, 1983, 1987; Verleur and Wanders, 1993) and channel activities have been detected in yeast and mammalian peroxisomes (Labarca *et al*, 1986; Lemmens *et al*, 1989; Douma *et al*, 1990; Sulter *et al*, 1993). It could not be conclusively excluded, however, that these effects were due to ruptured peroxisomal membranes or to mitochondrial contaminations. Genes of peroxisomal porin channels from fungi and mammals have not been cloned yet.

In leaf peroxisomes from *Spinacia oleracea* L., a porin-like channel could be characterized by electrophysiological means and is thought to mediate the diffusion of photorespiratory intermediates across the membrane (Reumann *et al*, 1995; (Figures 1, 4).

The channel is strongly anion selective and is, as indicated by single channel conductance analysis using different electrolytes, permeable to a variety of structurally different anions, such as chloride, nitrate, formate, and propionate (Reumann *et al*, 1995, 1998). Due to the relatively broad permeability for anions of low molecular weight, this channel is referred to as a "porin-like channel" until the primary structure and the postulated secondary structure of amphiphilic β-strands will be resolved.

Figure 4: Model of the diffusion of photorespiratory intermediates through the porin-like channel of leaf peroxisomes.

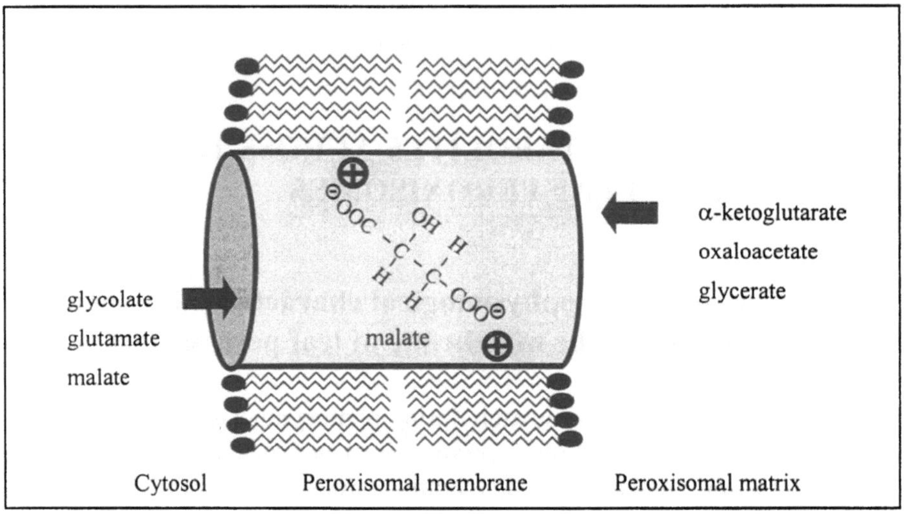

As deduced from the anion permeability, the channel diameter has been estimated to be a rather small size of about 0.6 nm. The permeability for the porin-like channel for intermediates of photorespiratory metabolites could be investigated by two complementary methods, single channel analysis for monovalent anions (for glycolate, glycerate, and glutamate) and macroscopic channel analyses in competitive binding studies with chloride (for oxaloacetate, malate, α-ketoglutarate, succinate, citrate, isocitrate). By single channel analyses it was demonstrated that the porin-like channel was permeable to the photorespiratory intermediates glycolate, glycerate, and glutamate (Reumann *et al*, 1995). By macroscopic channel analyses the porin-like channel was shown to contain an internal binding site for dicarboxylic acids, such as the photorespiratory intermediates malate, oxaloacetate, and α-ketoglutarate, and also for citrate and isocitrate (Reumann *et al*, 1998). It is noteworthy that reconstitution of the peroxisomal channel for the binding studies *in vitro* depended on a high ion concentrations of the electrolyte (1 M KCl). Under these conditions, the chloride ions competed with the dicarboxylic acids for the binding site, thereby causing an increase of the apparent half saturation concentrations of the dicarboxylic acids. The apparent half saturation concentrations in the millimolar range were estimated to be lowered by a factor of about 100 under physiological chloride concentrations (Reumann *et al*, 1998). It should be noted that a porin-like channel of almost identical electrophysiological properties has been characterized in glyoxysomes from *Ricinus communis* L. (Table 1, Reumann *et al*, 1997) and is thought to mediate the diffusion of malate, succinate, citrate, isocitrate, aspartate, oxaloacetate, and glutamate across the membrane of glyoxysomes during fatty acid β-oxidation (for review, see Reumann, 2000).

Table 1: Comparison of the electrophysiological properties of the porin-like channels from leaf peroxisomes (Spinacia oleracea L.) and glyoxysomes (Ricinus communis L.)

Organelle	Single channel conductance (in 1 M KCl) [nS]	Channel diameter [nm]	Selectivity	Binding site	Regulation of conductance	Reference
Leaf peroxisomes	0.35	0.6 nm	Strong anion	Present	Voltage dependence in low ion concentration	Reumann *et al*, 1995, 1998
Glyoxysomes	0.33	n.d.	Stron anion	Indicated	Voltage dependence in low ion concentration	Reumann *et al*, 1997

In summary, the electrophysiological properties indicate that the porin-like channel of plant peroxisomes is not a large general diffusion pore, as exemplified by VDAC and OmpF of the outer membrane of mitochondria (Colombini, 1979; Zalman *et al*, 1980; Benz, 1994a) and Gram negative bacteria (Benz *et al*, 1978; Nikaido, 1993; Benz, 1994b), respectively. Functionally, the porin-like channel of leaf peroxisomes belongs to the group of so-called 'specific porins', which have first been described in Gram negative bacteria. They are induced for the uptake of specific classes of substances and also contain a binding site inside of the channel for their substrates (LamB of *E. coli*: Benz *et al*, 1987; Schirmer *et al*, 1995; Tsx of *E. coli*: Maier *et al*, 1988; OprP of *P. aeruginosa*: Hancock *et al*, 1982; Benz and Hancock, 1987). The function of such an internal binding site inside the porin channel is to enable efficient uptake and transport of substrates especially at low substrate concentrations.

The molecular identification of the porin-like channel from plant peroxisomes remains to be resolved. A VDAC homologue has recently been identified in a membrane fraction of glyoxysomes from *Cucumis sativus* L. by microsequence peptide information and electron microscopic immunogold analysis (Corpas *et al*, 2000). Homologues of the mitochondrial VDAC have been localized in several internal membranes (amyloplasts from pea: Fischer *et al*, 1994; plasmalemma of B lymphocytes: Jakob *et al*, 1995; sarcoplasmatic reticulum: Jürgens *et al*, 1995). On the other hand, several results argue against the idea that a VDAC homologue occurs also in plant peroxisomes. Homologues of VDAC and the porin-like channel from plant peroxisomes differ significantly in many electrophysiological properties, such as the degree of anion selectivity, voltage-dependence, single channel conductance, the dependence of the single channel conductance on the chloride concentration, the presence of a binding site, and the channel diameter (Reumann *et al*, 1995, 1997, 1998). Apart from a minor contamination by mitochondria, a VDAC homologue could not be detected in leaf peroxisomes from *Spinacia oleracea* L. or glyoxysomes from *Ricinus communis* L. despite the high sensitivity of the electrophysiological methods (Reumann, Heldt and Benz, unpublished). It seems unlikely that a peroxisomal VDAC homologue escaped from its detection by electrophysiological means due to a closed state of conductance. Whether a VDAC homologue is specific for certain types of plant peroxisomes or mediates, for instance, the passage of larger molecules, e.g. co-factors such as NAD^+ and ATP, are important questions that need to be answered by cloning of the corresponding gene, electrophysiological analyses in comparison with the mitochondrial homologue, and subcellular localization studies. In our laboratory, a putative porin gene has been cloned of which the identity still needs to be verified by expression in a heterologous system and functional reconstitution of the membrane protein (Reumann *et al*, unpublished).

Physiological significance of the porin-like channel

Due to its electrophysiological properties found in reconstitution experiments using planar lipid bilayer membranes, the porin-like channel is expected to mediate *in vivo* the diffusion of at least six out of eight photorespiratory metabolites across the membrane of leaf peroxisomes (Figures 1, 4). The permeability of the channel to neutral amino acids, such as serine and glycine, has not been demonstrated yet and seems unlikely, considering its strong anion selectivity. The high anion specificity of the porin-like channel is in the range of that of OprP from *Pseudomonas aeruginosa*, which has been demonstrated to be impermeable to protons (Benz and Hancock, 1987). This property may be physiologically significant because a ΔpH gradient exists across the peroxisomal membrane from yeast and human fibroblasts (Nicolay *et al*, 1987; Dansen *et al*, 1999).

The porin-like channel allows metabolites to pass the peroxisomal membrane by diffusion, which is the fastest mode of membrane passage. The diffusion flux of organic acids is determined by the concentration gradient across the membrane, i.e. between the peroxisomal matrix and the cytosol. A serious disadvantage of diffusion-mediated metabolite transport, however, is a low diffusion flux at low metabolite concentration gradients, e.g. low cytosolic substrate concentrations for a matrix enzyme (e.g. glycolate, malate, or α-ketoglutarate). The small diameter of the channel additionally reduces the diffusion flux of photorespiratory metabolites as compared to large general diffusion pores but enables cofactors, such as ATP, NAD, and CoA, and fatty acids being compartmentalized in peroxisomes. In this aspect, the internal binding site with high specificity for dicarboxylic acids is an essential property of the peroxisomal channel to guarantee high diffusion rates of photorespiratory intermediates. The binding site enhances the diffusion rate especially at low substrate concentrations to a maximum and results in an even higher diffusion flux at low substrate concentrations as compared to general diffusion pores (Benz *et al*, 1987). Thus, the internal binding site compensates for reduced permeability due to the smaller channel diameter. The binding site also allows discrimination with respect to the diffusion flux. Dicarboxylic acids with terminal carboxyl groups have the highest affinity for the binding site. Due to the internal binding site, the diffusion flux is saturable and binding to the internal binding site is competitive. Whether, indeed, all photorespiratory intermediates pass the membrane by one and the same protein or by homologues of slightly different properties, remains to be seen after cloning of the gene and analysis of the properties of putative homologues.

The fact that the porin-like channel is relatively unspecific and permeable to a large variety of anions, such as mono-, di- and tricarboxylic organic acids, with a size up to at least six carbon atoms, may allow diffusion of intermediates of the photorespiratory C_2 cycle, such as glyoxylate and hydroxypyruvate, out of the organelle and requires metabolic compartmentalization by a multienzyme complex.

ACKNOWLEDGEMENT

I am very grateful to Prof. H.W. Heldt for critical comments on the manuscript and helpful discussions. This work is funded by the Deutsche Forschungsgemeinschaft.

REFERENCES

Andersen, B.R., Jin, G., Chen, R., Ertl, J.R. and Chen C. (1996) Transcriptional regulation of hydroxypyruvate reductase gene expression by cytokinin in etiolated pumpkin cotyledons. *Planta.* **198**: 1-5.

Andrews, T.J. and Lorimer, G.H. (1978) Photorespiration – still unavoidable? *FEBS Lett.* **90**: 1-9.

Bajracharya, D., Bergfeld, R., Hatzfeld, W.-D., Klein, S. and Schopfer, P. (1987) Regulatory involvement of plastids in the development of peroxisomal enzymes in the cotyledons of mustard (*Sinapis alba* L.) seedlings. *J. Plant Physiol.* **126**: 421-436.

Baker, A. and Schatz, G. (1987) Sequences from a prokaryotic genome or the mouse dihydrofolate reductase gene can restore the import of a truncated precursor protein into yeast mitochondria. *Proc. Natl. Acad. Sci.* **84**: 3117-3121.

Banjoko, A. and Trelease, R.N. (1995) Development and application of an *in vivo* plant peroxisome import system. *Plant Physiol.* **107**: 1201-1208.

Belanger, F.C. and Ogren, W.L. (1987) Phosphoglycolate phosphatase: purification and preparation of antibodies. *Photosynth. Res.* **14**: 3-13.

Benz, R. (1994a) Permeation of hydrophilic solutes through mitochondrial outer membranes: Review on mitochondrial porins. *Biochim. Biophys. Acta.* **1197**: 167-196.

Benz, R. (1994b) Uptake of solutes through bacterial outer membranes. In *Bacterial Cell Walls* (Ghuysen, J.-M. and Hakenbeck, R. eds.). pp. 397-423. Amsterdam (The Netherlands): Elsevier Science B.V.

Benz, R. and Hancock R.E.W. (1987) Mechanism of ion transport through the anion-selective channel of the *Pseudomonas aeruginosa* outer membrane. *J. Gen. Physiol.* **89**: 275-295.

Benz, R., Janko, K., Boos, W. and Laeuger P. (1978) Formation of large, ion-permeable membrane channels by the matrix protein (porin) of *Escherichia coli. Biochim. Biophys. Acta.* **511**: 305-319.

Benz, R., Schmid, A. and Vos-Scheperkeuter, G.H. (1987) Mechanism of sugar transport through the sugar-specific LamB channel of *Escherichia coli* outer membrane. *J. Membr. Biol.* **100**: 21-29.

Berkemeyer, M., Scheibe, R. and Ocheretina O. (1998) A novel, non-redox-regulated NAD-dependent malate dehydrogenase from chloroplasts of *Arabidopsis thaliana* L. *J. Biol. Chem.* **273**: 27927-27933.

Bertoni, G.P. and Becker, W.M. (1996) Expression of the cucumber hydroxypyruvate reductase gene is down-regulated by elevated CO_2. *Plant Physiol.* **112**: 599-605.

Bertoni, G.P. and Becker, W.M. (1993) Effects of light fluence and wavelength on expression of the gene encoding cucumber hydroxypyruvate reductase. *Plant Physiol.* **103**: 933-941.

Bethards, L.A., Skadsen, R.W. and Scandalios, J.G. (1987) Isolation and characterization of a cDNA clone for the *Cat2* gene in maize and its homology with other catalases. *Proc. Natl. Acad. Sci.* **84**: 6830-6834.

Bölter, B. and Soll, J. (2001) Ion channels in the outer membranes of chloroplasts and mitochondria: open doors or regulated gates? *EMBO J.* **20**: 935-940.

Bölter, B., Soll, J., Hill, K., Hemmler, R. and Wagner, R. (1999) A rectifying ATP-regulated solute channel in the chloroplastic outer envelope from pea. *EMBO J.* **18**: 5505-5516.

Brickner, D.G., Harada, J.J. and Olsen, L.J. (1997) Protein transport into higher plant peroxisomes. *In vitro* import assay provides evidence for receptor involvement. *Plant Physiol.* **113**: 1213-1221.

Brickner, D.G. and Olsen L.J. (1998) Nucleotide triphosphates are required for the transport of glycolate oxidase into peroxisomes. *Plant Physiol.* **116**: 309-317.

Bunkelmann, J.R. and Trelease, R.N. (1996) Ascorbate peroxidase. A prominent membrane protein in oilseed glyoxysomes. *Plant Physiol.* **110**: 589-598.

Campbell, W.J. and Ogren, W.L. (1990) Glyoxylate inhibition of ribulosebisphosphate carboxylase/oxygenase activation in intact, lysed, and reconstituted chloroplasts. *Photosyn. Res.* **23**: 257-268.

Chen, Z., Ricigliano, J.W. and Klessig, D.F. (1993a) Purification and characterization of a soluble salicylic acid-binding protein from tobacco. *Proc. Natl. Acad. Sci.* **90**: 9533-9537.

Chen, Z., Silva, H. and Klessig, D.F. (1993b) Active oxygen species in the induction of plant systemic acquired resistance by salicylic acid. *Science.* **262**: 1883-1886.

Chistoserdova, L.V. and Lidstrom, M.E. (1994) Genetics of the serine cycle in *Methylobacterium extorquens* AM1: identification of sgaA and mtdA and sequences of sgaA, hprA, and mtdA. *J. Bacteriol.* **176**: 1957-1968.

Christellar, J.T. and Tolbert N.E. (1978) Phosphoglycolate phosphatase. *J. Biol. Chem.* **253**: 1780-1785.

Colombini, M. (1979) A candidate for the permeability pathway of the outer mitochondrial membrane. *Nature.* **279**: 643-645.

Cook, C.M., Mulligan, R.M. and Tolbert N.E. (1985) Inhibition and stimulation of ribulose-1,5-bisphosphate carboxylase/oxygenase by glyoxylate. *Arch. Biochem. Biophys.* **240**: 392-401.

Corpas, F.J., Sandalio, L.M., Brown, M.J., del Rio, L.A. and Trelease R.N. (2000) Identification of porin-like polypeptide(s) in the boundary membrane of oilseed glyoxysomes. *Plant Cell Physiol.* **41**: 1218-1228.

Daniel S.G. and Becker W.M. (1995) Transgeneic analysis of the 5'- and 3'-flanking regions of the NADH-dependent hydroxypyruvate reductase gene from *Cucumis sativus* L. *Plant Mol. Biol.* **28**: 821-836.

Dansen, T.B., Wirtz, K.W.A., Wanders, R.J.A. and Pap, E.H.W. (2000) Peroxisomes in human fibroblasts have a basic pH. *Nature Cell Biol.* **2**: 51-53.

Douce, R. and Heldt, H.W. (2000) Photorespiration. In *Photosynthesis: Physiology and metabolism* (Leegood, R.C., Sharkey, T.D. and von Cammerer, S. eds.). pp. 115-136. The Netherlands, Kluwer Academic Publishers.

Douce, R. and Neuburger, M. (1999) Biochemical dissection of photorespiration. *Curr. Opin. Plant Biol.* **2**: 214-222.

Douce, R., Bourguignon, J., Neuburger, M. and Rebeille, F. (2001) The glycine decarboxylase system: a fascinating complex. *Trends Plant Sci.* **6**: 167-176.

Douma, A.C., Veenhuis, M., Sulter, G.J., Waterham, H.R., Verheyden, K., Mannaerts, G.P. and Harder, W. (1990) Permeability properties of peroxisomal membranes from yeast. *Arch. Microbiol.* **153**: 490-495.

Drumm, H. and Schopfer, P. (1974) Effect of phytochrome on development of catalase activity and isoenzyme pattern in mustard (*Sinapis alba* L.) seedlings. A reinvestigation. *Planta.* **120**: 13-30.

Ebbighausen, H., Jia, C. and Heldt, H.W. (1985) Oxaloacetate translocator in plant mitochondria. *Biochim. Biophys. Acta.* **810**: 184-199.

Eising, R. and Gerhardt, B. (1989) Catalase synthesis and turnover during peroxisome transition in the cotyledons of *Helianthus annuus* L. *Plant Physiol.* **89**: 1000-1005.

Eising, R. and Gerhardt, B. (1987) Catalase degradation in sunflower cotyledons during peroxisome transition from glyoxysomal to leaf peroxisomal function. *Plant Physiol.* **84**: 225-232.

Eising, R., Trelease, R.N. and Ni, W. (1990) Biogenesis of catalase in glyoxysomes and leaf-type peroxisomes of sunflower cotyledons. *Arch. Biochem. Biophys.* **278**: 258-263.

Emanuelsson, O., Nielsen, H., Brunak, S. and von Heijne, G. (2000) Predicting subcellular localization of proteins based on their N-terminal amino acid sequence. *J. Mol. Biol.* **300**: 1005-1016.

Esaka, M., Yamada, N., Kitabayashi, M., Setoguchi, Y., Tsugeki, R., Kondo, M. and Nishimura, M. (1997) cDNA cloning and differential gene expression of three catalases in pumpkin. *Plant Mol. Biol.* **33**: 141-155.

Feierabend, J. (1975) Developmental studies on microbodies in wheat leaves. III. On the photocontrol of microbody development. *Planta.* **123**: 63-77.

Feierabend, J. and Engel, S. (1986) Photoinactivation of catalase *in vitro* and in leaves. *Arch. Biochem. Biophys.* **251**: 567-576.

Feierabend, J. and Schubert, B. (1978) Comparative investigation of the action of several chlorosis-inducing herbicides on the biogenesis of chloroplasts and leaf microbodies. *Plant Physiol.* **61**: 1017-1022.

Fischer, K., Weber, A., Brink, S., Arbinger, B., Schünemann, D., Borchert, S., Heldt, H.W., Popp, B., Benz, R., Link, T.A., Eckerskorn, C. and Flügge, U.I. (1994) Porins from plants: molecular cloning, heterologous expression and functional characterization of two members of the porin family. *J. Biol. Chem.* **269**: 25754-25760.

Flügge, U.-I. (2000) Transport in and out of plastids: does the outer envelope membrane control the flow? *Trends Plant. Sci.* **5**: 135-137.

Flügge, U.-I. (1999) Phosphate translocators in plastids. *Ann. Rev. Plant Physiol. Mol. Biol.* **50**: 27-45.

Flügge, U.-I. and Benz, R. (1984) Pore-forming activity in the outer membrane of the chloroplast envelope. *FEBS Lett.* **169**: 85-89.

Flügge, U.-I. and Heldt, H.W. (1991) Metabolite translocators of the chloroplast envelope. *Ann. Rev. Plant Physiol. Mol. Biol.* **42**: 129-144.

Flügge, U.-I., Freisl, M. and Heldt, H.W. (1980) The mechanism of the control of carbon fixation by the pH in the chloroplast stroma. *Planta.* **149**: 48-51.

Flynn, C.R., Mullen, R.T. and Trelease, R.N. (1998) Mutational analyses of a type 2 peroxisomal targeting signal that is capable of directing oligomeric protein import into tobacco BY-2 glyoxysomes. *Plant J.* **16**: 709-720.

Frugoli, J.A., Zhong, H.H., Nuccio, M.L., McCourt, P., McPeek, M.A., Thomas, T.L. and McClung, C.R. (1996) Catalase is encoded by a multigene family in *Arabidopsis thaliana* (L.) Heynh. *Plant Physiol.* **112**: 327-336.

Gebhardt, J.S., Wadsworth, G.J. and Matthews, B.F. (1998) Characterization of a single soybean cDNA encoding cytosolic and glyoxysomal isozymes of aspartate aminotransferase. *Plant Mol. Biol.* **37**: 99-108.

Gerhardt, B. (1987) Higher plant peroxisomes and fatty acid degradation. In: *Peroxisomes in biology and medicine.* (Fahimi, H.D. and Sies, H. eds.). pp. 141-151, Springer, Berlin.

Gerhardt, B. (1983) Localization of β-oxidation enzymes in peroxisomes isolated from non-fatty plant tissue. *Planta.* **159**: 238-246.

Gerhardt, B. (1981) Enzyme activities of the β-oxidation pathway in spinach leaf peroxisomes. *FEBS Lett.* **126**: 71-73.

Gietl, C. (1996) Protein targeting and import into plant peroxisomes. *Physiol. Plant.* **97**: 599-608.

Gietl, C. (1992a) Malate dehydrogenase isoenzymes: cellular locations and role in the flow of metabolites between the cytoplasm and cell organelles. *Biochim. Biophys. Acta.* **1100**: 217-234.

Gietl, C. (1992b) Partitioning of malate dehydrogenase isoenzymes into glyoxysomes, mitochondria, and chloroplasts. *Plant Physiol.* **100**: 557-559.

Gietl, C. (1990) Glyoxysomal malate dehydrogenase from watermelon is synthesized with an amino-terminal transit peptide. *Proc. Natl. Acad. Sci.* **87**: 5773-5777.

Gietl, C., Faber, K.N., van der Klei, I.J. and Veenhuis, M. (1994) Mutational analysis of the N-terminal topogenic signal of watermelon glyoxysomal malate dehydrogenase using the heterologous host *Hansenula polymorpha. Proc. Natl. Acad. Sci.* **91**: 3151-3155.

Givan, C.V. (1980) Aminotransferases in higher plants. In *The Biochemistry of Plants* (Miflin, B.J. ed.). Vol 5, pp. 329-357. New York: (USA): Academic Press.

Givan, C.V. and Kleczkowski L.A. (1992) The enzymic reduction of glyoxylate and hydroxypyruvate in leaves of higher plants. *Plant Physiol.* **100**: 552-556.

Glover, J.R., Andrews, D.W., Subramani, S. and Rachubinski, R.A. (1994) Mutagenesis of the amino targeting signal of *Saccharomyces cerevisiae* 3-ketoacyl-CoA thiolase reveals conserved amino acids required for import into peroxisomes *in vivo*. *J. Biol. Chem.* **269**: 7558-7563.

Gontero, B., Cardenas, M.L. and Ricard, J. (1988) A functional five-enzyme complex of chloroplasts involved in the Calvin cycle. *Eur. J. Biochem.* **173**: 437-443.

Gould, S.J., Keller, G.-A. and Subramani, S. (1988) Identification of peroxisomal targeting signals located at the carboxy terminus of four peroxisomal proteins. *J. Cell Biol.* **107**: 897-905.

Gould, S.J., Keller, G.-A. and Subramani, S. (1987) Identification of a peroxisomal targeting signal at the carboxy terminus of firefly luciferase. *J. Cell Biol.* **105**: 2923-2931.

Gould, S.J., Keller, G.-A., Hosken, N., Wilkinson, J. and Subramani, S. (1989) A conserved tripeptide sorts proteins to peroxisomes. *J. Cell Biol.* **108**: 1657-1664.

Gould, S.J., Keller, G.-A., Schneider, M., Howell, S.H., Garrard, L.J., Goodman, J.M., Distel, B., Tabak, H. and Subramani S. (1990) Peroxisomal protein import is conserved between yeast, plants, insects and mammals. *EMBO J.* **9**: 85-90.

Greenler, J.McC. and Becker, W.M. (1990) Organ specificity and light regulation of NADH-dependent hydroxypyruvate reductase transcript abundance. *Plant Physiol.* **94**: 1484-1487.

Greenler, J.McC., Sloan, J.S., Schwartz, B.W. and Becker, W.M. (1989) Isolation, characterization and sequence analysis of a full-length cDNA clone encoding NADH-dependent hydroxypyruvate reductase from cucumber. *Plant Mol. Biol.* **13**: 139-150.

Grotjohann, N., Janning, A. and Eising, R. (1997) *In vitro* photoinactivation of catalase isoforms from cotyledons of sunflower (*Helianthus annuus* L.). *Arch. Biochem. Biophys.* **346**: 208-218.

Gühnemann-Schäfer, K. and Kindl, H. (1995) The leaf peroxisomal form (MFP IV) of multifunctional protein functioning in fatty-acid β-oxidation. *Planta.* **196**: 642-646.

Guex, N., Henry, H., Flach, J., Richter, H. and Widmer, F. (1995) Glyoxysomal malate dehydrogenase and malate synthase from soybean cotyledons (*Glycine max* L.): enzyme association, antibody production and cDNA cloning. *Planta.* **197**: 369-375.

Hagishita, T., Yoshida, T., Izumi, Y. and Mitsunaga, T. (1996) Cloning and expression of the gene for serine-glyoxylate aminotransferase from an obligate methylotroph *Hyphomicrobium methylovorum* GM2. *Eur. J. Biochem.* **241**: 1-5.

Hall, N.P., Reggiani, R. and Lea, P.J. (1985) Molecular weights of glycolate oxidase from C_3 and C_4 plants determined during early stages of purification. *Phytochem.* **24**: 1645-1648.

Hancock, R.E.W., Poole, K. and Benz, R. (1982) Outer membrane protein P of *Pseudomonas aeruginosa*: regulation by phosphate deficiency and formation of small anion-specific channels in lipid bilayer membranes. *J. Bacteriol.* **150**: 730-738.

Hanning, I. and Heldt, H.W. (1993) On the function of mitochondrial metabolism during photosynthesis in spinach (*Spinacia oleracea* L.) leaves. Partitioning between respiration and export of redox equivalents and precursors for nitrate assimilation products. *Plant Physiol.* **103**: 1147-1154.

Hanning, I., Baumgarten, K., Schott, K. and Heldt H.W. (1999) Oxaloacetate transport into plant mitochondria. *Plant Physiol.* **119**: 1025-1032.

Hardy, P. and Baldy, P. (1986) Corn phosphoglycolate phosphatase: purification and properties. *Planta* **168**: 245-252.

Hatch, M.D., Dröscher, L., Flügge, U.-I. and Heldt, H.W. (1984) A specific translocator for oxaloacetate transport in chloroplasts. *FEBS Lett.* **178**: 15-19.

Hayashi, M., Aoki, M., Kondo, M. and Nishimura, M. (1997) Changes in targeting efficiencies of proteins to plant microbodies caused by amino acid substitutions in the carboxy-terminal tripeptide. *Plant Cell Physiol.* **38**: 759-768.

Hayashi, M., Tsugeki, R., Kondo, M., Mori, H. and Nishimura M. (1996) Pumpkin hydroxypyruvate reductases with and without a putative C-terminal signal for targeting to microbodies may be produced by alternative splicing. *Plant Mol. Biol.* **30**: 183-189.

Heber, U. and Krause, G.H. (1980) What is the physiological role of photorespiration? *Trends Biochem. Sci.* **5**: 32-34.

Heineke, D., Riens, B., Grosse, H., Hoferichter, P., Peter, U., Flügge, U.-I. and Heldt, H.W. (1991) Redox transfer across the inner chloroplast envelope membrane. *Plant Physiol.* **95**: 1131-1137.

Hertwig, B., Streb, P. and Feierabend, J. (1992) Light dependence of catalase synthesis and degradation in leaves and the influence of interfering stress conditions. *Plant Physiol.* **100**: 1547-1553.

Heupel, R. and Heldt H.W. (1994) Protein organization in the matrix of leaf peroxisomes. A multi-enzyme complex involved in photorespiratory metabolism. *Eur. J. Biochem.* **220**: 165-172.

Heupel, R., Markgraf, T., Robinson, D.G. and Heldt, H.W. (1991) Compartmentation studies on spinach leaf peroxisomes. Evidence for channelling of photorespiratory metabolites in peroxisomes devoid of intact boundary membrane. *Plant Physiol.* **96**: 971-979.

Holbrook, J.D., Birdsey, G.M., Yang, Z., Bruford, M.W. and Danpure, C.J. (2000) Molecular adaptation of alanine:glyoxylate aminotransferase targeting in primates. *Mol. Biol. Evol.* **17**: 387-400.

Hondred, D., Hunter, J.McC., Keith, R., Titus, D.E. and Becker, W.M. (1985) Isolation of serine:glyoxylate aminotransferase from cucumber cotyledons. *Plant Physiol.* **79**: 95-102.

Hondred, D., Wadle, D.-M., Titus, D.E. and Becker, W.M. (1987) Light-stimulated accumulation of the peroxisomal enzymes hydroxypyruvate reductase and serine:glyoxylate aminotransferase and their translatable mRNAs in cotyledons of cucumber seedlings. *Plant Mol. Biol.* **9**: 259-275.

Horng, J.-T., Behari, R., Burke, L.E.C.-A. and Baker, A. (1995) Investigation of the energy requirement and targeting signal for the import of glycolate oxidase into glyoxysomes. *Eur. J. Biochem.* **230**: 157-163.

Howitz, K.T. and McCarty, R.E. (1991) Solubilization, partial purification, and reconstitution of the glycolate/glycerate transporter from chloroplast inner envelope membranes. *Plant Physiol.* **96**: 1060-1069.

Howitz, K.T. and McCarty, R.E. (1988) Measurement of proton-linked transport activities in pyranine loaded chloroplast inner envelope vesicles. *Plant Physiol.* **86**: 999-1001.

Howitz, K.T. and McCarty, R.E. (1986) D-glycerate transport by the pea chloroplast glycolate carrier. *Plant Physiol.* **80**: 390-395.

Howitz, K.T. and McCarty, R.E. (1985a) Kinetic characteristics of the chloroplast envelope glycolate transporter. *Biochem.* **24**: 2645-2652.

Howitz, K.T. and McCarty, R.E. (1985b) Substrate specificity of the pea chloroplast glycolate transporter. *Biochem.* **24**: 3645-3650.

Huang, A.H.C., Liu, K.D.F. and Youle, R.J. (1976) Organelle-specific isozymes of aspartate-α-ketoglutarate transaminase in spinach leaves. *Plant Physiol.* **58**: 110-113.

Huang, A.H.C., Trelease, R.N. and Moore, T.S. Jr. (1983) *Plant Peroxisomes.* New York, London: Academic Press.

Husic, D.W., Husic, H.D. and Tolbert, N.E. (1987) The oxidative photosynthetic carbon cycle or C_2 cycle. *CRC Crit. Rev. Plant Sci.* **5**: 45-100.

Ireland, R.I. and Joy, K.W. (1983) Purification and properties of an asparagine aminotransferase from *Pisum sativum* leaves. *Arch. Biochem. Biophys.* **223**: 291-296.

Isaya, G., Kalousek, F., Fenton, W.A. and Rosenberg, L.E. (1991) Cleavage of percursors by the mitochondrial processing peptidase requires a compatible mature protein or an intermediate octapeptide. *J. Cell Biol.* **113**: 65-76.

Jakob, C., Götz, H., Hellmann, T., Hellmann, K.P., Reymann, S., Flörke, H., Thinnes, F.P. and Hilschmann, N. (1995) Studies on human porin: XIII. The type-1 VDAC 'Porin 31 HL' biotinylated at the plasmalemma of trypan blue excluding human B lymphocytes. *FEBS Lett.* **368**: 5-9.

Jin, G., Davey, M.C., Ertl, J.R., Chen, R., Yu, Z., Daniel, S.G., Becker, W.M. and Chen, C. (1998) Interaction of DNA-binding proteins to the 5'-flanking region of a cytokinin-responsive cucumber hydroxypyruvate reductase gene. *Plant Mol. Biol.* **38**: 713-724.

Jones, J.M., Morrell, J.C. and Gould, S.J. (2000) Identification and characterization of HAOX1, HAOX2, and HAOX3, three human peroxisomal 2-hydroxy acid oxidases. *J. Biol. Chem.* **275**: 12590-12597.

Jürgens, L., Kleineke, J., Brdiczka, D., Thinnes, F.P. and Hilschmann, N. (1995) Localization of type-1 porin channel (VDAC) in the sarcoplasmatic reticulum. *Biol. Chem. Hoppe-Seyler.* **376**: 685-689.

Kato, A., Takeda-Yoshikawa, Y., Hayashi, M., Kondo, M., Hara-Nishimura, I. and Nishimura, M. (1998) Glyoxysomal malate dehydrogenase in pumpkin: cloning of a cDNA and functional analysis of its presequence. *Plant Cell Physiol.* **39**: 186-195.

Kim, D.J. and Smith, S.M. (1994) Expression of a single gene encoding microbody NAD-malate dehydrogenase during glyoxysome and peroxisome development in cucumber. *Plant Mol. Biol.* **26**: 1833-1841.

Kleczkowski, L.A. and Randall D.D. (1988) Purification and characterization of a novel NADPH(NADH)-dependent hydroxypyruvate reductase from spinach leaves. Comparison of immunological properties of leaf hydroxypyruvate reducatases. *Biochem. J.* **250**: 145-152.

Kleczkowski, L.A., Edwards, G.E., Blackwell, R.D., Lea, P.J. and Givan, C.V. (1990) Enzymology of the reduction of hydroxypyruvate and glyoxylate in a mutant of barley lacking peroxisomal hydroxypyruvate reductase. *Plant Physiol.* **94**: 819-825.

Kleczkowski, L.A., Givan, C.V., Hodgson, J.M. and Randall, D.D. (1988) Subcellular localization of NADPH-dependent hydroxypyruvate reductase of leaf protoplasts of *Pisum sativum* L. and its role in photorespiratory metabolsim. *Plant Physiol.* **88**: 1182-1185.

Kleczkowski, L.A., Randall, D.D. and Edwards, G.E. (1991) Oxalate as a potent and selective inhibitor of spinach (*Spinacia oleracea*) leaf NADPH-dependent hydroxypyruvate reductase. *Biochem. J.* **276**: 125-127.

Kleff, S., Sander, S., Mielke, G. and Eising, R. (1997) The predominant protein in peroxisomal cores of sunflower cotyledons is a catalase that differs in primary structure from the catalase in the peroxisomal matrix. *Eur. J. Biochem.* **245**: 402-410.

Kohler, S.A., Menotti, E. and Kuehn, L.C. (1999) Molecular cloning of mouse glycolate oxidase. High evolutionary conservation and presence of an iron-responsive element-like sequence in the mRNA. *J. Biol. Chem.* **274**: 2401-2407.

Kohn, L.D., Warren, W.A. and Carrol, W.R. (1970) The structural properties of spinach leaf glyoxylic acid reductase. *J. Biol. Chem.* **245**: 3821-3830.

Krömer, S. (1995) Respiration during photosynthesis. *Ann. Rev. Plant Physiol. Plant. Mol. Biol.* **46**: 45-70.

Krömer, S. and Heldt, H.W. (1991a) Respiration of pea leaf mitochondria and redox transfer between the mitochondrial and extramitochondrial compartment. *Biochim. Biophys. Acta.* **1057**: 42-50.

Krömer, S. and Heldt, H.W. (1991b) On the role of mitochondrial oxidative phosphorylation in photosynthesis metabolism as studied by the effect of oligomycin on photosynthesis in protoplasts and leaves of barley (*Hordeum vulgare*). *Plant Physiol.* **95**: 1270-1276.

Labarca, P., Wolff, D., Soto, U., Necochea, C. and Leighton, F. (1986) Large cation-selective pores from rat liver peroxisomal membranes incorporated to planar lipid bilayers. *J. Mem. Biol.* **94**: 285-291.

Lemmens, M., Verheyden, K., van Veldhoven, P., Vereecke, J., Mannaerts, G.P. and Carmeliet, E. (1989) Single channel analysis of a large conductance channel in peroxisomes from rat liver. *Biochim. Biophys. Acta.* **984**: 351-359.

Liepman, A.H. and Olsen, L.J. (2001) Peroxisomal alanine:glyoxylate aminotransferase (AGT1) is a photorespiratory enzyme with multiple substrates in *Arabidopsis thaliana*. *Plant J.* **25**: 1-14.

Lin, X., Kaul, S., Rounsley, S., Shea, T.P., Benito, M.I. *et al.* (1999) Sequence and analysis of chromosome 2 of the plant *Arabidopsis thaliana*. *Nature*. **402**: 761-768.

Lindqvist, Y. (1989) Refined structure of spinach glycolate oxidase at 2Å resolution. *J. Mol. Biol.* **209**: 151-166.

Lindqvist, Y. and Bränden C-I. (1989) The active site of spinach glycolate oxidase. *J. Biol. Chem.* **264**: 3624-3628.

Lindqvist, Y. and Bränden, C-I. (1985) Structure of glycolate oxidase from spinach. *Proc. Natl. Acad. Sci.* **82**: 6855-6859.

Lindqvist, Y., Bränden, C-I., Mathews, F.S. and Lederer, F. (1991) Spinach glycolate oxidase and yeast flavocytochrome b2 are structurally homologous and evolutionarily related enzymes with distinctly different function and flavin mononucleotide binding. *J. Biol. Chem.* **266**: 3198-3207.

Liu, K.D.F. and Huang, A.H.C. (1977) Subcellular localization and developmental changes of aspartate-α-ketoglutarate transaminase isozymes in the cotyledons of cucumber seedlings. *Plant Physiol.* **59**: 777-782.

Liu, K.D.F. and Huang, A.H.C. (1976) Developmental studies of NAD-malate dehydrogenase isozymes in the cotyledon of cucumber seedlings grown in darkness and in light. *Planta.* **131**: 279-284.

Lorimer, G.H., and Andrews, T.J. (1981) The C_2 chemo- and photorespiratory carbon oxidation cycle. In *The Biochemistry of Plants* (Hatch, M.D. and Boardman, N.K. eds) Vol. 8, pp. 329-374. New York (U.S.A.): Academic Press.

Ludt, C. and Kindl, H. (1990) Characterization of a cDNA encoding *Lens culinaris* glycolate oxidase and developmental expression of glycolate oxidase mRNA in cotyledons and leaves. *Plant Physiol.* **94**: 1193-1198.

Maier, C., Bremer, E., Schmid, A. and Benz, R. (1988) Pore-forming activity of Tsx protein from the outer membrane of *Escherichia coli*: demonstration of a nucleoside-specific binding site. *J. Biol. Chem.* **263**: 2493-2499.

Mano, S., Hayashi, M., Kondo, M. and Nishimura, M. (1997) Hydroxypyruvate reductase with a carboxy-terminal targeting signal to microbodies is expressed in Arabidopsis. *Plant Cell Physiol.* **38**: 449-455.

Mano, S., Hayashi, M., Nishimura, M. (1999) Light regulates alternative splicing of hydroxypyruvate reductase in pumpkin. *Plant J.* **17**: 309-320.

Mayer, K., Schuller, C., Wambutt, R., Murphy, G., Volckaert, G. *et al* (1999) Sequence and analysis of chromosome 4 of the plant *Arabidopsis thaliana*. *Nature*. **402**: 769-777.

McPhalen, C.A., Vincent, M.G. and Jansonius, J.N. (1992) X-ray structure refinement and comparison of three forms of mitochondrial aspartate aminotransferase. *J. Mol. Biol.* **225**: 495-517.

Mekhedov, S., Martinez de Ilarduya, O. and Ohlrogge, J. (2000) Toward a functional catalogue of the plant genome. A survey of genes for lipid biosynthesis. *Plant Physiol.* **122**: 389-401.

Metha, P.K., Hale, T.I. and Christen, P. (1993) Aminotransferases: demonstration of homology and division into evolutionary subgroups. *Eur. J. Biochem.* **214**: 549-561.

Mettler, I.J. and Beevers, H. (1980) Oxidation of NADH in glyoxysomes by a malate-aspartate shuttle. *Plant Physiol.* **66**: 555-560.

Miles, E.W., Rhee, S., Davies, D.R. (1999) The molecular basis of substrate channelling. *J. Biol. Chem.* **274**: 12193-12196.

Mullen, R.T., Lee, M.S. and Trelease, R.N. (1997) Identification of the peroxisomal targeting signal for cottonseed catalase. *Plant J.* **12**: 313-322.

Murray, A.J.S., Blackwell, R.D., Lea, P.J. (1989) Metabolism of hydroxypyruvate in a mutant of barley lacking NADH-dependent hydroxypyruvate reductase, an important photorespiratory enzyme activity. *Plant Physiol.* **91**: 395-400.

Nakai, K. and Kanehisa, M. (1992) A knowledge base for predicting protein localization sites in eukaryotic cells. *Genomics.* **14**: 897-911.

Nakamura, Y. and Tolbert, N.E. (1983) Serine:glyoxylate, alanine:glyoxylate and glutamate:glyoxylate aminotransferase reactions in peroxisomes from spinach leaves. *J. Biol. Chem.* **258**: 7631-7638.

Neuburger, M., Bourguignon, J. and Douce, R. (1986) Isolation of a large complex from the matrix of pea leaf mitochondria involved in the rapid transformation of glycine into serine. *FEBS Lett.* **207**: 18-22.

Ni, W. and Trelease, R.N. (1991) Post-transcriptional regulation of catalase isozyme expression in cotton seeds. *Plant Cell.* **3**: 737-744.

Ni, W., Trelease, R.N. and Eising, R. (1990) Two temporally synthesized charge subunits interact to form the five isoforms of cottonseed (*Gossypium hirsutum*) catalase. *Biochem. J.* **269**: 233-238.

Nicolay, K., Veenhuis, M., Douma, A.C. and Harder, W. (1987) A ^{31}P NMR study of the internal pH of yeast peroxisomes. *Arch. Microbiol.* **147**: 37-41.

Nikaido, H. (1993) Transport across the bacterial outer membrane. *J. Bioenerg. Biomembr.* **25**: 581-589.

Noguchi, T. and Fujiwara, S. (1982) Development of glutamate:glyoxylate aminotransferase in the cotyledons of cucumber (*Cucumis sativus*) seedlings. *Biochem J.* **201**: 209-214.

Noguchi T. and Hayashi S. (1981) Plant leaf alanine:2-oxoglutarate aminotransferase. Peroxisomal localization and identity with glutamate:glyoxylate aminotransferase. *Biochem J.* **195**: 235-239.

Noguchi, T. and Hayashi S. (1980) Peroxisomal localization of properties of tryptophan aminotransferase in plant leaves. *J. Biol. Chem.* **255**: 2267-2269.

Oda, T., Funai, T. and Ichiyama, A. (1990) Generation from a single gene of two mRNAs that encode the mitochondrial and peroxisomal serine:pyruvate aminotransferase of rat liver. *J. Biol. Chem.* **265**: 7513-7519.

Ogren, W.L. (1984) Photorespiration: Pathways, regulation, and modification. *Ann. Rev. Plant Physiol.* **35**: 415-442.

Oliver, D.J. (1994) The glycine decarboxylase complex from plant mitochondria. *Ann. Rev. Plant Physiol. Plant Mol. Biol.* **45**: 323-337.

Osumi, T., Tsukamoto, T., Hata, S., Yokota, S., Miura, S., Fujiki, Y., Hijikata, M., Miyazawa, S. and Hashimoto, T. (1991) Amino-terminal presequence of the precursor of peroxisomal 3-ketoacyl-CoA thiolase is a cleavable signal peptide for peroxisomal targeting. *Biochem. Biophys. Res. Comm.* **181**: 947-954.

Ovadi, J. (1991) Physiological significance of metabolite channelling. *J. Theor. Biol.* **152**: 1-22.

Palmieri, F., Bisaccia, F., Capobianco, L., Dolce, V., Fiermonte, G., Iacobazzi, V., Indiveri, C. and Palmieri, L. (1996) Mitochondrial metabolite transporters. *Biochim. Biophys. Acta.* **1275**: 127-132.

Pohlmeyer, K., Soll, J., Grimm, R., Hill, K. and Wagner, R. (1998) A high-conductance solute channel in the chloroplastic outer envelope from Pea. *Plant Cell.* **10**: 1207-1216.

Pohlmeyer, K., Soll, J., Steinkamp, T., Hinnah, S. and Wagner, R. (1997) Isolation and characterization of an amino acid-selective channel protein present in the chloroplastic outer envelope membrane. *Proc. Natl. Acad. Sci.* **94**: 9504-9509.

Raghavendra, A.S., Reumann, S. and Heldt, H.W. (1998) Participation of mitochondrial metabolism in photorespiration: Reconstituted system of peroxisomes and mitochondria from spinach leaves. *Plant Physiol.* **116**: 1333-1337.

Rault, M., Giudici-Orticoni, M.-T., Gontero, B. and Ricard J. (1993) Structural and functional properties of a multi-enzyme complex from spinach chloroplasts. 1. Stoichiometry of the polypeptide chains. *Eur. J. Biochem.* **217**: 1065-1073.

Rehfeld, D.W. and Tolbert N.E. (1972) Aminotransferases in peroxisomes from spinach leaves. *J. Biol. Chem.* **247**: 4803-4811.

Reumann, S. (2000) The structural properties of plant peroxisomes and their metabolic significance. *Biol. Chem.* **381**: 639-648.

Reumann, S., Bettermann, M., Benz, R. and Heldt H.W. (1997) Evidence for the presence of a porin in the membrane of glyoxysomes of castor bean. *Plant Physiol.* **115**: 891-899.

Reumann, S., Heupel, R. and Heldt H.W. (1994) Compartmentation studies on spinach leaf peroxisomes; II. Evidence for the transfer of reductant from the cytosol to the peroxisomal compartment via a malate shuttle. *Planta.* **193**: 167-173.

Reumann, S., Maier, E., Benz, R. and Heldt, H.W. (1996) A specific porin is involved in the malate shuttle of leaf peroxisomes. *Biochem. Soc. Trans.* **24**: 754-757.

Reumann, S., Maier, E., Benz, R. and Heldt, H.W. (1995) The membrane of leaf peroxisomes contains a porin-like channel. *J. Biol. Chem.* **270**: 17559-17565.

Reumann, S., Maier, E., Heldt, H.W. and Benz R. (1998) Permeability properties of the specific porin of spinach leaf peroxisomes. *Eur. J. Biochem.* **251**: 359-366.

Richardson K.E. and Tolbert N.E. (1961) Phosphoglycolic acid phosphatase. *J. Biol. Chem.* **236**: 1285-1290.

Robinson, J.B.Jr. and Srere P.A. (1985) Organization of Krebs tricarboxylic acid cycle enzymes in mitochondria. *J. Biol. Chem.* **260**: 10800-10805.

Robinson, J.B.Jr., Inman, L., Sumegi, B. and Srere, P.A. (1987) Further characterization of the Krebs tricarboxylic acid cycle metabolon. *J. Biol. Chem.* **262**: 1786-1790.

Röhl, T., Motzkus, M. and Soll, J. (1999) The outer envelope protein OEP24 from pea chloroplasts can functionally replace the mitochondrial VDAC in yeast. *FEBS Lett.* **460**: 491-494.

Roise, D., Theiler, F., Horvath, S.J., Tomich, J.M., Richards, J.H., Allison, D.S. and Schatz, G. (1988) Amphiphilicity is essential for mitochondrial presequence function. *EMBO J.* **7**: 649-653.

Roughan, P.G. and Ohlrogge J.B. (1996) Evidence that isolated chloroplasts contain an integrated lipid-synthesizing assembly that channels acetate into long-chain fatty acids. *Plant Physiol.* **110**: 1239-1247.

Rumsby, G. and Cregeen D.P. (1999) Identification and expression of a cDNA for human hydroxypyruvate/glyoxylate reductase. *Biochim. Biophys. Acta.* **1446**: 383-388.

Salanoubat, M., Lemcke, K., Rieger, M., Ansorge, W., Unseld, M. *et al* (2000) Sequence and analysis of chromosome 3 of the plant *Arabidopsis thaliana.* European Union Chromosome 3 Arabidopsis Sequencing Consortium, The Institute for Genomic Research and Kazusa DNA Research Institute. *Nature.* **408**: 820-822.

Schirmer, T., Keller, T.A., Wang, Y.F. and Rosenbusch, J.P. (1995) Structural basis for sugar translocation through maltoporin channels at 3.1 Å resolution. *Science.* **267**: 512-514.

Schmitt, M.R. and Edwards, G.E. (1983) Provision of reductant for the hydroxypyruvate to glycerate conversion in leaf peroxisomes. *Plant Physiol.* **72**: 728-734.

Schnarrenberger, C., Flechner, A. and Martin W. (1995) Enzymatic evidence for a complete oxidative pentose phosphate pathway in chloroplasts and an incomplete pathway in the cytosol of spinach leaves. *Plant Physiol.* **108**: 609-614.

Schnarrenberger, C., Oeser, A. and Tolbert N.E. (1973) Two isoenzymes each of glucose-6-phosphate dehydrogenase and 6-phosphogluconate dehydrogenase in spinach leaves. *Arch. Biochem. Biophys.* **154**: 438-448.

Schnarrenberger, C., Oeser, A. and Tolbert N.E. (1971) Development of microbodies in sunflower cotyledons and castor bean endosperm during germination. *Plant Physiol.* **48**: 566-574.

Schopfer, P., Bajracharya, D. Bergfeld, R. and Falk H. (1976) Phytochrome-mediated transformation of glyoxysomes into peroxisomes in the cotyledons of mustard (*Sinapis alba* L.) seedlings. *Planta.* **133**: 73-80.

Schultz, C.J. and Coruzzi G.M. (1995) The aspartate aminotransferase gene family of Arabidopsis encodes isoenzymes localized to three distinct subcellular compartments. *Plant J.* **7**: 61-75.

Schultz, C.J., Hsu, M., Miesak, B. and Coruzzi, G.M. (1998) Arabidopsis mutants define an *in vivo* role for isoenzymes of aspartate aminotransferase in plant nitrogen metabolism. *Genetics.* **149**: 491-499.

Schwartz, B.W., Daniel, S.G. and Becker, W.M. (1992) Photooxidative destruction of chloroplasts leads to reduced expression of peroxisomal NADH-dependent hydroxypyruvate reductase in developing cucumber cotyledons. *Plant Physiol.* **99**: 681-685.

Schwartz, B.W., Sloan, J.S. and Becker, W.M. (1991) Characterization of genes encoding hydroxypyruvate reductase in cucumber. *Plant Mol. Biol.* **17**: 941-947.

Sharkey T.D. (1988) Estimating the rate of photorespiration in leaves. *Physiol. Plant.* **73**: 147-152.

Sloan, J.S., Schwartz, B.W. and Becker, W.M. (1993) Promotor analysis of a light-regulated gene encoding hydroxypyruvate reductase, an enzyme of the photorespiratory glycolate pathway. *Plant J.* **3**: 867-874.

Smith, I.K. (1973) Purification and characterization of serine:glyoxylate aminotransferase from kidney bean (*Phaseolus vulgaris). Biochim. Biophys. Acta.* **321**: 156-164.

Somerville, C.R. (2001) An early Arabidopsis demonstration. Resolving a few issues concerning photorespiration. *Plant Physiol.* **125**: 20-24.

Somerville, C.R. and Ogren, W.L. (1982a) Genetic modification of photorespiration. *Trends Biochem. Sci.* **7**: 171-174.

Somerville, C.R. and Ogren, W.L. (1982b) Mutants of the cruciferous plant *Arabidopsis thaliana* lacking glycine decarboxylase activity. *Biochem J.* **202**: 373-380.

Somerville, C.R. and Ogren, W.L. (1981) Photorespiration-deficient mutants of *Arabidopsis thaliana* lacking mitochondrial serine transhydroxymethylase activity. *Plant Physiol.* **67**: 666-671.

Somerville, C.R. and Ogren, W.L. (1980) Photorespiration mutants of *Arabidopsis thaliana* deficient in serine-glyoxylate aminotransferase activity. *Proc. Natl. Acad. Sci.* **77**: 2684-2687.

Somerville, C.R. and Ogren, W.L. (1979) A phosphoglycolate phosphatase-deficient mutant of Arabidopsis. *Nature.* **280**: 833-836.

Srere, P.A. (1987) Complexes of sequential metabolic enzymes. *Ann. Rev. Biochem.* **56**: 89-124.

Stenberg, K., Clausen, T., Lindqvist, Y. and Macheroux, P. (1995) Involvement of Tyr24 and Trp108 in substrate binding and substrate specificity of glycolate oxidase. *Eur. J. Biochem.* **228**: 408-416.

Sulter, G.J., Verheyden, K., Mannaerts, G., Harder, W. and Veenhuis, M. (1993) The *in vitro* permeability of yeast peroxisomal membranes is caused by a 31 kDa integral membrane protein. *Yeast.* **9**: 733-742.

Sumegi, B., Sherry, A.D. and Malloy, C.R. Channelling of TCA cycle intermediates in cultured *Saccharomyces cerevisiae. Biochem.* **29**: 9106-9110.

Swinkels, B.W., Gould, S.J., Bodnar, A.G., Rachubinski, R.A. and Subramani S. (1991) A novel, cleavable peroxisomal targeting signal at the amino-terminus of the rat 3-ketoacyl-CoA thiolase. *EMBO J.* **10**: 3255-3262.

Swinkels, B.W., Gould, S.J. and Subramani, S. (1992) Targeting efficiencies of various permutations of the consensus C-terminal tripeptide peroxisomal targeting signal. *FEBS Lett.* **305**: 133-136.

Tabata, S., Kaneko, T., Nakamura, Y., Kotani, H., Kato, T. *et al* (2000) Sequence and analysis of chromosome 5 of the plant *Arabidopsis thaliana*. The Kazusa DNA Research Institute, The Cold Spring Harbor and Washington University in St Louis Sequencing Consortium and The European Union Arabidopsis Genome Sequencing Consortium. *Nature.* **408**: 823-826.

Takada, Y., Kaneko, N., Esumi, H., Purdue, P.E. and Danpure, C.J. (1990) Human peroxisomal L-alanine:glyoxylate aminotransferase. Evolutionary loss of a mitochondrial targeting signal by point mutation of the initiation codon. *Biochem J.* **268**: 517-520.

Tanaka, T., Yamamoto, S., Moriya, T., Taniguchi, M., Hayashi, H., Kagamiyama, H. and Oi, S. (1994) Aspartate aminotransferase from a thermophilic formate-utilizing methanogen, *Methanobacterium thermoformicicum* strain SF-4: relation to serine and phosphoserine aminotransferases, but not to the aspartate aminotransferase family. *J. Biochem.* **115**: 309-317.

Tchang, F., Lecharny, A. and Mazliak, P. (1984a) Photostimulation of hydroxypyruvate reductase activity in peroxisomes of *Pharbitis nil* seedlings: I. Action spectrum. *Plant Cell Physiol.* **25**: 1033-1037.

Tchang, F., Lecharny, A. and Mazliak, P. (1984b) Photostimulation of hydroxypyruvate reductase activity in peroxisomes of *Pharbitis nil* seedlings: II. Photoreceptors in blue light. *Plant Cell Physiol.* **25**: 1039-1043.

The Arabidopsis Genome Initiative Analysis of the genome sequence of the flowering plant *Arabidopsis thaliana.* (2000) *Nature.* **408**, 796-815.

Theologis, A., Ecker, J.R., Palm, C.J., Federspiel, N.A., Kaul, S. *et al* (2000) Sequence and analysis of chromosome 1 of the plant *Arabidopsis thaliana. Nature.* **408**: 816-820.

Titus, D.E., Hondred, D. and Becker W.M. (1983) Purification and characterization of hydroxypyruvate reductase from cucumber cotyledons. *Plant Physiol.* **72**: 402-408.

Tolbert, N.E. (1997) The C_2 oxidative photosynthetic carbon cycle. *Ann. Rev. Plant Physiol. Plant Mol. Biol.* **48**: 1-25.

Tolbert, N.E. (1981) Metabolic pathways in peroxisomes and glyoxysomes. *Ann. Rev. Biochem.* **50**: 133-157.

Tolbert, N.E. (1980a) Microbodies – peroxisomes and glyoxysomes. In *The Biochemistry of Plants* (Stumpf, P.K. and Conn, E.E. eds.) Vol. 1, pp. 359-388. New York (USA): Academic Press.

Tolbert, N.E. (1980b) Photorespiration. In *The Biochemistry of Plants.* (Davies, D.D. ed.). Vol. 2, pp. 487-523. New York (USA): Academic Press.

Tolbert, N.E., Yamazaki, R.K. and Oeser, A. (1970) Localization and properties of hydroxypyruvate reductase and glyoxylate reductases in spinach leaf particles. *J. Biol. Chem.* **245**: 5129-5136.

Tsugeki, R., Hara-Nishimura, I., Mori, H. and Nishimura, M. (1993) Cloning and sequencing of cDNA for glycolate oxidase from pumpkin cotyledons and Northern blot analysis. *Plant Cell Physiol.* **34**: 51-57.

van Veldhoven, P., de Beer, L.J. and Mannaerts, G.P. (1983) Water- and solute-accessible spaces of purified peroxisomes. Evidence that peroxisomes are permeable to NAD^+. *Biochem J.* **210**: 685-693.

van Veldhoven, P.P., Just, W.W. and Mannaerts, G.P. (1987) Permeability of the peroxisomal membrane to cofactors of β-oxidation. Evidence for the presence of a pore-forming protein. *J. Biol. Chem.* **262**: 4310-4318.

Verleur, N. and Wanders, R.J.A. (1993) Permeability properties of peroxisomes in digitonin-permealized rat hepatocytes. Evidence for free permeability towards a variety of substrates. *Eur. J. Biochem.* **218**: 75-82.

Verleur, N., Elgersma, Y., Van Roermund, C.W., Tabak, H.F. and Wanders, R.J. (1997) Cytosolic aspartate aminotransferase encoded by the AspAT2 gene is targeted to the peroxisomes in oleate-grown *Saccharomyces cerevisiae*. *Eur. J. Biochem.* **247**: 972-980.

Volokita, M. (1991) The carboxy-terminal end of glycolate oxidase directs a foreign protein into tobacco leaf peroxisomes. *Plant J.* **1**: 361-366.

Volokita, M. and Somerville, C.R. (1987) The primary structure of spinach glycolate oxidase deduced from the DNA sequence of a cDNA clone. *J. Biol. Chem.* **262**: 15825-15828.

von Heijne, G., Steppuhn, J. and Herrmann, R.G. (1989) Domain structure of mitochondrial and chloroplast targeting peptides. *Eur. J. Biochem.* **180**: 535-545.

von Schaewen, A., Langenkämper, G., Graeve, K., Wenderoth, I. and Scheibe, R. Molecular characterization of the plastidic glucose-6-phosphate dehydrogenase from potato in comparison to its cytosolic counterpart. *Plant Physiol.* **109**: 1327-1335.

Wadsworth, G.J. (1997) The plant aspartate aminotransferase gene family. *Physiol. Plant.* **100**: 998-1006.

Walk, R.A. and Hock, B. (1977) Glyoxysomal and mitochondrial malate dehydrogenase of watermelon (*Citrullus vulgaris*) cotyledons. II. Kinetic properties of the purified isoenzymes. *Planta.* **136**: 221-228.

Walk, R.A., Michaeli, S. and Hock, B. (1977) Glyoxysomal and mitochondrial malate dehydrogenase of watermelon (*Citrullus vulgaris*) cotyledons. I. Molecular properties of the purified isoenzymes. *Planta.* **136**: 211-220.

Wilkie, S.E. and Warren, M.J. (1998) Recombinant expression, purification, and characterization of three isoenzymes of aspartate aminotransferase from *Arabidopsis thaliana*. *Protein Expr. Purif.* **12**: 381-389.

Willekens, H., Chamnongpol, S., Davey, M., Schraudner, M., Langebartels, C., van Montagu, M., Inze D. and van Camp, W. (1997) Catalase is a sink for H_2O_2 and is indispensable for stress defense in C_3 plants. *EMBO J.* **16**: 4806-4816.

Willekens, H., Villarroel, R., Van Montagu, M., Inze, D. and Van Camp, W. (1994) Molecular identification of catalases from *Nicotiana plumbaginifolia* (L.). *FEBS Lett.* **352**: 79-83.

Williams, E., Cregeen, D. and Rumsby, G. (2000) Identification and expression of a cDNA for human glycolate oxidase. *Biochim. Biophys. Acta.* **1493**: 246-248.

Williamson, J.D. and Scandalios, J.G. (1993) Response of the maize catalases and superoxide dismutases to cercosporin-containing fungal extracts: the pattern of catalase response in scutella is stage specific. *Physiol. Plant.* **88**: 159-166.

Wimmer, B., Lottspeich, F., van der Klei, I., Veenhuis, M. and Gietl, C. (1997) The glyoxysomal and plastid molecular chaperones (70 kDa heat shock protein) of watermelon cotyledons are encoded by a single gene. *Proc. Natl. Acad. Sci.* **94**: 13624-13629.

Winter, H., Robinson, D.G. and Heldt, H.W. (1994) Subcellular volumes and metabolite concentrations in spinach leaves. *Planta.* **193**: 530-535.

Yamaguchi, J., Nishimura, M. and Akazawa, T. (1984) Maturation of catalase precursor proceeds to a different extent in glyoxysomes and leaf peroxisomes of pumpkin cotyledons. *Proc. Natl. Acad. Sci.* **81**: 4809-4813.

Yamaguchi, K., Mori, H. and Nishimura, M. (1995) A novel isoenzyme of ascorbate peroxidase localized on glyoxysomal and leaf peroxisomal membranes in pumpkin. *Plant Cell Physiol.* **36**: 1157-1162.

Yamazaki, R.K. and Tolbert, N.E. (1970) Enzymic characterization of leaf peroxisomes. *J. Biol. Chem.* **245**: 5137-5144.

Yu, C., Liang, Z., and Huang, A.H.C. (1984) Glyoxylate transamination in intact leaf peroxisomes. *Plant Physiol.* **75**: 7-12.

Yu, C., Huang and A.H.C. Conversion of serine to glycerate in intact spinach leaf peroxisomes: Role of malate dehydrogenase. *Arch. Biochem. Biophys.* **245**: 125-133.

Zalman, L.S., Nikaido, H. and Kagawa, Y. (1980) Mitochondrial outer membrane contains a protein producing non-specific diffusion channels. *J. Biol. Chem.* **255**: 1771-1774.

Zhong, H.H. and McClung, C.R. (1996) The circadian clock gates expression of two Arabidopsis catalase genes to distinct and opposite circadian phases. *Mol. Gen. Genet.* **251**: 196-203.

Zhong, H.H., Young, J.C., Pease, E.A., Hangartner, R.P. and McClung, C.R. (1994) Interactions between light and the circadian clock in the regulation of *CAT2* expression in Arabidopsis. *Plant Physiol.* **104**: 889-898.

6

PEROXISOME BIOGENESIS IN ROOT NODULES AND ASSIMILATION OF SYMBIOTICALLY-REDUCED NITROGEN IN TROPICAL LEGUMES

Desh Pal S. Verma

Ohio State University, Columbus, USA

KEYWORDS

Bacteroids; amide synthesis; purine biosynthesis; ureides; uricase.

INTRODUCTION

Tropical legumes fix nitrogen in association with *Bradyrhizobium* species. The symbiotically-reduced nitrogen is assimilated by the host plant using a unique pathway that ensures efficient utilization of carbon. Tropical legumes such as soybean, cowpea and bean primarily assimilate and transport fixed-nitrogen as ureides, allantoin and allantoic acid, while temperate legumes such as pea and alfalfa are amide-transporters, producing L-glutamine and L-asparagine. Ureides are produced via *de novo* biosynthesis and oxidation of purines. The activities of enzymes involved in *de novo* purine biosynthesis and catabolism vary in effective and ineffective nodules and correlate with the status of the nodules in fixing nitrogen (Atkins, 1991). The ureide-producing nodules are usually determinate in their structure but not all determinate nodules are ureide producers.

A. Baker and I.A. Graham (eds.), Plant Peroxisomes, 191–220.

The genes involved in assimilation of symbiotically-reduced nitrogen are induced prior to and independent of the commencement of nitrogen fixation in nodules, but their level of induction increases in effective root nodules following the onset of nitrogen fixation (Verma, 1989). A number of host genes encoding nodule-specific proteins (nodulins) have been identified and some of them are involved in nitrogen assimilation and peroxisome proliferation in root nodules (Verma *et al*, 1986). The primary site for ureide production in tropical legumes is peroxisomes, where purine is oxidized. This organelle in root nodules is proliferated in uninfected cells in response to purines synthesized and transported from the infected cells. This suggests that nitrogen assimilation in tropical legume nodules is compartmentalized within the cell and segregated between the infected and uninfected cells. Our knowledge of peroxisome biogenesis for nitrogen assimilation is limited. This Chapter summarizes various aspects of peroxisome biogenesis for assimilation of reduced nitrogen in tropical legumes. This process consumes the minimum amount of carbon for the assimilation of symbiotically reduced nitrogen, which is an energy-intensive process.

ASSIMILATION OF SYMBIOTICALLY-REDUCED NITROGEN

Synthesis of amides in root nodules

Once the nodule structure is developed on legume roots and an infection zone is established within a nodule, bacteroids begin to fix nitrogen (Verma, 2000). This event takes about ten days in soybean following infection of the roots with *Bradyrhizobium japonicum*. Based on the observation that isolated bacteroids excrete NH_4^+ (Bergersen and Turner, 1967), ammonium is considered to be assimilated in the host cell cytoplasm. An ammonium transporter has recently been isolated from *Lotus japonicus* (Salvemini *et al*, 2001). This transporter may be responsible for the export of ammonia from symbiosomes into the host cell cytoplasm (Blumwald *et al*, 1985). An excess of ammonia is generally toxic to the cell and must be immediately assimilated in the host cell cytoplasm. Glutamine synthetase and glutamate synthase are two major enzymes involved in the assimilation of NH_4^+. Meeks *et al* (1978) demonstrated that NH_4^+ is first incorporated into the amide position of glutamine by the glutamine synthetase-catalyzed reaction. This step can be blocked by methionine sulphoximine, a glutamine synthetase inhibitor. The amide group is subsequently transferred to the 2-carbon of oxoglutarate by the activity of glutamate synthase. Figure 1 outlines the path of assimilation of symbiotically reduced nitrogen in tropical legumes.

Figure 1: Metabolic pathways for the assimilation of symbiotically-reduced nitrogen in tropical legume root nodules segregated between the infected and uninfected cells of the symbiotic zone in root nodules

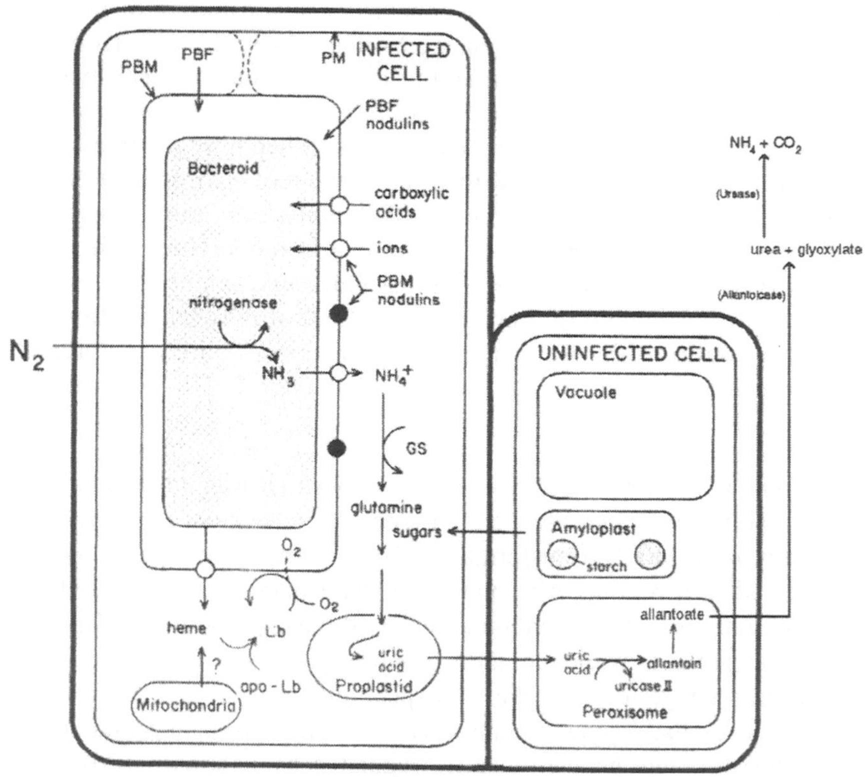

Figure 1. The conversion of NH_4^+ to purines occurs in the infected cells where bacteroids are present while the oxidation of purines and conversion of uric acid to allantoin occurs in the uninfected cells. This requires the presence of peroxisomes, proliferated in the uninfected cells of the root nodules. The ureides are then transported to shoot and converted back in to NH_4 via urea production to be assimilated in to various amino acids. PBM, peribacteroid membrane; PBF, peribacteroid fluid; PM, plasma membrane; Lb, leghaemoglobin; PBM nodulins, peribacteroid membrane nodulins; GS, glutamine synthetase.

The induction of glutamine synthetase activity in response to nitrogen fixation has long been observed (Robertson *et al*, 1975; Cook *et al*, 1990). The glutamine synthetase activity in the infected cells of cowpea nodules is primarily located in the cytoplasm (Groat and Schrader, 1982; Atkins *et al*, 1988; Atkins 1991). The cytoplasmic location of glutamine synthetase in the infected cells of soybean nodules was confirmed using immunogold labelling (Hirel *et al*, 1992). This observation established that the infected cells of determinate legume nodules are the primary site of NH_4^+ assimilation in root nodules. A more conclusive demonstration of glutamine synthetase induction was provided using Northern hybridization (Hirel *et al*, 1987) which showed that the availability of ammonium ions enhanced the expression of glutamine synthetase in root tissue within two hours, reaching a level similar to that in nodules by eight hours. When nitrogen fixation was prevented by replacing nitrogen with argon in the root environment or when nodules were formed by a Fix^- mutant of *B. japonicum*, the expression of glutamine synthetase did not increase over that in roots (Hirel *et al*, 1987). This observation led to the conclusion that glutamine synthetase is directly induced by the availablity of ammonia as a result of nitrogen fixation. This was further confirmed by isolating the glutamine synthetase gene promoter from soybean and fusing it with the β-glucuronidase (GUS) reporter gene (Miao *et al*, 1991).

The glutamine synthetase-GUS gene fusion construct was introduced into a legume (*Lotus corniculatus*) and a non-legume (tobacco) plant by *Agrobacterium*-mediated transformation. This chimaeric gene was found to be expressed in a root-specific manner in both plants (Miao *et al*, 1991). Treatment with exogenous ammonia increased the expression of GUS in the legume background, while no induction was observed in tobacco roots. These results suggested that the root-specificity of the soybean cytosolic glutamine synthetase gene is conserved in both non-legume and legume plants; however, this gene is directly regulated by externally-provided or symbiotically-fixed nitrogen only in the legume background (Miao *et al*, 1991). The localization of GUS activity was limited to the infected cells of the infection zone in transgenic root nodules, while the cortical cells showed no GUS activity (Figure 2). These data suggested that the symbiotic zone acts as a sealed compartment with a unique physiological environment that first acts as a sink for nitrogen (prior to the commencement of nitrogen fixation) and then serves as a source of nitrogen. Glutamine synthetase is required for both phases. In the first phase it assimilates nitrogen for the supply of growing nodule and in the second phase it assimilates symbiotically fixed nitrogen.

Figure 2: Expression of glutamine synthetase gene in transgenic Lotus japonicus root nodules as detected by glutamine synthetase promoter-GUS fusion experiments. Miao et al, 1991 Plant Cell 3:11; ©1991 American Society of Plant Biologists. Reprinted with permission.

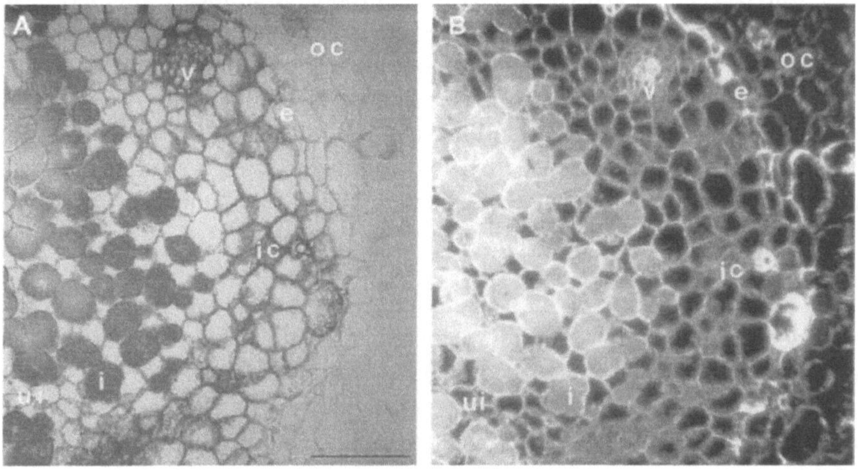

Figure 2. A, Light field micrograph of transgenic Lotus nodules stained with X-gluc. Bar 200μM. B. Darkfield micrograph of the nodule in A. e, endodermis; i, infected cell; ic, inner cortex; oc, outer cortex; v, vascular bundle. Note the presence of GS activity primarily in the infected cells of the symbiotic zone.

That tropical legumes operate two alternate nitrogen-assimilating systems is evident from the expression of the glutamine synthetase gene which is enhanced greatly by the availability of ammonium ions, while nitrate has no effect (Hirel *et al*, 1987; Miao *et al*, 1991). When plants are grown on nitrate-rich media, they reduce nitrate through nitrate reductase and utilize it as a major source of nitrogen, while in a nitrogen-limiting environment, they develop root nodules to fix nitrogen. Nitrate, in fact, inhibits the nodulation process and plant mutants able to nodulate in the presence of nitrate have been isolated (Carroll *et al*, 1985), but the gene responsible for this phenotype has not yet been identified. As shown above, the bacteroid-excreted ammonium acts as an inducer of the cytosolic form of glutamine synthetase and is incorporated into glutamine. Glutamine induces the expression of phosphoribosylpyrophosphate amidotransferase (PRAT) which catalyzes the first committed step of *de novo* purine biosynthesis. The induction of GS is developmentally controlled (Kim *et al*, 1995a; Walker and Coruzzi, 1989). The induction of other nitrogen assimilation enzymes during nodule development has also been documented (Robertson *et al*, 1975; Reynolds *et al*, 1982; Atkins *et al*, 1984).

In urease-negative plants supported by symbiotically-fixed nitrogen, xylem-borne nitrogen consisted predominantly of ureides (allantoin and allantoic acid), whereas plants grown on NH_4NO_3, produced amides (Schubert *et al*, 1986). Seed nitrogen yield was equal on either nitrogen regime, but the nitrogen-fixing plants accumulated about six times more leaf urea. To test whether urea accumulating in urease-negative seedlings was derived from ureides, seeds were treated with allopurinol, an inhibitor of ureide formation. Seedling ureides were decreased by 90%, but urea levels were unchanged in the leaves. This indicates that urea is accumulated via the amide pathway in plants grown on externally supplied nitrogen

During early development of root nodules of an effective cowpea *(Vigna unguiculata*), symbiosis rapidly increasing nitrogenase activity and leghaemoglobin content were accompanied by rapid increases in activities of glutamine synthetase, glutamate synthase and enzymes of *de novo* purine synthesis, xanthine oxidoreductase, urate oxidase and phosphoenolpyruvate carboxylase and there was increased export of ureides to the shoot of the host (Schubert *et al*, 1996). Nodulated plants maintained in the absence of nitrogen showed little increase in nitrogenase activity. As a result, in the absence of nitrogen fixation and consequent NH_3 production, there was no stimulation of activity of enzymes of NH_3 assimilation or of the synthesis of purines or ureides. Addition of nitrate relieved plant nitrogen deficiency but failed to increase levels of enzymes of nitrogen metabolism. These data suggest that ureides are not the precursors of urea in legume seeds and support the idea that there are two separate pathways for the assimilation of reduced nitrogen in legume plants.

De novo purine biosynthesis in root nodules

Most of the symbiotically reduced nitrogen in tropical legumes is assimilated via the ureide pathway. The ureides, allantoin and allantoic acid, are the most efficient forms of nitrogen transport compounds since they contain a higher molar ratio of nitrogen per carbon atom as compared to amides (Schubert, 1986). The production of ureides requires large amounts of purines, which are oxidatively catabolized to ureides (Atkins, 1981; Boland *et al*, 1982). Therefore, the rate of *de novo* purine biosynthesis in tropical legume nodules must be significantly increased in order to assimilate all the reduced nitrogen produced in association with rhizobia (Kohl *et al*, 1988).

In tropical legume root nodules, glutamine formed in the cytosol of the infected cells is not transported to the shoot, but rather is funnelled into the plastids of the infected cells where *de novo* purine biosynthesis commences. Although earlier studies had demonstrated that most of the enzymes involved in purine biosynthesis are localized in plastids (Boland and Schubert 1983), recent studies suggest that they may equally be distributed in mitochondria (Atkins *et al*, 1997). Glutamine is one of the substrates of PRAT which catalyzes the first step of the *de novo* purine biosynthesis pathway. It is likely that an organellar glutamine transporter is operative to take up glutamine from the cytosol and the export of glutamine is inhibited from the infected cells of the nodules. A glutamine transporter gene has been cloned from *E. coli* (Nohno *et al*, 1986), but no gene encoding this transporter has yet been isolated from plants.

A block in the activity of xanthine oxidoreductase by treatment with allopurinol results in the accumulation and transport of xanthine without affecting nitrogen fixation (Atkins *et al*, 1988), suggesting that purine biosynthesis is a committed step for the assimilation of reduced nitrogen during the development of root nodules in tropical legumes. This is confirmed from our recent studies on the inhibition of uricase activity as achieved by antisense regulation (see below).

The *de novo* purine biosynthesis pathway in plants consists of ten enzymatic reactions that convert the activated ribose precursor, phosphoribosyl pyrophosphate (PRPP), to the purine nucleotide, inosine 5'-monophosphate (IMP) (Ebbole and Zalkin, 1987; Henikoff, 1987; Neuhard and Nygaard, 1987; Rolfes and Zalkin, 1988). The first step in this pathway, involving the synthesis of 5-phosphoribosyl-1-amine from PRPP and glutamine, is catalyzed by PRAT. Figure 3 shows the general scheme of the *de novo* purine biosynthesis pathway and catabolism (oxidation) of purines into ureides. End product inhibition of the PRAT enzyme in bacteria (Messenger and Zalkin, 1979; Meyer and Switzer, 1979) and animals (Holmes *et al*, 1973) as well as in plants (Reynolds *et al*, 1984) plays a key role in regulating the metabolic flux of carbon and nitrogen through this pathway and, thus, controlling the level of purine synthesis. Multiple controls on this step suggest that PRAT is the key enzyme in this process and increasing the level of expression of this enzyme may enhance assimilation of reduced nitrogen through this pathway. However, feedback inhibition of this step by purines ensures that a toxic level of purines is not built up in the cell. This situation in root nodules is apparently alleviated by the export of purines out of the infected cell in to the neighbouring uninfected cells, where it is catabolized into ureides.

Figure 3: Pathway of de novo purine biosynthesis and its regulation in root nodules

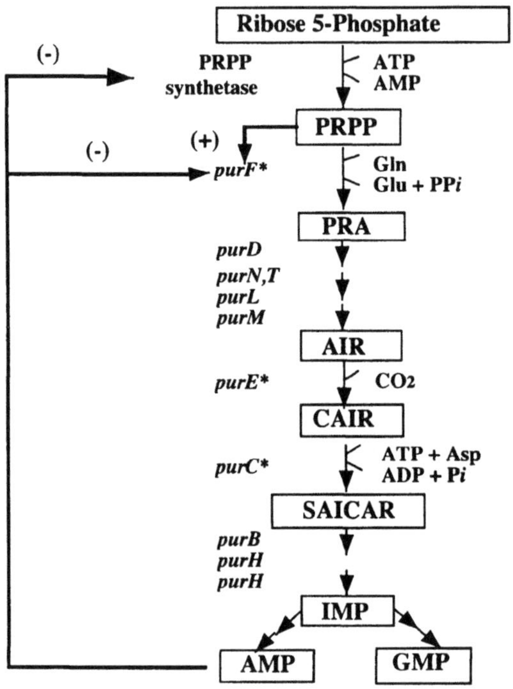

Figure 3. purF, gene encoding phosphoribosylpyrophosphate (PRPP) amidotransferase; purD, glycineamide ribonucleotide (GAR) synthetase; purN,T, GAR transformylase; purL, formylglycinamidine ribonucleotide (FGAM) synthetase; purM, aminoimidazole ribonucleotide (AIR) synthetase; purE, AIR carboxylase; purC, phosphoribosylamino-imidazole carboxamide (SAICAR) synthetase; purB, adenylosuccinate lyase; purH, aminoimidazole carboxaimde ribonucleotide (AICAR) transformylase/Inosine mono-phosphate (IMP) cyclohydrolase; PRA, phosphoribosylamine; CAIR, carboxyamino-imidazole ribonucleotide; AMP, adenosine monophosphate; GMP, guanosine monophosphate.

The *de novo* purine biosynthesis pathway may, in fact, be induced prior to the commencement of nitrogen fixation since uricase induction occurs several days prior to this event (Nguyen *et al*, 1985). Also PRAT is induced prior to the commencement of nitrogen fixation in nodules (Kim *et al*, 1995b). These data suggest that the pathway may be activated in response to high demand for purines during early nodule development, where most of the infected cells become polyploid (up to 64N, Mitchell, 1965) and this pathway is later used by the cell to assimilate reduced nitrogen, once nitrogen fixation commences. Consequently, the expression of all genes involved in purine biosynthesis and oxidation is further enhanced at that time in parallel to the increase in nitrogen fixation capacity of the nodules.

Several enzymes of the *de novo* purine biosynthesis pathway, PRAT, glycinamide ribonucleotide (GAR) synthetase, GAR transformylase and adenylosuccinate-AMP lyase, have been identified and partially purified from soybean and cowpea root nodules (Reynolds *et al*, 1982; 1984; Atkins, 1991) and the genes encoding most of these enzymes have been cloned from soybean and mothbean (*Vigna aconitifolia)* (Chapman *et al*, 1994; Kim *et al*, 1995a) as well as from Arabidopsis (Senecoff and Meagher, 1993; Ito *et al*, 1994; Schnorr *et al*, 1994). In nodules, most of these enzymes are apparently localized in the plastids of infected cells (Boland and Schubert, 1983). The amino acid sequence deduced from the soybean PRAT cDNA clone showed >85% homology to the PRAT sequence of mothbean and a significant homology to those of bacteria, yeast, chicken, rat and human (Kim *et al*, 1995a). The soybean PRAT clone encodes a protein with an N-terminal sequence resembling a plastid-targeting peptide consistent with the location of this enzyme in plastids. The mothbean cDNA encodes the putative propeptide and efficiently complements purine auxotrophy in an *Escherichia coli purF* mutant impaired in PRAT synthesis.

The soybean PRAT gene promoter was transcriptionally fused to GUS reporter and introduced into both tobacco and soybean by *A. tumefaciens*- and *A. rhizogenes*-mediated transformation, respectively. GUS expression was specifically detected in tobacco roots and soybean hairy roots. The transgenic roots from both plants provided with 10 mM L-glutamine showed higher GUS expression than control roots, indicating that glutamine, one of the substrates for PRAT, acts as an inducer of this gene. This result confirmed the previous work (Kim *et al*, 1995a) and is in accordance with the idea that the ureide biosynthesis in tropical legume root nodules is tightly regulated by the availability of glutamine formed by symbiotically-reduced nitrogen. The transcriptional data show that the PRAT gene is induced prior to the commencement of nitrogen fixation (Kim *et al*, 1995a), indicating a developmental control on the induction of this pathway. Accordingly, the levels of PRAT mRNA in soybean and cowpea nodules increase steadily as the nodule matures and the rate of nitrogen fixation increases. PRAT mRNA was not detectable in uninfected root tissue, but a low level of transcript was detected in leaves. Treatment of uninfected root with L-glutamine induced the PRAT mRNA transcript, suggesting that glutamine produced as a result of assimilation of fixed nitrogen induces *de novo* purine biosynthesis in the infected cells of root nodules (Kim *et al*, 1995a).

Functional complementation of *E. coli purE* and *purC* mutants with mothbean cDNA library yielded two cDNA clones, encoding 5-aminoimidazole ribonucleotide (AIR) carboxylase and 5-aminoimidazole-4-*N*-succinocarboxamide ribonucleotide (SAICAR) synthetase, involved in *de novo* purine biosynthesis (Chapman *et al*, 1994). Sequencing of these clones revealed that the two enzymes are distinct proteins in mothbean, unlike in animals, where both activities are associated with a single bifunctional polypeptide. The mothbean AIR carboxylase has an N-terminal domain homologous to the eubacterial *purK* gene product. This PurK-like domain appears to facilitate the binding of CO_2 and is dispensable in the presence of high CO_2. Availability of all the genes of *de novo* purine biosynthesis pathway may now allow determination of the rate-limiting step(s) in this pathway. The regulation of this pathway may change under the microaerobic environment which exists in the symbiotic zone of root nodules.

Export of purines from infected cells and their oxidation to ureides

The assimilation of nitrogen is a highly compartmentalized process even in normal plant tissues (Suzuki *et al*, 1981). That uninfected cells of the symbiotic zone of root nodules primarily participate in ureide production was suggested from the observation that these cells accumulate peroxisomes (Newcomb and Tanden, 1981). The compartmentation of the purine and ureide biosynthetic pathways in different organelles of both infected and uninfected cells is important for maintaining metabolite channelling, as well as levels of key intermediates and effectors (Schubert and Boland, 1990). Oxygen appears to play an important role in the intercellular compartmentation of the enzymes of the ureide biosynthesis pathway. For example, uricase requires more oxygen than is available in the infected cells and, accordingly, is located in the uninfected cells, whereas PRAT is oxygen sensitive and is located in the infected cells that are very low in oxygen (Hu 1991). Thus, different steps of this pathways are divided between the infected and uninfected cells of the root nodules (Verma, 1989). Moreover, purines inhibit PRAT and uricase activities (Hurst *et al*, 1985) and purine catabolism is facilitated in tissues with low oxygen tension (Mohamedali *et al*, 1993). In fact, the catabolism of purines is sensitive to oxygen and urate oxidase has been suggested to be a sensor for oxygen levels (Sies, 1977).

The plasma membrane has low permeability to nucleotides (Muller *et al*, 1982), which must be translocated from infected cells to the uninfected cells by an active transport system (Schubert and Boland, 1990). However, ATP and ADP are efficiently translocated from both mitochondria and chloroplasts (Heldt, 1969; Robinson and Wiskich, 1977; Robinson, 1985). This suggests the possibility that IMP, the final product of *de novo* purine biosynthesis, may be transported from these organelles in to the cytosol (Schubert and Boland, 1990) and then transported out of the infected cells to uninfected cells. Considering the location of xanthine dehydrogenase or xanthine oxidoreductase, which has been localized in the cytosol of uninfected cells (Nguyen *et al*, 1986; Datta *et al*, 1991), we can assume that XMP, xanthosine or xanthine, converted from IMP, may also be transported from plastids. High concentrations of purines are generally toxic to the cell and, thus, may have to be removed from the infected cells as they are made.

Little is known about the purine transporters in plants. *Aspergillus nidulans* can utilize purines as a sole nitrogen source (Gorfinkiel *et al*, 1993). Two purine permease genes have been cloned from *Aspergillus* (Gorfinkiel *et al*, 1993; Diallinas *et al*, 1995). One encodes uric acid-xanthine permease (UapA) and is a highly hydrophobic protein with twelve to fourteen putative transmembrane segments. This protein has no amino acid sequence homology to other permeases (Gorfinkiel *et al*, 1993). The transcription of this gene is inducible by 2-thiouric acid, and is highly repressed by ammonium. Another type of purine permease in *Aspergillus* is a wide specificity purine permease (UapC), which contributes partially to uric acid and hypoxanthine transport (Diallinas *et al*, 1995). UapC is also hydrophobic with several putative transmembrane segments. The UapC protein shows high similarity to the UapA protein and some similarity to several bacterial transporters. Both *uapA* and *uapC* genes are regulated by UaY, the specific regulator of the purine utilization pathway in *A. nidulans* (Suarez *et al*, 1995). These studies may be useful in investigating the regulation of purine transport in legume plants. The *Aspergillus* mutants impaired in purine transport may provide tools for complementation studies for isolation of plant purine transporter genes.

After transport to the uninfected cells, the xanthine is catabolized to uric acid by xanthine dehydrogense (XDH). This enzyme is localized in cytosol in root nodules (Nguyen *et al*, 1986; Datta *et al*, 1991). Uric acid is further oxidized to allantoin by uricase, which is localized in the peroxisome (Nguyen *et al*, 1985). Ureides account for up to 90% of the total nitrogen transported through the xylem of nodulated tropical legumes (Pate *et al*, 1980). In the presence of nitrate in the rooting medium, ureide synthesis is reduced and asparagine synthesis is enhanced, switching the plant from an ureide to an amide-transporter (Schubert and Boland, 1990). Non-nodulated roots of tropical legumes are capable of synthesizing and transporting ureides but they constitute a very small fraction of the total nitrogen in the xylem sap of nodulated plants (Pate *et al*, 1980). Several plant genes encoding uricases including soybean nodulin-35; (Nguyen *et al*, 1985); *Vigna* nodulin-35 (Lee *et al*, 1993) and *Phaseolus* (bean) nodulin 35 (Capote-Mainez and Sanchez, 1997) have been cloned and characterized. The ureides are eventually converted in the shoot to urea which gives rise to ammonia. Thus, the assimilation of symbiotically reduced nitrogen via ureide pathway consumes less energy than the amide synthesis transport system. This may be one of the resons why tropical legumes are able to produce high protein seeds.

ANTISENSE CONTROL OF URICASE AND ITS EFFECT ON NITROGEN ASSIMILATION IN ROOT NODULES

The role of uricase in ureide biosynthesis in root nodules was deciphered using an antisense approach (Delauney *et al*, 1988). For obtaining transgenic nodules expressing antisense uricase (nodulin-35) gene on the ureide-producing plants, a chimaeric system was used where nodules were formed on transgenic hairy roots arising from infection of *V. aconitifolia* with *Agrobactrium rhizogenes* (Cheon *et al*, 1993). The 5'-region of *Vigna* nodulin-35 cDNA linked in antisense orientation to the 35S promoter, was introduced into a binary vector pBI121. Since pBI121 contains the GUS gene attached to the 35S promoter, GUS expression could be used to select transgenic hairy roots. *V. aconitifolia* seedlings were inoculated with *A. rhizogenes* containing the *Vigna* antisense N-35 construct or pBI121 and the hairy roots formed were examined for GUS activity. The main root system and all but one GUS-positive hairy roots were removed, and the plants were fertilized with NH_4NO_3 to support growth. Gus-positive hairy roots were then infected by *Bradyrhizobium* (cowpea*)* strain to obtain root nodules (Cheon *et al*, 1993; Lee *et al*, 1993).

Nodules developed on transgenic roots were smaller in size as compared to the wild type roots. While all of the nodulated control plants were healthy with green leaves and grew well throughout the experiments, most of the nodulated plants containing antisense uricase RNA showed chlorosis. The chlorosis in the antisense plants was reversed by application of NH_4NO_3 to the watering medium (Lee *et al*, 1993) indicating that it was due to nitrogen defficiency. The specific activity of uricase in the antisense nodules was reduced to about half of that observed in control nodule. Although the uricase activity increased in older nodules, the relative level of activity between the antisense nodules and the control nodules remained the same. On a per nodule basis, the level of uricase activity was about 20% of that in control nodules and this was in proportion to the size of the nodules. PCR amplification of nodule RNA yielded a 230 bp DNA fragment, the identity of which was confirmed to be uricase, suggesting that the reduction in uricase activity was a result of antisense control. Recently, uricase has been detected in *Lotus japonicus*, an amide-producing temperate legume root nodules (Takane *et al*, 2000).

Ultrastructure analysis and immunogold labelling using antibodies against soybean nodulin-35 (Nguyen *et al*, 1985) showed that peroxisomes in uninfected cells of the transgenic antisense uricase nodules were smaller than those of the control nodules (Figure 4C, D) (Lee *et al*, 1993). On the basis of the relative concentration of gold particles in the peroxisomes, no difference in the uricase levels per unit area of peroxisomes between the control and the antisense *Vigna* N-35 RNA-expressing nodules was found. The density of gold particles in *Vigna* and soybean peroxisomes was about the same (Figure 4A, B), suggesting that the packing density of this protein is very similar and does not change with the size of the nodules.

These results indicated that uricase packaging density in peroxisomes was not altered. Since most of the protein in nodule peroxisomes is uricase (Legocki and Verma, 1979), lack of synthesis and accumulation of uricase apparently retarded peroxisome development. Although the specific activity of uricase in the nodules expressing antisense *Vigna* N-35 RNA was reduced by only 50%, the total activity was only 20% of that in the control nodules on a per nodule basis. The high specific activity of uricase in the antisense nodules is, in part, due to the lack of leghaemoglobin accumulation which accounts for almost 20% of the cellular protein in nodules and its accumulation is directly inhibited in ineffective nodules (Fuller and Verma 1984). Furthermore, the inhibition of gene expression by antisense RNA is usually not complete (Delauney *et al*, 1988). Moreover, uricase is a very stable protein (Suzuki and Verma, 1991) and once produced, it is translocated to and accumulates in, the peroxisomes of uninfected nodule cells where it is protected from the normal turnover of proteins (Nguyen *et al*, 1985; Van den Bosch *et al*, 1985). We observed that most of the antisense-uricase plants, regardless of the level of uricase activity, showed severe nitrogen-deficiency. This result is consistent with the finding by Atkins *et al* (1988) that a block of ureide biosynthesis by allopurinol caused severe chlorosis in tropical legume plants. Since allopurinol is taken up and transported through the xylem to the shoots of the plant (Atkins *et al*, 1988), it might inhibit ureide metabolism in other parts of the plant. By specifically blocking the uricase activity in nodules by the antisense approach, the possibility of side effects which may be generated using a metabolic inhibitor were eliminated. To also eliminate the possibility that uric acid accumulated as a result of decreased uricase activity, might be toxic to the plant or inhibit nodule function, uric acid was applied to the plants. Growth of plants watered with uric acid, whether nodulated or not, was not affected in this study (Lee *et al*, 1993).

These observations suggest that when ureide biosynthesis is blocked, amides and other nitrogenous compounds formed from reduced nitrogen are not sufficient to support the nitrogen requirements of the tropical legume plant.

This study also suggested that the development of peroxisomes is tightly coupled with the synthesis of uricase which is the major protein in these organelles in tropical legume nodules (Bergmann *et al*, 1983; Legocki and Verma, 1979). Furthermore, packaging of uricase inside the peroxisomes appears to determine their size since reduction in synthesis of uricase in antisence uricase expressing nodules proportionally reduced the size of peroxisomes (Figure 4B).

Figure 4: Localization of uricase in peroxisomes of Vigna control and antisense uricase-expressing plants (see Lee et al, 1993)

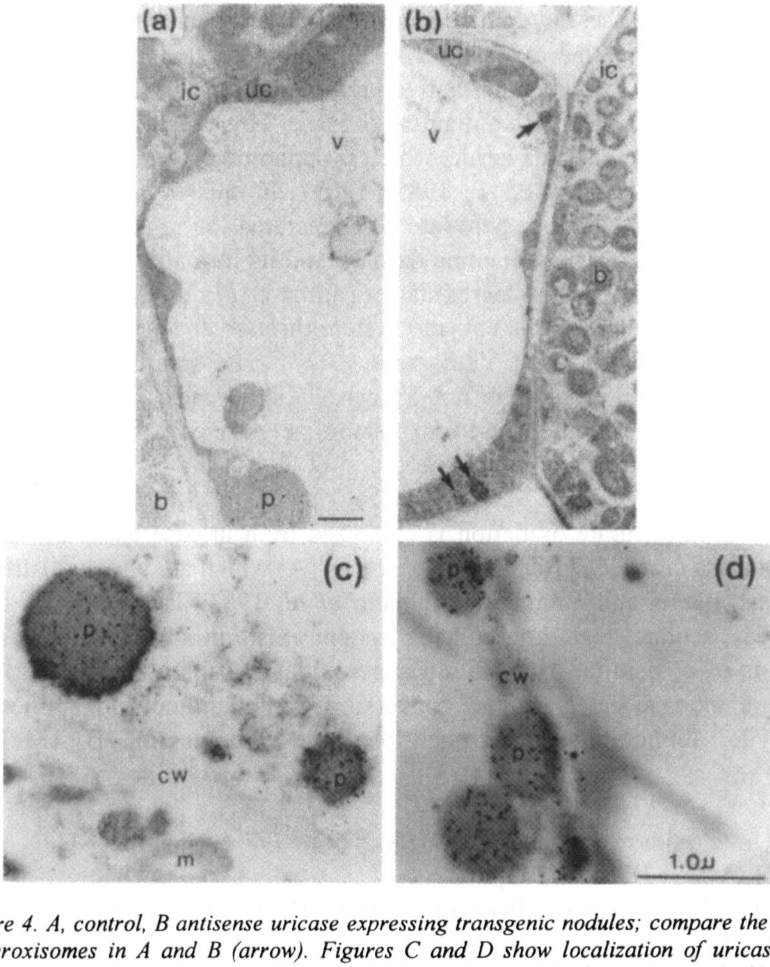

Figure 4. A, control, B antisense uricase expressing transgenic nodules; compare the size of peroxisomes in A and B (arrow). Figures C and D show localization of uricase in soybean root nodules. Section in C is embedded with Epon while that in D is embedded in Lowicryl resin (see Nguyen et al, 1985). b, bacteroid; ic, infected cell; m, mitochondria; p, peroxisome; uc, uninfected cell; v, vacuole.

Catabolism of uric acid can occur in the absence of peroxisomes

Generally, urate oxidation into allantoin is catalyzed by uricase, a peroxisomal matrix protein with high K_m for uric acid. Uricase is highly induced during nitrogen fixation and nodule formation in tropical legume nodules. However, a horseradish peroxidase has also been reported to be able to catalyze the oxidation of uric acid (Paul and Avi-dorl, 1954). Young roots lack the uricase, but contain a urate-degradation enzyme system, which is defined as a diamine oxidase-peroxidase urate-degrading enzyme system (Tajima *et al*, 1985). This alternative urate oxidation is catalyzed by a low K_m peroxidase coupled with a diamine oxidase. This reaction is enhanced by addition of a diamine, cadaverine, which is apparently required as a co-factor in this process (Tajima *et al*, 1985; 1983). In this system, it has been suggested that hydrogen peroxide is generated as a by-product from cadaverine oxidation by a diamine oxidase and its formation is coupled with uric acid degradation by a peroxidase (Tajima *et al*, 1983). Wounding has been shown to enhance this enzyme system through an increase in the level of cadaverine (Tajima and Yamamoto, 1977). The significance of ureide generation by this system is not known. It is possible that under stress conditions, this system is used to catabolize purines to supply nitrogen consuming a reduced amount of energy.

In root nodules, urate oxidation is catalyzed by a uricase using the uricase and catalase enzyme system in peroxisomes (A) which are widely distributed in yeast, plants and animals (Bergmann *et al*, 1983; Tanaka *et al*, 1977). Whereas, a urate degrading enzyme system exists in soybean radicles (B), consists of a diamine oxidase and a peroxidase (Tajima *et al*, 1985). The diamine oxidase catalyzes the conversion of polyamine and amines into aldehyde, ammonia and hydrogen peroxide (see reaction B) (McIntire, 1993).

(A) Urate oxidase + Catalase: Urate + $2H_2O + O_2 \rightarrow$ Allantoin + $H_2O_2 + CO_2$
$H_2O_2 \rightarrow H_2O + 1/2\ O_2$
(B) Diamine Oxidase + Peroxidase: $RH_2NH_2 + H_2O \rightarrow RCHO + NH_3 + H_2O_2$;
Urate + $H_2O_2 \rightarrow$ Allantoin + H_2O

The peroxidase thus degrades uric acid by using the hydrogen peroxide as an oxidant. Therefore, the expression of the H_2O_2-dependent peroxidase may confer the ability of the cells to grow on urate, even in the absence of functional peroxisomes. This has allowed rescue of a non-uric acid-utilizing mutant (see below) when grown on uric acid as a sole source of nitrogen.

A yeast mutant *spb1* developed in our laboratory (Wu, 1998) was shown to have lower levels of catalase, uricase and peroxidase activities than that of the wild-type strain. This mutant was found to be defective in peroxisome proliferation due to a mutation in a transcription factor (Wu, 1998), which is likely to inhibit peroxisome proliferation, as well as repressing the induction of some of the peroxisomal proteins (Kos *et al*, 1995). We cloned a novel peroxisomal cytochrome P450 from a soybean (*Glycine max*) nodule cDNA library by functional complementation of the *spb1* yeast mutant (Wu 1998). Functional complementation of this mutant with a soybean cDNA allowed the cells to grow on uric acid as the sole source of nitrogen, while the mutant did not utilize uric acid due to the lack of peroxisomes. This complementation did not occur at the gene level because yeast cells were not mutated in the gene coding for cytochrome P450, but likely occurred at the enzyme level through a functional 'bypass' which allowed them to utilize ureic acid. Therefore, we believe that the degradation of ureic acid was not catalyzed by classical uricase which requires the presence of peroxisomes, but rather occurred via an alternate pathway. We postulate that this cytochrome P450, referred to as P450W, catalyzes uric acid by degrading H_2O_2 by the scheme out lined in equation B. Accordingly, the cells expressing P450W are shown to be more tolerant to H_2O_2. Induction of the peroxisomal enzyme by a substrate or a proliferator is accomplished via different mechanisms (Kos *et al*, 1995). Without the catalase activity (a peroxisomal enzyme), normal uricase activity could be compromised by H_2O_2 or cells may die due to the high levels of H_2O_2 generated by urate oxidation (A). The cytochrome P450W appears to function in ureide metabolism demonstrating the existence of a urate-degrading diamine oxidase and peroxidase enzyme system. Such a system may be involved in cell defense and early seedling growth and in nitrogen assimilation in nodules when plants are under stress conditions. P450W shows homology with many cytochrome P450s and contains the characteristic haem-binding domain which is highly conserved among known P450s. Many cytochrome P450s have been shown to contain peroxidase activities (Marnett *et al*, 1986).

Northern blot data suggested that P450W is a constitutive protein existing in cotyledons, roots, leaves and nodules where peroxisomes are abundant. P450W could be a membrane protein in cotyledon glyoxysomes, leaf peroxisomes and nodule peroxisomes. P450W has a molecular weight of 57 kDa and previous studies have shown that a 56-57 kDa membrane protein exists in plant peroxisomes and glyoxysomes (del Rio *et al*, 1992), but its identity has not been directly confirmed. P450W contains a putative peroxisome targeting signal residing near the C-terminus of the protein suggesting that it is a membrane-bound protein with the C-terminus inside the peroxisomes. However, PTS1 type signals do not generally target membrane proteins. A C-terminus tripeptide, acting as peroxisome targeting signal, has been defined by Gould *et al* (1990). The tripetide signal Ser-Lys-Leu (SKL) is degenerate and conforms to the consensus sequence S/A/C/-K/H/R-L/I which occurs in a variety of peroxisomal proteins at or near the C-terminus (Gould, 1990; Hayashi *et al*, 1997). SKL is also found in the soybean uricase C-terminus (Nguyen *et al*, 1985; Ganzalez, 1991) and this sequence has been shown to be involved in the import of urate oxidase in peroxisomes (Miura *et al*, 1994). This targeting signal at an internal position in many peroxisomal proteins has been suggested to act as a topogenic signal for protein translocation into peroxisomes. The carboxyl terminal 27 amino acids of human catalase, containing an internal Ser-His-Leu (Subramani, 1993), is able to direct a heterologous protein into the microbody, although, subsequently, human catalase was shown to have a non-standard PTS1 KANL-COOH which is necessary and sufficient to act as a targeting signal (Purdue and Lazarow 1996), as does plant catalase (Mullen *et al*, 1997). The features of P450W suggest that its C-terminus may guide it in to the peroxisomes, and its hydrophobic N-terminus may serve as a membrane anchor. However, a definite conclusion must await direct evidence by *in vivo*/*in vitro* import experiments or immunolocalization studies.

Ureide formation by the alternate pathway may facilitate nitrogen assimilation under stress conditions

Allantoin is formed in the roots of many plants (Mothes, 1961; Tajima, 1977), though its biosynthetic pathway is unclear. Biosynthesis of allantoin in root tissues is apparently accomplished by a diamine oxidase and peroxidase enzyme system (Tajima and Yamamoto, 1977). Under stress conditions, allantoin generation in soybean root radicles is catalyzed through an alternative pathway which is regulated by a cadaverine. This enzyme system is activated only when catalase activity is low (Tajima *et al*, 1985). Catalase is the key enzyme for scavenging H_2O_2 in peroxisomes. Catalase activity is inhibited in plants under stress conditions such as exposure to heavy metals, salinity, and pathogen attack (for review, see del Rio *et al*, 1992). Interestingly, catalase activity remains low during soybean seedling growth (Tajima *et al*, 1985). The presence of a unique uricase system in soybean radicles suggests that an alternate pathway for ureic acid degradation may exist in other tissues and be used under specific physiological conditions. A human diamine oxidase-peroxidase enzyme system has been found to be involved in anti-virus processes (Klebanoff and Kazazi, 1995).

The presence of a diamine oxidase-peroxidase enzyme system within plant peroxisomes suggests a mechanism which may be involved in controlling cellular toxicity caused by H_2O_2, oxidative stress, and plant cell defense during seedling growth and which may be secondarily utilized for nitrogen assimilation via the ureide pathway.

Xanthine may act as a peroxisome proliferator in root nodules

Peroxisome proliferators are a group of structurally diverse compounds that cause a dramatic increase in both number and size of peroxisomes, as well as the induction of peroxisomal enzymes (Arnaiz *et al*, 1995; Veenhuis *et al*, 1987; Osumi, 1990). For example, catalase and uricase can be induced by some peroxisome proliferators in rodents (Arnaiz *et al*, 1995). When *S. cerevisiae* is grown on oleic acid as the sole C-source, a remarkable proliferation in peroxisomes occurs. As a result, there is a considerable induction of peroxisomal enzymes in yeast. Oleic acid is thus far considered to be the only peroxisome proliferator in *S. cerevisiae* (Veenhuis *et al*, 1987).

The use of xanthine as the sole nitrogen source in yeast studies, resulted in a three and a half- to four-fold increase in catalase and uricase activities (Wu, 1998). These enzyme activities were induced at about the same time and to about the same level and were present in the microsomal fraction containing peroxisomes. This suggested that the increased enzyme activities may be accounted for by the increase in peroxisome number and/or size of individual peroxisomes because the packaging density of the enzyme inside the peroxisomes is the same irrespective of the size of the peroxisomes (see Figure 3) (Arnaiz *et al*, 1995; Osumi *et al*, 1990). Our results demonstrated that xanthine can act as an inducer of peroxisome proliferation in *S. cerevisiae*, as shown by an organelle associated increases in uricase activity. In tropical legumes, purine biosynthesis is highly induced during nodule formation and nitrogen assimilation (Kim, 1996) suggesting the possibility that xanthine produced as a result of *de novo* purine biosynthesis in root nodules may function as a peroxisome proliferator in soybean nodules during nodule development.

As a result of purine catabolism H_2O_2 concentration can become very high inside peroxisomes. The excess H_2O_2 and O_2^- permeate through peroxisome membranes into the cytosol (Heupel *et al*, 1991) and together they may cause high oxidative stress inside the cell. This oxidative stress, in turn, may cause peroxisome proliferation (Palma *et al*, 1991; Lopez-Huertas *et al*, 2000).

POSSIBLE INVOLVEMENT OF NODULINS IN *DE NOVO* PURINE BIOSYNTHESIS IN ROOT NODULES

Considering the unique physiological environment in root nodules and the requirements of *de novo* purine biosynthesis and oxidation, it is likely that some of the enzymes involved in purine biosynthesis and catabolism are nodule specific (nodulins). The enzymes involved in the pathway for purine synthesis are contained primarily in the plastids. This group of enzymes includes phosphoribosylpyrophosphate (PRPP) amidotransferase (PRAT), PRPP synthetase, aspartate aminotransferase (AAT), NADH-glutamate synthase, asparagine synthetase, phosphoserine aminotransferase, serine hydroxymethyl transferase, phosphoglycerate dehydrogenase, methylene tetrahydrofolate oxidoreductase and triose-phosphate isomerase. PRAT catalyzes the first committed step of *de novo* purine biosynthsis, using PRPP as one of the substrates.

It is possible that a new isoform of this enzyme is induced during *de novo* purine biosynthesis. Different PRAT genes have been shown to be expressed in different tissues in Arabidopsis (Ito *et al*, 1994). Phosphoserine aminotransferase, serine hydroxymethyltransferase, phosphoglycerate dehydrogenase, methylene tetrahydrofolate oxidoreductase and triose phosphate isomerase are other key enzymes providing carbon source to the *de novo* purine biosynthetic pathway, and activity of these enzymes is readily detected in the proplastid fractions of the ureide-transporting legumes. Four genes of the *de novo* purine biosynthesis pathway, PRAT, glycinamide ribonucleotide (GAR) synthetase, GAR transformylase and adenylosuccinate-AMP lyase, have been isolated from soybean (Chapman *et al*, 1994; Schnorr *et al*, 1994; Kim *et al*, 1995a) and the activities encoded by these genes have been shown to be induced in root nodules (Reynolds *et al*,

1982; 1984). Sequencing of these clones revealed that the two enzymes are distinct proteins in cowpea, unlike in animals where both activities are associated with a single bifunctional polypeptide and therefore is a nodulin. Nodulin 35 was the first nodulin identified in root nodules (Legocki and Verma, 1979) and since then many proteins of the *de novo* purine biosynthesis and degradation have been identified and shown to be nodulins since these enzymes function in a unique microaerobic environment of root nodules (Larsen and Jochimsen, 1987).

PEROXISOME BIOGENESIS IN ROOT NODULES

As the topic of peroxisome biogenesis is discussed in great detail in Chapters 10-12, only the proliferation of these organelles in the context of nitrogen assimilation is briefly discussed here. Plants possess several types of peroxisomes formed at distinct stages of plant development and serve different metabolic functions (Kindl, 1982; Olsen, 1998). Accordingly, many metabolic enzymes are found to be located in these organelles. Acyl-CoA oxidase, uricase, catalase, malate synthase, glutamate-glyoxylate aminotransferase, alanine: 2-oxoglutarate aminotransferase, glycolate oxidase, delta 2, delta 3-enoyl-CoA isomerase, superoxide dismutase and thiolase have been localized in these organelles and the proteins have been purified (Huang *et al*, 1983; Pistelli *et al*, 1995; Riezman *et al*, 1980; Suzuki and Verma, 1991). Ureidoglycolate amidohydrolase, an enzyme involved in allantoin catabolism has also been found to be associated with peroxisomes in tropical legumes (Wells and Lees, 1991). Thus, peroxisomes in plants are involved in photorespiration, fatty acid oxidation (see Chapters 2 and 5) and nitrogen assimilation.

It has been demonstrated that the induction of peroxisome biogenesis genes in both plant and animal cells occurs by the universal stress signal, hydrogen peroxide (Brickner, 1999; Lopez-Huertas *et al*, 2000). Various stress conditions that lead to the production of hydrogen peroxide can induce peroxisome proliferation. In the case of ureide biosynthesis in root nodules, hydrogen peroxide is generated during the oxidation of purines and, thus, peroxisomes proliferate to catabolize purines.

Although it is believed that peroxisomes, including those in root nodules, are formed from pre-existing microbodies (Trelease, 1984; Lazarow and Fujiki, 1985; Borst, 1989) in rat liver *de novo* proliferation of peroxisomes has been suggested. This is currently an area of active research and is dealt with in detail in Chapter 10. This is initiated by membranous attachments onto the surface of pre-existing peroxisomes. These membranous structures may provide the appropriate lipid environment for the incorporation of peroxisomal membrane proteins and subsequently become the sites for import of the newly synthesized matrix proteins (Lueres *et al*, 1993; Fahimi *et al*, 1993). The fact that peroxisomes can accumulate different enzymes that are needed for the catabolism of various xenobiotic agents, suggest that the peroxisome serves as a subcellular compartment to carry out any reaction that is generally toxic to the cell cytoplasm, such as those producing H_2O_2. Our knowledge of peroxisome biogenesis in plants is very limited, however, using yeast genetics and molecular tools it may become possible to address this important topic in more detail and understand how this organelle is proliferated in different organs of plant in response to a variety of compounds.

PERSPECTIVE

Since assimilation of reduced nitrogen is more energy efficient using the ureide pathway, it may be desirable to convert an amide producing crop plant into a ureide producer. This requires a clear understanding of the key regulatory steps involved in funnelling of carbon and nitrogen into the *de novo* purine biosynthesis pathway and then inducing peroxisome proliferation. Since *de novo* purine biosynthesis is facilitated in hypoxic environments such as infected nodule cells, this pathway may not function optimally in the normal root cells and hence production of purines may become rate limiting for the assimilation of reduced nitrogen in a normal cell through this route. A nitrogen sensor needs to be identified and the regulation of transport of glutamine needs to be understood before such steps can be taken. With this knowledge, the conversion of amide-producing legumes to ureide producers, may be possible and desirable to save energy for nitrogen assimilation. For this, the rate-limiting steps in *de novo* purine biosynthesis and catabolism need to be identified by under- and overexpression of respective genes in transgenic nodules. The latter can be simplified using the chimaeric system as we have demonstrated (Cheon *et al*, 1993; Lee *et al*, 1993). Clearly, more studies are needed to understand the nitrogen regulatory circuits in the plant cell in order to modify nitrogen assimilation pathways.

ACKNOWLEDGEMENT

Figure 4C and D reproduced with permission from Nguyen et al, 1985 ©Blackwell Publishing, Oxford, UK.

REFERENCES

Arnaiz, S.L., Travacio, M., Llesuy, S. and Boveris, A. (1995) Hydrogen peroxide metabolism during peroxisome proliferation by fenofibrate. *Biochim. Biophys. Acta.* **1272**: 175-80.

Atkins, C.A. (1991) Ammonia assimilation and export of nitrogen from the legume nodule. In *Biology and Biochemistry of Nitrogen Fixation*, (Dilworth, M. and Glenn, A. eds). pp. 293-319. Elsevier, Amsterdam.

Atkins, C.A. (1981) Metabolism of purine nucleotides to form ureides in nitrogen-fixing nodules of cowpea (*Vigna unguiculata L. Walp*). *FEBS Lett.* **125**: 89-93.

Atkins, C.A., Shelp, B.J., Storer, P.J. and Pate, J.S. (1984) Nitrogen nutrition and the development of biochemical functions associated with nitrogen fixation and ammonia assimilation of nodules on cowpea seedlings. *Planta.* **162**: 327-333.

Atkins, C.A., Smith, P.M.C. and Storer, P.J. (1997) Re-examination of the intracellular localization of *de novo* purine synthesis in cowpea nodules. *Plant Physiol.* **113**: 127-135.

Atkins, C.A., Storer, P.J. and Pate, J.S. (1988) Pathways of nitrogen assimilation in cowpea nodules studied using $^{15}N_2$ and allopurinol. *Plant Physiol.* **86**: 204-207.

Bergersen, F.J. and Turner, G.L. (1967) Nitrogen fixation by the bacteroid fraction of breis of soybean root nodules. *Biochim. Biophys. Acta.* **141**: 507-515.

Bergmann, H., Preddie, E. and Verma, D.P.S. (1983) Nodulin-35: a subunit of specific uricase (uricase II) induced and localized in the uninfected cells of soybean nodules. *EMBO J.* **2**: 2333-2339.

Blumwald, E., Fortin, M., Rea, P.A.,Verma, D.P.S. and Poole, R.J. (1985) Presence of plasma membrane type H^+-ATPase in the membrane envelope enclosing the bacteroids in soybean root nodules. *Plant Physiol.* **78**: 665-672.

Boland, M.J. and Schubert, K.R. (1983) Biosynthesis of purines by a proplastid fraction from soybean nodules. *Arch. Biochem. Biophys.* **220**: 179-187.

Boland, M.J., Hanks, J.F., Reynolds, P.H.S., Blevins, D.G., Tolbert, N.E. and Schubert, K.R. (1982) Subcellular organization of ureide biogenesis from glycolytic intermediates and ammonium in nitrogen-fixing soybean nodules. *Planta.* **155**: 45-51.

Borst, P. (1989) Peroxisome biogenesis revisited. *Biochem Biophys Acta.* **1008**: 1-13.

Brickner, D.G. (1999) Signals, receptors and protein targeting: determining the molecular mechanisms for peroxisome biogenesis in higher plants. Ph.D. Thesis. The University of Michigan, USA.

Capote-Mainez, N. and Sanchez, F. (1997) Characterization of the common bean uricase II and its expression in organs other than nodules *Plant Physiol.* **115**: 1307-1317.

Carroll, B.J., McNeil, D.L. and Gresshoff, P.M. (1985) Isolation and properties of soybean [*Glycine max* (L.) Merr.] mutants that nodulate in the presence of high nitrate concentrations. *Proc. Natl. Acad. Sci. USA*. **82**: 4162-4166.

Chapman, K.A., Delauney, A.J., Kim, J.H. and Verma, D.P.S. (1994) Structural characterization of *de novo* purine biosynthesis enzymes in plants: 5-aminoimidazole ribonucleotide carboxylase and 5-aminoimidazole-4-*N*-succinocarboxamide ribonucleotide synthetase cDNAs from *Vigna aconitifolia. Plant Mol. Biol.* **24**: 389-395.

Cheon C.-I., Lee, N.-G., Siddique, A. B. M., Bal, A. K. and Verma, D. P. S. (1993) Roles of plant homologs of Rab1p and Rab7p in biogenesis of peribacteroid membrane, a subcellular compartment formed *de novo* during root nodule symbiosis. *EMBO J.* **12**: 4125-4135.

Cock, J.M., Mould, R.M., Bennett, M.J. and Cullimore, J.V. (1990) Expression of glutamine synthetase genes in roots and nodules of *Phaseolus vulgaris* following changes in ammonium supply and infection with various *Rhizobium* mutants. *Plant Mol. Biol.* **14**: 549-560.

Datta, D.B., Triplett, E.W. and Newcomb, E.H. (1991) Localization of xanthine dehydrogenase in cowpea root nodules: Implications for the interaction between cellular compartments during ureide biogenesis. *Proc. Natl. Acad. Sci. USA*. **88**: 4700-4702.

Delauney, A.J., Z. Tabaeizadeh and D.P.S. Verma (1988) A stable bifunctional antisense transcript inhibiting gene expression in transgenic plants. *Proc. Nat. Acad. Sci. USA*. **85**: 4300-4304.

del Río, L.A., Sandalio, L.M., Palma, J.M., Bueno, P., and Corpas, F.J. (1992) Metabolism of oxygen radicals in peroxisomes and cellular implications. *Free Rad. Biol. Med.* **13**: 557-580.

Diallinas, G., Gorfinkiel, L., Arst,H.N., Cecchetto, G. and Scazzocchio, C. (1995) Genetic and molecular characterization of a gene encoding a wide specificity purine permease of *Aspergillus nidulans* reveals a novel family of transporters conserved in prokaryotes and eukaryotes. *J. Biol. Chem.* **270**: 8610-8622.

Ebbole, D.J. and Zalkin, H. (1987) Cloning and characterization of a 12 gene cluster from *Bacillus subtilis* encoding nine enzymes for *de novo* purine nucleotide synthesis. *J. Biol. Chem.* **262**: 8274-8287.

Fahimi, H.D., Baumgart, E. and Volkl, A. (1993) Ultrastructural aspects of the biogenesis of peroxisomes in rat liver. *Biochimie.* **75**: 201-208.

Fuller, F. and Verma, D.P.S. (1984). Accumulation of nodulin mRNAs during the development of effective root nodules of soybean. *Plant Mol. Biol.* **3**: 21-28.

Gonzalez E. (1991) The C-terminal domain of plant catalases. Implications for a glyoxysomal targeting sequence. *Eur. J. Biochem.* **199**: 211-215.

Gorfinkiel, L., Diallinas, G. and Scazzocchio, C. (1993) Sequence and regulation of the *uapA* gene encoding a uric acid-xanthine permease in the fungus *Aspergillus nidulans. J. Biol. Chem.* **268**: 23376-23381.

Gould, S.J., Keller, G.A., Schneider, M., Howell, S.H., Garrard, L.J., Goodman, J.M., Distel, B., Tabak, H. and Subramani, S. (1990) Peroxisomal protein import is conserved between yeast, plants, insects and mammals. *EMBO J.* **9**: 85-90.

Groat, R.G. and Schrader, L.E. (1982) Isolation and immunochemical characterization of plant glutamine synthetase in alfalfa (*Medicago sativa* L.) nodules. *Plant Physiol.* **70**: 1759-1761.

Hayashi, M., Aoki, M., Kondo, M. and Nishimura, M. (1997) Changes in targeting efficiencies of proteins to plant microbodies caused by amino acid substitutions in the carboxy-terminal tripeptide. *Plant Cell Physiol.* **38**(6): 759-768.

Heldt, H.W. (1969) Adenine nucleotide translocation in spinach chloroplasts. *FEBS Lett.* **5**: 11-14.

Henikoff, S. (1987) Multi-functional polypeptides for purine *de novo* synthesis. *BioEssays.* **6**: 8-13.

Heupel, R., Markgraf, T., Robinson, D.G., Heldt, H.W. (1991) Compartmentation studies on spinach leaf peroxisomes. Evidence for channelling of photorespiratory metabolites in peroxisomes devoid of intact boundary membrane. *Plant Physiol.* **96**: 971-979.

Hirel, B., Bouet, C., King, B., Layzell, D., Jacobs, F. and Verma, D.P.S. (1987) Glutamine synthetase genes are regulated by ammonia provided externally or by symbiotic nitrogen fixation. *EMBO J.* **6**: 1167-1171.

Hirel, B., Marsolier, M.C., Hoarau, A., Hoarau, J. Bargeon, J., Schafer, R. and Verma, D.P.S. (1992) Forcing expression of a soybean root glutamine synthetase gene in tobacco leaves induces a native gene encoding cytosolic enzyme. *Plant Mol. Biol.* **20**: 207-218.

Holmes, E.W., McDonald, J.A., McCord, J.M., Wyngaarden, J.B. and Kelley, W.N. (1973) Human glutamine phosphoribosylpyrophosphate amidotransferase. *J. Biol. Chem.* **248**: 144-150.

Hu, C.-A. A. (1991) M.Sc. Thesis. Ohio State University, Columbus OH, USA.

Huang, A.H.C., Trelease, R.N. and Moore, Jr., T.S. (1983) *Plant Peroxisomes.* Academic Press, New York.

Hurst, D.T., Griffiths, E. and Vayianos, C. (1985) Inhibition of uricase by pyrimidine and purine drugs. *Clin. Biochem.* **18**: 247-251.

Ito, T., Shiraishi, H., Okada, K. and Shimura, Y. (1994) Two amidophosphoribosyltransferase genes of *Arabidopsis thaliana* expressed in different organs. *Plant Mol. Biol.* **26**: 529-533.

Kim, J.H., Delauney, A.J. and Verma, D.P.S. (1995b) Control of *de novo* purine biosynthesis genes in ureide-producing legumes: Induction of glutamine phosphoribosylpyrophosphate amidotransferase gene and characterization of its cDNA from soybean and *Vigna*. *Plant J.* **7**: 77-86.

Kim, J.H., Humphreys, J.M. and Verma, D.P.S. (1995a) Regulation of ureide biosynthesis genes in soybean root nodules. *Plant Physiol.* **108S**: 72.

Kim, J.K. (1996) Regulation of ureide biosynthesis genes in tropical legume root nodules. Ph.D. Thesis. The Ohio State University, USA.

Kindl, H. (1982) The biogenesis of microbodies (peroxisomes glyoxysomes) *Int. Rev Cyto.* **80**: 193-229.

Klebanoff, S.J. and Kazazi, F. (1995) Inactivation of human immunodeficiency virus type 1 by the amine oxidase-peroxidase system. *Clin. Microbiol.* **33**: 2054-2057.

Kohl, D.H., Schubert, K.R., Carter, M.B., Hagedorn, C.H. and Shearer, G. (1988) Proline metabolism in N_2-fixing root nodules: Energy transfer and regulation of purine synthesis. *Proc. Natl. Acad. Sci. USA*. **85**: 2036-2040.

Kos, W., Kal, A., Van Wilpe, S. and Tabak, H. (1995) Expression of genes encoding peroxisomal proteins in *Saccharomyces cerevisiae* is regulated by different circuits of transcriptional control. *Biochim. Biophys. Acta*. **1264**: 79-86.

Larsen, K. and Jochimsen, B.U. (1987) Appearance of purine-catabolizing enzymes in Fix^+ and Fix^- root nodules on soybean and effect of oxygen on the expression of the enzymes in callus tissue. *Plant Physiol.* **85**: 452-456.

Lazarow, P. B. and Fujiki. (1985) Biogenesis of peroxisomes. *Ann. Rev. Cell Biol.* **1**: 489-530.

Lee, N.G., Stein, B., Suzuki, H. and Verma, D.P.S. (1993) Antisense expression of nodulin-35 RNA in *Vigna aconitifolia* root nodules retards peroxisome development and the availability of nitrogen to the plant. *Plant J.* **3**: 599-606.

Legocki, R.P. and D.P.S. Verma (1979). A nodule-specific plant protein (nodulin-35) from soybean. *Science*. **205**: 190-193.

López-Huertas, E., Charlton, W.L., Johnson, B., Graham, I.A. and Baker, A. (2000) Stress induces peroxisome biogenesis genes. *EMBO J.* **19**(24): 6770-6777.

Luers, G., Hashimoto, T., Fahimi, H.D. and Volkl, A. (1993) Biogenesis of peroxisomes: isolation and characterization of two distinct peroxisomal populations from normal and regenerating rat liver. *J. Cell Biol.* **121**: 1271-80.

Marnett, L., Weller, P. and Battista, J. (1986) Comparison of the peroxidase activity of Haem proteins and cytochrome P-450. In: *Cytochrome P-450. Structure, Mechanism and Biochemistry*. (Ortiz de Montellano, P. ed.) pp 29-76, Plenum Press, New York.

McIntire, W.S. and Hartmann, C. (1993) Copper-containing amine oxidase. In: *Principle and application of quinoproteins*. (Davidson, V.L. ed.). pp 97-171. Marcel Dekker Inc., New York.

Meeks, J.C., Wolk, C.P., Schilling, N., Shaffer, P.W., Avissar, Y. and Chien, W.S. (1978) Initial organic products of fixation of [^{13}N]dinitrogen by root nodules of soybean (*Glycine max*). *Plant Physiol.* **61**: 980-983.

Messenger, L.J. and Zalkin, H. (1979) Glutamine phosphoribosyl pyrophosphate amidotransferase from *Escherichia coli*. *J. Biol. Chem.* **254**: 3382-3392.

Meyer, E. and Switzer, R.L. (1979) Regulation of *Bacillus subtilis* glutamine phosphoribosyl-pyrophosphate amidotransferase activity by end products. *J. Biol. Chem.* **254**: 5397-5402.

Miao, G.H., Hirel, B., Marsolier, M.C., Ridge, R.W. and Verma, D.P.S. (1991) Ammonia-regulated expression of a soybean gene encoding cytosolic glutamine synthetase in transgenic *Lotus corniculatus*. *Plant Cell.* **3**: 11-22.

Mitchell, J.P. (1965) The DNA content of nuclei in pea root nodules. *Annals Bot.* **29**: 371-376.

Miura, S., Oda, T., Funai, T., Ito, M., Okada, Y. and Ichiyama, A. (1994) Urate oxidase is imported into peroxisomes recognizing the C-terminal SKL motif of proteins. *Eur. J. Biochem.* **223**: 141-146.

Mohamedali, K.A., Guicherit, O.M., Kellems, R.E. and Rudolph, F.B. (1993) The highest levels of purine catabolic enzymes in mice are present in the proximal small intestine. *J. Biol. Chem.* **268**: 23728-23733.

Mothes, K. K. (1961) The metabolism of urea and ureides. *Can. J. Bot.* **39**: 1785-1807.

Mullen, R.T., Lee, M.S. and Trelease, R.N. (1997) Identification of the peroxisomal targeting signal for cottonseed catalase. *Plant J.* **12**:313-322.

Muller, M., Kraupp, M., Chiba, P. and Rumpold, H. (1982) Regulation of purine uptake in normal and neoplastic cells. *Adv. Enzyme Regul.* **21**: 239-256.

Neuhard, J. and Nygaard, P. (1987) Purines and pyrimidines. In *Escherichia coli and Salmonella typhimurium*: *Cellular and Molecular Biology*. Vol. 1 (Neidhardt, F.C., Ingraham, J.L., Low, K.B., Magasanik, B., Schaechter, M., Umbarger, H.E., eds.). pp. 445-473: American Society for Microbiology, Washington D.C., USA.

Newcomb, E.H. and Tandon, S.R. (1981) Uninfected cells of soybean root nodules: Ultrastructure suggests key role in ureide production. *Science.* **212**: 1394-1396.

Nguyen J., Machal, L., Vidal, J., Perrot-Rechenmann, C. and Gadal, P. (1986) Immunochemical studies on xanthine dehydrogenase of soybean root nodules. *Planta.* **167**: 190-195.

Nguyen, T., Zelechowska, M., Foster, V., Bergmann, H. and Verma, D.P.S. (1985) Primary structure of the soybean nodulin-35 gene encoding uricase II localized in the peroxisomes of uninfected cells of nodules. *Proc. Natl. Acad. Sci. USA.* **82**: 5040-5044.

Nohno, T., Saito, T. and Hong, J.S. (1986) Cloning and complete nucleotide sequence of the *Escherichia coli* glutamine permease operon (*glnHPQ*). *Mol. Gen. Genet.* **205**: 260-269.

Olsen, L.J. (1998) The surprising complexity of peroxisome biogenesis. *Plant Mol Biol.* **1-2**: 163-189.

Osumi, T., Yokota, S., Hashimoto, T. (1990) Proliferation of peroxisomes and induction of peroxisomal β-oxidation enzymes in rat hepatoma H4IIEC3 by ciprofibrate. *J. Biochem.* (Tokyo). **108**: 614-621.

Palma, J.M., Garrido, M., Rodriguez-Garcia, M.I. and del Rio, L.A. (1991) Peroxisome proliferation and oxidative stress mediated by activated oxygen species in plant peroxisomes. *Arch. Biochem. Biophys.* **287**: 68-74.

Pate, S.J., Atkins, C.A., White, S.T., Rainbird, R.M. and Woo, K.C. (1980) Nitrogen nutrition and Xylem transport of nitrogen in ureide-producing grain legumes. *Plant Phyisol.* **65**: 961-965.

Paul, K.G. and Avi-dorl, Y. (1954) The oxidation of uric acid with horseradish peroxidase. *Acta. Chem. Scand.* **8**: 637-648.

Pistelli, L., De Bellis, L. and Alpi, A. (1995) Evidences of glyoxylate cycle in peroxisomes of senescent cotyledons. *Plant Sci.* **109**: 13-21.

Purdue, P.E. and Lazarow, P.B. (1996) Targeting of human catalase to peroxisomes is dependent upon a novel COOH-terminal peroxisomal targeting sequence. *J. Cell Biol.* **134**(4): 849-62.

Reynolds, P.H.S., Blevins, D.G. and Randall, D.D. (1984) 5-phosphoribosylpyrophosphate amidotransferase from soybean root nodules: Kinetic and regulatory properties. *Arch. Biochem. Biophys.* **229**: 623-631.

Reynolds, P.H.S., Boland, M.J., Blevins, D.G., Schubert, K.R. and Randall, D.D. (1982) Enzymes of amide and ureide biogenesis in developing soybean nodules. *Plant Physiol.* **69**: 1334-1338.

Riezman, H., Weir, E., Leaver, C., Titus, D.E. and Becker, W.M. (1980) Regulation of glyoxysomal enzymes during germination of cucumber 3; *in vitro* translation of four glyoxysomal enzymes. *Plant Physiol.* **64**: 40-46.

Robertson, J.G., Warburton, M. and Farnden, K.J.F. (1975) Induction of glutamate synthase during nodule development in lupin. *FEBS Lett.* **55**: 33-37.

Robinson, S.P. (1985) The involvement of stromal ATP in maintaining the pH gradient across the chloroplast envelope in the light. *Biochim. Biophys. Acta.* **806**: 187-194.

Robinson, S.P. and Wiskich, J.T. (1977b) Uptake of ATP analogs by isolated pea chloroplasts and their effect in CO_2 fixation and electron transport. *Biochim. Biophys. Acta.* **461**: 131-140.

Rolfes, R.J. and Zalkin, H. (1988) *Escherichia coli* gene *purR* encoding a repressor protein for purine nucleotide synthesis. *J. Biol. Chem.* **263**: 19653-19661.

Salvemini, F., Marini, A.-M., Riccio, A., Patriarca, E. J. and Chiurazzi, M. (2001) Functional characterization of an ammonium transporter gene from *Lotus japonicus. Gene.* **270**: 237-243.

Schnorr, K.M., Nygaard, P. and Laloue, M. (1994) Molecular characterization of *Arabidopsis thaliana* cDNAs encoding three purine biosynthetic enzymes. *Plant J.* **6**: 113-121.

Schubert, K.R. (1986) Products of biological nitrogen fixation in higher plants: Synthesis, transport, and metabolism. *Ann. Rev. Plant Physiol.* **37**: 539-574.

Schubert, K.R. and Boland, M.J. (1990) The ureides. In: *Biochemistry of Plants.* **16**: 197-281.

Senecoff, J.F. and Meagher, R.B. (1993) Isolating the *Arabidopsis thaliana* genes for *de novo* purine synthesis by suppression of *Escheria coli* mutants. *Plant Physiol.* **102**: 387-399.

Sies, H. (1977) Cytochrome oxidase and urate oxidase as intracellular O_2 indicator in studies of O_2 gradient during hypoxia in liver. *Adv. Exp. Med. Biol.* **94**: 561-566.

Suarez, T., de Queiroz, M.V., Oestreicher, N. and Scazzocchio, C. (1995) The sequence and binding specificity of UaY, the specific regulator of the purine utilization pathway in *Aspergillus nidulans*, suggest an evolutionary relationship with the PPR1 protein of *Saccharomyces cerevisiae. EMBO J.* **14**: 1453-1467.

Subramani, S. (1993) Protein import into peroxisomes and biogenesis of the organelle. *Ann. Rev. Cell Biol.* **9**: 445-478.

Suzuki, A., Gadal, P. and Oaks, A. (1981) Intracellular distribution of enzymes associated with nitrogen assimilation in roots. *Planta.* **151**: 457-461.

Suzuki, H. and Verma, D.P.S. (1991) Soybean nodule-specific uricase (Nodulin-35) is expressed and assembled into a functional tetrameric holoenzyme in *Escherichia coli. Plant Physiol.* **95**: 384-389.

Tajima, S. and Yamamoto, Y., 1977, Regulation of uricase activty in developing roots of *Glycine max*, non-nodulation variety A62-2. *Plant Cell Physiol.* **18**: 247-253.

Tajima, S., Kanazawa, T., Takeuchi, E., Yamamoto, Y. (1985) Characteristics of a urate-degrading diamine oxidase-peroxidase enzyme system in soybean radicles. *Plant Cell Physiol.* **26**: 787-795.

Tajima, S., Kato, N., and Yamamoto, Y., (1983) Cadaverine involved in urate degrading activity (uricase activity) in soybean radicles. *Plant Cell Physiol.* **24**: 247-253.

Tanaka, A., Yamamura, M., Kawamoto, S. and Fukui, S. (1977) Production of uricase by *Candida tropicalis* using n-alkane as substrate. *Appl. Env. Microbiol.* **34**: 342-346.

Tanake, K., Tajima, S. and Kouchi, H. (2000) Structure and expression analysis of uricase mRNA from *Lotus japonicus*. *Mol. Plant-Microbe Interact.* **13**: 1156-1160.

Trelease, R.N. (1984) Biogenesis of glyoxysomes. *Ann. Rev. Plant Physiol.* **35**: 321-347.

Van den Bosch, K.A., Noel, K.D., Kaneko, Y. and Newcomb, E.H. (1985) Nodule initiation elicited by non-infective mutants of *Rhizobium phaseoli*. *J. Bacteriol.* **162**: 950-959.

Veenhuis, M., Mateblowski, M., Kunau, W.H., and Harder, W. (1987) Proliferation of microbodies in *Saccharomyces cerevisiae*. *Yeast.* **3**(2): 77-84.

Verma D.P.S. (2000) Nodulins: Nodule-specific host gene products, their induction and function in root nodule symbiosis. In *Prokaryotic Nitrogen Fixation: A Model System for Analysis of Biological Process.* (Tripton, ed.). Horizon Sci. Press, Wymondham, UK.

Verma, D.P.S. (1989) Plant genes involved in carbon and nitrogen assimilation in root nodules. In *Plant nitrogen metabolism*, (Poulton, J.E., Romeo, J.T. and Conn, E.E., eds.). pp 43-63. New York: Plenum Pub. Corp.

Verma, D.P.S., Fortin, M.G., Stanley, J., Mauro, V.P., Purohit, S. and Morrison, N. (1986) Nodulins and nodulin genes of *Glycine max*. *Plant Mol. Biol.* **7**: 51-61.

Walker, E.L. and Coruzzi, G.M. (1989) Developmentally regulated expression of the gene family for cytosolic glutamine synthetase in *Pisum sativum*. *Plant Physiol.* **91**: 702-708.

Wells, X. E. and Lees, E.M. (1991) Ureidoglycolate amidohydrolase from developing French bean fruits (*Phaseolus vulgaris* [L.]). *Arch. Biochem. Biophys.* **287**: 151-159.

Wu, T. (1998) Assimilation of symbiotically-reduced nitrogen in tropical legumes: Regulation of peroxisome proliferation and ureide production. Ph.D. Thesis, The Ohio State University, Columbus, Ohio. USA.

7

PEROXISOMES, REACTIVE OXYGEN METABOLISM, AND STRESS-RELATED ENZYME ACTIVITIES

Luis A. del Río, Luisa M. Sandalio, José M. Palma, Francisco J. Corpas, Eduardo López-Huertas[1], María C. Romero-Puertas and Iva McCarthy

Estación Experimental del Zaidín, CSIC, Granada, Spain
[1]Puleva Biotech, Granada, Spain

KEYWORDS

Abiotic stress; nitric oxide; peroxisomes; ROS; signalling.

INTRODUCTION

The main functions described for peroxisomes in plant cells are the oxidative photosynthetic carbon cycle of photorespiration, fatty acid β-oxidation, the glyoxylate cycle, and the metabolism of ureides (Huang *et al*, 1983; del Río *et al*, 1992; Reumann, 2000). A characteristic property of peroxisomes is their metabolic plasticity since their enzymatic content can vary depending on the organism, cell/tissue-type and environmental conditions. An illustrative example of the inducible nature of peroxisomal metabolism is the light-induced transition of glyoxysomes, the specialized peroxisomes of oilseeds, to leaf-type peroxisomes (Huang *et al*, 1983; Mullen and Trelease, 1996). Likewise, during the senescence of leaves the reverse process is observed, that is the conversion of leaf peroxisomes into glyoxysomes (Vicentini and Matile, 1993; Nishimura *et al*, 1996; del Río *et al*, 1998a). On the other hand, the cellular population of peroxisomes can proliferate in plants during senescence and under different abiotic stress conditions

A. Baker and I.A. Graham (eds.), Plant Peroxisomes, 221–258.

(del Río *et al*, 1996, 1998a, 2001; Corpas *et al*, 2001). The induction of peroxisome biogenesis genes (*PEX*) by H_2O_2 has been demonstrated in both plant and animal cells, indicating that the signal molecule H_2O_2 is responsible for the proliferation of peroxisomes (López-Huertas *et al*, 2000).

Another important characteristic of peroxisomes is that they have an essentially oxidative type of metabolism. It is well-known that an important part of the H_2O_2 produced in plant cells originates in peroxisomes, and that catalase is responsible for the removal of toxic H_2O_2 (Huang *et al*, 1983; del Río *et al*, 1992). However, the production of superoxide radicals ($O_2^{\cdot-}$), another reactive oxygen species (ROS), has been demonstrated in plant peroxisomes, as well as the presence of proteases and a battery of antioxidative enzymes, including superoxide dismutases (SODs), the components of the ascorbate-glutathione cycle, several NADP-dependent dehydrogenases, and the nitric oxide-generating enzyme nitric oxide synthase (NOS) (Corpas *et al*, 2001). This has suggested the existence of new cellular functions for peroxisomes related to ROS metabolism which appear to be particularly relevant in the process of leaf senescence and in different abiotic stress conditions. In this review, the production of superoxide radicals, the different antioxidative enzymes and proteases in peroxisomes and the response of these oxidative organelles to different plant stress situations will be analyzed in the light of these new ROS-mediated functions of plant peroxisomes.

REACTIVE OXYGEN SPECIES

Reactive oxygen species (ROS) is a collective term used to include mainly superoxide radicals ($O_2^{\cdot-}$), hydroxyl radicals ($^{\cdot}OH$), hydroperoxyl radicals ($HO^{\cdot}_2$), alkoxyl radicals ($RO^{\cdot}$), peroxyl radicals ($ROO^{\cdot}$), and also some non-radical derivatives of O_2, such as hydrogen peroxide (H_2O_2), singlet oxygen ($^1O_2$ or $^1\Delta g$), ozone (O_3), hypochlorous acid (HOCl) and peroxynitrite ($ONOO^-$) (Halliwell and Gutteridge, 2000). Some of these ROS, particularly hydroxyl radicals ($^{\cdot}OH$), are very strong oxidising species than can rapidly attack biological membranes and all types of biomolecules, including DNA and proteins, leading to irreparable metabolic dysfunction and cell death (Halliwell and Gutteridge, 2000). However, plants possess enzymatic and nonenzymatic antioxidative defence systems distributed in different cell compartments. Superoxide dismutases (SODs), catalases, peroxidases, and the ascorbate-glutathione cycle enzymes are examples of antioxidative enzymes. The nonenzymatic antioxidants are mainly ascorbate (vitamin C), glutathione (GSH), α-tocopherol (vitamin E), β-carotene, and flavonoids, which are distributed chiefly in chloroplasts but also in other cellular compartments such as mitochondria and peroxisomes. Under normal conditions, the antioxidative defence system of plants provides adequate cellular protection against ROS, but when the generation of ROS overcomes the defence provided by the cellular antioxidant systems, then oxidative stress results.

Production of oxygen radicals in peroxisomes

In plant cells, the production of ROS has been demonstrated mainly in chloroplasts, mitochondria, plasma membrane and the apoplastic space (Bolwell, 1999). In peroxisomes from pea leaves and glyoxysomes from watermelon cotyledons, both biochemical and electron spin resonance (ESR) methods demonstrated the existence of at least two sites of $O_2^{\cdot -}$ generation: one in the organelle matrix, in which the generating system was identified as xanthine oxidase (XOD), and another site in the peroxisomal membranes dependent on NAD(P)H (Sandalio *et al*, 1988; del Río *et al*, 1998b) (Figure 1). Xanthine oxidase catalyses the oxidation of xanthine and hypoxanthine to uric acid and is a well-known producer of superoxide radicals (Fridovich, 1986). The presence of xanthine and uric acid, and allantoin, the product of the urate oxidase reaction, was detected in leaf peroxisomes by HPLC analysis (Corpas *et al*, 1997). This indicates a cellular role for these organelles in the catabolism of xanthine produced as a result of the turnover of nucleotides, RNA and DNA (Corpas *et al*, 1997; del Río *et al*, 1998b).

The other site of $O_2^{\cdot -}$ production is the peroxisomal membrane, where a small electron chain similar to that reported in glyoxysomal membranes from castor bean endosperm (Fang *et al*, 1987) and peroxisomal membranes from potato tuber appears to be involved (Struglics *et al*, 1993). This electron transport chain is composed of a flavoprotein NADH:ferricyanide reductase of about 32 kDa and a Cyt *b* (Fang *et al*, 1987). The integral peroxisomal membrane polypeptides (PMPs) of pea leaf peroxisomes were identified by SDS-PAGE (López-Huertas *et al*, 1995) and three of these membrane polypeptides, with molecular masses of 18, 29 and 32 kDa have been characterized and demonstrated to be responsible for $O_2^{\cdot -}$ generation (López-Huertas *et al*, 1997, 1999; del Río *et al*, 1998b).

Figure 1: Model proposed for the function of the ascorbate-glutathione cycle in leaf peroxisomes

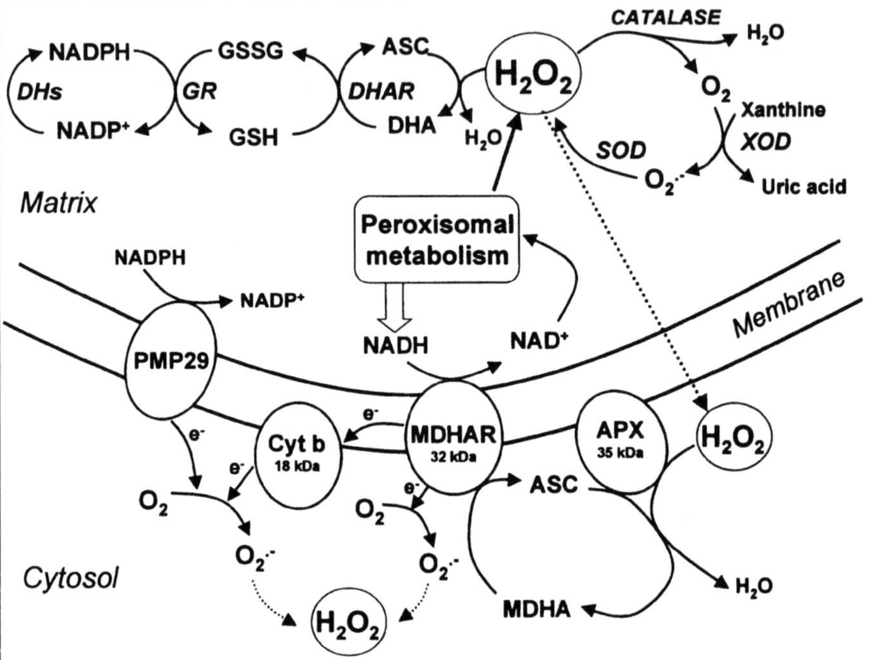

***Figure 1**. The model is based on results described by Jiménez et al. (1997) and those reported on the characterization of PMPs from pea leaves (López-Huertas et al, 1997) and the NADH:MDHAR of glyoxysomal membranes from castor bean endosperm (Bowditch and Donaldson, 1990). ASC, ascorbate, reduced form; DHA, ascorbate, oxidised form (dehydroascorbate). DHs, NADP-dehydrogenases. PMP29, 29-kDa peroxisomal membrane polypeptide. Plant Physiol. 116: 1198; 1998. Copyright 1998 American Society of Plant Biologists, Rockville, MD.*

The main producer of superoxide radicals in the peroxisomal membrane was the 18 kDa PMP which was proposed to be a cytochrome possibly belonging to the *b*-type group (López-Huertas *et al*, 1997). While the 18- and 32-kDa PMPs use NADH as electron donor for $O_2^{\cdot-}$ production, the 29-kDa PMP was clearly dependent on NADPH, and was also able to reduce cytochrome *c* with NADPH as electron donor (López-Huertas *et al*, 1997, 1999; del Río *et al*, 1998b). On the basis of its biochemical and immunochemical properties, the PMP32 very probably corresponds to the monodehydroascorbate reductase (MDHAR) (López-Huertas *et al*, 1999) whose activity was previously detected in pea leaf peroxisomal membranes by Jiménez *et al* (1997). This indicates the participation of MDHAR in $O_2^{\cdot-}$ production by peroxisomal membranes. Very recently, Miyake *et al* (1998) reported that chloroplast MDHAR mediates the production of $O_2^{\cdot-}$ in spinach thylakoid membranes. The third $O_2^{\cdot-}$-generating polypeptide, PMP29, is strictly dependent on NADPH as electron donor, and could be related to the peroxisomal NADPH:cytochrome P-450 reductase.

Superoxide production by peroxisomal membranes may be an obligatory consequence of NADH re-oxidation by the peroxisomal electron-transport chain, in order to regenerate NAD^+ to be re-utilized in peroxisomal metabolic processes (del Río *et al*, 1992; del Río and Donaldson, 1995; Jiménez *et al*, 1997). Under normal metabolic conditions, the $O_2^{\cdot-}$ production by peroxisomal membranes is not dangerous to the cell, which is adequately protected against these radicals. However, under certain conditions of stress, the release of peroxisomal membrane-generated $O_2^{\cdot-}$ radicals into the cytosol can be enhanced (del Río *et al*, 1996), producing cellular oxidative stress situations mediated by ROS. Since $O_2^{\cdot-}$ radicals have a short half-life under physiological conditions and are rapidly converted into H_2O_2 and O_2, the final result of these stress conditions will be an increase of H_2O_2 in the cell (del Río *et al*, 1996).

Antioxidant systems in peroxisomes

Superoxide dismutases

Superoxide dismutases (SODs; EC 1.15.1.1) are a family of metalloenzymes that catalyse the disproportionation of $O_2^{\cdot-}$ radicals into H_2O_2 and O_2, and play an important role in protecting cells against the toxic effects of superoxide radicals produced in different cellular compartments (Fridovich, 1986; Halliwell and Gutteridge, 2000). SODs are distributed in different cell compartments, mainly chloroplasts, cytosol, and mitochondria (Fridovich, 1986; Halliwell and Gutteridge, 2000; del Río *et al*, 1992; Bowler *et al*, 1994), but the presence of SOD in peroxisomes was demonstrated for the first time in plant tissues (del Río *et al*, 1983; Sandalio *et al*, 1987). Since then, the occurrence of SODs in isolated plant peroxisomes has been reported in at least eight different plant species (del Río *et al*, 2001), and in four of these plants the presence of SOD in peroxisomes has been confirmed by immunogold electron microscopy (Sandalio *et al*, 1997; Corpas *et al*, 1998a) (Figure 2). Results obtained concerning the presence of SOD in plant peroxisomes were extended years later to human and animal cells, which were found to contain CuZn-SOD in peroxisomes and, more recently, also

Mn-SOD (for a review see del Río *et al*, 2001). Three SODs of peroxisomal origin have been purified and characterized, a CuZn-SOD and a Mn-SOD from watermelon cotyledons (Bueno *et al*, 1995; Pastori *et al*, 1996) and a Mn-SOD from pea leaves (Palma *et al*, 1998).

Ascorbate-glutathione cycle

The ascorbate-glutathione cycle, also called Foyer-Halliwell-Asada cycle, is an efficient way for plant cells to dispose of H_2O_2 in certain cellular compartments where this metabolite is produced and no catalase is present (Halliwell and Gutteridge, 2000). This cycle makes use of the nonenzymic antioxidants ascorbate and glutathione in a series of reactions catalyzed by four antioxidative enzymes, and has been demonstrated in chloroplasts, the cytosol and root nodule mitochondria (Foyer *et al*, 1997). In peroxisomes and mitochondria purified from pea leaves, the presence of the enzymes of the ascorbate-glutathione cycle was investigated. The four enzymes of the cycle, ascorbate peroxidase (APX; EC 1.11.1.11), monodehydroascorbate reductase (MDHAR; EC 1.6.5.4), dehydroascorbate reductase (DHAR; EC 1.8.5.1.), and glutathione reductase (GR; EC 1.6.4.2) were present in peroxisomes (Jiménez *et al*, 1997) (Figure 1). Likewise, in intact peroxisomes and mitochondria, the presence of reduced ascorbate (ASC) and glutathione (GSH), and their oxidised forms DHA and GSSG, respectively, was demonstrated by HPLC analysis (Jiménez *et al*, 1997). The presence of the ascorbate-glutathione cycle enzymes has also been recently reported in peroxisomes of leaf and root cells of two species of tomato (Mittova *et al*, 2000). The intraperoxisomal distribution of the four enzymes was studied and a model for the function of the ascorbate-glutathione cycle in leaf peroxisomes is shown in Figure 1. DHAR and GR were found in the soluble fraction of peroxisomes, whereas APX activity was bound to the cytosolic side of the peroxisomal membrane. By Western blot analysis with a specific antibody against pumpkin APX, an integral membrane polypeptide of 35 kDa was identified as the APX of pea leaf peroxisomes (López-Huertas *et al*, 1999). These results agree with previous findings of an APX isoenzyme in membranes of pumpkin and cotton peroxisomes (Yamaguchi *et al*, 1995; Bunkelmann and Trelease, 1996).

Figure 2: Immunolocalization of SOD in plant peroxisomes

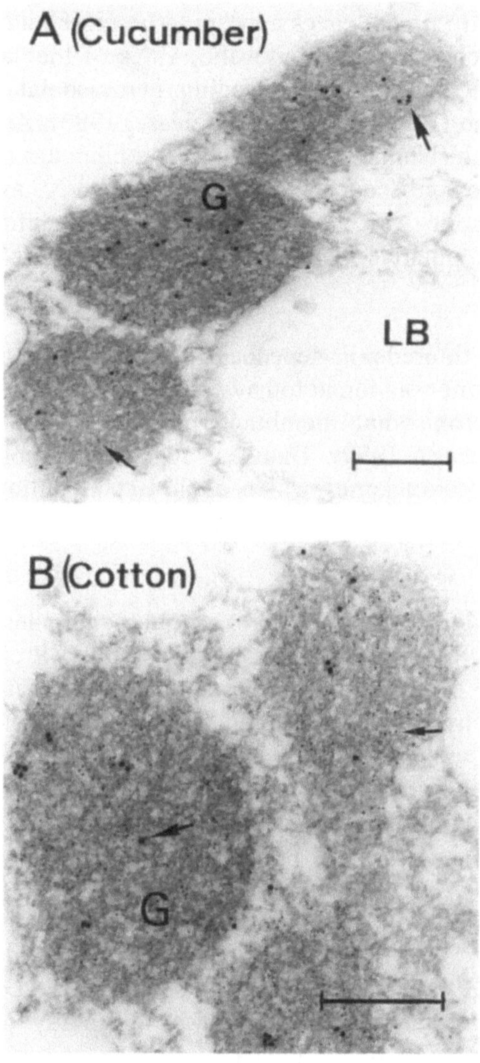

Figure 2. *The electron micrographs are representative of thin sections of cucumber (a) and cotton (b) cotyledons showing specific immunolocalizations of CuZn-SOD and malate synthase using indirect, double immunogold labelling. The 15-nm gold particles (large arrows) show the matrix localization of CuZn-SOD using antibody against the °peroxisomal CuZn-SOD of watermelon (Bueno et al., 1995). The 5-nm gold particles (small arrows) illustrate the site of malate synthase, a characteristic enzyme of oilseed peroxisomes, using IgG to cotton malate synthase. G, peroxisomes; LB, lipid body. Bars = 0.5 μm. (Corpas et al., 1998a).*

The substrate specificity and stability of the membrane-bound peroxisomal APX was studied (Jiménez *et al*, 1998a). Considering its low stability, the peroxisomal APX from pea leaves appears to be more related to chloroplast APX than to mitochondrial and cytosolic APXs of the same plant species (Jiménez *et al*, 1998a). cDNAs encoding peroxisomal APX have been isolated from cotton (Bunkelmann and Trelease, 1996), Arabidopsis (Zhang *et al*, 1997), spinach (Ishikawa *et al*, 1998), and pumpkin (Nito *et al*, 2001). The deduced amino acid sequence of peroxisomal APX has a high degree of identity with cytosolic APX but it has a C-terminal amino acid extension containing a single, putative membrane-spanning region (Mullen *et al*, 1999; Nito *et al*, 2001).

An *E. coli* mutant thioredoxin-dependent peroxidase has been expressed in yeast, and the enzyme was found to have high homology with the C-terminal sequence of a peroxisomal membrane protein (PMP20) from *Candida boidini* (Verdoucq *et al*, 1999). This has suggested the possible presence of peroxiredoxins in peroxisomes which could act as antioxidative enzymes together with catalase and APX in the control of the peroxisomal H_2O_2 concentration.

MDHAR was also localized in the peroxisomal membranes (Figure 1). It has been proposed that the *trans*-membrane protein MDHAR can oxidise NADH on the matrix side of the peroxisomal membrane and transfer the reducing equivalents as electrons to the acceptor MDHA on the cytosolic side of the membrane (Luster and Donaldson, 1987; Bowditch and Donaldson, 1990). In this process, O_2 could also act as an electron acceptor, with the concomitant formation of $O_2^{\cdot-}$ radicals (López-Huertas *et al*, 1999).

The presence of APX and MDHAR in leaf peroxisomal membranes suggests a dual complementary role in peroxisomal metabolism of these membrane-bound antioxidative enzymes. The first function could be to reoxidise endogenous NADH to maintain a constant supply of NAD^+ for peroxisomal metabolism (Figure 1), an idea which was originally postulated for the membrane-bound NADH dehydrogenase of glyoxysomes from castor bean endosperm (Fang *et al*, 1987; Luster and Donaldson, 1987; Bowditch and Donaldson, 1990). A second function of the membrane antioxidative enzymes could be to protect against H_2O_2 leaking from peroxisomes. H_2O_2 can easily permeate the peroxisomal membrane, but APX would degrade leaking H_2O_2, as well as the H_2O_2 that is being continously formed by dismutation of the $O_2^{\cdot-}$ generated in the NADH-dependent electron transport system of the peroxisomal membrane (Figure 1). This membrane scavenging of H_2O_2 could prevent an increase in the cytosolic H_2O_2 concentration during normal metabolism and under certain plant stress situations, when the level of H_2O_2 produced in peroxisomes can be substantially enhanced (del Río *et al*, 1992, 1996).

NADP-dependent dehydrogenases

In recent years, different studies have shown that the NADPH-generating glucose-6-phosphate dehydrogenase (G6PDH; EC 1.1.1.49) has a relevant role in the mechanism of protection against oxidative stress of bacteria, yeast and mammalian cells (see Corpas *et al*, 1998b). In plant cells, NADPH has an important role in the protection systems against oxidative stress due to its involvement in the ascorbate-glutathione cycle of chloroplasts (Foyer and Halliwell, 1976; Asada, 2000). This evidence supports the notion of G6PDH as an antioxidative enzyme which can be included in the group of catalase, SOD, APX and GR/peroxidase (Martini and Ursini, 1996). In higher plants, only two isoforms of G6PDH have been reported, which are localized in the cytosol and the plastidic stroma (Schnarrenberger *et al*, 1995).

In work carried out in our laboratory, the presence of three NADP-dehydrogenases in leaf peroxisomes purified from pea leaves was investigated. The dehydrogenases included: G6PDH, 6-phosphogluconate dehydrogenase (6PGDH; EC 1.1.1.44), and isocitrate dehydrogenase (ICDH; EC1.1.1.42). The only NADP-dependent dehydrogenase that had been previously detected in peroxisomes was the ICDH, reported by Tolbert (1981) as present in trace amounts in spinach leaf peroxisomes, and by Donaldson (1982) in castor bean glyoxysomes where the occurrence of NADP(H) was also demonstrated. The three dehydrogenases were found to be present in the matrix of leaf peroxisomes and showed a typical Michaelis-Menten kinetic saturation curve (Corpas *et al*, 1998b, 1999a). By isoelectric focusing, four isoforms of ICDH, three isoforms of G6PDH, and one isoform of 6PGDH were detected in peroxisomal matrices (Corpas *et al*, 1998b, 1999a). The presence of G6PDH and ICDH in leaf peroxisomes was also checked by immunoblot analysis of peroxisomal matrices and was confirmed by immunogold electron microscopy using antibodies against G6PDH and pea ICDH (Corpas *et al*, 1998b, 1999a).

In peroxisomes from senescent leaves, it was found that the K_m of ICDH decreased almost 11-fold compared to young leaves, which resulted in a catalytic efficiency approximately twelve times higher for peroxisomal ICDH from senescent leaves (Corpas *et al*, 1999a). The physiological significance of this K_m change of ICDH is probably two-fold: (i) To compete with isocitrate lyase, which is present in peroxisomes from senescent leaves, for the intracellular pool of isocitrate; and (ii) to provide a higher and constant supply of NADPH to eliminate, by the ascorbate-glutathione cycle, the excess of H_2O_2 produced during senescence (Pastori and del Río, 1997).

The presence in peroxisomes of the two oxidative enzymes of the pentose phosphate pathway, G6PDH and 6PGDH, and ICDH implies that these organelles have the capacity to reduce $NADP^+$ to NADPH for its re-utilization in their metabolism. NADPH is necessary for the function of the NADPH:cytochrome *P*-450 reductase, whose presence has been detected in the membranes of castor bean peroxisomes (Donaldson and Luster, 1991). On the other hand, one of the $O_2^{\cdot -}$-generating polypeptides of peroxisomal membranes, the PMP29, was clearly dependent on NADPH and was also able to reduce cytochrome *c* with NADPH as electron donor (López-Huertas *et al*, 1999). NADPH has been reported to protect catalase from oxidative damage (Kirkman *et al*, 1999) and is also required for the maintenance of the ascorbate-glutathione cycle operativity. In this cycle, NADPH is necessary for the glutathione reductase activity which recycles reduced glutathione (GSH) from its oxidised form (GSSG) to be used for the scavenging of H_2O_2 in the cycle (Foyer and Halliwell, 1976; Halliwell and Gutteridge, 2000). An additional function for NADPH in peroxisomes could be related to the mechanism of protein import into these organelles. It was recently shown that the NADPH to $NADP^+$ ratio is important in peroxisomal protein import (Pool *et al*, 1998).

PEROXISOMAL PROTEASES

To our knowledge, the first report on the occurrence of protease activity in peroxisomes is that of Gray *et al* (1970) who demonstrated the existence of a neutral protease activity in peroxisomes purified from rat liver. This evidence passed practically unnoticed for many years perhaps due to the widely accepted dogma, among researchers in the animal field, that peroxisomes lacked proteolytic activity because their proteins were imported in a mature form without processing. Further research demonstrated that several peroxisomal proteins from different origins, including thiolases, malate dehydrogenases and non-specific lipid transfer proteins (nsLTPs), were synthesised as larger precursors and were proteolytically processed after import into peroxisomes (Fujiki *et al*, 1989; Gietl, 1990; Gietl *et al* 1997; Tsuboi *et al*, 1992).

In plants, the presence of proteases in different cell compartments, including vacuoles, chloroplasts, mitochondria, Golgi apparatus, cell wall and cytosol, is well documented (see Distefano *et al*, 1997), but in recent years the occurrence of proteolytic activity in plant peroxisomes was also reported. A list of the different proteases characterized in plant peroxisomes is shown in Table 1. The presence of endo- and exo-proteolytic activity in plant peroxisomes was demonstrated for the first time in cell organelles purified from pea leaves (Corpas *et al*, 1993a). By native PAGE using different exopeptidase substrates (L-aa-βNA), one leucine aminopeptidase (AP) was found in peroxisomes. The peroxisomal AP by its sensitivity to different protease inhibitors was characterized as a serine protease and had essential thiol groups and metal(s) in its active centre. The enzyme had a maximal activity at pH 7.5, a molecular mass of 56.8 kDa and a pI of 5.3, and was mainly present in the soluble fraction of peroxisomes (Corpas *et al*, 1993a).

Table 1. Proteases characterized in plant peroxisomes

Isoenzyme Reference	**Plant source**	**Molecular mass (kDa)**	**Type of protease**
Exo-peptidase	Pea	57	Leucine-aminopeptidase Corpas *et al* (1993a)
Endo-peptidases			
EP-1	Pea	220	Serine-proteinase Distefano *et al* (1997)
EP-2	Pea	88	Cysteine-proteinase
EP-3	Pea	76	Serine-proteinase
EP-4	Pea	64	Serine-proteinase
EP-5	Pea	50	?
EP-6	Pea	46	Cysteine-proteinase
EP-7	Pea	34	Metallo-proteinase

The endo-proteolytic activity of pea leaf peroxisomes was also characterized and the effect of senescence on the pattern of the peroxisomal endoprotease (EP) isoenzymes was also reported (Distefano *et al*, 1997). Results obtained from the incubation of peroxisomal matrices with azocasein showed a much higher endo-proteolytic activity in senescent leaves than in young leaves. In gelatin-containing SDS/polyacrylamide-gradient gels three EP isoenzymes (EP2, EP4 and EP5) were found in young plants, but four additional isoenzymes (EP1, EP3, EP6 and EP7) were detected in senescent plants (Distefano *et al*, 1997). By using different class-specific inhibitors, the electrophoretically separated EP isoenzymes were characterized as indicated in Table 1. The serine-proteinase isoenzymes (EP1, EP3 and EP4) represent approximately 70% of the total EP activity of the organelle, and they have a notable thermal stability (Distefano *et al*, 1997).

The peroxisomal serine-proteinases may have an important role during senescence, when the most abundant isoenzymes of this class, EP3 and EP4, are present. The peroxisomal enzymes glycolate oxidase (EC 1.1.3.1), catalase (EC 1.11.1.6) and glucose-6-phosphate dehydrogenase (EC 1.1.1.49) were susceptible to proteolytic degradation by peroxisomal endoproteases, and this cleavage was prevented, to some extent, by incubation with 2 mM phenylmethylsulfonylfluoride (PMSF), an inhibitor of serine-proteinases (Distefano *et al*, 1999). On the contrary, peroxisomal manganese superoxide dismutase (EC 1.15.1.1) was not endoproteolytically degraded. Ribulose-1,5-bisphosphate carboxylase/oxygenase (EC 4.1.1.39) from spinach and urease (EC 3.5.1.5) from jackbean were strongly degraded in the presence of peroxisomal matrices, indicating that endogenous EPs are not specific for peroxisomal proteins (Distefano *et al*, 1999).

The conversion of glyoxysomes into leaf peroxisomes during germination of oilseeds, as well as the opposite process converting leaf peroxisomes into glyoxysomes during senescence, implies the degradation of the organelle's pre-existing proteins (Nishimura *et al*, 1996). Some indirect results were obtained indicating that glyoxysomal malate synthetase might specifically be degraded in glyoxysomes during greening of pumpkin cotyledons (Mori and Nishimura, 1989). On the other hand, the decrease in catalase activity, observed under some stress conditions (Hertwig *et al*, 1992; Pastori and del Río, 1997) suggested that some proteolytic degradation of this enzyme could occur. However, up until now, in senescent plants there was no evidence showing that peroxisomal proteins might be cleaved by their own organellar proteases.

Xanthine oxidoreductase catalyzes the conversion of the purines hypoxanthine and xanthine into uric acid with the concomitant formation of either NADH or $O_2^{\cdot-}$ (Nishino, 1994). The enzyme has been found as two interconvertible forms (D and O) (Nishino, 1994). The conversion of xanthine dehydrogenase (XDH; form D; EC 1.2.1.27) into xanthine oxidase (XOD; form O; EC 1.2.3.2) can be carried out by either reversible or irreversible pathways. In the irreversible step, form D is converted into form O by a proteolytic cleavage (Nishino, 1994; Saksela *et al*, 1999). The effect of incubating peroxisomal matrices from pea leaves with microbial xanthine oxidoreductase was studied and results obtained suggested that peroxisomal endoproteases could potentially carry out the irreversible conversion of xanthine dehydrogenase into the superoxide-generating xanthine oxidase (Distefano *et al*, 1999). A similar protease-mediated conversion of XDH into XOD has recently been reported in animal mitochondria (Saksela *et al*, 1999).

These results indicate that peroxisomal EPs from pea leaves may play an important role in the turnover of peroxisomal proteins during senescence. This means that the organelle's own proteolytic machinery could participate in the mechanism of the senescence-induced conversion of leaf peroxisomes into glyoxysomes with a concomitant degradation of pre-existent proteins, as is known to occur in different senescent tissues. Also, peroxisomal EPs might be involved in the turnover of proteins located in other cell compartments during advanced stages of senescence, when deterioration of membranes and leakage of the organelle's soluble fractions takes place. Moreover, results obtained with XDH/XOD assays suggest that peroxisomal EPs could take part in a regulated modification of proteins which do not necessarily imply full degradation of proteins.

Peroxisomal proteases could also be involved in the posttranslational processing of cytoplasmically synthesized peroxisome polypeptides. The topogenic signal PTS2 resides in a cleavable N-terminal presequence and it is known that several plant peroxisomal enzymes, including thiolases, malate dehydrogenase and citrate synthase, are targeted by PTS2 into peroxisomes (Gietl *et al*, 1997). The peroxisomal proteases could have a function in the proteolytic cleavage of the amino-terminal presequences of these proteins upon their translocation into peroxisomes.

RESPONSE OF LEAF PEROXISOMES TO DIFFERENT PLANT STRESS SITUATIONS

In plants, reactive oxygen species such as $O_2^{\cdot -}$ and H_2O_2 are formed in different cell compartments as normal by-products of aerobic metabolism, and their level is conveniently regulated by different enzymatic and non-enzymatic antioxidants (Boiwell, 1999; Halliwell and Gutteridge, 2000; del Río *et al*, 2001). However, in most biotic and abiotic stress conditions an overproduction of ROS has been demonstrated and these species are thought to be responsible for the oxidative damage associated with plant stress. Conditions which induce the generation of ROS in plants include pathogen infection, exposure to high light intensities, drought and salt stress, low and high temperature exposure, heavy metals, UV radiation, air pollutants, and physical and mechanical wounding (del Río *et al*, 1999; Dat *et al*, 2000).

In plant cells, as in most eukaryotic organisms, peroxisomes are probably the major sites of intracellular H_2O_2 production, as a result of the essentially oxidative type of metabolism of these organelles. Superoxide radicals generated in peroxisomes are finally converted into hydrogen peroxide, but in these organelles H_2O_2 is also produced in the photorespiration glycolate oxidase reaction, fatty acid β-oxidation, and the enzymatic reaction of flavin oxidases (Huang *et al*, 1983; del Río *et al*, 1992). Under normal physiological conditions, the production by peroxisomes of the ROS H_2O_2 and $O_2^{\cdot -}$ should be adequately controlled by catalase and ascorbate peroxidase, and SOD, respectively, which are present in peroxisomes. However, the risk of serious damage can arise when, under stress conditions the peroxisomal generation of ROS is enhanced and the protective antioxidative systems of the organelle are depressed. This alteration of the pro-oxidant/antioxidant balance can bring about extensive oxidative damage in the plant cell. The metabolism of ROS in leaf peroxisomes from pea plants subjected to different types of stress, including xenobiotics, heavy metals, salinity, and also during leaf senescence, will be discussed in this section.

Xenobiotics

Clofibrate

In animals, a variety of xenobiotics, mainly hypolipidemic drugs such as clofibrate, and certain herbicides and phthalate-ester plasticizers, induce the proliferation of the peroxisomal population as well as the activity of the H_2O_2-producing acyl-CoA oxidase in liver (Reddy *et al*, 1987; Reddy and Chu, 1996). In mussels, the proliferation of peroxisomes has been proposed as a specific biomarker of pollution by xenobiotics in marine and estuarine environments (Cancio and Cajaraville, 2000).

The existence of a relationship between clofibrate-induced peroxisome proliferation and oxidative stress mediated by reactive oxygen species was studied in leaf peroxisomes purified from pea plants (Palma *et al*, 1991). Incubation of leaves with 1 mM clofibrate (ethyl-α-*p*-

chlorophenoxyisobutyrate) produced a remarkable increase in the peroxisomal activity of H_2O_2-producing acyl-CoA oxidase and, to a lesser extent, of $O_2^{\cdot-}$-producing xanthine oxidase. There was a complete loss of catalase activity and a decrease in Mn-SOD. Electron microscopy studies of intact leaves showed that clofibrate induced a five- and two-fold proliferation of the peroxisomal and mitochondrial populations, respectively, as well as considerable alterations in the ultrastructure of cells (Palma *et al*, 1991). These results showed that the hypolipidemic drug clofibrate produced similar effects in plant tissues to those described in animals, namely proliferation of peroxisomes and mitochondria (Reddy *et al*, 1987; Reddy and Chu, 1996; Goel *et al*, 1986).

In peroxisomal membranes, clofibrate produced a large decrease in the content of the 43- and 32-kDa PMPs and an increase in the content of PMP29 (López-Huertas *et al*, 1995). This means that this xenobiotic strongly depresses the activity of MDHAR (PMP32) which, as indicated in a previous section, is an important antioxidative enzyme of the ascorbate-glutathione cycle. On the other hand, the increase in the content of PMP29 suggests that this polypeptide may be related to the NADPH:cytochrome P-450 which has been found in plant peroxisomes and is involved in processes of xenobiotic detoxification (Donaldson and Luster, 1991). The internal H_2O_2 concentration of leaf peroxisomes, the NADH-induced generation of $O_2^{\cdot-}$ radicals and the lipid peroxidation of peroxisomal membranes were increased by treatment with clofibrate. Considering the superoxide-producing electron transport chain of peroxisomal membranes, this would imply that clofibrate promotes the extrusion of membrane-generated $O_2^{\cdot-}$ radicals into the cytosol, which would join the H_2O_2 leaking out of peroxisomes as a result of the catalase activity depletion. This would pose a very serious situation not only for peroxisomes, but also for other cell organelles, such as mitochondria, nuclei and chloroplasts, due to the risk of generation of the highly reactive hydroxyl radicals ($^{\cdot}OH$) (Halliwell and Gutteridge, 2000). This effect is accompanied by an inhibition of catalase and Mn-SOD, two important peroxisomal enzymatic defences against H_2O_2 and $O_2^{\cdot-}$, respectively. These results indicate that in the toxicity of clofibrate, at the level of peroxisomes, an oxidative stress mechanism mediated by ROS is involved. Although there is very little information on the generation of superoxide radicals in purified animal peroxisomes, it seems reasonable that a mechanism similar to the one found in plant peroxisomes could also be operative in peroxisomes of those animals in which clofibrate and other xenobiotics have been demonstrated to induce the proliferation of hepatic peroxisomes and be carcinogenic (Goel *et al*, 1986; Reddy and Chu, 1996).

Herbicides

Chlorinated phenoxyacid herbicides are used in agriculture for the control of unwanted weeds. These compounds comprise a broad family of phytotoxic substances including 2,4,5-trichlorophenoxyacetic acid (2,4,5-T) and 2,4-dichlorophenoxyacetic acid (2,4-D). The effect of 2,4,5-T on some oxidative enzymes of the peroxisomal metabolism was studied in germinating Norway spruce (Segura-Aguilar *et al*, 1995). In crude peroxisomal fractions from cotyledons of seedlings treated with 200 μM 2,4,5-T, large increases in fatty acyl-CoA oxidase and catalase activities were found and, to a lesser extent, in SOD activity. These results suggested the involvement of oxidative stress in the effect of 2,4,5-T on Norway spruce seedlings (Segura-Aguilar *et al*, 1995). However, organelle counting of tissue sections by electron microscopy did not show any marked increase in the number of peroxisomes in seedlings treated with 2,4,5-T (Segura-Aguilar *et al*, 1995).

The herbicide 2,4-D (2,4-dichlorophenoxyacetic acid) also has auxinic properties and in cell cultures is used at low concentrations as stimulator of somatic embryogenesis (Bronsema *et al*, 1998). In animal tissues it has been reported that 2,4-D produces changes in the oxidative metabolism of peroxisomes and induces the proliferation of these organelles (Abdellatif *et al*, 1990). The study of the effect of 2,4-D on the oxidative metabolism of leaf peroxisomes purified from pea plants has been recently carried out in our laboratory. The herbicide was applied either by spraying leaves with a single dose of 22.6 mM 2,4-D or by growing plants in nutrient solutions containing 45.2 mM 2,4-D, and leaf peroxisomes were purified from the treated pea plants. In sprayed leaves, 2,4-D brought about a decrease of catalase (activity and protein) but, on the contrary, important enhancements were observed in the activities of $O_2^{\cdot-}$-producing XOD and H_2O_2-producing acyl-CoA oxidase. The activity of the glyoxylate cycle enzyme, isocitrate lyase, was augmented two to five times in peroxisomes of herbicide-treated plants and, to a lesser extent (50-70%), the activities of glutathione reductase and Mn-SOD. The effect of the herbicide on the peroxisomal population of pea leaves was studied but no significant changes were found in the number of peroxisomes per cell section of 2,4-D-treated plants compared to control ones.

A herbicide that was demonstrated to affect the peroxisomal population of leaves is N,N-dimethyl-N'-[4-(1-methylethyl)phenyl]urea (isoproturon) (de Felipe *et al*, 1988). This is a phenylurea herbicide used for the control of weeds in cereal crops. In plants of *Lolium rigidum* which were grown with nutrient solutions containing 34 μM isoproturon, it was found that the number of peroxisomes in mesophyll cells increased nearly two-fold. This herbicide also affected the normal position of peroxisomes which, in the control mesophyll cells appeared apressed to the chloroplast membrane, as is common in most photosynthetic cells, but the application of isoproturon drastically reduced the frequency of the peroxisome-chloroplast associations detectable in electron micrographs which are common in leaf cells (de Felipe *et al*, 1988). However, no studies were conducted on the effect of isoproturon on the metabolism of ROS in peroxisomes.

Metal toxicity

Ions of transition metals, such as iron, copper and manganese, are involved in many free radical reactions that often lead to the generation of very reactive ROS, like ·OH, from less reactive ones (Halliwell and Gutteridge, 2000). Experimental evidence linking plant peroxisomes with metal toxicity situations have been presented, mainly dealing with copper and cadmium.

Copper

The effect *in vivo* of high nutrient levels of copper (240 μM) on the activity of different metalloenzymes, including SOD isozymes, distributed in chloroplasts, peroxisomes, and mitochondria, was studied in leaves of two cultivars of pea plants with different sensitivity to copper (Palma *et al*, 1987). The activity of mitochondrial SOD isozymes was very similar in Cu-tolerant and Cu-sensitive plants, and chloroplastic SOD activity was the same in the two varieties. In these two cellular organelles the production of superoxide radicals, and the induction of SOD isozymes in response to enhanced rates of intracellular $O_2^{\cdot -}$ production, has been described in different organisms (del Río *et al*, 1991; Bowler *et al*, 1992; Halliwell and Gutteridge, 2000). The fact that the activity levels of mitochondrial and chloroplastic SODs are very similar to those of pea plants grown under optimum copper nutrition suggests that, in the two cultivars studied, Cu toxic levels apparently did not bring about $O_2^{\cdot -}$-derived induction of SOD isozymes. In contrast, the peroxisomal Mn-SOD was considerably higher in Cu-tolerant than in Cu-sensitive plants, and the activity of catalase was also increased in peroxisomes of Cu-tolerant plants (Palma *et al*, 1991).

The higher activities of the peroxisomal enzymes (Mn-SOD and catalase) found in Cu-tolerant plants suggests: (i) the involvement of reactive oxygen intermediates ($O_2^{\cdot -}$, ·OH) in the mechanism of Cu lethality in peroxisomes;

and (ii) that peroxisomes, and reactive oxygen-related enzymes in particular, might have a function in the molecular mechanisms responsible for the plant tolerance to Cu toxicity. An increase in the peroxisomal concentration of copper could, under appropriate conditions, originate the generation of $^{\cdot}OH$ radicals (Halliwell and Gutteridge, 2000). Therefore, Cu-tolerant plants could have evolved a protection mechanism against the production in peroxisomes of $O_2^{\cdot-}$-dependent toxic species by high levels of copper by inducing the peroxisomal Mn-SOD and catalase activities. In this way, $O_2^{\cdot-}$ radicals and H_2O_2 could be effectively removed, avoiding the eventual formation of the highly toxic $^{\cdot}OH$ radicals.

Cadmium

Cadmium is a toxic heavy-metal for humans, animals and plants, and is one of the widespread trace pollutants with a long biological half-life (Wagner, 1993). In recent years, diverse studies mainly in animal tissues have suggested that oxidative stress could be involved in the toxicity of cadmium (Stochs and Bagchi, 1995; Shaw, 1995). To investigate the possible involvement of peroxisomes in the oxidative stress induced by cadmium, the effect of growing pea plants with 50 μM $CdCl_2$ was studied in whole plants (Sandalio *et al*, 2001) and in organelles isolated from pea leaves. The responses of different enzymes of the peroxisomal metabolism, antioxidative enzymes, proteases, and ROS levels in peroxisomes were analyzed (Romero-Puertas *et al*, 1999; McCarthy *et al*, 2001). In peroxisomes from plants treated with cadmium, an increase in the total protein concentration and in the content of H_2O_2 in these organelles was found, but in peroxisomal membranes no significant effect on the NADH-dependent $O_2^{\cdot-}$ production was observed. The activity of glycolate oxidase, which generates H_2O_2, was significantly enhanced by the exposure to Cd, and the protein content of this enzyme, measured by Western blotting, was also higher in metal-treated than in control plants. The activity of xanthine oxidase, which is responsible for $O_2^{\cdot-}$ production in peroxisomal matrices, was slightly increased by exposure to Cd.

In pea leaf peroxisomes, five isoforms of catalase (CAT 1-5) have been detected by isoelectric focusing (Corpas *et al*, 1999b). The activity bands corresponding to CAT3, CAT4 and CAT5 were more intense in peroxisomes from Cd-treated plants. The determination of enzymes involved in the ascorbate-glutathione cycle of pea leaf peroxisomes showed a significant increment of ascorbate peroxidase and glutathione reductase as a result of Cd treatment. Cadmium also produced an increase of about two-fold in the activity of G6PDH and 6PGDH in the peroxisomal matrix, and also of ICDH. In the case of G6PDH and ICDH, the activity enhancements were due to an increase in their proteins contents, as checked by Western blotting (Romero-Puertas *et al*, 1999). The glyoxylate cycle enzymes, malate synthase and isocitrate lyase, whose activity is normally very low in leaf peroxisomes, were also enhanced by Cd treatment. The activity of the endogenous proteases of leaf peroxisomes (aminopeptidase and endopeptidase activity) was also increased in Cd-treated plants (McCarthy *et al*, 2001).

Taken together, these results indicate that the main effect of cadmium in leaf peroxisomes is an enhancement of the H_2O_2 concentration of these organelles. Peroxisomes respond to Cd toxicity by increasing the activity of antioxidative enzymes involved in the ascorbate-glutathione cycle and the NADP-dependent dehydrogenases located in these organelles. The enhancement in the activity of the glyoxylate cycle enzymes suggests that Cd induces senescence symptoms in leaf peroxisomes and, probably, a metabolic transition of leaf peroxisomes into glyoxysomes, with a participation of the peroxisomal proteases in all these Cd-induced metabolic changes (McCarthy *et al*, 2001). An important conclusion is that the antioxidative enzymes of peroxisomes appear to be a good choice to design molecular strategies directed to improve the tolerance of plants to heavy-metals.

The ultrastructural analysis of pea leaves showed that exposure to Cd resulted in disorganisation of the chloroplast structure, the formation of vesicles in the vacuoles, and structures containing peroxisomes and mitochondria near the plasma membrane (Scandalio *et al*, 2001). Moreover, cadmium treatment produced a slight increase in the number of peroxisomes per cell (Romero-Puertas *et al*, 1999). This suggests that the increase of peroxisomal population could contribute to the toxicity of cadmium by inducing oxidative stress as a result of the increased H_2O_2 production, such as was demonstrated in the yeast *Candida albicans* (Chen *et al*, 1995).

Salinity

It is known that in plants salinity can induce alterations in the metabolism of proteins and nucleic acids and can affect important processes such as photosynthesis and respiration (see Hernández *et al*, 1993, 1995). The effect of salt stress on the metabolism of reactive oxygen was studied in peroxisomes purified from leaves of two cultivars of pea plants with different sensitivity to NaCl (Corpas *et al*, 1993b). Catalase activity was inhibited by NaCl in both cultivars of *Pisum sativum* L., and salinity brought about a significant decrease of urate oxidase and hydroxypyruvate reductase activities in peroxisomes from NaCl-sensitive plants. Conversely, the activity of glycolate oxidase was stimulated by NaCl in salt-tolerant plants. Since glycolate oxidase is a key enzyme of the glycolate pathway of photorespiration (Huang *et al*, 1983), this seems to be an indication that this process is affected by salinity. When the CO_2 concentration inside the chloroplast decreases, there is also a lower availability of $NADP^+$ to accept electrons from photosystem I so that they can be transferred to O_2 with the concomitant generation of reactive oxygen species (Halliwell, 1982). It has been suggested that the function of the photorespiratory pathway in C_3 plants is to prevent the photooxidative damage that would result from excessive oxygen-radical production, by continually recycling CO_2 so that chloroplasts are never depleted of this molecule (Halliwell, 1982; Halliwell and Gutteridge, 2000). This would explain the significant increase observed in the glycolate oxidase activity of these plants compared to the salt-sensitive cultivars.

The peroxisomal Mn-SOD and XOD activity and the NADH-dependent generation of $O_2^{\cdot-}$ radicals by peroxisomal membranes were not altered by salinity (Corpas *et al*, 1993b). This contrasts with experiments carried out with leaf chloroplasts and mitochondria from pea plants grown with NaCl. In NaCl-tolerant plants, salt stress brought about the induction of CuZn-SOD and APX activity in chloroplasts (Hernández *et al*, 1995), as well as increases in the mitochondrial production of $O_2^{\cdot-}$ radicals and a decrease of Mn-SOD activity in NaCl-sensitive plants (Hernández *et al*, 1993). The internal concentration of H_2O_2 in leaf peroxisomes was decreased by NaCl in both salt-tolerant and salt-sensitive plants, in spite of the inhibition of the catalase activity and the increase in the H_2O_2-producing glycolate oxidase activity. These results suggest that H_2O_2 could diffuse out into the cytosol as a result of NaCl-induced leakiness of the peroxisomal membrane. In pea plants it was demonstrated that salt-stress produces a reduction in the 43-kDa PMP and an increase in the 10-kDa PMP (López-Huertas *et al*, 1995). The decrease of the 43-kDa peroxisomal membrane polypeptide could be partly responsible for the enhancement of the peroxisomal membrane permeability produced by salt stress.

On the other hand, the inhibition of urate oxidase activity by NaCl in NaCl-sensitive plants is indicative of a salinity effect on the metabolism of ureides (Corpas *et al*, 1993b). Urate oxidase catalyses the oxidation of uric acid by O_2 with production of the ureide allantoin and hydrogen peroxide (Huang *et al*, 1983). The inhibitory effect of salinity on urate oxidase in salt-sensitive plants indicates that the production of allantoin, at the level of peroxisomes, is depressed by NaCl, and NaCl-tolerant plants could have evolved molecular strategies of defence in order to keep active the biogenesis of allantoin in leaf peroxisomes. More studies are necessary to understand the function of ureide allantoin produced inside leaf peroxisomes, in the metabolism of leaf cells .

Senescence

During leaf senescence dramatic increases in lipid peroxidation and membrane leakiness occur and these changes are mainly due to the strong enhancement in the generation of reactive oxygen species that takes place during this physiological stage (Thompson *et al*, 1987; see del Río *et al*, 1998a). In these conditions, the generation of reactive oxygen species overcomes the defence provided by the cellular antioxidant systems, and oxidative stress is produced.

In senescent leaves strong changes were observed in many enzymes of peroxisomes purified from pea leaves (Pastori and del Río, 1994a, 1994b). Glycolate oxidase and hydroxypyruvate reductase decreased markedly and were hardly detectable when senescence was advanced. The glyoxylate cycle enzymes, malate synthase and isocitrate lyase, which could not be detected in young leaves, increased drastically (Pastori and del Río 1994a, 1997). A transition of leaf peroxisomes into glyoxysomes during leaf senescence has been observed by different authors (De Bellis *et al*, 1990; Landolt and Matile, 1990; Vicentini and Matile, 1993; Nishimura *et al*, 1993, 1996; Pastori and del Río, 1994a; Pistelli *et al*, 1996). The results obtained in peroxisomes from senescent pea leaves on the induction of malate synthase and isocitrate lyase (Pastori and del Río, 1994a, 1997), agree with results reported for peroxisomes from different senescent plants (Pistelli *et al*, 1996, and refs. therein) and support the idea that leaf senescence is associated with the reverse transition of leaf peroxisomes to glyoxysomes, with the channeling of acetyl-CoA through the glyoxylate cycle. The enhancement by senescence of the fatty acid β-oxidation and the glyoxylate cycle activity of leaf peroxisomes could be a means of converting thylakoidal lipids into sugars to be used as building blocks or for respiration in younger leaves or storage tissues (Landolt and Matile, 1990; del Río *et al*, 1998a).

Ultrastructural studies of senescent pea leaves show that whereas chloroplasts are gradually altered and degraded, peroxisomes remain intact, and their population, together with that of mitochondria, increases about four and five times, respectively, compared with young leaves (Pastori and del Río, 1994a). The proliferation of the peroxisomal population during senescence was first demonstrated in carnation petals (Droillard and Paulin, 1990) and is also induced by different abiotic stress conditions (del Río *et al*, 1996, 1998a, 2001; Corpas *et al*, 2001).

Senescent pea leaves contain two populations of peroxisomes with different equilibrium densities (Pastori and del Río, 1997). The characteristic glyoxysomal enzymes malate synthase and isocitrate lyase are found in both populations. When analysed by electron microscopy, peroxisomes from the first population have the typical size but a lower matrix electron density than peroxisomes from young leaves. In contrast, peroxisomes from the higher-density peak are smaller and have a higher matrix electron density (Pastori and del Río, 1997). In animals, several cases of peroxisomal heterogeneity have been reported as a result of different treatments (for references see Pastori and del Río, 1997, and refs. therein; Völkl *et al*, 1999). As to the peroxisomal SOD isozymes, the constitutive Mn-SOD activity of leaf peroxisomes increased significantly in senescent leaves and two new CuZn-SODs were apparently induced, one of which was recognized by an antibody against glyoxysomal CuZn-SOD (Pastori and del Río, 1994b, 1997). The constitutive Mn-SOD was present in the lower density peroxisomal peak, whereas the new CuZn-SODs occurred predominantly in the higher density peroxisomal peak (Pastori and del Río, 1997).

In conclusion, senescence brings about important alterations in the oxidative metabolism, SOD isozymes, and ascorbate-glutathione cycle of peroxisomes, as well as in the quantity and quality of the peroxisomal population. The senescence-induced changes in the activated oxygen metabolism of peroxisomes are mainly characterized by the disappearance of catalase activity and an overproduction of $O_2^{\cdot-}$ and H_2O_2 (Pastori and del Río, 1994a, 1997). This accumulation of ROS can only be partly counteracted by the peroxisomal ascorbate-glutathione cycle, since this is also negatively affected by senescence. In dark-induced senescent leaves the peroxisomal APX and MDHAR activities were notably decreased, but DHAR was considerably enhanced (Jiménez *et al*, 1998b). With regard to the peroxisomal antioxidants, whereas the ascorbate content was only slightly increased by senescence, the total glutathione content augmented about 20 times. The predominant form of glutathione during senescence was GSSG, resulting in a forty three-fold increase in the GSSG/GSH ratio in peroxisomes (Jiménez *et al*, 1998b). This suggests that glutathione and part of the ascorbate-glutathione cycle can have a protective role for the elimination of H_2O_2 in peroxisomes as senescence proceeds and catalase disappears. It is well known that glutathione is very important for antioxidant defence, redox balance and the regulation of gene expression in plants (Foyer *et al*, 1997).

Since $O_2^{\cdot -}$ radicals under physiological conditions quickly dismutate into H_2O_2 and O_2, the final result of senescence is a build-up in leaf peroxisomes of the more stable metabolite H_2O_2, which can diffuse into the cytosol. This represents a serious situation not only for peroxisomes, but also for other cell organelles such as mitochondria, nuclei and chloroplasts, because of the possible formation of the strongly oxidising $^{\cdot}OH$ radicals by the metal-catalyzed reaction of H_2O_2 with $O_2^{\cdot -}$. In other words, peroxisomes appear to have a ROS-mediated role in the oxidative reactions characteristic of senescence. On the other hand, the increase in peroxisomal xanthine oxidase and urate oxidase activities in senescent leaves suggests a function for leaf peroxisomes in the catabolism of purines resulting from senescence-induced nucleic acid degradation (Corpas *et al*, 1997; del Río *et al*, 1998b).

Other stress situations

The role of catalase in the effect of hydrogen peroxide stress (1-100 mM H_2O_2) in plants has been demonstrated in transgenic tobacco plants with about 10% of wild-type catalase activity (Willekens *et al*, 1997). Catalase was found to be crucial for maintaining the redox balance during oxidative stress by H_2O_2 (Willekens *et al*, 1997). Catalase-deficient plants were subjected to other stress conditions and were found to be susceptible to paraquat, salt and ozone, demonstrating that catalase was critical for the cellular defence against these H_2O_2-dependent stresses (Willekens *et al*, 1997).

The induction of peroxisome biogenesis genes (*PEX*) by hydrogen peroxide has been recently demonstrated in both plant and animal cells (López-Huertas *et al*, 2000). Using *Arabidopsis thaliana* plants transformed with a *PEX*1 promoter-luciferase reporter, rapid local and systemic induction of *PEX*1-luciferase could be demonstrated *in vivo* in response to 1 mM H_2O_2 concentrations. *PEX*1-luciferase was also induced *in vivo* in naturally occurring stress situations such as wounding and infection with an avirulent pathogen. On the basis of these results a model was proposed whereby diverse stresses that generate H_2O_2 as a signalling molecule result in peroxisome proliferation via the up-regulation of *PEX* genes required for biogenesis of the organelle and import of proteins (López-Huertas *et al*,

2000). This model would predict that other stress conditions which generate H_2O_2 would also result in *PEX* gene induction and peroxisome proliferation. According to this model, the peroxisome proliferation by the signalling molecule H_2O_2 might be a common mechanism of protection against oxidative stress and restoration of the cellular redox balance, by making use of the enzymatic and non-enzymatic antioxidants of peroxisomes (López-Huertas *et al*, 2000).

PEROXISOMES AS A SOURCE OF ROS AND NITRIC OXIDE SIGNAL MOLECULES

Nitric oxide ($NO^{\cdot}$) is a widespread intra- and inter-cellular messenger in vertebrates with a broad spectrum of regulatory functions in the central nervous, cardiovascular, and immune systems (Moncada *et al*, 1991; Knowles and Moncada, 1994). Research on $NO^{\cdot}$ in plants has gained considerable attention in recent years and there is increasing evidence of a role of this molecule as an endogenous plant growth regulator as well as a signal molecule in the transduction pathways leading to the induction of defence responses against pathogens and in damage leading to cell death (Van Camp *et al*, 1998; Delledonne *et al*, 1998; Durner and Klessig, 1999; McDowell and Dangl, 2000; Grant and Loake, 2000; Leshem, 2000; Pedroso *et al*, 2000). In addition, low concentrations of $NO^{\cdot}$ have also been reported to have antioxidant properties in cytotoxic processes mediated by ROS in plant tissues (Beligni and Lamattina, 1999). As a consequence of the physiological importance of the free radical $NO^{\cdot}$, numerous studies have been focused on the enzyme responsible for its endogenous production. Nitric oxide synthase (NOS; EC 1.14.13.39) catalyses the oxygen- and NADPH-dependent oxidation of L-arginine to $NO^{\cdot}$ and citrulline in a complex reaction requiring FAD, FMN, and tetrahydrobiopterin, and in some cases also Ca^{2+} and calmodulin (Knowles and Moncada, 1994). In plant extracts, NOS-like activity has been detected mainly during the interaction *Rhyzobium*-legume (Cueto *et al*, 1996) and fungi-plants (Ninnemann and Maier, 1996), in soybean cell suspensions and *Arabidopsis* (Delledonne *et al*, 1998), and in tobacco leaves (Durner *et al*, 1998). The presence of NOS in the cytosol and nucleus of maize cells has been reported (Ribeiro *et al*, 1999).

The presence of NOS has been demonstrated in peroxisomes from leaves of pea plants (Barroso *et al*, 1999). For this localization of NOS in plant peroxisomes four complementary approaches were used: (i) distribution of NOS activity in leaf peroxisomes purified by sucrose density-gradient centrifugation, (ii) sensitivity of NOS activity to different well characterised inhibitors of mammalian NOSs, (iii) cross-reactivity of peroxisomes on Western blot with antibodies against murine iNOS, and (iv) immunogold electron microscopy analysis of intact plant tissue. The peroxisomal NOS had a subunit molecular mass of about 130-kDa, was calcium-dependent, constituvely expressed, and immunorelated with the mammalian inducible NOS (Barroso *et al*, 1999).

The presence of NOS in peroxisomes suggests that these organelles are a cellular source of $NO^{\cdot}$. Taking together this fact and the superoxide radical generating systems and diverse antioxidants of peroxisomes, a model has been designed of peroxisomal metabolism (Figure 3) that shows that these organelles can release several signal molecules, such as H_2O_2, $O_2^{\cdot -}$ and $NO^{\cdot}$, to the cytosol. Nitric oxide produced by the enzymatic reaction of NOS can react with $O_2^{\cdot -}$ radicals generated in the peroxisomal matrix by xanthine oxidase (XOD) to form the powerful oxidant peroxynitrite ($ONOO^-$), which according to Sakuma *et al* (1997) can regulate the conversion of xanthine dehydrogenase (XDH) into the superoxide-generating XOD. On the other hand, $NO^{\cdot}$ in the presence of O_2 can react with reduced glutathione, also present in peroxisomes (Jiménez *et al*, 1997), to form *S*-nitrosoglutathione (GSNO), a reactive nitrogen oxide species (Wink *et al*, 1996). In animal systems, GSNO has been reported to function as an inter- and intracellular $NO^{\cdot}$ carrier, and in plants GSNO was found to be a powerful inducer of defence genes (Durner *et al*, 1998). As hypothesised by Durner and Klessig (1999), GSNO could function as a long distance signal molecule, transporting glutathione-bound $NO^{\cdot}$ throughout the plant. In this mechanism, leaf peroxisomes could participate through the endogenous production of GSNO which could diffuse to the cytosol.

$NO^{\cdot}$ can also diffuse through the peroxisomal membrane to the cytosol, where it could react with $O_2^{\cdot -}$ produced in the cytosolic side of the membrane thus generating the oxidant peroxynitrite in the cytosol. However, a modulation by $NO^{\cdot}$ of the endogenous antioxidant enzymes of peroxisomes cannot be ruled out. In animal systems, catalase and glutathione peroxidase activity are down-regulated by $NO^{\cdot}$, whereas the activity of the peroxisomal H_2O_2-producing β-oxidation is enhanced by $NO^{\cdot}$ (Dobashi *et al*, 1997).

Figure 3: Model for the function of NOS in leaf peroxisomes and the role of these organelles in the generation of the signal molecules hydrogen peroxide (H_2O_2), superoxide radicals ($O_2^{\cdot-}$), nitric oxide ($NO^{\cdot}$), and possibly S-nitrosoglutathione (GSNO)

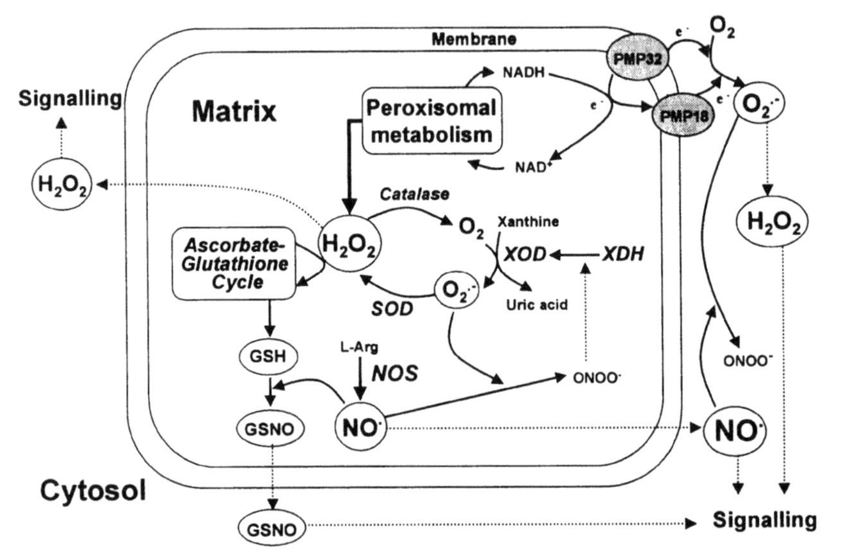

***Figure 3**. PMP32, 32-kDa peroxisomal membrane polypeptide. PMP18, 18-kDa peroxisomal membrane polypeptide. XOD, xanthine oxidase. XDH, xanthine dehydrogenase. L-Arg, L-arginine. $ONOO^-$, peroxynitrite. GSH, reduced glutathione. GSNO, S-nitrosoglutathione.*

Under normal physiological conditions, the peroxisomal level of the ROS H_2O_2 and $O_2^{\cdot-}$ should be adequately controlled by catalase and APX, and SOD, respectively, which are present in peroxisomes. However, catalase is known to be inactivated by light and different stress conditions that suppress protein synthesis (Schäfer and Feierabend, 2000). Superoxide radicals also inhibit catalase activity (Kono and Fridovich, 1982) and it has recently been reported that $NO^{\cdot}$ and peroxynitrite inhibit catalase and APX activity in tobacco plants (Clark *et al*, 2000). In addition, an enhanced synthesis of nitric oxide was found to increase the peroxisomal H_2O_2-producing β-oxidation in animal cells (Dobashi *et al*, 1997). Taken together, these data indicate that if, under any type of plant stress an induction of the peroxisomal production of $O_2^{\cdot-}$ and $NO^{\cdot}$ radicals takes place, this can lead to the inhibition of catalase and APX activities and possibly to an increase in the H_2O_2 level. This breakdown of the peroxisomal antioxidant defences would finally originate an overproduction of H_2O_2 in peroxisomes, leading to oxidative damage and, possibly, cell death.

However, the rate of ROS and $NO^{\cdot}$ generation in plant cells have an ambivalent effect. While a high cellular production of these active molecules can bring about extensive oxidative damage (Halliwell and Gutteridge, 2000), low levels of $NO^{\cdot}$ and ROS, like H_2O_2 and $O_2^{\cdot -}$, are involved as signal molecules in many physiological processes and plant stresses. Hydrogen peroxide has been described as a diffusible transduction signal in plant-pathogen interactions (Levine *et al*, 1994; Alvarez *et al*, 1998; Chamnongpol *et al*, 1998; López-Huertas *et al*, 2000), response to wounding (Rea *et al*, 1998; López-Huertas *et al*, 2000), ABA-mediated guard cell closure (Pei *et al*, 2000), osmotic stress (Guan *et al*, 2000), and excess light stress (Karpinski *et al*, 1999). In all these cases, a function for hydrogen peroxide as a signal molecule leading tc the induction of genes encoding different cellular protectants has been demonstrated. Apart from H_2O_2, $NO^{\cdot}$ and $O_2^{\cdot -}$ are key mediators of pathogen-induced programmed cell death in plants and appear to function as part of a signal transduction pathway leading to the induction of defence responses against pathogens and cell death (Delledonne *et al*, 1998; Van Camp *et al*, 1998; Durner and Klessig, 1999; McDowell and Dangl, 2000; Grant and Loake, 2000).

Accordingly, peroxisomes should be considered as cellular compartments with the capacity to generate and release into the cytosol important signal molecules such as $O_2^{\cdot -}$, H_2O_2, $NO^{\cdot}$, and possibly GSNO, which can contribute to a more integrated communication among cell compartments. The study of the molecules that mediate the inter-organellar communication within the cell is an important emerging area of plant organelle research, which can supply more information on how peroxisomal, mitochondrial and nuclear fractions are coupled throughout plant development and under different stress conditions (Mackenzie and McIntosh, 1999). The signal-producing function of plant peroxisomes is still more relevant from a physiological viewpoint considering that the population of these oxidative organelles can proliferate in plants during senescence and under different stress conditions (del Río *et al*, 2001).

CONCLUSIONS

The existence of reactive oxygen metabolism in plant peroxisomes and the presence in these organelles of a complex battery of antioxidative enzymes apart from catalase, such as SOD, the components of the ascorbate-glutathione cycle, NADP-dehydrogenases, and the $NO^{\cdot}$-generating enzyme nitric oxide synthase, emphasizes the importance of these organelles in cellular oxidative metabolism. The new enzymatic and non-enzymatic proteins recently found in plant peroxisomes evidence the importance of these organelles as a pool of metabolites shared with other cell compartments. This supports the idea postulated by Tolbert *et al* (1987) on the effect of peroxisomal metabolism on metabolic pathways in other cell compartments. The fluxes of metabolites from and to peroxisomes are facilitated by pore-forming proteins (porins) in the peroxisomal membranes, whose existence has been demonstrated in plant peroxisomes in recent years (Chapter 5, this volume, Reumann *et al*, 1995; Reumann, 2000; Corpas *et al*, 2000).

Plant peroxisomes have a ROS-mediated metabolic function in leaf senescence and in abiotic stress conditions induced by xenobiotics (clofibrate and some herbicides) and heavy metals (Cd and Cu), and probably in other types of abiotic stress. Until recent years, chloroplasts and mitochondria were considered to be almost exclusively responsible for the intracellular oxidative damage induced by different stresses. However, results currently available indicate that peroxisomes have two roles in cells which can be seen as being mutually opposed, the production of harmful reactive oxygen species and the production of signal molecules which coordinate defences against these and other stresses. In the cell biology of plant stress more consideration should be given to peroxisomes in their two antagonistic roles, as oxidative stress-generators and as a source of ROS and nitric oxide signal molecules in transduction pathways inducing defence gene expression.

There are still many unanswered questions relating to the role of peroxisomes in the metabolism of plant cells. Further research on the cloning of genes for the biosynthetic enzymes of peroxisomal ROS metabolism and endogenous proteases, and the characterization of mutants that are defective in stress responses, will throw more light on the physiological significance of peroxisomes in cellular metabolism and their function as a source of signalling molecules. This molecular information could be valuable in helping to design molecular strategies directed to improve the tolerance of plants to different biotic and abiotic stresses (see also Chapter 13).

It seems reasonable to think that a ROS, $NO^{\cdot}$ and GSNO signal molecule-producing function similar to that postulated for plant peroxisomes could also be performed by human, animal and yeast peroxisomes. The importance of human peroxisomes in biomedicine because of their association with several important genetic diseases caused by peroxisomal dysfunction (Poll-The *et al*, 1998), suggests that important research advances for human health could be derived from these studies.

ACKNOWLEDGEMENTS

The authors are grateful to the Ministry of Education and Science (DGESIC grants PB95-0004-01, PB98-0493-01, and 1FD97-0889-02), the European Union (projects CHRX-CT94-0605 and RTN1-1999-00174) and the *Junta de Andalucía* (group CVI 0192), for financial support.

REFERENCES

Abdellatif, A.G., Préat, V., Vamecq, J., Nilsson, R. and Roberfroid, M. (1990) Peroxisome proliferation and modulation of rat liver carcinogenesis by 2,4-dichlorophenoxyacetic acid, 2,4,5-trichlorophenoxyacetic acid, perfluorooctanoic acid and nafenopin. *Carcinogenesis.* **11**: 1899-1902.

Alvarez, M.E., Pennell, R.I., Meier, P-J., Ishikawa, A., Dixon, R.A. and Lamb, C. (1998) Reactive oxygen intermediates mediate a systemic signal network in the establishment of plant immunity. *Cell.* **92**: 773-784.

Asada, K. (2000) The water-water cycle as alternative photon and electron sinks. *Phil. Trans. R. Soc.* Lond. B. **355**: 1419-1431.

Barroso, J.B., Corpas, F.J., Carreras, A., Sandalio, L.M., Valderrama, R., Palma, J.M., Lupiáñez, J.A. and del Río, L.A. (1999) Localization of nitric-oxide synthase in plant peroxisomes. *J Biol Chem.* **274**: 36729- 36733.

Beligni, M.V. and Lamattina, L. (1999) Nitric oxide counteracts cytotoxic processes mediated by reactive oxygen species in plant tissues. *Planta.* **208**: 337-344.

Bolwell, G.P. (1999) Role of reactive oxygen species and NO in plant defence responses. *Curr. Opin. Plant Biol.* **2**: 287-294.

Bowditch, M.Y. and Donaldson, R.P. (1990) Ascorbate free-radical reduction by glyoxysomal membranes. *Plant Physiol.* **94**: 531-537.

Bowler, C., Van Camp, W., Van Montagu, M. and Inzé, D. (1994) Superoxide dismutase in plants. *Crit. Rev. Plant Sci.* **13**: 199-218.

Bowler, C., Van Montagu, M. and Inzé, D. (1992) Superoxide dismutase and stress tolerance. Ann. Rev. *Plant. Physiol. Plant Mol. Biol.* **43**: 83-116.

Bronsema, F.B.F, van Oostveen, W.J.F., Prinsen, E. and van Lammeren, A.A.M. (1998) Distribution of [^{14}C]dichlorophenoxyacetic acid in cultured zygotic embryos of *Zea mays* L. *J. Plant Growth Regul.* **17**: 81-88.

Bueno, P., Varela, J., Giménez Gallego, G. and del Río, L.A. (1995) Peroxisomal copper, zinc superoxide dismutase: Characterization of the isoenzyme from watermelon cotyledons. *Plant Physiol.* **108**: 1151-1160.

Bunkelmann, J. and Trelease, R.N. (1996) Ascorbate peroxidase: a prominent membrane protein in oilseed glyoxysomes. *Plant Physiol.* **110**: 589-598.

Cancio, I. and Cajaraville, M.P. (2000) Cell Biology of peroxisomes and their characteristics in aquatic organisms. *Int. Rev. Cytol.* **199**: 201-293.

Chamnongpol, S., Willekens, H., Moeder, W., Langebartels, C., Sandermann, H., Van Montagu, M., Inzé, D. and Van Camp, W. (1998) Defense activation and enhanced pathogen tolerance induced by H_2O_2 in transgenic tobacco. *Proc. Nat. Acad. Sci. USA.* **95**: 5818-5823.

Chen, T., Li, W., Schulz, P.J., Furst, A. and Chien, P.K. (1995) Induction of peroxisome proliferation and increase of catalase activity in yeast, *Candida albicans*, by cadmium. *Biol. Trace Element Res.* **50**: 125-133.

Clark, D., Durner, J., Navarre, D.A. and Klessig, D.F. (2000) Nitric oxide inhibition of tobacco catalase and ascorbate peroxidase. *Mol. Plant-Microbe Inter.* **13**: 1380-1384.

Corpas, F.J., Barroso, J.B. and del Río, L.A. (2001) Peroxisomes as a source of reactive oxygen species and nitric oxide signal molecules in plant cells. *Trends Plant Sci.* **6**: 145-150.

Corpas, F.J., Barroso, J.B., Sandalio, L.M., Palma, J.M., Lupiáñez, J.A. and del Río, L.A. (1999a) Peroxisomal NADP-dependent isocitrate dehydrogenase. Characterization and activity regulation during natural senescence. *Plant Physiol.* **121**: 921-928.

Corpas, F.J., Barroso, J.B., Sandalio, L.M., Distefano, S., Palma, J.M., Lupiáñez, J.A. and del Río, L.A. (1998b) A dehydrogenase-mediated recycling system of NADPH in plant peroxisomes. *Biochem J.* **330**: 777-784.

Corpas, F.J., de la Colina, C., Sánchez-Rasero, F. and del Río, L.A. (1997) A role for leaf peroxisomes in the catabolism of purines. *J. Plant Physiol.* **151**: 246-250.

Corpas, F.J., Gómez, M., Hernández, J.A. and del Río, L.A. (1993b) Metabolism of activated oxygen in peroxisomes from two *Pisum sativum* L. cultivars with different sensitivity to sodium chloride. *J. Plant Physiol.* **141**: 160-165.

Corpas, F.J., Palma, J.M. and del Río, L.A. (1993a) Evidence for the presence of proteolytic activity in peroxisomes. *Eur. J. Cell Biol.* **61**: 81-85.

Corpas, F.J., Palma, J.M., Sandalio, L.M., López-Huertas, E., Romero-Puertas, M.C., Barroso, J.B. and del Río, L.A. (1999b) Purification of catalase from pea leaf peroxisomes: Identification of five different isoforms. *Free Rad. Res.* **31**: S235-241.

Corpas, F.J., Sandalio, L.M., Brown, M.J., del Río, L.A. and Trelease, R.N. (2000) Identification of porin-like polypeptide(s) in the boundary membrane of oilseed glyoxysomes. *Plant Cell Physiol.* **41**: 1218-1228.

Corpas, F.J., Sandalio, L.M., del Río, L.A. and Trelease, R.N. (1998a) Copper-zinc superoxide dismutase is a constituent enzyme of the matrix of peroxisomes in the cotyledons of oilseed plants. *New Phytol.* **138**: 307-314.

Cueto, M., Hernández-Perera, O., Martin, R., Bentura, M.L., Rodrigo, J., Lamas, S. and Golvano, M.P. (1996) Presence of nitric oxide synthase activity in roots and nodules of *Lupinus albus*. *FEBS Lett.* **398**: 159-164.

Dat, J., Vandenabeele, S., Vranová, E., Van Montagu, M., Inzé, D. and Van Breusegem, F. (2000) Dual action of the active oxygen species during plant stress responses. *Cell Mol. Life Sci.* **57**: 779-795.

de Bellis, L., Picciarelli, P., Pistelli, L. and Alpi, A. (1990) Localization of glyoxylate-cycle enzymes in peroxisomes of senescent leaves and green cotyledons. *Planta.* **180**: 435-439.

de Felipe, M.R., Lucas, M.M. and Pozuelo, J.M. (1988) Cytochemical study of catalase and peroxidase in the mesophyll of *Lolium rigidum* plants treated with isoproturon. *J. Plant Physiol.* **132**: 67-73.

Delledonne, M., Xia, Y.J., Dixon, R.A. and Lamb, C. (1998) Nitric oxide functions as a signal in plant disease resistance. *Nature.* **394**: 585-588.

del Río, L.A. and Donaldson, R.P. (1995) Production of superoxide radicals in glyoxysomal membranes from castor bean endosperm. *J. Plant Physiol.* **146**: 283-287.

del Río, L.A., Corpas, F.J., Sandalio, L.M., Palma, J.M., Gómez, M. and Barroso, J.B. (2001) Reactive oxygen species, antioxidant systems and nitric oxide in peroxisomes. *J. Exp. Bot.* In press.

del Río, L.A., Lyon, D.S., Olah, I., Glick, B. and Salin, M.L. (1983) Immunocytochemical evidence for a peroxisomal localization of manganese superoxide dismutase in leaf protoplasts from a higher plant. *Planta.* **158**: 216-224.

del Río, L.A., Palma, J.M., Sandalio, L.M. and Navari-Izzo, F. (eds.). (1999) Winter Meeting 1998 of the Society for Free Radical Research-Europe on *Oxygen, Free Radicals and Oxidative Stress in Plants*. *Free Rad. Res.* **31**: S1-256.

del Río, L.A., Palma, J.M., Sandalio, L.M., Corpas, F.J., Pastori, G.M., Bueno, P. and López-Huertas E. (1996) Peroxisomes as a source of superoxide and hydrogen peroxide in stressed plants. *Biochem. Soc. Trans.* **24**: 434-438.

del Río, L.A., Pastori, G.M., Palma, J.M., Sandalio, L.M., Sevilla, F., Corpas FJ, Jiménez, A., López-Huertas, E. and Hernández, J.A. (1998a) The activated oxygen role of peroxisomes in senescence. *Plant Physiol.* **116**: 1195-1200.

del Río, L.A., Sandalio, L.M., Corpas, F.J., López-Huertas, E., Palma, J.M. and Pastori, G.M. (1998b) Activated oxygen-mediated metabolic functions of leaf peroxisomes. *Physiol. Plant.* **104**: 673-680.

del Río, L.A., Sandalio, L.M., Palma, J.M., Bueno, P. and Corpas, F.J. (1992) Metabolism of oxygen radicals in peroxisomes and cellular implications. *Free Rad. Biol. Med.* **13**: 557-580.

del Río, L.A, Sevilla, F., Sandalio, L.M. and Palma, J.M. (1991) Nutritional effect and expression of SODs: induction and gene expression; diagnostics; prospective protection against oxygen toxicity. *Free Rad Res.* **12-13**: 819-827.

Distefano, S., Palma, J.M., Gómez, M. and del Río, L.A. (1997) Characterization of endopeptidases from plant peroxisomes. *Biochem J.* **327**: 399-405.

Distefano, S., Palma, J.M., McCarthy, I. and del Río, L.A. (1999) Proteolytic cleavage of plant proteins by peroxisomal endoproteases from senescent pea leaves. *Planta.* **209**: 308-313.

Dobashi, K., Pahan, K., Chahal, A. and Singh, I. (1997) Modulation of endogenous antioxidant enzymes by nitric oxide in rat C_6 glial cells. *J. Neurochem.* **68**: 1896-1903.

Donaldson, R.P. (1982) Nicotinamide cofactors (NAD and NADP) in glyoxysomes, mitochondria and plastids isolated from castor bean endosperm. *Arch. Biochem. Biophys.* **215**: 274-279.

Donaldson, R.P. and Luster, D. (1991) Multiple forms of plant cytochromes P_{450}. *Plant Physiol.* **96**: 669- 674.

Droillard, M-J. and Paulin, A. (1990) Isozymes of superoxide dismutase in mitochondria and peroxisomes isolated from petals of carnation (*Dianthus caryophyllus*) during senescence. *Plant Physiol.* **94**: 1187-1192.

Durner, J. and Klessig, D.F. (1999) Nitric oxide as a signal molecule in plants. *Curr. Opin. Plant Biol.* **2**: 369-374.

Durner, J., Wendehenne, D. and Klessig, D.F. (1998) Defence gene induction in tobacco by nitric oxide, cyclic GMP, and cyclic ADP-ribose. *Proc. Natl. Acad. Sci. USA* **95**: 10328-10333.

Fang, T.K., Donaldson, R.P. and Vigil, E.L. (1987) Electron transport in purified glyoxysomal membranes from castor bean endosperm. *Planta.* **172**: 1-13.

Foyer, C.H. and Halliwell B. (1976) Presence of glutathione and glutathione reductase in chloroplasts: A proposed role in ascorbic acid metabolism. *Planta.* **133**: 21-25.

Foyer, C.H., López-Delgado, H., Dat, J.F. and Scott, I.M. (1997) Hydrogen peroxide- and glutathione-associated mechanisms of acclimatory stress tolerance and signalling. *Physiol. Plant.* **100**: 241-254.

Fridovich, I. (1986) Superoxide dismutases. *Adv. Enzymol. Rel. Areas Mol. Biol.* **58**: 61-97.

Fujiki, Y., Tsuneoka, M. and Tahiro, Y. (1989) Biosynthesis of non-specific lipid transfer protein (sterol carrier protein 2) on polyribosomes as a larger precursor in rat liver. *J. Biochem.* **106**: 1126-1131.

Gietl, C. (1990) Glyoxysomal malate dehydrogenase from watermelon is synthesized with an amino-terminal transit peptide. *Proc. Natl. Acad. Sci. USA.* **87**: 5733-5777.

Gietl, C., Wimmer, B., Adamec, J. and Kalousek, F. (1997) A cysteine endopeptidase isolated from castor bean endosperm microbodies processes the glyoxysomal malate dehydrogenase precursor protein. *Plant Physiol.* **113**: 863-871.

Goel, S.K., Lalwani, N.D. and Reddy, J.K. (1986) Peroxisome proliferation and lipid peroxidation in rat liver. *Cancer Res.* **46**: 1324-1330.

Grant, J.J. and Loake, G.J. (2000) Role of reactive oxygen intermediates and cognate redox signalling in disease resistance. *Plant Physiol.* **124**: 21-29.

Gray, R.W., Arsensis, C. and Jeffay, H. (1970) Neutral protease activity associated with the rat liver peroxisomal fraction. *Biochim. Biophys. Acta.* **222**: 627-636.

Guan, L.M., Zhao, J. and Scandalios, J. (2000) *Cis* elements and *trans*-factors that regulate expression of the maize Cat 1 antioxidant gene in response to ABA and osmotic stress: H_2O_2 is the likely intermediary signalling molecule for the response. *Plant J.* **22**: 87-95.

Halliwell, B. (1982) The toxic effects of oxygen on plant tissues. In *Superoxide Dismutase.* (Oberley, L.W. ed.). Vol. 1. pp. 89-123. Boca Ratón, FL. CRC Press.

Halliwell, B. and Gutteridge, J.M.C. (2000) *Free Radicals in Biology and Medicine.* Oxford University Press, Oxford.

Hernández, J.A., Corpas, F.J., Gómez, M., del Río, L.A. and Sevilla, F. (1993) Salt-induced oxidative stress mediated by activated oxygen species in pea leaf mitochondria. *Physiol. Plant.* **89**: 103-110.

Hernández, J.A., Olmos, E., Corpas, F.J., Sevilla, F. and del Río, L.A. (1995) Salt-induced oxidative stress in chloroplasts of pea plants. *Plant Sci.* **105**: 151-167.

Hertwig, B., Streb, P. and Feierabend, J. (1992) Light dependence of catalase synthesis and degradation in leaves and the influence of interfering stress conditions. *Plant Physiol.* **100**: 1547-1553.

Huang, A.H.C., Trelease, R.N. and Moore, T.S., Jr. (1983) *Plant Peroxisomes.* New York: Academic Press.

Ishikawa, T., Yoshimura, K., Sakai, K., Tamoi, M., Takeda, T. and Shigeoka, S. (1998) Molecular characterization and physiological role of glyoxysome-bound ascorbate peroxidase from spinach. *Plant Cell Physiol.* **39**: 23-34.

Jiménez, A., Hernández, J.A., del Río, L.A. and Sevilla, F. (1997) Evidence for the presence of the ascorbate-glutathione cycle in mitochondria and peroxisomes of pea leaves. *Plant Physiol.* **114**: 275-284.

Jiménez, A., Hernández, J.A., Pastori, G.M., del Río, L.A. and Sevilla, F. (1998b) Role of the ascorbate-glutathione cycle of mitochondria and peroxisomes in the senescence of pea leaves. *Plant Physiol.* **118**: 1327-1335

Jiménez, A., Hernández, J.A., Ros Barceló, A., Sandalio, L.M., del Río, L.A. and Sevilla, F. (1998a) Mitochondrial and peroxisomal ascorbate peroxidase of pea leaves. *Physiol. Plant.* **104**: 687-692.

Karpinski, S., Reynolds, H., Karpinska, B., Wingsle, G., Creissen, G. and Mullineaux, P.M. (1999) Systemic signalling and acclimation in response to excess excitation energy in Arabidopsis. *Nature.* **284**: 654-657.

Kirkman, H.N., Rolfo, M., Ferraris, A.M. and Gaetani, G.F. (1999) Mechanisms of protection of catalase by NADPH. Kinetics and stoichiometry. *J. Biol. Chem.* **275**: 13908-13914.

Knowles, R.G. and Moncada, S. (1994) Nitric oxide synthases in mammals. *Biochem. J.* **298**: 249-258.

Kono, Y. and Fridovich, I. (1982) Superoxide radical inhibits catalase. *J. Biol. Chem.* **257**: 5751-5754.

Landolt, R. and Matile, P. (1990) Glyxoxysome-like microbodies in senescent spinach leaves. *Plant Sci.* **72**: 159-163.

Leshem, Y.Y. (2000) *Nitric Oxide in Plants. Occurrence, Function and Use.* Dordrecht, The Netherlands: Kluwer Academic Publishers.

Levine, A., Tenhaken, R., Dixon, R. and Lamb, C. (1994) H_2O_2 from the oxidative burst orchestrates the plant hypersensitive disease resistance response. *Cell.* **79**: 583-593.

López-Huertas, E., Charlton, W.L., Johnson, B., Graham, I.A. and Baker, A. (2000) Stress induces peroxisome biogenesis genes. *EMBO J.* **19**: 6770-6777.

López-Huertas, E., Corpas, F.J., Sandalio, L.M. and del Río, L.A. (1999) Characterization of membrane polypeptides from pea leaf peroxisomes involved in superoxide radical generation. *Biochem. J.* **337**: 531-536.

López-Huertas, E., Sandalio, L.M. and del Río, L.A. (1995) Integral membrane polypeptides of pea leaf peroxisomes: characterization and response to plant stress. *Plant Physiol. Biochem.* **33**: 295-302.

López-Huertas, E., Sandalio, L.M., Gómez, M. and del Río, L.A. (1997) Superoxide radical generation in peroxisomal membranes: evidence for the participation of the 18 kDa integral membrane polypeptide. *Free Rad. Res.* **26**: 497-506.

Luster, D.G. and Donaldson, R.P. (1987) Orientation of electron transport activities in the membrane of intact glyoxysomes isolated from castor bean endosperm. *Plant Physiol.* **85**: 796-800.

Mackenzie, S. and McIntosh, L. (1999) Higher plant mitochondria. *Plant Cell.* **11**: 571-585.

Martini, G. and Ursini, M.V. (1996) A new lease of life for an old enzyme. *Bioessays.* **18**: 631-637.

McCarthy, I., Romero-Puertas, M.C., Palma, J.M., Sandalio, L.M., Corpas, F.J., Gómez, M. and del Río, L.A. (2001) Cadmium induces senescence symptoms in leaf peroxisomes of pea plants. *Plant Cell Environ.* In press.

McDowell, J.M. and Dangl, J.L. (2000) Signal transduction in the plant immune response. *Trends Plant Sci.* **290**: 79-82.

Mittova, V., Volokita, M., Guy, M. and Tal, M. (2000) Activities of SOD and the ascorbate-glutathione cycle enzymes in subcellular compartments in leaves and roots of the cultivated tomato and its wild salt-tolerant relative *Lycopersicon pennellii. Physiol. Plant.* **110**: 42-51.

Miyake, C., Schreiber, U., Hormann, H., Sano, S. and Asada, K. (1998) The FAD-enzyme monodehydroascorbate radical reductase mediates the photoproduction of superoxide radicals in spinach thylakoid membranes. *Plant Cell Physiol.* **39**: 821-829.

Moncada, S., Palmer, R.M.J. and Higgs, E.A. (1991) Nitric oxide: physiology, pathophysiology and pharmacology. *Pharmacol. Rev.* **43**: 109-142.

Mori, H. and Nishimura, M. (1989) Glyoxysomal malate synthase is specifically degraded in microbodies during greening of pumpkin cotyledons. *FEBS Lett.* **244**: 163-166.

Mullen, R.T. and Trelease, R.N. (1996) Biogenesis and membrane properties of peroxisomes: does the boundary membrane serve and protect? *Trends Plant Sci.* **1**: 389-394.

Mullen, R.T., Lisenbee, C.S., Miernyk, J.A. and Trelease, R.N. (1999) Peroxisomal membrane ascorbate peroxidase is sorted to a membranous network that resembles a subdomain of the endoplasmic reticulum. *Plant Cell.* **11**: 2167-2185.

Ninnemann, H. and Maier, J. (1996) Indications for the occurrence of nitric oxide synthases in fungi and plants, and the involvement in photoconidation of *Neurospora crassa*. *Photochem. Photobiol.* **64**: 393-398.

Nishimura, M., Hayashi, M., Kato, A., Yamaguchi, K. and Mano, S. (1996) Functional transformation of microbodies in higher plant cells. *Cell Struct. Funct.* **21**: 387-393.

Nishimura, M., Takeuchi, Y., De Bellis, L. and Hara-Nishimura, I. (1993) Leaf peroxisomes are directly transformed to glyoxysomes during senescence. *Protoplasma.* **175**: 131-137.

Nishino, T. (1994) The conversion of xanthine dehydrogenase to xanthine oxidase and the role of the enzyme in reperfusion injury. *Biochem J.* **116**: 1-6.

Nito, K., Yamaguchi, K., Kondo, M., Hayashi, M. and Nishimura, M. (2001) Pumpkin peroxisomal ascorbate peroxidase is localized on peroxisomal membranes and unknown membranous structures. *Plant Cell Physiol.* **42**: 20-27.

Palma, J.M., Garrido, M., Rodríguez-García, M.I. and del Río, L.A. (1991) Peroxisome proliferation and oxidative stress mediated by activated oxygen species in plant peroxisomes. *Arch. Biochem. Biophys.* **287**: 68-74.

Palma, J.M., Gómez, M., Yáñez, J. and del Río, L.A. (1987) Increased levels of peroxisomal active oxygen-related enzymes in copper-tolerant pea plants. *Plant Physiol.* **85**: 570-574.

Palma, J.M., López-Huertas, E., Corpas, F.J., Sandalio, L.M., Gómez, M. and del Río, L.A. (1998) Peroxisomal manganese superoxide dismutase: Purification and properties of the isozyme from pea leaves. *Physiol. Plant.* **104**: 720-726.

Pastori, G.M. and del Río, L.A. (1997) Natural senescence of pea leaves: an activated oxygen-mediated function for peroxisomes. *Plant Physiol.* **113**: 411-418.

Pastori, G.M. and del Río, L.A. (1994a) An activated-oxygen-mediated role for peroxisomes in the mechanism of senescence of pea leaves. *Planta.* **193**: 385-391.

Pastori, G.M. and del Río, L.A. (1994b) Activated oxygen species and superoxide dismutase activity in peroxisomes from senescent pea leaves. *Proc. R. Soc. Edinb. Sec. B. Biol.* **102B**: 505-509.

Pastori, G.M., Distefano, S., Palma, J.M. and del Río, L.A. (1996) Purification and characterization of peroxisomal and mitochondrial Mn-superoxide dismutase from watermelon cotyledons. *Biochem. Soc. Trans.* **24**: 196S.

Pedroso, M.C., Magalhaes, J.R. and Durzan, D. (2000) A nitric oxide burst precedes apoptosis in angiosperm and gymnosperm callus cells and foliar tissues. *J. Exp. Bot.* **51**: 1027-1036.

Pei, Z-M., Murata, Y., Benning, G., Thiomine, S., Klüsener, B., Allen, G.J., Grill, E. and Schroeder, J.I. (2000) Calcium channels activated by hydrogen peroxide mediate abscisic acid signalling in guard cells. *Nature.* **406**: 731-734.

Pistelli, L., Nieri, B., Smith, S.M., Alpi, A. and De Bellis, L. (1996) Glyoxylate cycle enzyme activities are induced in senescent pumpkin fruits. *Plant Sci.* **119**: 23-29.

Poll-The, B.T., de Koning, T.J., Dorland, L. and Duran, M. (1998) Peroxisomal disorders. *Neurosci. Res. Comm.* **22**: 63-71.

Pool, M.R., López-Huertas, E., Horng, J.T. and Baker, A. (1998) NADPH is a specific inhibitor of protein import into glyoxysomes. *Plant J.* **15**: 1-14.

Rea, G., Laurenzi, M., Tranquili, E., D'Ovidio, R., Federico, R. and Angelini, R. (1998) Developmentally and wound-regulated expression of the gene encoding a cell wall copper amine oxidase in chickpea seedlings. *FEBS Lett.* **437**: 177-182.

Reddy, J.K. and Chu, R. (1996) Peroxisome proliferator-induced pleiotropic responses: Pursuit of a phenomenon. *Ann. New York Acad. Sci.* **804**: 176-201.

Reddy, J.K., Rao, M.S., Lalwani, N.D., Reddy, M.K., Nemali, M.R. and Alvares, K. (1987) Induction of hepatic peroxisome proliferation by xenobiotics. In *Peroxisomes in Biology and Medicine* (Fahimi, H.D. and Sies, H. eds.). pp. 254-262. Berlin-Heidelberg: Springer-Verlag.

Reumann, S. (2000) The structural properties of plant peroxisomes and their metabolic significance. *Biol. Chem.* **381**: 639-648.

Reumann, S., Maier, E., Benz, R. and Heldt, H.W. (1995) The membrane of leaf peroxisomes contains a porin-like channel. *J. Biol. Chem.* **270**: 17559-17565.

Ribeiro, E.A. Cunha, F.Q., Tamashiro, W.M.S.C. and Martins, I.S. (1999) Growth phase-dependent subcellular localization of nitric oxide synthase in maize cells. *FEBS Lett.* **445**: 283-286.

Romero-Puertas, M.C., McCarthy, I., Sandalio, L.M., Palma, J.M., Corpas, F.J., Gómez, M. and del Río, L.A. (1999) Cadmium toxicity and oxidative metabolism of pea leaf peroxisomes. *Free Rad. Res.* **31**: S25-31.

Saksela, M., Lapatto, R. and Raivio, K.O. (1999) Irreversible conversion of xanthine dehydrogenase into xanthine oxidase by a mitochondrial protease. *FEBS Lett.* **443**: 117-120.

Sakuma, S., Fujimoto, Y., Sakamoto, Y., Uchiyama, T., Yoshioka, K., Nishida, H. and Fujita, T. (1997) Peroxynitrite induces the conversion of xanthine dehydrogenase to oxidase in rabbit liver. *Biochem. Biophys. Res. Comm.* **230**: 476-479.

Sandalio, L.M., Dalurzo, H.C., Gómez, M., Romero-Puertas, M.C. and del Río, L.A. (2001) Cadmium-induced changes in the growth and oxidative metabolism of pea plants. *J. Exp. Bot.* **52**. In press.

Sandalio, L.M., Fernández, V.M., Rupérez, F.L. and del Río, L.A. (1988) Superoxide free radicals are produced in glyoxysomes. *Plant Physiol.* **87**: 1-4.

Sandalio, L.M., Lopez-Huertas, E., Bueno, P. and del Río, L.A. (1997) Immunocytochemical localization of copper, zinc superoxide dismutase in peroxisomes from watermelon (*Citrullus vulgaris Schrad.*). *Free Rad. Rēs.* **26**: 187-194.

Sandalio, L.M., Palma, J.M. and del Río, L.A. (1987) Localization of manganese superoxide dismutase in peroxisomes isolated from *Pisum sativum* L. *Plant Sci.* **51**: 1-8.

Schäfer, L. and Feierabend, J. (2000) Photoinactivation and protection of glycolate oxidase *in vitro* and in leaves. *Z. Naturforsch.* **55c**: 361-372.

Schnarrenberger, C., Flechner, A. and Martin, W. (1995) Enzymatic evidence for a complete oxidative pentose phosphate pathway in chloroplasts and an incomplete pathway in the cytosol of spinach leaves. *Plant Physiol.* **108**: 609-614.

Segura-Aguilar, J., Hakman, I. and Rydström, J. (1995) Studies on the mode of action of the herbicidal effect of 2,4,5-trichlorophenoxyacetic acid on germinating Norway spruce. *Environ. Exp. Bot.* **35**: 309-319.

Shaw, B.P. (1995) Effect of mercury and cadmium on the activities of antioxidative enzymes in the seedling of *Phaseolus aureus*. *Biol. Plant.* **37**: 587-596.

Stochs, S.J. and Bagchi, D. (1995) Oxidative mechanism in the toxicity of metal ions. *Free Rad. Biol. Med.* **18**: 321-336.

Struglics, A., Fredlund, K.M., Rasmusson, A.G. and Møller, I.M. (1993) The presence of a short redox chain in the membrane of intact potato tuber peroxisomes and the association of malate dehydrogenase with the peroxisomal membrane. *Physiol Plant.* **88**: 19-28.

Tolbert, N.E. (1981) Metabolic pathways in peroxisomes and glyoxysomes. *Ann. Rev. Biochem.* **50**: 133-157.

Tolbert, N.E., Gee, R., Husic, D.W. and Dietrich, S. (1987) Peroxisomal glycolate metabolism and the C_2 oxidative photosynthetic carbon cycle. In: *Peroxisomes in Biology and Medicine,* (Fahimi, H.D. and Sies, H. eds.). pp. 213-222. Berlin: Springer-Verlag.

Thompson, J.E., Ledge, R.L. and Barber, R.F. (1987) The role of free radicals in senescence and wounding. *New Phytol.* **105**: 317-344.

Tsuboi, T., Osafune, T., Tsugeki, R., Nishimura, M. and Yamada, M. (1992) Non-specific lipid transfer protein in castor bean cotyledons cells: Subcellular localization and a possible role in lipid metabolism. *J. Biochem.* **111**: 500-508.

Van Camp, W., Van Montagu, M. and Inzé, D. (1998) H_2O_2 and NO: redox signals in disease resistance. *Trends Plant Sci.* **3**: 330-334.

Verdoucq, L., Vignols, F., Jacquot, J-P., Chartier, Y. and Meyer, Y. (1999) *In vivo* characterization of a thioredoxin h target protein defines a new peroxiredoxin family. *J. Biol. Chem.* **274**: 19714-19722.

Vicentini, F. and Matile, P. (1993) Gerontosomes, a multi-functional type of peroxisomes in senescent leaves. *J. Plant Physiol.* **142**: 50-56.

Völkl, A., Mohr, H. and Fahimi, H.D. (1999) Peroxisome sub-populations of the rat liver: Isolation by immune free flow electrophoresis. *J. Histochem. Cytochem.* **47**: 1111-1117.

Wagner, G.J. (1993) Accumulation of cadmium in crop plants and its consequences to human health. *Adv. Agron.* **51**: 173-212.

Willekens, H., Chamnongpol, S., Davey, M., Schraudner, M., Langebartels, C., Van Montagu, M., Inzé, D. and Van Camp, W. (1997) Catalase is a sink for H_2O_2 and is indispensable for stress defence in C_3 plants. *EMBO J.* **16**: 4806-4816.

Wink, D.A., Hanbauer, I., Grisham, M.B., Laval, F., Nims, R.W., Laval, J., Cook, J., Pacelli, R., Liebmann, J., Krishna, M., Ford, P.C. and Mitchell, J.B. (1996) Chemical biology of nitric oxide: Regulation and protective and toxic mechanisms. *Curr. Top. Cell. Regul.* **34**: 159-187.

Yamaguchi, K., Mori, H. and Nishimura, M. (1995) A novel isoenzyme of ascorbate peroxidase localized on glyoxysomal and leaf peroxisomal membranes in pumpkin. *Plant Cell Physiol.* **36**: 1157-1162.

Zhang, H., Wang, J., Nickel, U., Allen, R.D. and Goodman, H.M. (1997) A novel isoenzyme of ascorbate peroxidase localized on glyoxysomal and leaf peroxisomal membranes in pumpkin. *Plant Cell Physiol.* **36**: 1157-1162.

8

PEROXISOMAL MEMBRANE ENZYMES

Robert P. Donaldson

George Washington University, Washington DC , USA

KEYWORDS

Membrane permeability; membrane composition; electron transport; ROS metabolism; peroxidases.

INTRODUCTION

Peroxisomes are enclosed by a single membrane that separates the contents of the organelles from the cytosol and defines them as distinct intracellular entities. This membrane appears to be indiscriminately porous under some conditions while in other situations the membrane is more restrictive and can maintain a pH gradient. The understanding of the functions of this organelle requires knowledge of the properties and enzymatic activities of its membrane. This membrane has a variety of integral proteins that consume NADH and conduct activities that generate superoxide and hydrogen peroxide, as well as activities that participate in ascorbate metabolism. There are two types of peroxide scavengers, ascorbate peroxidase and thiol peroxidase, in the peroxisomal membranes of plants, yeasts, and mammalian cells. In addition, some of the membrane proteins orchestrate the utilization of ATP and GTP to promote protein import and lipid processing in ways that are not well understood. Lipid metabolism in glyoxysomes of seeds and cotyledons involves the passage of fatty acids through the membranes. The resulting gluconeogenic intermediates must exit from the organelles. Likewise, photorespiratory metabolism in leaf peroxisomes requires that metabolites be transported through the membrane. The entry and egress of

A. Baker and I.A. Graham (eds.), Plant Peroxisomes, 259–278.

some metabolites may be through a pore-forming protein, porin, which may be responsible for variations in membrane permeability.

One fascinating view of the peroxisomal membrane that emerges from the literature is that it is not a barrier at all. Folded, oligomeric proteins are able to pass through the membranes of peroxisomes and glyoxysomes as discussed in Chapter 11. Even when the membrane is not intact some metabolic containment can be observed suggesting that the compartmentation of metabolic activities can be a consequence of enzyme assemblages or organized aggregates (Heupel and Heldt, 1994 and Chapter 5). The image of the membrane is of a porous basket or sieve that holds larger molecules together, but does not prevent the passage of metabolites or protein dimers. However, this does not mean that molecules pass through the membrane unfettered. The presence of a variety of enzyme activities in the membranes suggests that the membrane can be somewhat selective about the molecules that pass through it. Unlike the mitochondria, chloroplasts or nuclei, peroxisomes have a single membrane rather than two separate lipid bilayers. Entry and exit from peroxisomes would seem to be simpler; however, this single membrane may be more versatile. Under some conditions the membrane may allow passage of large molecules, protein complexes, and unrestricted passage of cofactors (NAD, CoA) along with metabolites. In other circumstances the membrane may become quite selective to the extent that a pH gradient maybe established while metabolites and proteins are confined.

A more traditional view of the membrane is that molecules, metabolites and proteins that pass through it require specific carriers or selective channels to allow them through. The presence of a porin and peroxins in the membrane suggests that there is some selectivity. The two views are not exclusive. Under some conditions the membrane may be freely permeable to many metabolites and proteins. Yet specific metabolites and proteins would be sequestered in the organelle because there are binding sites for them in the matrix. Thus, some molecules would experience a net flux into the organelle because they are being consumed or bound therein. Specific proteins would accumulate in the matrix as they join in aggregates with other proteins within. Under other conditions the membrane may become more selective, pores and channels may become more restrictive.

What goes on in the membranes relates to the processes occurring inside plant peroxisomes and glyoxysomes. For example, peroxisomes of photosynthetic cells are actively involved in photorespiration, the metabolism of glycolate. Elsewhere, when lipids are being metabolized, as they are in cotyledons or endosperm of seeds, the glyoxysomes are responsible for fatty acid oxidation and conversion of acetate to gluconeogenic intermediates. Thus, the necessary enzymes need to be sequestered within the organelles and requisite metabolites will be transported through the membrane. Both the metabolism of glycolate and of fatty acid begin with oxidations that create hydrogen peroxide (H_2O_2) which can be damaging to cells. It would seem that the membrane would be somewhat resistant to allowing H_2O_2 to escape the organelle. Also, both metabolic processes involve NADH, suggesting that reducing equivalents need mechanisms to pass through the membranes.

This chapter will consider the proteins and enzymes in the membranes of plant peroxisomes (including glyoxysomes) that deal with reactive oxygen species (ROS) such as H_2O_2 and superoxide, as well as enzymes that utilize NADH or ATP. Also, proteins that may conduct traffic of metabolites through the membrane will be discussed. What little is known about the mechanism(s) by which peroxisomal membrane proteins are sorted to the correct membrane is discussed in Chapter 10. Membrane proteins involved in the import of matrix proteins are discussed in Chapters 11 and 13. The enzyme activities of peroxisomal membranes in yeasts and mammalian cells will also be considered in this chapter. Comparisons among the various types of peroxisomes will lead to ideas about other activities that might be found in the membranes of plant peroxisomes.

DEFINING MEMBRANE PROTEINS OF PEROXISOMES

It is operationally difficult to define which proteins are associated with the membranes of peroxisomes. The tendency of peroxisomal matrix proteins to aggregate causes them to sediment under conditions used to centrifuge down membranes (Heupel and Heldt, 1994). The technique that has been widely accepted to characterize integral membrane proteins involves treating the membranes with sodium carbonate, pH 11.5 (Fang *et al*, 1987). This removes visible aggregates from the inside of castor bean glyoxysomal membranes and releases 98% to 100% of each of the matrix enzyme activities such as catalase, glyoxylate cycle enzymes and fatty acid oxidation activities. Freeze fracture electron microscope images of the membranes show particles that may represent pores, carrier proteins, docking proteins of the import apparatus, or H_2O_2/NADH scavengers. The resulting membrane ghosts consist of about 5% of the total glyoxysomal protein and possess activities such as NADH: ferricyanide reductase and cytochrome c reductase.

However, this treatment destroys some enzyme activities and removes legitimate peripheral membrane proteins.

The specific association of some proteins with peroxisomal membranes has been clearly demonstrated by immunogold electron microscopy (Bunkelmann and Trelease, 1996). The orientation of a protein in the membrane can be defined by treatment of intact isolated organelles with a protease or by the use of domain specific antibodies. Also, latency has been used as an indication that a membrane enzyme has active sites on the matrix side of the membrane. However, if an activity is not latent and is measurable in intact organelles a possible interpretation is that the membrane is permeable to the substrates involved. The membranes may become more permeable in the process of isolation and then many activities would appear to be accessible to external substrates.

Certain enzymes that are predominantly located in the matrix of the organelle may also be associated with the membrane to some extent. It is possible that this represents a technical uncertainty when attempting to separate matrix proteins from the membranes. However, there may be a limited number of sites in the membrane for the attachment of some matrix proteins and considering the volume of the matrix relative to the surface area of the membrane, those sites could be saturated. The molecules of an enzyme population that happen to be associated with the membrane may have a distinctive role in metabolism. A dehydrogenase, for example, could contribute NADH directly to enzymes residing in the membrane.

NADH-DRIVEN ELECTRON TRANSPORT

The first enzymes to be discovered in the membranes of peroxisomes were NADH dehydrogenase activities such as NADH: ferricyanide reductase and cytochrome c reductase (Hicks and Donaldson, 1982). Also, b-type cytochrome was observed. The reductase activities were found to be associated with a 32 kDa protein that was extracted from castor bean endosperm glyoxysomal membranes with detergents and purified by affinity chromatography (Luster *et al*, 1988). These reductases are probably flavoproteins that accept electrons from NADH and then donate the electrons to cytochrome c or ferricyanide, which are not considered to be physiological electron acceptors for peroxisomal activities. More recently it has been shown that oxygen or oxidized ascorbate can accept electrons from the membrane dehydrogenases as discussed below (Lopez-Huertas *et al*, 1999). Some of these activities are depicted in Figure 1.

Figure 1: Peroxisomal antioxidant systems

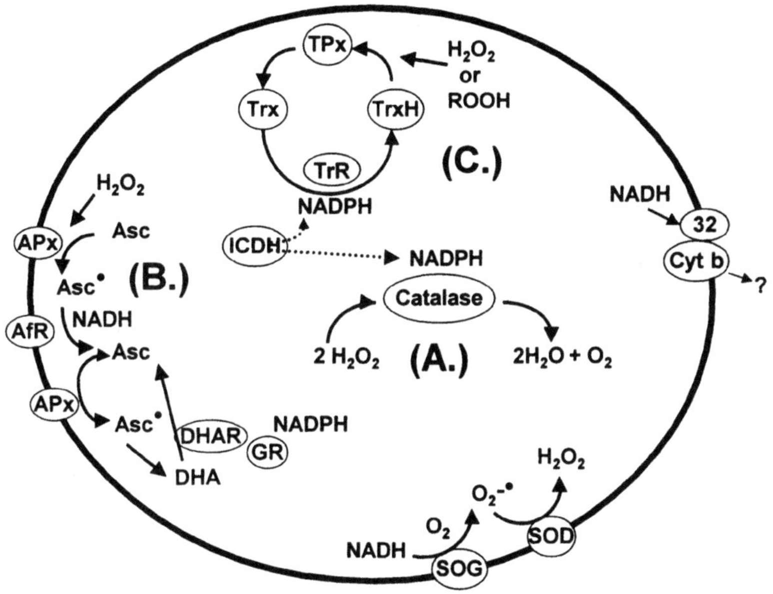

Figure 1. (A.) Catalase is the most active scavenger of H_2O_2, but at physiological (μM) concentrations of H_2O_2 it does not function near maximum velocity. Also, if H_2O_2 is limited catalase is likely to be inactivated unless NADPH is available to prevent it (Kirkman et al, 1999). This inactivation and NADPH reactivation have not been investigated in peroxisomes. (B.) Ascorbate Peroxidase (APx) uses ascorbate (Asc) as an H_2O_2 scavenger resulting in ascorbate free radical (Asc•), which can be recycled to ascorbate by an NADH dependent Ascorbate Free Radical Reductase (AfR), also known as MDHA reductase. Ascorbate free radical can also be non-enzymatically converted to fully oxidized dehydroascorbate (DH) which is reduced to ascorbate by a glutathione dependent Dehydroascorbate Reductase (DHAR). Glutathione is re-reduced by Glutathione Reductase (GR) which uses NADPH. All of these components have been found in plant peroxisomes. (C.) The thioredoxin system consists of an NADPH thioredoxin Reductase (TrR), thioredoxin (Trx), and thioredoxin peroxidase (TPx). TPx has been recently found in peroxisomes indicating that TrR and Trx should be there also. TPx uses Trx to reduce H_2O_2 or lipid peroxides (ROOH). NADPH can be supplied to the antioxidant systems by Isocitrate Dehydrogenase (ICDH) or other enzymes (Corpas et al, 1998 and 1999). NADH contributes electrons to superoxide generating proteins (SOG) and superoxide dismutase (SOD) produces H_2O_2. NADH can also donate electrons to a 32 kDa protein, which reduces a cytochrome b (Cyt b) or an artificial acceptor such as ferricyanide.

Peroxisomal membranes from pea leaves and potato tubers also have similar electron transport systems (Lopez-Huertas *et al*, 1999; Struglics *et al*, 1993). The activity is distinguishable from that in the endoplasmic reticulum (ER) in that it is specific for the beta-hydrogen of NADH whereas the ER activity uses the alpha-hydrogen. The activity of the matrix enzyme, malate dehydrogenase, is latent indicating that the transport of NADH or oxaloacetic acid is restricted. The NADH:ferricyanide reductase activity is also latent and thus the active site is on the matrix side of the membrane. This is also the case for glyoxysomes isolated from castor bean (Luster and Donaldson, 1987). These observations of latency show that the membranes of the isolated organelles can be selective.

The matrix activities, β-oxidation and the glyoxylate cycle, can be coupled to NADH:cytochrome c and ferricyanide reductases in glyoxysomes (Donaldson and Fang, 1987). The membrane activities have the capacity to oxidize NADH at rates equivalent to the matrix enzymes that generate the NADH. However, the activity of this system *in vivo* would depend on what physiological electron acceptors are available.

SUPEROXIDE METABOLISM

Superoxide (O_2-•) is produced when electrons are transferred from NADH to O_2 (del Rio and Donaldson, 1995). Superoxide production has been attributed to an 18 kDa peroxisomal membrane protein in pea leaf peroxisomes (Lopez-Huertas *et al*, 1997). This 18 kDa protein was recognized by an antibody to human cytochrome b5 from red blood cells, and difference spectroscopy also indicated the presence of this type of cytochrome in the peroxisomal membranes. The isolated 18 kDa protein was shown to conduct an NADH dependent production of superoxide radicals. However, this is somewhat unexpected since cytochrome b5, for example, does not usually accept electrons directly from NADH but requires an NADH:cytochrome b5 reductase, a flavoprotein, to mediate the transfer. In another report this same group presented evidence that the 32 kDa protein from pea leaf peroxisomes would generate O_2-• using NADH as the source of electrons (Lopez-Huertas *et al*, 1999). This protein appeared to be a monodehydroascorbate (MDHA) reductase since it was recognized by an antibody made against a cucumber MDHA reductase. This 32 kDa protein would also donate electrons to cytochrome c or ferricyanide. This is likely to be a flavoprotein, which can accept electrons from NADH and in the absence of another electron acceptor the electrons could go to oxygen resulting in superoxide. The 32 kDa protein found in pea leaf peroxisomes is probably similar to the 32 kDa membrane protein characterized as an NADH:cytochrome c reductase from castor bean glyoxysomes (Luster *et al*, 1988). Similarly, a 29 kDa membrane protein from the pea leaf peroxisomes was found to use NADPH as a source of electrons for a cytochrome c reductase activity or to produce O_2-• (Lopez-Huertas *et al*, 1997). Once again, cytochrome c is not likely to be a normal substrate for this protein.

In general any protein in the membrane which can take electrons from NADH (or other donors) has the potential to donate electrons to oxygen creating O_2-• if the appropriate electron acceptor is not available. This inadvertent production of superoxide has the potential to damage membrane lipids and proteins. Thus, protective mechanisms would be expected to be present in the membranes.

Superoxide radical generation activity is usually accompanied by superoxide dismutase (SOD) activity that converts superoxide to H_2O_2. SODs are present in the matrix and membranes of peroxisomes. For example, a Cu,Zn-SOD is found in the matrix together with a Mn-SOD in the membranes of watermelon cotyledon peroxisomes (Sandalio *et al*, 1997) and castor bean glyoxysomes (del Rio and Donaldson, 1995). The Mn-SOD is a peripheral protein that can be removed by sodium carbonate washing.

ASCORBATE AND H_2O_2 METABOLISM

Perhaps one of the most interesting discoveries in peroxisomes in the past decade is the enzymes that conduct scavenging of radicals and H_2O_2 at the membranes. Initially an NADH:MDHA reductase was characterized in the membranes of castor bean endosperm glyoxysomes (Bowditch and Donaldson, 1990). This enzyme has also been referred to as ascorbate free radical reductase (AfR). This activity has not been well characterized, but it may represent the 32 kDa protein that has been reported to accept electrons from NADH (Luster *et al*, 1988; Lopez-Huertas *et al*, 1999). Subsequently, pumpkin seed and leaf peroxisomes were found to have an ascorbate peroxidase (APX) activity (Yamaguchi *et al*, 1995). Shortly after that, a membrane protein of cotton seed glyoxysomes was found to have an amino acid sequence that was similar to ascorbate peroxidases found in chloroplasts and cytosol (Bunkelmann and Trelease, 1996). This APX was shown by immunocytochemistry to be located in the membrane. An antibody made to recognize this APX also recognizes a 34 kDa protein isolated from castor bean glyoxysomes. The purified protein has APX activity (Karyotou *et al*, 1997). The glyoxysomal APX has a hydrophobic domain near the

C-terminus that would likely function as the membrane anchor, as is the case for thylakoid APX. APX uses two molecules of ascorbate to convert H_2O_2 to water resulting in ascorbate free-radical, that is MDHA.

The sequence of the cotton glyoxysomal membrane protein predicted that it would have APX activity. This was demonstrated for the spinach glyoxysomal APX by expressing the protein in *E. coli* and showing that the protein had APX activity (Ishikawa *et al*, 1998). The spinach glyoxysomal membrane APX is a 286 amino acid, 31.5 kDa protein and represents one of five APX isozymes found in the spinach. MDHA reductase is also present in spinach glyoxysomal membranes. Both enzymes were detectable in intact, unbroken isolated glyoxysomes suggesting that the active sites of the enzymes are on the outside, that is the cytosolic side, of the membrane. In other words the activities were not latent. However, this experimental observation could also mean that the substrates for the enzymes, H_2O_2, MDHA, and NADH permeate the membrane at rates greater than the turnover number for these particular enzymes.

Pea leaf peroxisomal APX is also membrane bound, but not latent (Jimenez *et al*, 1997). However, the MDHA activity in the peroxisomes is latent and thus appears to be on the matrix side of the membrane.

Arabidopsis has seven types of APX, two soluble in cytosol, one in the stroma of chloroplasts, another in the thylakoids, two of unknown locations and one in the membranes of peroxisomes (Jespersen *et al*, 1997). The latter three types have C-terminal extensions including sufficient spans of hydrophobic residues to suggest membrane association. The charged residues that are responsible for electrostatic interactions between dimers are present in the peroxisomal APXs but not in the chloroplast forms. The peroxisomal and cytosolic APXs are less specific for ascorbate than the chloroplast forms, which may be determined by the peptide segments around the active site entrance channel approaching the edge of the haem.

APX is an effective scavenger of H_2O_2. The K_m for H_2O_2 is in the range of 70 μM (Ishikawa *et al*, 1998). Catalase is less active than APX at this concentration of H_2O_2 (Klapheck *et al*, 1990). Also, the K_m for ascorbate is around 2 mM which is the concentration of ascorbate found in castor bean during germination (Klapheck *et al*, 1990). Thus, as long as the ascorbate concentration is around this level then it can effectively support H_2O_2 scavenging. Under conditions of oxygen stress when ascorbate becomes oxidized this system may be less effective. A pool of ascorbate appears to be present in peroxisomes (Jimenez *et al*, 1997). The peroxisomal APX is less specific than some peroxidases and will oxidize pyrogallol in addition to ascorbate.

Although APX is anchored in the membrane, the protein probably does not have transmembrane activity. An ascorbate reducible b-cytochrome is known to have transmembrane activity and to transfer electrons from one side of a membrane to the other in mammalian cell compartments such as chromaffin granules or clathrin-coated vesicles (Rubinstein, 1994). Nevertheless, if its active sites are facing the cytosol they may have access to H_2O_2 as it passes out through the membrane. As discussed in Chapter 10, the information that directs the peroxisomal APX to the membrane resides in a hydrophilic sequence of eight amino acids at the C-terminus. Before taking up residence in the peroxisomal membrane the APX is found in an ER-like membrane referred to as a peroxisomal endoplasmic reticulum (Mullen and Trelease, 2000; Nito *et al*, 2001).

MDHA reductase (also referred to as AfR), like APX, is anchored in the membrane and these two enzymes can function together to scavenge H_2O_2 in an NADH dependent manner: APX consumes H_2O_2 and produces MDHA; then MDHA reductase uses NADH to recycle the ascorbate (Karyotou *et al*, 1997). The physiological role of the peroxisomal APX appears to be to protect the membrane or the cell from oxidative damage that could be caused by H_2O_2. For example, the overexpression of the *Arabidopsis* peroxisomal APX in tobacco protects the leaves from H_2O_2 generated by peroxisomes (Wang *et al*, 1999). Environmental factors such as exposure to cold, UV light or increased H_2O_2 cause changes in the APX mRNA levels.

If ascorbate free radical created by APX is not quickly reduced back to ascorbate it will disproportionate to dehydroascorbate (DHA) which may be reduced back to ascorbate by a DHA reductase which uses glutathione as the source of reductant. Thus, NADPH glutathione reductase should also be present in the peroxisomes for this system to function. If these activities take place within peroxisomes then all of the attendant enzymes and substrates would be needed inside the organelle. However, (some of) these processes could occur on the outer surface of the membrane and utilize cytosolic substrates. Pea leaf peroxisomes have been reported to have membrane-bound APX. These peroxisomes also have MDHA reductase together with glutathione reductase and DHA reductase within the matrix (Jimenez *et al*, 1998). During senescence of the leaves the membrane enzyme activities are diminished while the matrix DHA reductase is increased along with an increase in the glutathione content of the peroxisomes.

The question that arises from the consideration of the APX is the extent of H_2O_2 permeation through the membrane. While the membrane may be permeable to H_2O_2 under some conditions, the scavenging activities of APX in the membrane may be sufficient to trap the H_2O_2 before it can escape.

THIOL PEROXIDASES

Other peroxide scavenging activities that have been recently identified in peroxisomes are thiol peroxidases. These peroxidases do not contain haem but rather have cysteines their active sites to transfer hydrogens from thiols such as thioredoxin (Trx) or glutathione to peroxides. Such peroxidases are also referred to as peroxiredoxins or thiol specific antioxidants. For example, *Saccharomyces cerevisiae* has a thioredoxin peroxidase (TPX) which has a peroxisomal targeting sequence and prefers lipid peroxides as substrates rather than H_2O_2 (Lee *et al*, 1999, Jeong *et al*, 1999). A similar TPX is found in human and mouse peroxisomes (Yamashita *et al*, 1999). A thiol peroxidase appears to be associated with the inside of the peroxisomal membrane in the yeast *Candida*. The *Candida* activity uses glutathione rather than thioredoxin as the source of reductant. This glutathione peroxidase functions especially to decompose lipid hydroperoxides that are created in the membrane (Horiguchi *et al*, 2001).

The TPX system consists of three components, thioredoxin reductase, thioredoxin, and TPX, but only the peroxidase has been described in peroxisomes to date. Genes for a TPX protein and several thioredoxins have been described in the Arabidopsis genome. The Arabidopsis TPX protein was expressed in *Saccharomyces*, which had been mutated for TPX, and consequently the yeast regained the ability to tolerate H_2O_2 and lipid peroxides (Verdoucq *et al*, 1999). The Arabidopsis gene was also expressed in *E. coli* and shown to have TPX activity. The second component of this system, Arabidopsis thioredoxin, AtTrx3 was shown to interact with a *Saccharomyces* TPX. AtTrx3 is one of the five genes for cytosolic, h-type thioredoxins in Arabidopsis in addition to the m-type and f-type thioredoxins found in the chloroplasts (Rivera-Madrid *et al*, 1995). Based on the activities of the Arabidopsis TPX and AtTrx3 when expressed in yeast, these two proteins might also be expected to interact with each other within Arabidopsis peroxisomes and, thus, AtTrx3 might be found in peroxisomes. However, of the three components that comprise this system (Figure 1C), only the peroxidase, TPX, has been found in peroxisomes. The other two components, thioredoxin (Trx) and NADPH:thioredoxin reductase (TrR) have not been found in peroxisomes of any species. None of the Arabidopsis thioredoxins have peroxisomal targeting sequences. However, thioredoxins are small molecules, around 10 kDa, and since a thioredoxin may associate with the peroxisomal TPX it may be imported into peroxisomes by 'piggy back'.

The *Candida* TPX protein has been referred to as a peroxisomal membrane protein, PMP20, but is not anchored in the membrane by a hydrophobic domain. Rather than typical peroxisomal membrane targeting information the peroxisomal TPXs have a PTS1 sequence at the C-terminus that is characteristic of proteins that are transported from cytosol, through the membrane into the matrix. This does not preclude an association with the membrane, however. The Arabidopsis TPX is a 20 kDa protein and like the *Candida* protein it may be associated with the membrane. However, this Arabidopsis TPX is not the same protein as the integral PMP22 protein previously described (Tugal *et al*, 1999).

The existence of the thiol peroxidase system in peroxisomes poses some interesting considerations with respect to activities that may take place in the membranes. Lipid peroxide metabolism may be a significant process in some circumstances. TPX functions within peroxisomes could involve an influx of reduced thioredoxin.

ATPase AND GTPase ACTIVITIES

The import of proteins into isolated, intact peroxisomes is energy dependent. The import of matrix enzymes such as glycolate oxidase (Horng *et al*, 1995) and isocitrate lyase (Behari and Baker, 1993) has been shown to be ATP dependent. The process of transporting proteins through the membrane involves the hydrolysis of ATP or GTP, but it is not understood how nucleotide triphosphate is used in this process (Brickner and Olsen, 1998). There are indications that a membrane potential is generated by these activities since ionophores have some effect on import. The proteins responsible for these activities have not been identified. GTP-binding proteins have also been described in peroxisomal membranes from rat liver (Verheyden *et al*, 1992).

The chaperone protein, DnaJ is located in the membranes of cucumber glyoxysomes. Immunologically detected DnaJ is removed from intact glyoxysomes by protease treatment. Thus it appears to be on the outside of the membrane. (Preisig-Muller *et al*, 1994). The presence of DnaJ in glyoxysomal membranes was confirmed along with a report of an Hsp70

(Corpas and Trelease, 1997). The glyoxysomal DnaJ protein has been shown to specifically interact with the cytosolic isoform 1 of Hsp70. This interaction promotes the ATPase activity of the Hsp70 (Diefenbach and Kindl, 2000). The purpose of these interactions may be to facilitate the translocation of proteins through the membrane and may be responsible for the energy requirement for the import of proteins into plant peroxisomes. There is also evidence in *Saccharomyces* of a DnaJ being involved in protein import (Hettema *et al*, 1998).

Peroxisomes from the yeasts, *Candida* and *Hansenula* have an ATPase activity in their membranes, which may be responsible for generating a pH gradient (Douma, *et al*, 1987). The peroxisomes in these yeasts have an acidic interior as indicated by a weak base that can be visualized within the peroxisomes by immunogold electronmicroscopy (Waterham *et al*, 1990) and by ^{31}P nuclear magnetic resonance (NMR) spectroscopy (Nicolay *et al*, 1987). In contrast to the yeasts, the peroxisomes of human fibroblasts appear to have a basic pH. A pH-sensitive fluorescent reporter linked to a PTS1 containing heptapeptide is imported into peroxisomes and indicates a pH of 8.2 (Dansen *et al*, 2000). Interestingly, this pH gradient is not diminished by ATP depletion, suggesting that an ATP-driven proton pump is not involved. Perhaps electron transport activities in the membrane could generate a pH gradient. Also, a mutation in an import receptor, PEX7, results in a pH of 6.5 rather than 8.2, one explanation being that this PTS2 receptor induces a pore in the membrane. Thus yeast and human peroxisomal membranes are normally impermeable to protons, but some conditions may cause the membrane to become more porous.

As discussed in the next section, ATP-dependent activities in the peroxisomal membranes of yeast and mammals are involved in the processing of fatty acids.

LIPID PROCESSING IN THE MEMBRANES

In germinating oil seeds there should be a considerable amount of lipid traffic through the membranes of glyoxysomes as fatty acids move from lipid bodies into the glyoxysomal matrix where the enzymes of β-oxidation and the glyoxylate cycle are located. Free fatty acids are present in glyoxysomal membranes in the endosperm of germinating castor bean, (Donaldson and Beevers, 1977). Triacylglycerols and free fatty acids are also present in the enlarging glyoxysomes of cotton cotyledons (Chapman and

Trelease, 1991). Unusual, unidentified fatty acids are predominant, perhaps representing metabolites or lipid peroxides.

An AMP binding protein was found to be present in the membranes of *Saccharomyces cerevisiae* peroxisomes (Blobel and Erdmann, 1996). This is probably an enzyme that joins free fatty acids to CoA prior to β-oxidation, an acyl-CoA synthetase. This 60 kDa peripheral membrane protein has a C-terminal SKL sequence typical of proteins that are imported from the cytosol into the matrix. However, it also has a membrane-spanning domain and the protein is found both associated with the membrane and in the matrix. Medium chain free fatty acids are transported into peroxisomes and then combined with CoA (Hettema *et al*, 1996, Verleur *et al*, 1997, Wanders *et al*, 2001). Long chain fatty acids such as C18:1, however, are attached to CoA in the cytosol and then transported through the membrane by a specific, ATP-binding-cassette (ABC) transporter. A similar ABC transporter, a 70 kDa protein, has been described in mammalian peroxisomal membranes (Imanaka *et al*, 2000, Roerig *et al*, 2001). The peroxisomes of the yeast *Candida boidinii* possess a membrane protein, PMP47 which is necessary for the oxidation of laureate (Nakagawa *et al*, 2000). Mutations and complementation of this protein indicate that it may be involved in the transport of ATP or CoA into the peroxisomes to support the conversion of the laurate to lauroyl-CoA. The acyl-CoA synthetases and transporters have yet to be described for the fatty acids that are metabolized in plant peroxisomes.

Irrespective of the permeability properties of the membrane, it seems likely that fatty acids and other hydrophobic molecules would traverse the lipid phase of the membrane in the process of entering the organelle.

TRANSPORT OF METABOLITES

The metabolic processes that occur in plant peroxisomes and glyoxysomes require the passage of a variety of metabolites through the membrane. Photorespiration involves glycolate and serine entry into leaf peroxisomes and the egress of glycine and glycerate. Also, NADH or malate would need to enter the organelle to supply the reducing equivalents required by the hydroxypyruvate reductase to produce glycerate. Fatty acids enter glyoxysomes to be oxidized to acetyl-CoA which is then converted to succinate via the glyoxylate cycle enzymes. One of the enzymes of this cycle, aconitase, is located in the cytosol rather than within the glyoxysomes (Courtois-Verniquet and Douce, 1993). Also, some of the reducing equivalents generated by β-oxidation and the glyoxylate cycle may be transported out of glyoxysomes by a malate shuttle. These metabolic processes would require the transport of citrate, isocitrate, malate, succinate, and possibly aspartate through the glyoxysomal membrane (Escher and Widmer, 1997). The question is whether the flow of metabolites is conducted through passive channels in an unrestricted manner or if specific carriers are required.

A porin protein may accommodate the transport of metabolites through the membranes of peroxisomes and glyoxysomes. Porin activity was detected in detergent extracts from spinach leaf peroxisomes (Reumann *et al*, 1995 and 1998 and Chapter 5 of this volume). When this protein was reconstituted into a membrane electrophysiological measurements indicated the presence of a channel with a diameter of about 0.6 nm, with specificity for binding C4 dicarboxylates, and with greater selectivity than the mitochondrial-type porin. Similar single-channel analysis of an activity from rat liver peroxisomes indicated a diameter of 1.7 nm (Lemmens *et al*, 1989) and a 31 kDa protein from *Hansenula polymorpha* peroxisomal membranes was shown to have pore forming activity (Sulter *et al*, 1993). Meanwhile, a 36 kDa protein found in the membranes of plant peroxisomes and glyoxysomes has sequence similarities to porins found in mitochondria and chloroplasts (Corpas *et al*, 2000). This protein was localized to peroxisomal membranes by immunogold electron microscopy analysis. Therefore, what is known at this point is that plant peroxisomes have a porin-like activity and a protein that has sequence similarities to a porin.

This porin may be responsible for the permeability properties of peroxisomal membranes. The opening of the channel may vary depending on the physiological conditions in a cell. Under certain conditions it may be more open. Under other conditions the pore may close and allow a gradient of ions and metabolites to build up.

SUMMARY: MEMBRANE/MATRIX INTERACTIONS

The membranes of plant peroxisomes (including glyoxysomes) have a variety of activities that contribute to the processes that take place in the matrix of the compartment. The membrane provides for the transport of the appropriate proteins and metabolites that are used in peroxisomes. Although much is known about the various receptors and membrane docking proteins that are involved in the recruiting of matrix proteins to be brought into the organelle, the process of translocating a protein through the membrane is mysterious (Olsen, 1998). Hence, it is not clear how (ATPase, GTPase, dehydrogenase) activities in the membrane might contribute to the translocation process. Free fatty acids and/or acyl-CoAs pass through the membrane to be metabolized within, but very little is known about membrane proteins that might be responsible for this in plants. The observation of the unusual free fatty acids in the membranes suggest that some fatty acid may be translocated and processed in the lipid phase (Chapman and Trelease, 1991). Smaller, hydrophilic metabolites may pass through the membrane porin that has been identified in peroxisomes of plants as well as yeasts and humans. The selectivity of this channel varies among species. Also, it is possible that some more specific carriers for certain metabolites will eventually be found. Since the porin can be somewhat non-specific about which metabolites are allowed passage, some metabolic compartmentation may depend on physical associations of matrix enzymes that channel metabolites directly from one matrix enzyme to the next (Heupel and Heldt, 1994). The membrane pores may open and close under various physiological conditions. When the organelles are isolated the pores may be opened and the membranes may become very porous. A peroxisome could be viewed as a reaction vessel, which can be opened to allow ingredients to be added. Then the vessel could be closed so that the ingredients can react together efficiently and by-products such as reactive oxygen species are kept under control. The use of *in vivo* fluorescent probes may enlighten our understanding of the physiological properties of the membrane.

The oxidases within the matrix can generate copious amounts of H_2O_2 as hydrogens are extracted from the incoming substrates such as fatty acyl-CoA and glycolate. Also, $O_2^{-}\bullet$ can be produced in the membrane as a consequence of NAD(P)H oxidation. Thus, there is considerable potential for oxidative damage to proteins and lipids in peroxisomal matrix and membranes. Peroxisomes have a multiplicity of protective mechanisms. Catalase is abundant in the matrix and SODs are present in the membranes and the matrix. The membrane has two facilities for H_2O_2 scavenging, APX and thiol peroxidase. The APX system uses ascorbate and NADH while the thiol peroxidase systems use thioredoxin or glutathione and NADPH. The membranes of peroxisomes have some unique activities and characteristics that will provide investigators with some stimulating problems to solve in the coming years.

REFERENCES

Behari, R. and Baker, A. (1993) The carboxyl terminus of isocitrate lyase is not essential for import into glyoxysomes in an *in vitro* system. *J. Biol. Chem.* **268**: 7315-7322.

Blobel, F., Erdmann, R. (1996) Identification of a yeast peroxisomal member of the family of AMP-binding proteins. *Eur. J. Biochem.* **240**: 468-476.

Bowditch, M.Y. and Donaldson, R.P. (1990) Ascorbate free-radical reduction by glyoxysomal membranes. *Plant Physiol.* **94**: 531-537.

Brickner, D.G. and Olsen, L.J. (1998) Nucleotide triphosphates are required for the transport of glycolate oxidase into peroxisomes. *Plant Physiol.* **116**: 309-17.

Bunkelmann, J.R. and Trelease, R.N. (1996) Ascorbate peroxidase. A prominent membrane protein in oilseed glyoxysomes. *Plant Physiol.* **110**: 589-98.

Chapman, K.D. and Trelease, R.N. (1991) Acquisition of membrane lipids by differentiating glyoxysomes: role of lipid bodies. *J. Cell Biol.* **115**: 995-1007.

Corpas, F.J. and Trelease, R.N. (1997) The plant 73 kDa peroxisomal membrane protein (PMP73) is immunorelated to molecular chaperones. *Eur. J. Cell Biol.* **73**: 49-57.

Corpas, F.J., Barroso, J.B., Sandalio, L.M., Distefano, S., Palma, J.M., Lupianez, J.A. and del Rio, L.A. (1998) A dehydrogenase-mediated recycling system of NADPH in plant peroxisomes. *Biochem. J.* **330**: 777-84.

Corpas, F.J., Bunkelmann, J. and Trelease, R.N. (1994) Identification and immunochemical characterization of a family of peroxisome membrane proteins (PMPs) in oilseed glyoxysomes. *Eur. J. Cell Biol.* **65**: 280-90.

Corpas, F.J., Sandalio, L.M., Brown, M.J., Rio, L.A. and Trelease, R.N. (2000) Identification of porin-like polypeptide(s) in the boundary membrane of oilseed glyoxysomes. *Plant Cell Physiol.* **41**: 1218-1228.

Courtois-Verniquet, F. and Douce, R. (1993) Lack of aconitase in glyoxysomes and peroxisomes. *Biochem. J.* **294**: 103-7.

Dansen, T.B., Wirtz, K.W., Wanders, R.J. and Pap, E.H. (2000) Peroxisomes in human fibroblasts have a basic pH. *Nat. Cell Biol.* **2**, 51-3.

del Rio, L.A. and Donaldson, R. (1995) Production of superoxide radicals in glyoxysomal membranes from castor bean endosperm. *J. Plant Physiol.* **146**: 283-287.

Diefenbach, J. and Kindl, H. (2000) The membrane-bound DnaJ protein located at the cytosolic site of glyoxysomes specifically binds the cytosolic isoform 1 of Hsp70, but not other Hsp70 species. *Eur. J. Biochem.* **267**: 746-54.

Donaldson, R.P. and Beevers, H. (1977) Lipid composition of organelles from germinating castor bean endosperm. *Plant Physiol.* **59**: 259-263.

Donaldson, R.P. and Fang, T.K. (1987) β-oxidation and glyoxylate cycle coupled to NADH:cytochrome c and ferricyanide reductases in glyoxysomes. *Plant Physiol.* **85**: 792-795.

Douma, A.C., Veenhuis, M., Sulter, G.J. and Harder, W. (1987) A proton-translocating adenosine triphosphatase is associated with the peroxisomal membrane of yeasts. *Arch. Microbiol.* **147**: 42-7.

Escher, C.L. and Widmer, F. (1997) Lipid mobilization and gluconeogenesis in plants: do glyoxylate cycle enzyme activities constitute a real cycle? A hypothesis. *Biol. Chem.* **378**: 803-13

Fang, T.K., Donaldson, R.P. and Vigil, E.L. (1987) Electron transport in purified glyoxysomal membranes from castor bean endosperm. *Planta.* **172**: 1-13.

Hettema, E.H., Ruigrok, C.C., Koerkamp, M.G., van den Berg, M., Tabak, H.F., Distel, B. and Braakman, I. (1998) The cytosolic DnaJ-like protein djp1p is involved specifically in peroxisomal protein import. *J. Cell Biol.* **142**: 421-34.

Hettema, E.H., van Roermund, C.W., Distel, B., van den Berg, M., Vilela, C., Rodrigues-Pousada, C., Wanders, R.J. and Tabak, H.F. (1996) The ABC transporter proteins Pat1 and Pat2 are required for import of long-chain fatty acids into peroxisomes of *Saccharomyces cerevisiae*. *EMBO J.* **15**: 3813-22.

Heupel, R. and Heldt, H.W. (1994) Protein organization in the matrix of leaf peroxisomes. A multi-enzyme complex involved in photorespiratory metabolism. *Eur. J. Biochem.* **220**: 165-72

Hicks, D.B. and Donaldson, R.P. (1982) Electron transport in glyoxysomal membranes. *Arch. Biochem. Biophys.* **215**: 280-8.

Horiguchi, H., Yurimoto, H., Kato, N. and Sakai, Y. (2001) Antioxidant system within yeast peroxisome: biochemical and physiological characterization of CbPmp20 in the methylotrophic yeast *Candida boidinii*. *J. Biol. Chem.* **276**: 14279-88.

Horng, J.T., Behari, R., Burke, L.E. and Baker, A. (1995) Investigation of the energy requirement and targeting signal for the import of glycolate oxidase into glyoxysomes. *Eur. J. Biochem.* **230**: 157-63.

Imanaka, T., Aihara, K., Suzuki, Y., Yokota, S. and Osumi, T. (2000) The 70-kDa peroxisomal membrane protein (PMP70), an ATP-binding cassette transporter. *Cell Biochem. Biophys.* **32** Spring: 131-8.

Ishikawa, T., Yoshimura, K., Sakai, K., Tamoi, M., Takeda, T. and Shigeoka, S. (1998) Molecular characterization and physiological role of a glyoxysome-bound ascorbate peroxidase from spinach. *Plant Cell Physiol.* **39**: 23-34.

Jeong, J.S., Kwon, S.J., Kang, S.W., Rhee, S.G. and Kim, K. (1999) Purification and characterization of a second type thioredoxin peroxidase (type II TPx) from *Saccharomyces cerevisiae*. *Biochem.* **38**, 776-83.

Jespersen, H.M., Kjaersgard, I.V., Ostergaard, L. and Welinder, K.G. (1997) From sequence analysis of three novel ascorbate peroxidases from *Arabidopsis thaliana* to structure, function and evolution of seven types of ascorbate peroxidase. *Biochem. J.* **326**: 305-10.

Jimenez, A., Hernandez, J.A., Pastori, G., del Rio, L.A. and Sevilla, F. (1998) Role of the ascorbate-glutathione cycle of mitochondria and peroxisomes in the senescence of pea leaves. *Plant Physiol.* **118**: 1327-35.

Jimenez, A., Hernandez, J.A., del Rio, L.A. and Sevilla, F (1997) Evidence for the presence of the ascorbate-glutathione cycle in mitochondria and peroxisomes of pea leaves. *Plant Physiol.* **114**: 275-284.

Karyotou, K., Bushey, J.L., Hu, L., Donaldson, R.P. (1997) NADH oxidation dependent on ascorbate and hydrogen peroxide in glyoxysomal membranes. *Plant Physiol.* **114**: 200.

Kirkman, H.N., Rolfo, M., Ferraris, A.M. and Gaetani, G.F. (1999) Mechanisms of protection of catalase by NADPH. Kinetics and stoichiometry. *J. Biol. Chem.* **274**, 13908-14.

Klapheck, S., Zimmer, I. and Cosse, H. (1990) Scavenging of hydrogen peroxide in the endosperm of *Ricinus communis* by ascorbate peroxidase. *Plant Cell Physiol.* **31**: 1005-1013.

Lee, J., Spector, D., Godon, C., Labarre, J. and Toledano, M.B. (1999) A new antioxidant with alkyl hydroperoxide defense properties in yeast. *J. Biol. Chem.* **274**: 4537-44.

Lemmens, M., Verheyden, K., Van Veldhoven, P., Vereecke, J., Mannaerts, G.P. and Carmeliet, E. (1989) Single-channel analysis of a large conductance channel in peroxisomes from rat liver. *Biochim. Biophys. Acta.* **984**: 351-9.

López-Huertas, E., Corpas, F.J., Sandalio, L.M. and Del Rio, L.A. (1999) Characterization of membrane polypeptides from pea leaf peroxisomes involved in superoxide radical generation. *Biochem. J.* **337**: 531-6.

López-Huertas, E., Sandalio, L.M., Gomez, M. and Del Rio, L.A. (1997) Superoxide radical generation in peroxisomal membranes: evidence for the participation of the 18 kDa integral membrane polypeptide. *Free Radic. Res.* **26**: 497-506.

Luster, D.G. and Donaldson, R.P. (1987) Orientation of electron transport components in the glyoxysomal membrane. *Plant Physiol.* **85**: 796-800.

Luster, D.G., Bowditch, M.I., Eldridge, K.M. and Donaldson, R.P. (1988) Characterization of membrane-bound electron transport enzymes from castor bean glyoxysomes and endoplasmic reticulum. *Arch. Biochem. Biophys.* **265**: 50-61.

Mullen, R.T. and Trelease, R.N. (2000) The sorting signals for peroxisomal membrane-bound ascorbate peroxidase are within its C-terminal tail. *J. Biol. Chem.* **275**: 16337-44.

Nakagawa, T., Imanaka, T., Morita, M., Ishiguro, K., Yurimoto, H., Yamashita, A., Kato, N. and Sakai, Y. (2000) Peroxisomal membrane protein Pmp47 is essential in the metabolism of middle-chain fatty acid in yeast peroxisomes and Is associated with peroxisome proliferation. *J. Biol. Chem.* **275**: 3455-61.

Nicolay, K., Veenhuis, M., Douma, A.C. and Harder, W. (1987) A 31P NMR study of the internal pH of yeast peroxisomes. *Arch. Microbiol.* **147**: 37-41.

Nito, K., Yamaguchi, K., Kondo, M., Hayashi, M. and Nishimura, M. (2001) Pumpkin peroxisomal ascorbate peroxidase is localized on peroxisomal membranes and unknown membranous structures. *Plant Cell Physiol.* **42**: 20-7.

Olsen, L.J. (1998) The surprising complexity of peroxisome biogenesis. *Plant Mol. Biol.* **38**: 163-89.

Preisig-Müller, R., Muster, G. and Kindl, H. (1994) Heat shock enhances the amount of prenylated DnaJ protein at membranes of glyoxysomes. *Eur. J. Biochem.* **219**: 57-63.

Reumann, S., Maier, E., Benz, R. and Heldt, H.W. (1995) The membrane of leaf peroxisomes contains a porin-like channel. *J. Biol. Chem.* **270**: 17559-65.

Reumann, S., Maier, E., Heldt, H.W. and Benz, R. (1998) Permeability properties of the porin of spinach leaf peroxisomes. *Eur. J. Biochem.* **251**: 359-66.

Rivera-Madrid, R., Mestres, D., Marinho, P., Jacquot, J.P., Decottignies, P., Miginiac-Maslow, M. and Meyer, Y. (1995) Evidence for five divergent thioredoxin h sequences in *Arabidopsis thaliana*. *Proc. Natl. Acad. Sci. USA.* **92**: 5620-4.

Roerig, P., Mayerhofer, P., Holzinger, A. and Gartner, J. (2001) Characterization and functional analysis of the nucleotide binding fold in human peroxisomal ATP binding cassette transporters. *FEBS Lett.* **492**: 66-72.

Rubinstein, B. (1994) The action of ascorbate in vesicular systems. *J. Bioenerg. Biomembr.* **26**: 385-92

Sandalio, L.M., Lopez-Huertas, E., Bueno, P. and Del Rio, L.A. (1997) Immunocytochemical localization of copper, zinc superoxide dismutase in peroxisomes from watermelon (*Citrullus vulgaris Schrad.*) cotyledons. *Free Rad. Res.* **26**: 187-94.

Struglics, A., Fredlund, K.M. and Rasmusson, A.G. (1993) The presence of a short redox chain in the membrane of intact potato tuber peroxisomes and the association of malate dehydrogenase with the peroxisomal membrane. *Physiol. Plantarum.* **88**: 19-28.

Sulter, G.J., Harder, W. and Veenhuis, M. (1993) Structural and functional aspects of peroxisomal membranes in yeasts. *FEMS Microbiol. Rev.* **11**: 285-96.

Tugal, H.B., Pool, M. and Baker, A. (1999) Arabidopsis 22-kDa peroxisomal membrane protein. Nucleotide sequence analysis and biochemical characterization. *Plant Physiol.* **120**: 309-20.

Verdoucq, L., Vignols, F., Jacquot, J.P., Chartier, Y. and Meyer, Y. (1999) *In vivo* characterization of a thioredoxin h target protein defines a new peroxiredoxin family. *J. Biol. Chem.* 274: 19714-22.

Verheyden, K., Fransen, M., Van Veldhoven, P.P. and Mannaerts, G.P. (1992) Presence of small GTP-binding proteins in the peroxisomal membrane. *Biochim. Biophys. Acta.* **1109**: 48-54.

Verleur, N., Hettema, E.H., van Roermund, C.W., Tabak, H.F. and Wanders, R.J. (1997) Transport of activated fatty acids by the peroxisomal ATP-binding-cassette transporter Pxa2 in a semi-intact yeast cell system. *Eur. J. Biochem.* **249**: 657-61.

Wanders, R.J., Vreken, P., Ferdinandusse, S., Jansen, G.A., Waterham, H.R., Van Roermund, C.W. and Van Grunsven, E.G. (2001) Peroxisomal fatty acid α- and β-oxidation in humans: enzymology, peroxisomal metabolite transporters and peroxisomal diseases. *Biochem. Soc. Trans.* **29**: 250-67.

Wang, J., Zhang, H. and Allen, R.D. (1999) Overexpression of an Arabidopsis peroxisomal ascorbate peroxidase gene in tobacco increases protection against oxidative stress. *Plant Cell Physiol.* **40**: 725-32.

Waterham, H.R., Keizer-Gunnink, I., Goodman, J.M., Harder, W. and Veenhuis, M. (1990) Immunocytochemical evidence for the acidic nature of peroxisomes in methylotrophic yeasts. *FEBS Lett.* **262**: 17-9.

Yamaguchi, K., Mori, H. and Nishimura, M. (1995) A novel isoenzyme of ascorbate peroxidase localized on glyoxysomal and leaf peroxisomal membranes in pumpkin. *Plant Cell Physiol.* **36**: 1157-62.

Yamashita, H., Avraham, S., Jiang, S., London, R., Van Veldhoven, P.P., *et al.* (1999) Characterization of human and murine PMP20 peroxisomal proteins that exhibit antioxidant activity *in vitro*. *J. Biol. Chem.* **274**: 29897-904.

9

GENETIC APPROACHES TO UNDERSTAND PLANT PEROXISOMES

Makoto Hayashi and Mikio Nishimura

National Institute for Basic Biology, Okazaki, Japan

KEYWORDS

PED; PEX; β-oxidation; glycolate pathway; glyoxylate cycle.

INTRODUCTION

Peroxisomes in higher plant cells are known to differentiate in function depending on the cell type. Because of the functional differentiation, plant peroxisomes are sub-divided into at least three different classes, namely glyoxysomes, leaf peroxisomes and unspecialized peroxisomes (Beevers, 1979). As already described in the previous chapters, extensive studies of each class of plant peroxisomes, especially glyoxysomes and leaf peroxisomes, had revealed that they contain special sets of enzymes that lead to a differentiated metabolic role, although they share features common to peroxisomes.

These peroxisomes are known to interconvert in their functions with each other during certain cellular processes. The functional transformation of plant peroxisomes has been most extensively studied using oil seed plants. For example, reversible interconversion between glyoxysomes and leaf peroxisomes is observed during greening and senescence of the cotyledonary cells (Titus and Becker, 1985; Nishimura *et al*, 1986; Nishimura *et al*, 1993).

A. Baker and I.A. Graham (eds.), Plant Peroxisomes, 279–303.

When the seeds germinate, seedlings start to grow using the seed reserve substances in etiolated cotyledons. After the seedlings grow and are then irradiated, the etiolated cotyledons become green and produce energy by photosynthesis. To support the drastic metabolic change, glyoxysomes are directly transformed into leaf peroxisomes during the greening process of the cotyledons. Once seedlings expand leaves, green cotyledons gradually undergo senescence. Along with the process of senescence, reverse transformation from leaf peroxisomes to glyoxysomes occurs in the cotyledonary cells. Induction of leaf peroxisomes is also found in cells of other senescent organs, such as leaves and petals. Glyoxysomes appearing in these senescent organs may be responsible for the recovery of carbon from membrane lipids in the senescent organs before they die. Although the plasticity for the peroxisomal function is a remarkable feature of higher plant peroxisomes, the mechanism involved in the regulation of these peroxisomal function in the plant cells is still obscure.

In the past ten years, significant progress in the study of peroxisome biogenesis has been made using yeast mutants. Analyses of these mutants allowed the identification of over twenty *PEX* genes (Tabak *et al*, 1999). These genes have been assigned numbers in accordance with a unified nomenclature. Proteins encoded by these *PEX* genes are called peroxins. The functions of these peroxins are discussed in detail in Chapters 11 and 12. Characterization of the peroxins indicated that more than twenty peroxins are involved in the process of peroxisome protein import, biogenesis, proliferation, and inheritance. These studies clearly indicated the usefulness of a genetical approach to study the regulatory mechanism(s) for the functions of plant peroxisomes. It is worth mentioning that the first plant mutant with a defect in peroxisomal function had been isolated almost a decade before the studies of the yeast mutants (Somerville and Ogren, 1982). This pioneering work allowed isolation of the Arabidopsis mutant called*sat* (Table 1). Recent progress in molecular genetical tools and whole genome sequencing of Arabidopsis has accelerated isolation and detailed analyses of a wide variety of Arabidopsis mutants. Some of them are known to have defects in peroxisomal functions, and their defective genes have been determined (Table 1). In this chapter, genetic studies of these mutants will be discussed in relation to the function of plant peroxisomes.

Table 1: A list of peroxisome-related mutants described in this chapter

Mutant	Screening	Defective protein	Defect related to peroxisomes
aiml	Abnormal inflorescence	The multifunctional protein (enoyl-CoA hydrataseβ-hydroxyacyl-CoA dehydrogenase)	Fatty acid β-oxidation
chyl	Indole-3-butyric acid	β-hydroxyisobutyryl-CoA hydrolase	Fatty acid β-oxidation
Delayed dehiscence1 (opr3)	Male-sterile	12-oxophytodienic acid reductase	N.D.[1]
ped1	2,4-dichlorophenoxybutyric acid	3-keotacyl CoA thiolase	Fatty acid β-oxidation Morphology of glyoxysome
ped2	2,4-dichlorophenoxybutyric acid	*At*Pex14p	Fatty acid β-oxidation Photorespiration Morphology of all peroxisomes
ped3	2,4-dichlorophenoxybutyric acid	N.D.[1]	Fatty acid β-oxidation
pex5	Indole-3-butyric acid	*At*Pex5p	Fatty acid β-oxidation
sat	CO_2 concentration	Serine-glyoxylate aminotransferase	Photorespiration
sse1	Shrunken seed phenotype	*At*Pex16p	N.D.[1]

[1]N.D.: Not determined

MUTANTS WITH DEFECTS IN LEAF PEROXISOMAL FUNCTION

Leaf peroxisomes are widely found in cells of photosynthetic organs, such as green cotyledons and leaves (Tolbert, 1982). In C3 plants, these organs have a light-dependent O_2 uptake and CO_2 release called photorespiration. This physiological phenomenon is initiated by the oxygenase reaction of ribulose bisphosphate carboxylase/oxygenase (RuBisCO; a key enzyme for CO_2 fixation in photosynthesis) that depends on the O_2 concentration and light intensity. Two phosphoglycolates, byproducts of the oxygenase reaction, are converted to produce one phosphoglycerate, an intermediate of the Calvin-Benson cycle, and one CO_2 by the photorespiratory glycolate pathway. This pathway involves many enzymatic reactions located in leaf peroxisomes, chloroplasts and mitochondria (Chapter 5). Within the entire photorespiratory glycolate pathway, leaf peroxisomes possess glycolate oxidase, hydroxypyruvate reductase and some aminotransferases (Figure 1). By the combination of these enzymes, leaf peroxisomes convert glycolate to glycine and serine to glycerate.

Figure 1: Glycolate pathway of photorespiration in photosynthetic tissue of C3 plants

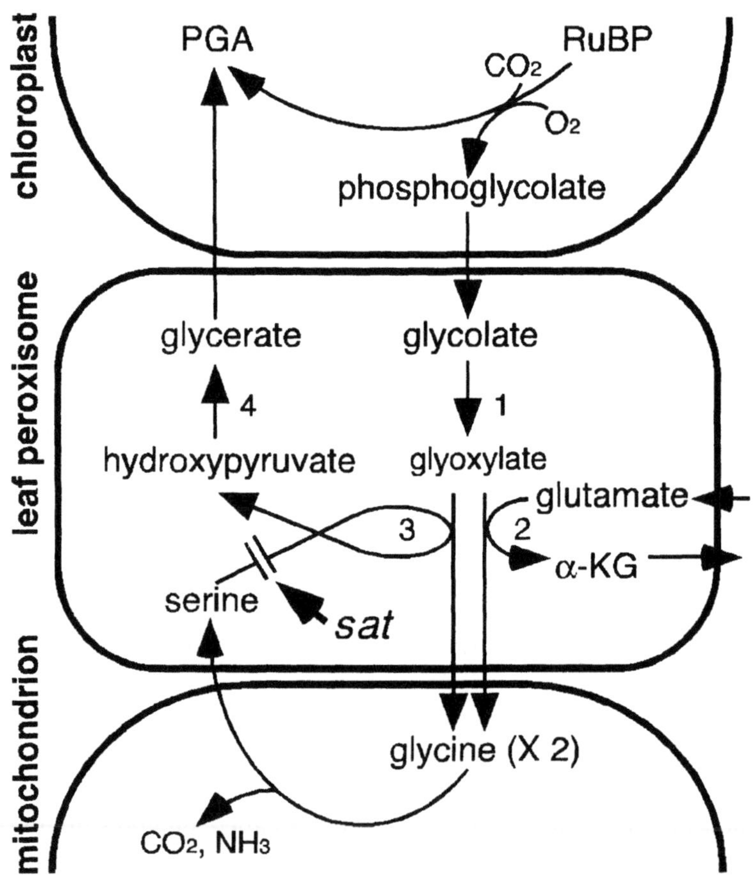

Figure 1. Within the entire photorespiratory glycolate pathway, leaf peroxisome converts glycolate to glycine and serine to glycerate. The enzymes involved in this metabolism are; 1, glycolate oxidase; 2, glutamate-glyoxylate aminotransferase; 3, serine-glyoxylate aminotransferase; 4, hydroxypyruvate reductase. The sat mutant with a defect in the serine-glyoxylate aminotransferase has been characterized.

sat mutant

In 1980, *sat*, the first plant mutant with a defect in peroxisomal function, was identified from the mutant collection with defects in the photorespiratory glycolate pathway (Somerville and Ogren, 1980). These mutants are viable in an atmosphere enriched to 1% CO_2, but inviable in a normal atmosphere (approximately 0.036% CO_2). The selection procedure was based on the observation that plants still show normal or enhanced growth even in an atmosphere containing a relatively high CO_2/O_2 ratio that inhibits photorespiration. This observation turned out to be an expectation that the mutants with reduced activity of photorespiration show conditional lethal phenotypes depending on the CO_2 concentration of the atmosphere. This screening procedure was fruitful, and seven different alleles were identified (Somerville and Ogren, 1982). Six alleles (*sat*, *pcoA*, *gluS*, *glyD*, *stm*, *dct*) have been known to induce biochemical lesions in one of the enzymatic activities for the photorespiratory glycolate pathway. Identification and analyses of these mutants provided the first experimental evidence that the photorespiratory glycolate pathway plays an important role, i.e. it removes the excess reducing power produced under certain photosynthetic conditions such as low CO_2/O_2 ratio and high light irradiation (Somerville, 2001).

Of these alleles, *sat* is the only one that reduces the activity of the leaf peroxisomal enzyme (Figure 1). The *sat* mutant is viable and exhibited normal photosynthesis under conditions that suppressed photorespiration, but they were non-viable and photosynthesized at greatly reduced rates under conditions that promoted photorespiration. In the mutants, serine and glycine accumulated as end products of photosynthesis, mostly at the expense of starch and sucrose. Because photorespiration is the major source of serine and glycine in photosynthesizing tissue, the accumulation of these amino acids in the mutants suggested a block in the photorespiratory glycolate pathway immediately after serine, in a reaction that is catalyzed by the peroxisomal enzyme, serine-glyoxylate aminotransferase (Figure 1). Indeed, analysis of leaf extracts indicated that the mutants lose the enzyme activity, while they have normal glutamate-glyoxylate and alanine-glyoxylate aminotransferase activities. By using a similar screening procedure, mutants defective in serine-glyoxylate aminotransferase were also identified in tobacco (*Nicotiana sylvestris*) and barley (*Hordeum vulgare*) (Murray *et al*, 1987; McHale *et al*, 1988).

Although these biochemical analyses suggested that the conditional lethal phenotype is due to the defect in serine-glyoxylate aminotransferase, the gene encoding the enzyme was not identified until recently. About twenty years after the isolation of the *sat* mutant, a cDNA encoding alanine-glyoxylate aminotransferase (AGT1) was isolated from Arabidopsis (Liepman and Olsen, 2001). Analyses of the AGT1 revealed that this enzyme exists in leaf peroxisomes, and catalyzes the transamination reaction using serine-glyoxylate combination as the preferred substrates, although alanine-glyoxylate and serine-pyruvate can be used. Genomic sequences of AGT1 from the *sat* mutant line revealed the presence of a single mutation that is predicted to result in a proline-to-leucine substitution at amino acid position 251 of AGT1. Overall, data indicate that *sat* mutant has a defective AGT1 gene encoding leaf peroxisomal serine-glyoxylate aminotransferase that specifically functions when the photorespiratory glycolate pathway is active (Figure 1). Because of this defect, *sat* mutants cannot metabolize phosphoglycolate by the photorespiratory glycolate pathway, and are unable to maintain sufficient activity of the Calvin-Benson cycle under a normal atmosphere. The reduced activity of the Calvin-Benson cycle may cause the conditional lethal phenotype of the *sat* mutant recognized under certain CO_2 concentrations.

MUTANTS WITH DEFECTS IN GLYOXYSOMAL FUNCTION

In higher plants, peroxisomes are the sole, or at least major, site of fatty acid β–oxidation. As described earlier, peroxisomes functionally specialized to degrade fatty acids are called glyoxysomes. Glyoxysomes are present in cells of storage organs, such as endosperms and cotyledons, during post-germinative growth of oil-seed plants, as well as in cells of senescent organs (Beevers, 1982; Nishimura *et al*, 1996). They play an important role in lipid metabolism (Figure 2). In dry seeds, large amounts of triacylglycerols accumulate as seed lipid reserves in organelles called either lipid bodies or spherosomes. In general, triacylglycerols in oil seeds mainly contain long-chain fatty acids such as palmitic acid (C16:0), stearic acid (C18:0), oleic acid (C18:1), linoleic acid (C18:2) and linolenic acid (C18:3) (Trelease and Doman, 1984). During the post-germinative growth of the seedlings, fatty acids released from the triacylglycerols are metabolized to produce sucrose. Sucrose is then transferred to shoot and root apical meristems, and provides a carbon source that is necessary for growth before the plants start photosynthesis. Within the entire gluconeogenic pathway, the conversion of fatty acids to succinate takes place in the glyoxysomes via fatty acid β-oxidation and the glyoxylate cycle.

Figure 2: Gluconeogenesis from seed lipid reserves (TGA; triacylglycerol) during germination

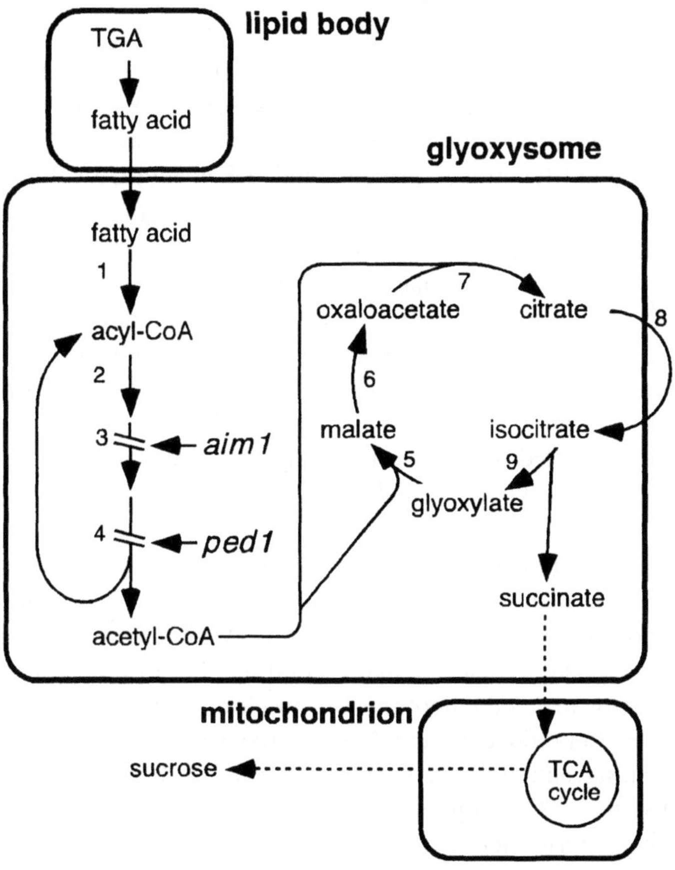

Figure 2. Within the entire gluconeogenic pathway, the conversion of fatty acids to succinate takes place in the glyoxysomes via fatty acid β-oxidation and the glyoxylate cycle. The enzymes involved in these pathways are; 1, acyl-CoA synthetase; 2, acyl-CoA oxidase; 3, the multifunctional protein possessing enoyl-CoA hydratase and 3-hydroxyacyl-CoA dehydrogenase activities; 4, 3-ketoacyl-CoA thiolase; 5, malate synthase; 6, malate dehydrogenase; 7, citrate synthase; 8, aconitase; 9, isocitrate lyase. Aconitase is the only enzyme that is not localized within the glyoxysome. Mutants with defects in the multifunctional protein (aim1) and 3-ketoacyl-CoA thiolase (ped1) have been characterized.

ped1 mutant

In 1998, Arabidopsis mutants with defects in the glyoxysomal fatty acid β–oxidation were reported (Hayashi *et al*, 1998b). These mutants were primarily identified by their resistance to a toxic level of 2,4-dichlorophenoxybutyric acid (2,4DB). The concept of the screening is based on the bioactivation at 2,4 DB to 2,4D an auxin by β-oxidation. During germination, fatty acids released from the seed reserved lipids are first activated to fatty acyl-coenzyme A (CoA) by acyl-CoA synthetase. Fatty acyl-CoA (Cn) is converted to truncated fatty acyl-CoA (Cn-2) and acetyl-CoA by a combination of four enzymatic reactions (Figure 3). They are acyl-CoA oxidase, multifunctional protein, and 3-ketoacyl-CoA thiolase. The multifunctional protein is an enzyme involved in the fatty acid β-oxidation, that possesses two enzymatic activities, enoyl-CoA hydratase and 3-hydroxyacyl-CoA dehydrogenase, in addition to two other activities, i.e. 3-hydroxyacyl-CoA epimerase and enoyl-CoA isomerase. The truncated fatty acyl-CoA (Cn-2) becomes a substrate for the same reactions. Because of this spiral pathway, fatty acyl-CoA produced from the seed lipid reserves is completely metabolized to produce acetyl-CoA. The acetyl-CoAs are subsequently converted to sucrose by many enzymatic reactions located in the glyoxysome, mitochondrion and cytosol.

The existence of fatty acid β-oxidation in plant cells was first elucidated by demonstrating the plant growth-regulating activities of a homologous series of 2,4-dicholorophenoxyalkylcarboxylic acids (Wain and Wightman, 1954). It has been demonstrated clearly that an odd number of aliphatic side chain methylene groups in 2,4-dichlorophenoxyalkylcarboxylic acids (n = 3, 5 and 7) are degraded to produce 2,4-dichlorophenoxyacetic acid (2,4D), and show growth-regulating activity. For example, when 2,4-dichlorophenoxy*butyric acid* (2,4DB; n = 3) is supplied to wild-type plants, the two methylene groups of the butyric side chain in 2,4DB are removed by the action of glyoxysomal fatty acid β-oxidation to produce the acetic side chain of a herbicide, 2,4-D (Figure 3). Because 2,4D inhibits root elongation of Arabidopsis at an early stage of seedling growth, one can assume that 2,4-DB also inhibits the root elongation of wild-type Arabidopsis by its conversion to 2,4-D, whereas the mutants that have defects in glyoxysomal fatty acid β–oxidation would no longer produce a toxic level of 2,4-D from 2,4-DB. Indeed, when wild-type Arabidopsis seeds were placed on growth media containing an appropriate concentration of 2,4-DB, the roots of the seedlings do not elongate properly (Figure 4, WT/2,4-DB). In contrast, the mutants that have defects in glyoxysomal fatty acid β–oxidation, such as *ped1*, have elongated roots in the presence of 2,4-DB (Figure 4, *ped1*/2,4-DB). These mutants are obviously different from auxin-related mutants, since 2,4-D itself inhibits their growth.

Figure 3: Fatty acid β-oxidation and the conversion of 2,4-dicholorophenoxybutyric acid (2,4-DB) to 2,4-dicholorophenoxyacetic acid (2,4-D) in the glyoxysome

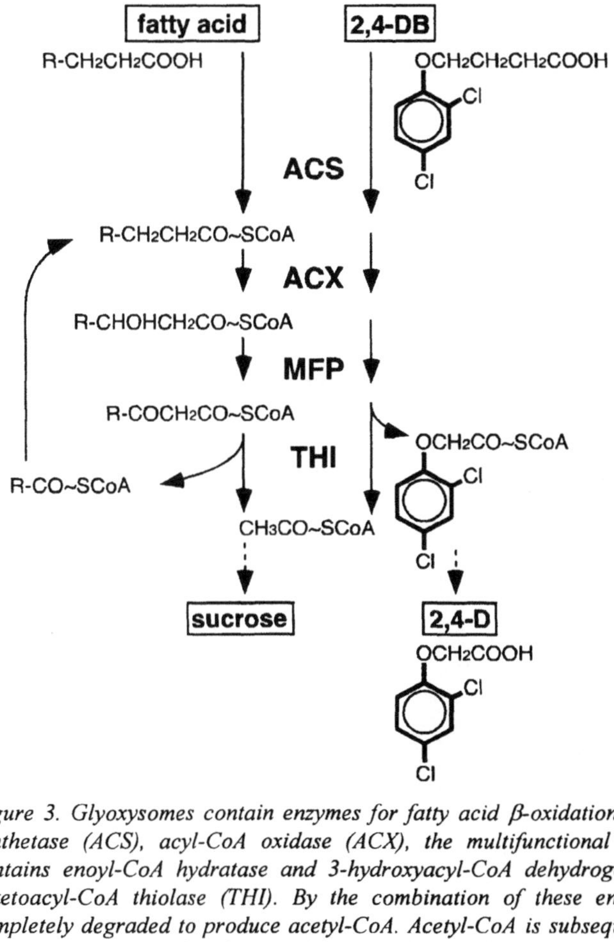

Figure 3. Glyoxysomes contain enzymes for fatty acid β-oxidation. They are acyl-CoA synthetase (ACS), acyl-CoA oxidase (ACX), the multifunctional protein (MFP) that contains enoyl-CoA hydratase and 3-hydroxyacyl-CoA dehydrogenase activities, and 3-ketoacyl-CoA thiolase (THI). By the combination of these enzymes, fatty acid is completely degraded to produce acetyl-CoA. Acetyl-CoA is subsequently metabolized in the glyoxysome, mitochondrion and the cytosol to produce sucrose. Exogenously supplied 2,4-DB is also metabolized by the same pathway, and finally converted to 2,4-D.

Twelve mutants that show resistance specifically to 2,4-DB have been identified. Of these mutant lines, four mutants are strongly inhibited by the absence of sucrose in the growth medium (cf Figure 4, *ped1*/-sucrose). Genetic analyses of these four mutant lines revealed that they can be classified as carrying alleles at three independent loci, which were designated *ped1*, *ped2* and *ped3* (Hayashi *et al*, 1998b). These mutants could grow only when sucrose was supplied to the growth medium, while the wild-type Arabidopsis seeds germinated and grew normally, regardless of the presence or absence of sucrose (Figure 4, WT/-sucrose). Once these mutants start photosynthesis, they no longer require sucrose for growth. The requirement of sucrose only for post-germinative growth of the mutants indicates that these mutants are defective in glyoxysomal fatty acid β-oxidation, and that the lack of fatty acid β-oxidation activity in glyoxysomes prevents the conversion of the seed lipid reserves into sucrose that is required for heterotrophic growth at an early stage of seedling growth. This indicates that fatty acids derived from the seed lipid reserves are predominantly degraded in glyoxysomes, and that the fatty acid β-oxidation is essential for post-germinative growth before the plant starts photosynthesis.

Figure 4: Effect of 2,4-DB and sucrose on the growth of wild-type and ped1 seedlings

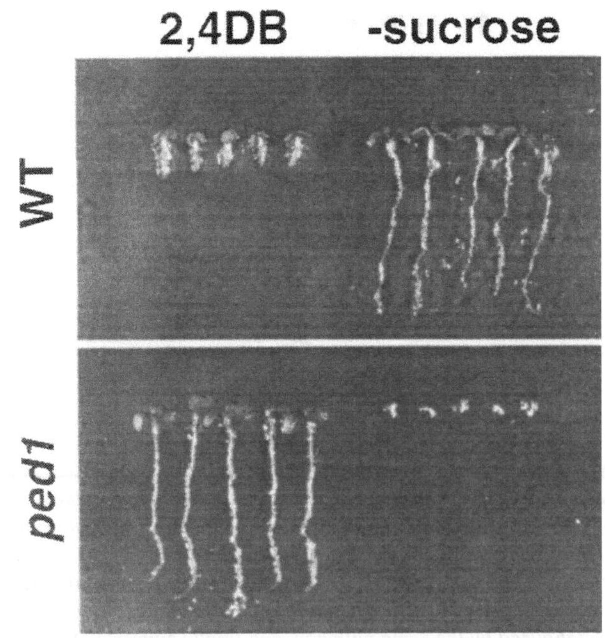

Figure 4. Wild-type Arabidopsis (WT) and ped1 mutant (ped1) seeds were grown for seven days on growth medium containing 2,4-dichlorophenoxybutyric acid (2,4-DB) or growth medium without sucrose (-sucrose) under constant illumination at 22°C.

Molecular genetical analyses revealed that one of the mutants, which is designated as *ped1,* contains a frame-shift mutation in the fourth exon of a gene for 3-ketoacyl-CoA thiolase, an enzyme for the final step of fatty acid β-oxidation (Figure 2). Immunoblot analysis indicated that the *ped1* mutant lacks a detectable amount of 3-ketoacyl-CoA thiolase in glyoxysomes. Evidently, the lack of 3-ketoacyl-CoA thiolase in glyoxysomes is the reason for the loss of fatty acid β-oxidation activity in the *ped1* mutant. Because of this defect, the *ped1* mutant cannot produce 2,4-D from 2,4-DB, and shows resistance to 2,4-DB. At the same time, it does not expand green cotyledons and leaves in the absence of sucrose, because the lack of the 3-ketoacyl-CoA thiolase prevents the production of sucrose from seed lipid reserves.

Electron microscopic analyses of the *ped1* mutant indicate that the loss of 3-ketoacyl-CoA thiolase also affects the morphology of the glyoxysomes, while leaf peroxisomes and unspecialized peroxisomes in the mutant have normal structures (Hayashi *et al*, in press). This suggests that the morphological defect is limited to the organelle that actively metabolizes fatty acids. The glyoxysomes in cotyledonary cells of wild-type Arabidopsis are approximately 0.5 μm in diameter, and have a round or oval shape with a uniform matrix. In contrast, glyoxysomes in the *ped1* mutant become enlarged organelles, probably because of accumulation of a metabolic intermediate for fatty acid β-oxidation by the lack of 3-ketoacyl-CoA thiolase. The largest glyoxysomes in the *ped1* mutant are more than 2 μm in diameter. Surprisingly, these enlarged glyoxysomes contain many electron-translucent, tubular structures surrounded by membranes. Serial sections revealed that these tubular structures are connected to each other to form several branched tubes inside the enlarged glyoxysome. These tubes are finally connected to the glyoxysomal membrane to make several holes on the surface of the glyoxysome. It is worth mentioning the following two points: (i) the hole always borders the lipid body; and (ii) the tubes often contain small vesicles whose membranes appear to be independent of the glyoxysomal membrane. These morphological observations suggest a direct interaction between glyoxysomes and lipid bodies. The small vesicles found in the tube may be derived from the lipid body and reflect an unknown mechanism for incorporating the fatty acids from the lipid body to glyoxysomes as substrates for fatty acid β-oxidation. Further analyses of the *ped1* mutant may give us a clue to resolve this problem.

aim1 mutant

Another *Arabidopsis* mutant called *abnormal inflorescence meristem 1* (*aim1*) is also known to have a defect in fatty acid β-oxidation (Richmond and Bleecker, 1999). This mutant was originally identified by its abnormal inflorescence and floral development from a T-DNA tagged mutant lines. The *aim1* mutant shows a wide range of morphological phenotypes associated with reproductive development. For example, the inflorescence meristem sometimes does not produce any recognizable floral structures, but rather produces only a small mass of undifferentiated tissue that subsequently ceases to develop altogether. More often, the inflorescence meristems will produce a few to several floral structures in what appears to be a normal spiral pattern before terminating. There are often many floral organs missing from each flower, and there are often homeotic conversions. Although many of these phenotypes are similar to the phenotypes produced by the defects of previously identified genes regulating the morphology of reproductive organs in Arabidopsis, genetic analyses revealed that the *AIM1* gene does not coincide with any of these genes, but encodes a protein showing extensive similarity to the cucumber multifunctional protein. Indeed, a recombinant protein produced from the *AIM1* cDNA has enoyl-CoA hydratase activity, one of the activities known to be involved in the multifunctional protein. These data suggest that the *aim1* mutant has a defective multifunctional protein involved in the fatty acid β-oxidation (Figure 2). This conclusion is supported by the results showing that the *aim1* mutant is resistant to a toxic level of 2,4-DB, and has altered fatty acid composition of leaf cells. However, the *aim1* mutant shows a sucrose-independent germination phenotype, although the *ped1* mutant requires sucrose for germination. One of the explanations for this phenotype is that *AIM1* shares a role in fatty acid β-oxidation with another isozyme of the multi-functional protein, *At*MFP2, existing in the Arabidopsis genome.

The role of *AIM1* in determining the morphology of the reproductive organs is unclear. There are several possibilities but the most attractive explanation is that *aim1* blocks the production of a lipid signaling molecule that is necessary for intercellular communication in the reproductive meristems (Richmond and Bleecker, 1999). The possible candidates for fatty acid-derived signal molecules are jasmonic acid, the traumatin family and related alkenals, highly oxygenated fatty acid derivatives and lipo-oligosaccharides. It is worth mentioning analyses of two independently identified male-sterile Arabidopsis mutants, *delayed dehiscence1* and *opr3* (Sanders *et al*, 2000;

Stintzi and Browse, 2000). The male-sterile phenotype can be rescued by exogenous application of jasmonic acid. These T-DNA-mutagenized lines have T-DNA insertion, within a gene encoding a third isoenzyme of 12-oxophytodienoic acid reductase (OPR3). Comparison of enzymatic properties between OPR3 and other two isoenzymes, OPR1 and OPR2, indicated that OPR3 is the isoenzyme relevant for jasmonate biosynthesis (Schaller *et al*, 2000). The presence of peroxisomal targeting signal (PTS1; see below) at the carboxyl terminus of OPR3 indicates that the protein is probably resident in the peroxisome (Stintzi and Browse, 2000). There is no such signal in OPR1 and OPR2. Although the subcellular localization of OPR3 still needs experimental confirmation, it is most likely that the peroxisome is the site of reaction catalyzed by OPR3 and the subsequent fatty acid β-oxidation required for jasmonic acid biosynthesis. Future analyses of these mutants may give us new insight into unidentified function(s) for plant peroxisomes.

chy1 mutant

Indole-3-butyric acid (IBA) is an endogenous auxin. Recently, IBA was used for the isolation of IBA-resistant *Arabidopsis* mutants (Zolman *et al*, 2000). As is the case for 2,4-DB, two methylene groups of the butyric side chain in IBA are removed by the action of fatty acid β-oxidation to produce the acetic side chain in a plant hormone, indole-3-acetic acid (IAA). These IBA-resistant mutants fall into four classes. One of these classes contains six independent loci. Mutations in these loci appear to induce defects in the utilization of seed lipid reserves. Mapping of these loci revealed that none of them coincides with the defective genes in the previously isolated peroxisome-defective mutants, such as *sat*, *ped1*, *ped2*, *ped3*, *aim1*, and *sse1*.

The *chy1* mutant was identified from the collection of the IBA-resistant mutants. Poor growth in the absence of supplemental sucrose suggests that the *chy1* mutant has a defect in fatty acid β-oxidation. Molecular genetic analyses revealed that *CHY1* encodes a peroxisomal protein that is 43% identical to a mammalian valine catabolic enzyme, β-hydroxy-isobutyryl-CoA hydrolase (Zolman *et al*, 2001). Interestingly, the mammalian orthologue of this enzyme is localized in mitochondria. The branched amino acids including valine, leucine, and isoleucine can be broken down to generate energy during germination, senescence or carbon starvation. The subcellular localization of branched chain amino acid catabolism in plants remains controversial. The existence of peroxisomal β-hydroxyisobutyryl-CoA hydrolase strongly supports the idea that valine catabolism occurs at least partially in peroxisomes of higher plants. Although the relationship between valine catabolism and fatty acid β-oxidation is unclear, it is hypothesized that the loss of β-hydroxyisobutyryl-CoA hydrolase in the *chy1* mutant indirectly disrupts both fatty acid β-oxidation and the conversion of IBA to IAA because the toxic intermediate, methacryl-CoA, accumulates in the glyoxysomes.

Mutants obtained by reverse genetics

The availability of the complete genome sequence of *Arabidopsis thaliana* and large populations of plants with randomly located T-DNA or transposon insertions has combined to make reverse genetics a very powerful strategy for obtaining mutants in peroxisome function. Several mutants have been successfully isolated using this procedure. For example, the *kat2* mutant produced by a T-DNA insertion in a gene for 3-ketoacyl CoA thiolase was identified (Germain *et al*, 2001). This is the same gene responsible for the *ped1* mutant described above. The *kat2* mutant showed an identical phenotype to the *ped1* mutant. This result clearly indicates the potential of the reverse genetic approach for future research into plant peroxisomes. Similarly, mutants have been isolated for acyl CoA oxidase (Eastmond *et al*, 2000b), isocitrate lyase and malate synthase (Eastmond *et al*, 2000a). These mutants are discussed in detail in Chapters 2 and 3.

MUTANTS WITH DEFECTS IN *PEX* ORTHOLOGUES

Remarkable progress in recent studies on peroxisomes has been made by molecular genetical approaches using the yeast *pex* mutants. These mutants led to the rapid identification of over twenty yeast *PEX* genes through functional complementation cloning, and the identification of their products, the peroxins, that play an important role in the process of peroxisome protein import, biogenesis, proliferation, and inheritance. As illustrated in Figure 5, some of the *PEX* genes, such as *PEX5*, *PEX7* and *PEX14*, are known to directly involved in the peroxisomal protein import machinery (Subramani *et al*, 2000). Information on the yeast *PEX* genes gave a strong influence on identification of *PEX* orthologues from other organisms, such as human *PEX* genes, whose defects cause human peroxisome biogenesis disorders, such as Zelweger cerebrohepatorenal syndrome (Sacksteder and Gould, 2000).

ped2 mutant

There are few Arabidopsis mutants so far that have defects in plant *PEX* gene orthlogs. One of them is the *ped2* mutant (Hayashi *et al*, 2000). *ped2* was identified from the Arabidopsis mutants that show resistance to a toxic level of 2,4-DB (Hayashi *et al*, 1998b). As is the case for *ped1* described above, *ped2* has reduced activity for fatty acid β-oxidation. Detailed analyses revealed that the *ped2* mutant has defects in intracellular transport of peroxisomal matrix proteins.

Figure 5: Hypothetical model for import of plant peroxisomal matrix proteins

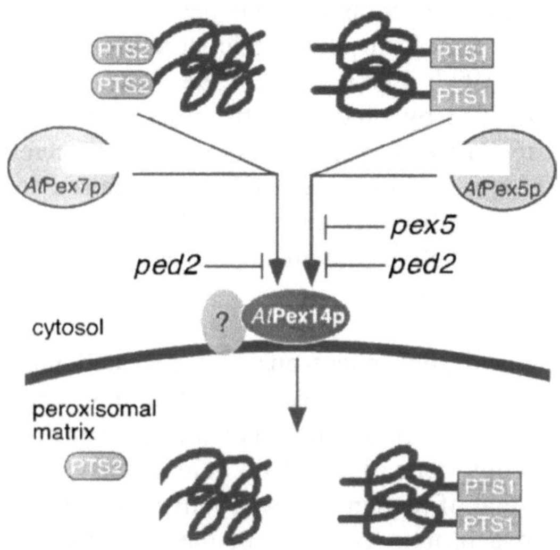

Figure 5. Nascent polypeptides are synthesized and folded in the cytosol. The oligomeric proteins with either carboxy-terminal PTS1 or amino-terminal PTS2 are recognized by a PTS1 receptor (AtPex5p) and a PTS2 receptor (AtPex7p), respectively. These PTS receptors bind to their cargo, and are then recognized by AtPex14p located in the peroxisomal membrane. AtPex14p may form receptor complex with other proxins as indicated by question mark. Finally, the oligomeric proteins are translocated into the peroxisomal matrix. Amino-terminal presequences of PTS2-containing proteins are processed after the translocation within the peroxisomes. The ped2 mutant has defects in the intracellular transport pathways for both PTS1-containing and PTS2-containing proteins. The pex5 mutant may have a defect at least in the intracellular transport of PTS1-containing proteins.

Figure 5 illustrates the molecular mechanism for the import of plant peroxisomal matrix proteins. The peroxisomal matrix proteins are translated on free polysomes in the cytosol, and function after their post-translational transport into peroxisomes. Most of the plant peroxisomal enzymes have been shown to contain one of two peroxisome targeting signals within their amino acid sequences (Figure 5). One of the targeting signals is a unique tripeptide sequence found in the carboxy-terminus of the proteins (Hayashi *et al*, 1996a; Trelease *et al*, 1996). The tripeptide sequence is designated as PTS1. The permissible combinations of tripeptide sequence for plant PTS1 are [C/A/S/P]-[K/R]-[I/L/M] (Hayashi *et al*, 1997). Another type of targeting signal is involved in a cleavable amino-terminal presequence (Gietl *et al*, 1994). The amino-terminal presequences contain a consensus sequence, [R]-[L/Q/I]-X5-[H]-[L] (X stands for any amino acid), designated as PTS2 (Kato *et al*, 1996a; Kato *et al*, 1998). These proteins are synthesized as precursor proteins that show a higher molecular mass due to the amino-terminal presequence. The amino-terminal presequence is processed to form the mature protein after its transport into peroxisomes. Proteins containing either PTS1 or PTS2 may form oligomeric structures after translation in the cytosol (Lee *et al*, 1997; Flynn *et al*, 1998; Kato *et al*, 1999), and are subsequently recognized by binding with either a PTS1 receptor or PTS2 preceptor, that are called Pex5p and Pex7p, respectively. In Figure 5, Arabidopsis orthologues of these receptors are referred to as *At*Pex5p and *At*Pex7p, respectively (Brickner *et al*, 1998; Schumann *et al*, 1999). A detailed discussion of matrix protein import and function of *PEX* genes is presented in Chapters 11 and 12.

Positional cloning and subsequent analyses revealed that the *PED2* gene encodes *At*Pex14p, a 75-kDa peroxisomal membrane protein (Figure 5). Similarity search of the Arabidopsis genome sequence revealed that the *PED2* is a single copy gene. The amino acid sequence of the *At*Pex14p is 29.6% identical to that of human Pex14p (Will *et al*, 1999). Pex14p has been elucidated to form a protein complex in the peroxisomal membrane with other peroxins exist in peroxisomal membrane, such as Pex13p and Pex17p (Albertini *et al*, 1997; Brocard *et al*, 1997; Komori *et al*, 1997; Huhse *et al*, 1998; Girzalsky *et al*, 1999). This protein complex is sometimes called a docking protein complex, since Pex14p binds with both Pex5p and Pex7p following the initial recognition of PTS1-containing and PTS2-containing proteins in the cytosol. Thus, Pex14p is believed to be a point of convergence of the PTS1-dependent and PTS2-dependent peroxisomal protein import pathway. Indeed, loss of *At*Pex14p in the *ped2* mutant reduces the activity for import of both PTS1-containing and PTS2-containing proteins into peroxisomes, and induces accumulation of these peroxisomal proteins in the cytosol. Consistent with this an antibody against human Pex14p blocks binding of matrix, but not membrane proteins to isolated plant peroxisomes (Lopez-Huertas *et al*, 1999). In higher plant cells, all enzymes involved in fatty acid β-oxidation and the glyoxylate cycle except cytosolic aconitase are predominantly localized in the glyoxysome (Hayashi, 2000). Among these enzymes, short-chain acyl-CoA oxidase (Hayashi *et al*, 1999), the multifunctional protein (Preisig-Muller *et al*, 1994), malate synthase (Mori *et al*, 1991), and isocitrate lyase (Mano *et al*, 1996) are PTS1-containing proteins, whereas long-chain acyl-CoA oxidase (Hayashi *et al*, 1998a), 3-ketoacyl-CoA thiolase (Kato *et al*, 1996b), malate dehydrogenase (Kato *et al*, 1998) and citrate synthase (Kato *et al*, 1995) are PTS2-containing proteins. The defect in the *PED2* gene resulted in reduced levels of enzymes in glyoxysomes, which reduced the ability of glyoxysomes to carry out fatty acid β-oxidation. The loss of glyoxysomal matrix proteins also reduced the density of the glyoxysomes.

Electron microscopic analyses indicated that not only glyoxysomes, but also leaf peroxisomes and unspecialized peroxisomes have abnormal shrunken structures in the *ped2* mutant. In addition, the *ped2* mutant had a defect in the physiological function of leaf peroxisomes, photorespiration. Measurements of maximal quantum yield of photosystem II suggested that the activity of photorespiration in the *ped2* mutant is lower than that in wild-type plants but higher than that in *stm*, a mutant with a defect in mitochondrial serine transhydroxymethylase that is involved in the photorespiratory glycolate pathway. Leaf peroxisomes contain enzymes for photorespiration such as hydroxypyruvate reductase, glycolate oxidase and serine-glyoxylate aminotransferase. These enzymes contain PTS1 at their carboxy-termini (Tsugeki *et al*, 1993; Hayashi *et al*, 1996b; Mano *et al*, 1997; Mano *et al*, 1999; Liepman and Olsen, 2001). Since the *ped2* mutant has a weakened ability to import PTS1-containing proteins, leaf peroxisomes in the mutant also contain reduced amounts of these enzymes. Although photorespiration consists of many enzymatic reactions located not only in leaf peroxisomes, but also in chloroplasts and mitochondria (Tolbert, 1982), reduced amounts of leaf peroxisomal enzymes diminish the overall activity of photorespiration in the *ped2* mutant. Because of this, *ped2* grows normally in an atmosphere enriched to 1% CO_2, but shows a growth defect and has yellowish leaves in a normal atmosphere. The phenotype caused by the defect in photorespiration is weaker, but similar to the photorespiration-defective mutants mentioned above.

These observations strongly suggest that *At*Pex14p is a key component of the peroxisomal matrix protein import machinery which maintains physiological functions of all types of plant peroxisomes by binding with *At*Pex5p and *At*Pex7p (Figure 5). However, the detailed molecular mechanisms of *At*Pex14p in peroxisomal protein import in conjunction with the function of *At*Pex5p and *At*Pex7p remain to be clarified.

pex5 mutant

The Arabidopsis *pex5* mutant was identified from the collection of IBA-resistant mutants described above (Zolman *et al*, 2000). Detailed analyses revealed that the loci corresponds to the *PEX5* orthologue, which encodes a PTS1 receptor (Brickner *et al*, 1998; Kragler *et al*, 1998; Wimmer *et al*, 1998). However, the phenotype of the Arabidopsis *pex5* mutant in relation to the peroxisomal functions and intracellular transport of peroxisomal proteins have not yet been characterized.

Analyses of the *pex5* mutant in several organisms indicated that the recognition of PTS1-containing proteins begins by the binding of PTS1 and Pex5p in the cytosol (Figure 5). *PEX5* was first identified in *Pichia pastoris*, and the mutant was notable for its specific defect in PTS1-mediated import (McCollum *et al*, 1993). Differing from the PTS1-mediated import pathway, PTS2-containing proteins are recognized by a PTS2 receptor, called Pex7p (Figure 5). *PEX7* was first identified in *Saccharomyces cerevisiae* (Marzioch *et al*, 1994). The *pex7* mutant has a defect in PTS2-mediated import, but has no defect in PTS1-mediated import defect, indicating that *PEX7* and *PEX5* are functionally independent in these yeast. By contrast, the human cell line with null alleles of *PEX5* has a severe defect in both PTS1- and PTS2-mediated import, while a cell line with a defective *PEX7* gene has a PTS2-specific import defect (Dodt *et al*, 1995). These data suggest that, in mammalian cells, *PEX7* is dependent on *PEX5* for proper function, while *PEX5* is not dependent on *PEX7*.

The Arabidopsis *pex5* mutant may also have defects at least in intracellular transport of PTS1-containing proteins (Figure 5), as is the case of *pex5* mutants in other organisms. In contrast, import competency of PTS2-containing proteins in the mutant is obscure. This is because of the different behaviors of the *PEX7*-dependent import pathway in different organisms. Whether the Arabidopsis *pex5* mutant is defective only in the PTS1-mediated import pathway or has defects both in PTS1- and PTS2-mediated import pathways remains to be clarified. It would be interesting to determine the functions of peroxisomes in the Arabidopsis *pex5* mutant. The present data indicate the mutant has a defect at least in glyoxysomal function. Does

*At*Pex5p maintain physiological function(s) of only glyoxysomes or all types of plant peroxisomes? There remain many interesting questions to be answered.

sse1 mutant

A *shrunken seed 1* (*sse1*) mutant was selected by its shrunken seed phenotype from a collection of Arabidopsis T-DNA tagged mutant lines (Lin *et al*, 1999). The *sse1* seeds shrink upon dessication (a likely consequence of insufficient deposition of storage molecules), whereas the wild-type seeds are dessication-tolerant. The homozygous *sse1* plant produces 90% shrunken seeds and 10% normal round seeds. The shrunken seeds are not viable. However, plants grown from the round seeds produce 90% shrunken seeds. Genetical analyses revealed that *sse1* behaves as a typical single-recessive Mendelian gene. The *SSE1* gene and its cDNA were cloned. The *SSE1* cDNA encodes a protein similar to Pex16p, a membrane-associated peroxin identified in the yeast, *Yarrowia lipolytica* (Eitzen *et al*, 1997). Pex16p in the yeast is a glycosylated 45.5 kD protein residing mainly on the matrix face of the peroxisome, but transiently in the endoplasmic reticulum (Titorenko and Rachubinski, 1998). Overexpression of the protein causes accumulation of enlarged peroxisomes. The yeast *pex16* mutant can not grow on oleic acid as a sole carbon source, but *SSE1* partially complements the phenotype of the yeast mutant. This indicates that at least a part of the functions of the*SSE1* is identical to that of the *PEX16* in yeast cells. These data suggest that the *sse1* mutant has defective peroxisomes. However, detailed analyses of peroxisomes in the *sse1* mutant have not yet been reported.

In Arabidopsis, proteins and lipids are the major reserves in mature seeds. These storage compounds are deposited into specific organelles, i.e. protein bodies and lipid bodies, respectively, in the storage organs, such as cotyledonary cells, of dry seeds. The *sse1* mutant alters this seed storage profile, accumulating starch over proteins and lipids. The storage organs of the *sse1* seed contain no recognizable protein bodies and few lipid bodies. By contrast, the *sse1* cells contain starch granules, membrane stacks, vesicles and vacuoles, in spite of the fact that these structures are absent in the wild-type cells. The lipid bodies in *sse1* contain more electron-dense substances than the wild-type cells. This indicates that *SSE1* is required for the formation of protein bodies and lipid bodies. The reason why the defect in *SSE1* results in accumulation of excess starch over lipids and proteins in seed is unclear. The functional similarity to Pex16p, however, excludes the possibility that *SSE1* suppresses starch synthesis in wild-type seeds.

Detailed analyses of the *sse1* mutant in relation to the peroxisomal functions may help us to understand the molecular mechanisms underlying the biogenesis and differentiation of not only peroxisomes but also other organelles, such as protein bodies, lipid bodies and amyloplasts.

CONCLUSION

The recent progress in molecular genetical tools as well as the whole genome sequence of Arabidopsis allows us to access plant mutants easily. Indeed, the number of Arabidopsis mutants related to the peroxisomal functions has rapidly increased. Recent findings of new mutants suggest that there are still more mutants to be identified. Indeed, there are twenty three *pex* mutants identified in yeast, and fourteen complementation groups for peroxisome biogenesis disorder (PBD) human cell lines (Sacksteder and Gould, 2000). There is still a strong possibility to design more extensive screens and identify more plant mutants. As described above, these mutants may be useful in elucidating the unidentified functions, biogenesis, mechanism of protein trafficking and mechanism of functional differentiation of the plant peroxisomes. Careful and detailed analyses of these mutants in relation to the peroxisomal functions at the molecular level are also indispensable.

ACKNOWLEDGEMENT

The authors deeply thank all co-workers in our laboratory for their excellent techniques and critical discussions. This work was supported, in part, by a grant from the Research for the Future Program of the Japanese Society for the Promotion of Science (JSPS-RFTF96L00407), grants-in-aid for scientific research from the Ministry of Education, Science and Culture of Japan (12440231 to M.N. and 12640625 to M.H.), and CREST of JST (Japan Science and Technology) to M.H.

REFERENCES

Albertini, M., Rehling, P., Erdmann, R., Girzalsky, W., Kiel, J.A.K.W., Veenhuis, M. and Kunau, W.H. (1997) Pex14p, a peroxisomal membrane protein binding both receptors of the two PTS-dependent import pathways. *Cell.* **89**: 83-92.

Beevers, H. (1982) Glyoxysomes in higher plants. *Annals NY Acad. Sci.* **386**: 243-251.

Beevers, H. (1979) Microbodies in higher plants. *Annu. Rev. Plant Physiol.* **30**: 159-193.

Brickner, D.G., Brickner, J.H. and Olsen, J.J. (1998) Sequence analysis of a cDNA encoding Pex5p, a peroxisomal targeting signal type 1 receptor from Arabidopsis. *Plant Physiol.* **118**: 330.

Brocard, C., Lametschwandtner, G., Koudelka, R. and Hartig, A. (1997) Pex14p is a member of the protein linkage map of Pex5p. *EMBO J.* **16**: 5491-5500.

Dodt, G., Braverman, N., Wong, C., Moser, A., Moser, H.W., Watkins, P., Valle, D. and Gould, S.J. (1995) Mutations in the PTS1 receptor gene, *PXR1*, define complementation group 2 of the peroxisome biogenesis disorders. *Nature Genet.* **9**: 115-125.

Eastmond, P.J., Germain, V., Lange, P.R., Bryce, J.H., Smith, S.M. and Graham, I.A. (2000a) Post-germinative growth and lipid catabolism in oilseeds lacking the glyoxylate cycle. *Proc. Nat. Acad. Sci. USA.* **97**: 5669-5674.

Eastmond, P.J., Hooks, M.A., Williams, D., Lange, P., Bechtold, N., Sarrobert, C., Nussaume, L. and Graham, I.A. (2000b) Promoter trapping of a novel medium-chain acyl-CoA oxidase, which is induced transcriptionally during Arabidopsis seed germination. *J. Biol. Chem.* **275**: 34375-34381.

Eitzen, G.A., Szilard, R.K. and Rachubinski, R.A. (1997) Enlarged peroxisomes are present in oleic acid-grown *Yarrowia lipolytica* overexpressing the *PEX16* gene encoding an intraperoxisomal peripheral membrane peroxin. *J. Cell. Biol.* **137**: 1265-1278.

Flynn, C.R., Mullen, R.T. and Trelease, R.N. (1998) Mutational analyses of a type 2 peroxisomal targeting signal that is capable of directing oligomeric protein import into tobacco BY-2 glyoxysomes. *Plant J.* **16**: 709-720.

Germain, V., Rylott, E., Larson, T.R., Sherson, S.M., Bechtold, N., Carde, J-P., Bryce, J.H., Graham, I. and Smith, M.S. (2001) Requirement for 3-ketoacyl-CoA thiolase-2 in peroxisome development, fatty acid β-oxidation and breakdown of triacylglycerol in lipid bodies of Arabidopsis seedlings. *Plant J.* in press.

Gietl, C., Faber, K.N., Vanderklei, I.J. and Veenhuis, M. (1994) Mutational analysis of the N-terminal topogenic signal of watermelon glyoxysomal malate dehydrogenase using the heterologous host *Hansenula polymorpha*. *Proc. Natl. Acad. Sci. USA.* **91**: 3151-3155.

Girzalsky, W., Rehling, P., Stein, K., Kipper, J., Blank, L., Kunau, W.H. and Erdmann, R. (1999) Involvement of Pex13p in Pex14p localization and peroxisomal targeting signal 2-dependent protein import into peroxisomes. *J. Cell Biol.* **144**: 1151-1162.

Hayashi, M. (2000) Plant peroxisomes: molecular basis of the regulation of their functions. *J. Plant Res.* **113**: 103-109.

Hayashi, M., Aoki, M., Kondo, M. and Nishimura, M. (1997) Changes in targeting efficiencies of proteins to plant microbodies caused by amino acid substitutions in the carboxy-terminal tripeptide. *Plant Cell Physiol.* **38**: 759-768.

Hayashi, M., Aoki, M., Kato, A., Kondo, M. and Nishimura, M. (1996a) Transport of chimaeric proteins that contain a carboxy-terminal targeting signal into plant microbodies. *Plant J.* **10**: 225-234.

Hayashi, H., De Bellis, L., Ciurli, A., Kondo, M., Hayashi, M. and Nishimura, M. (1999) A novel acyl-CoA oxidase that can oxidize short-chain acyl-CoA in plant peroxisomes. *J. Biol. Chem.* **274**: 12715-12721.

Hayashi, H., De Bellis, L., Yamaguchi, K., Kato, A., Hayashi, M. and Nishimura, M. (1998a) Molecular characterization of a glyoxysomal long chain acyl-CoA oxidase that is synthesized as a precursor of higher molecular mass in pumpkin. *J. Biol. Chem.* **273**: 8301-8307.

Hayashi, Y., Hayashi, H., Hayashi, M., Hara-Nishimura, I. and Nishimura, M. (2001) Direct interaction between glyoxysomes and lipid bodies in cotyledons of the *Arabidopsis thaliana ped1* mutant. *Protoplasma*, in press.

Hayashi, M., Nito, K., Toriyama-Kato, K., Kondo, M., Yamaya, T. and Nishimura, M. (2000) *At*Pex14p maintains peroxisomal functions by determining protein targeting to three kinds of plant peroxisomes. *EMBO J.* **19**: 5701-5710.

Hayashi, M., Toriyama, K., Kondo, M. and Nishimura, M. (1998b) 2,4-dichlorophenoxybutyric acid-resistant mutants of Arabidopsis have defects in glyoxysomal fatty acid β-oxidation. *Plant Cell.* **10**: 183-195.

Hayashi, M., Tsugeki, R., Kondo, M., Mori, H. and Nishimura, M. (1996b). Pumpkin hydroxypyruvate reductases with and without a putative C-terminal signal for targeting to microbodies may be produced by alternative splicing. *Plant Mol. Biol.* **30**: 183-189.

Huhse, B., Rehling, P., Albertini, M., Blank, L., Meller, K. and Kunau, W.H. (1998) Pex17p of *Saccharomyces cerevisiae* is a novel peroxin and component of the peroxisomal protein translocation machinery. *J. Cell Biol.* **140**: 49-60.

Kato, A., Hayashi, M. and Nishimura, M. (1999) Oligomeric proteins containing N-terminal targeting signals are imported into peroxisomes in transgenic Arabidopsis. *Plant Cell Physiol.* **40**: 586-591.

Kato, A., Hayashi, M., Kondo, M. and Nishimura, M. (1996a) Targeting and processing of a chimaeric protein with the N-terminal presequence of the precursor to glyoxysomal citrate synthase. *Plant Cell.* **8**: 1601-1611.

Kato, A., Hayashi, M., Mori, H. and Nishimura, M. (1995) Molecular characterization of a glyoxysomal citrate synthase that is synthesized as a precursor of higher molecular mass in pumpkin. *Plant Mol. Biol.* **27**: 377-390.

Kato, A., Hayashi, M., Takeuchi, Y. and Nishimura, M. (1996b) cDNA cloning and expression of a gene for 3-ketoacyl-CoA thiolase in pumpkin cotyledons. *Plant Mol. Biol.* **31**: 843-852.

Kato, A., Takeda-Yoshikawa, Y., Hayashi, M., Kondo, M., Hara-Nishimura, I. and Nishimura, M. (1998) Glyoxysomal malate dehydrogenase in pumpkin: Cloning of a cDNA and functional analysis of its presequence. *Plant Cell Physiol.* **39**: 186-195.

Komori, M., Rasmussen, S.W., Kiel, J.A.K.W., Baerends, R.J.S., Cregg, J.M., Van der Klei, I.J. and Veenhuis, M. (1997). The *Hansenula polymorpha PEX14* gene encodes a novel peroxisomal membrane protein essential for peroxisome biogenesis. *EMBO J.* **16**: 44-53.

Kragler, F., Lametschwandtner, G., Christmann, J., Hartig, A. and Harada, J.J. (1998) Identification and analysis of the plant peroxisomal targeting signal 1 receptor NtPEX5. *Proc. Natl. Acad. Sci. USA*. **95**: 13336-13341.

Lee, M.S., Mullen, R.T. and Trelease, R.N. (1997) Oilseed isocitrate lyases lacking their essential type 1 peroxisomal targeting signal are piggy-backed to glyoxysomes. *Plant Cell*. **9**: 185-197.

Liepman, A.H. and Olsen, L.J. (2001) Peroxisomal alanine: glyoxylate aminotransferase (AGT1) is a photorespiratory enzyme with multiple substrates in *Arabidopsis thaliana*. *Plant J*. **25**: 487-498.

López-Huertas, E., Oh, J.S. and Baker, A. (1999). Antibodies against Pex14p block ATP-independent binding of matrix proteins to peroxisomes *in vitro*. *FEBS Lett*. **459**: 227-229.

Lin, Y., Sun, L., Nguyen, L.V., Rachubinski, R.A. and Goodman, H.M. (1999) The Pex16p homolog *SSE1* and storage organelle formation in Arabidopsis seeds. *Science*. **284**: 328-330.

Mano, S., Hayashi, M. and Nishimura, M. (1999). Light regulates alternative splicing of hydroxypyruvate reductase in pumpkin. *Plant J*. **17**: 309-320.

Mano, S., Hayashi, M., Kondo, M. and Nishimura, M. (1997) Hydroxypyruvate reductase with a carboxy-terminal targeting signal to microbodies is expressed in Arabidopsis. *Plant Cell Physiol*. **38**: 449-455.

Mano, S., Hayashi, M., Kondo, M. and Nishimura, M. (1996) cDNA cloning and expression of a gene for isocitrate lyase in pumpkin cotyledons. *Plant Cell Physiol*. **37**: 941-948.

Marzioch, M., Erdmann, R., Veenhuis, M. and Kunau, W.H. (1994) *PAS7* encodes a novel yeast member of the WD-40 protein family essential for import of 3-oxoacyl-CoA thiolase, a PTS2-containing protein, into peroxisomes. *EMBO J*. **13**: 4908-4918.

McCollum, D., Monosov, E. and Subramani, S. (1993) The *pas8* mutant of *Pichia pastoris* exhibits the peroxisomal protein import deficiencies of Zellweger syndrome cells. The *PAS8* protein binds to the COOH-terminal tripeptide peroxisomal targeting signal, and is a member of the TPR protein family. *J. Cell Biol*. **121**: 761-774.

McHale, N.A., Havir, E.A. and Zelitch, I. (1988) A mutant of *Nicotiana sylvestris* deficient in serine: glyoxylate aminotransferase activity. *Theor. Appl. Genet*. **76**: 71-75.

Mori, H., Takeda-Yoshikawa, Y., Hara-Nishimura, I. and Nishimura, M. (1991) Pumpkin malate synthase: cloning and sequencing of the cDNA and Northern blot analysis. *Eur. J. Biochem*. **197**: 331-336.

Murray, A.J.S., Blackwell, R.D., Joy, K.W. and Lea, P.J. (1987) Photorespiratory N donors, aminotransferase specificity and photosynthesis in a mutant of barley deficient in serine: glyoxylate aminotransferase activity. *Planta*. **172**: 106-113.

Nishimura, M., Hayashi, M., Kato, A., Yamaguchi, K. and Mano, S. (1996) Functional transformation of microbodies in higher plant cells. *Cell Struct. Funct*. **21**: 387-393.

Nishimura, M., Takeuchi, Y., De Bellis, L. and Hara-Nishimura, I. (1993) Leaf peroxisomes are directly transformed to glyoxysomes during senescence of pumpkin cotyledons. *Protoplasma.* **175**: 131-137.

Nishimura, M., Yamaguchi, J., Mori, H., Akazawa, T. and Yokota, S. (1986) Immunocytochemical analysis shows that glyoxysomes are directly transformed to leaf peroxisomes during greening of pumpkin cotyledons. *Plant Physiol.* **80**: 313-316.

Preisig-Müller, R., Gühnemann-Schäfer, K. and Kindl, H. (1994) Domains of the tetrafunctional protein acting in glyoxysomal fatty acid β-oxidation. Demonstration of epimerase and isomerase activities on a peptide lacking hydratase activity. *J. Biol. Chem.* **269**: 20475-20481.

Richmond, T.A. and Bleecker, A.B. (1999) A defect in β-oxidation causes abnormal inflorescence development in Arabidopsis. *Plant Cell.* **11**: 1911-1923.

Sacksteder, K.A. and Gould, S. (2000) The genetics of peroxisome biogenesis. *Annu. Rev. Genet.* **34**:

Sanders, P.M., Lee, P.Y., Biesgen, C., Boone, J.D., Beals, T.P., Weiler, E.W. and Goldberg, R.B. (2000) The Arabidopsis delayed *dehiscence1* gene endoces an enzyme in the jasmonic acid synthesis pathway. *Plant Cell.* **12**: 1041-4061.

Schaller, F., Biesgen, C., Müssig, C., Altmann, T. and Weiler, E.W. (2000) 12-Oxophytodienoate reductase 3 (OPR3) is the isoenzyme involved in jasmonate biosynthesis. *Planta.* **210**: 979-984.

Schumann, U., Gietl, C. and Schmid, M. (1999) Sequence analysis of a cDNA encoding Pex7p, a peroxisomal targeting signal 2 receptor from *Arabidopsis thaliana. Plant Physiol.* **120**: 339.

Somerville, C.R. (2001) An early Arabidopsis demonstration. Resolving a few issues concerning photorespiration. *Plant Physiol.* **125**: 20-24.

Somerville, C.R. and Ogren, W.L. (1982) Genetic modification of photorespiration. *Trends Biochem. Sci.* **7**: 171-174.

Somerville, C.R. and Ogren, W.L. (1980) Photorespiration mutants of *Arabidopsis thaliana* deficient in serine-glyoxylate aminotransferase activity. *Proc. Natl. Acad. Sci. USA.* **77**: 2684-2687.

Stintzi, A. and Browse, J. (2000) The Arabidopsis male-sterile mutant, *opr3*, lacks the 12-oxophytodienoic acid reductase required for jasmonate synthesis. *Proc. Natl. Acad. Sci. USA.* **97**: 10625-10630.

Subramani, S., Koller, A. and Snyder, W.B. (2000) Import of peroxisomal matrix and membrane proteins. *Ann. Rev. Biochem.* **69**: 399-418.

Tabak, H.F., Braakman, I. and Distel, B. (1999) Peroxisomes: simple in function but complex in maintenance. *Trends Cell Biol.* **9**: 447-453.

Titorenko, V.I. and Rachubinski, R.A. (1998) Mutants of the yeast *Yarrowia lipolytica* defective in protein exit from the endoplasmic reticulum are also defective in peroxisome biogenesis. *Mol. Cell. Biol.* **18**: 2789-2803.

Titus, D.E. and Becker, W.M. (1985) Investigation of the glyoxysome-peroxisome transition in germinating cucumber cotyledons using double-label immunoelectron microscopy. *J. Cell. Biol.* **101**: 1288-1299.

Tolbert, N.E. (1982) Leaf peroxisomes. *Annals NY Acad. Sci.* **386**: 254-268.

Trelease, R.N. and Doman, D.C. (1984) Mobilization of oil and wax reserves. In *Seed Physiology* Vol. 2. *Germination and reserve mobilization.* (Murray, D.R., ed.). pp. 201-245. New York: Academic Press.

Trelease, R.N., Lee, M.S., Banjoko, A. and Bunkelmann, J. (1996) C-terminal polypeptides are necessary and sufficient for *in vivo* targeting of transiently-expressed proteins to peroxisomes in suspension-cultured plant cells. *Protoplasma.* **195**: 156-167.

Tsugeki, R., Hara-Nishimura, I., Mori, H. and Nishimura, M. (1993) Cloning and sequencing of cDNA for glycolate oxidase from pumpkin cotyledons and Northern blot analysis. *Plant Cell Physiol.* **34**: 51-57.

Wain, R.L. and Wightman, F. (1954) The growth-regulating activity of certain β-substituted alkyl carboxylic acids in relation to their β-oxidation within the plant. *Proc. Roy. Soc. Lond. Biol. Sci.* **142**: 525-536.

Will, G.K., Soukupova, M., Hong, X., Erdmann, K.S., Kiel, J., Dodt, G., Kunau, W.H. and Erdmann, R. (1999) Identification and characterization of the human orthologue of yeast pex14p. *Mol. Cell. Biol.* **19**: 2265-2277.

Wimmer, C., Schmid, M., Veenhuis, M. and Gietl, C. (1998) The plant PTS1 receptor: similarities and differences to its human and yeast counterparts. *Plant J.* **16**: 453-464.

Zolman, B.K., Yoder, A. and Bartel, B. (2000) Genetic analysis of indole-3-butyric acid responses in *Arabidopsis thaliana* reveals four mutant classes. *Genetics.* **156**: 1323-1337.

Zolman, B.K., Monroe-Augustus, M., Thompson, B., Hawes, J.W., Krukenberg, K.A., Matsuda, S.P.T. and Bartel, B. (2001) *Chy1*, an Arabidopsis mutant with impared β-oxidation, is defective in a peroxisomal β-hydroxyisobutyryl-CoA hydrolase. *J. Biol. Chem.* **276**: 31037-31046.

10

PEROXISOMAL BIOGENESIS AND ACQUISTION OF MEMBRANE PROTEINS

Richard N. Trelease

Arizona State University, Tempe, USA

KEYWORDS

Endoplasmic reticulum; growth and division; membrane composition; membrane protein targeting; targeting signals.

ORGANELLE BIOGENESIS – ORIGINATION AND DIFFERENTIATION

In this chapter, peroxisomal biogenesis is considered in a comprehensive view that includes at least these two major components: (a) *origination,* i.e. the site and means by which peroxisomes (pre-peroxisomes or mature peroxisomes) proliferate in number, and (b) *differentiation*, i.e. the acquisition (targeting, sorting, uptake) of matrix and membrane molecules by peroxisomes (pre-peroxisomes or pre-existing peroxisomes). Morphological changes (e.g. pleiomorphic enlargement) are a manifestation of the latter. Since peroxisomes do not possess their own DNA, nuclear-encoded genes mediate all aspects of these processes.

The mechanisms of biogenesis have been a matter of continuous discussion for many years, due in part to the variations among organisms that are attributable to the incredible metabolic and morphological plasticity of peroxisomes. Lopez-Huertas and Baker (1999) presented a schematic view (their Figure 9) of the evolution of peroxisomal biogenesis models. In the

A. Baker and I.A. Graham (eds.), Plant Peroxisomes, 305–337.

1970s, peroxisomes (microbodies) were thought to originate and acquire virtually all of their lipids and proteins directly from rough ER according to a so-called 'ER-vesiculation' model (Beevers, 1979). A key feature of this model was the *co-translational* acquisition of matrix and membrane proteins. Proliferation in the number of peroxisomes per cell was thought to occur via continued budding of nascent peroxisomes from specialized, smooth (polysome-free) regions of the rough ER. Differentiation of these newly formed peroxisomes generally was not considered as part of the model. In a more comprehensive scheme, however, Huang *et al* (1983) presented several possible alternatives (monograph Figure 5.13); interestingly, some of these proposed alternatives are supported by current information.

This ER-vesiculation model lost favor in the early 1980s for several reasons. One main reason was that close ultrastructural associations of rough ER and peroxisomes that often were cited as supportive evidence for the model did not reveal a direct connection between the lumen of authentic rough ER and nascent peroxisomes. In mammalian liver and kidney cells, the putative ER portion of such associations was found to be part of a 'peroxisomal reticulum', which is characteristic of the dynamic proliferation/ differentiation of these animal peroxisomes. In other eukaryotic cells, particularly common in plant cells, the cited "close associations" were peroxisomes juxtaposed to rough ER whose polysomes were present only on the ER surface away from the peroxisomes (e.g. Vigil, 1973; Huang *et al*, 1983; Frederick *et al*, 1968). Figure 1 shows examples in several different plant tissues. Although these views do not support the ER vesiculation model, they certainly illustrate a provocative relationship between ER and peroxisomes, especially in the context of renewed interest in ER involvement in peroxisomal biogenesis (see next section). It is quite interesting to me that a functional relationship has never been shown for this rather commonly observed association. Addition of lipid or other components to 'pre-existing' peroxisomes to facilitate peroxisomal differentiation seems attractive (see Figure 3, later). Perhaps these views are morphological manifestations of data showing that phospholipid-synthesizing enzymes are in ER (not in peroxisomes), and that radiolabelled phospholipids move from ER to peroxisomal membranes (Moore, 1982; Lord and Roberts, 1983). Indirect evidence for this postulate comes from recent data obtained with *S. cerevisiae* where subfractions of ER closely associated with mitochondria and plasma membrane were characterized by a high capacity to synthesize common membrane phospholipids (Gaigg *et al*, 1995; Pichler *et al*, 2001).

Figure 1: Electron micrographs illustrating the commonly-observed 'close association' between rough ER and pre-existing peroxisomes in various plant cells and organisms

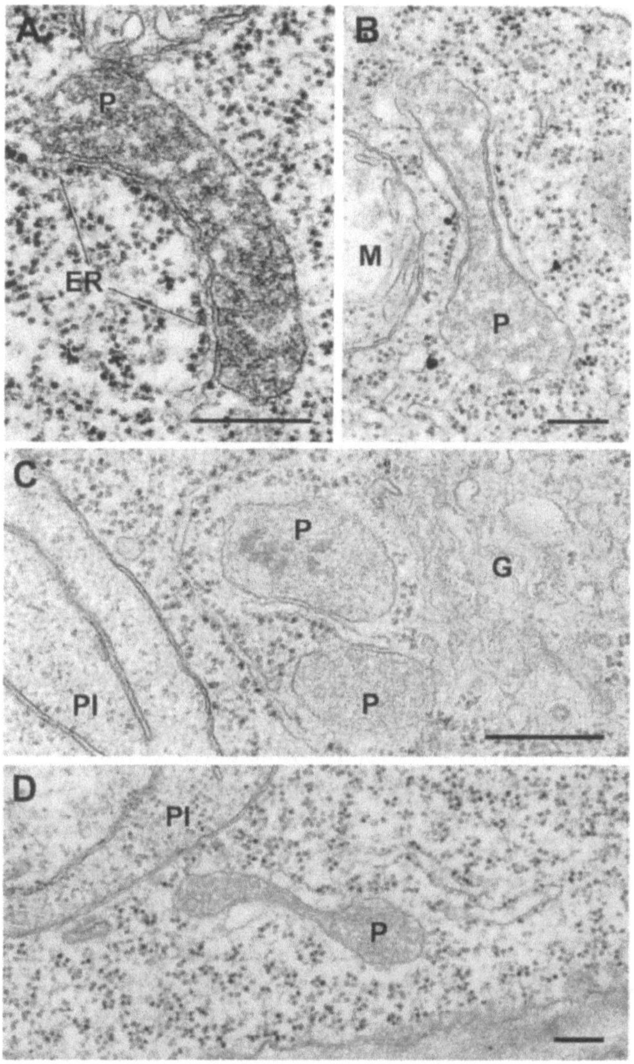

Figure 1. In all of these examples, the surface of the ER that is adjacent to the peroxisome lacks bound polyribosomes, whereas the opposite ER surface has bound polyribosomes. A. Elongate peroxisome within the cortex of a tobacco root cell (from Vigil, 1973) ; B. Dumbbell-shaped peroxisome in a meristematic cell of a bean root tip (from Frederick et al, 1968); C. Peroxisomes in a parenchyma cell of a wheat coleoptile (from Trelease in Huang et al, 1983); D. Elongate peroxisome in a palisade cell of an achlorophyllous bean leaf cell (from Gruber in Huang et al, 1983). ER – endoplasmic reticulum; G – Golgi body; M – mitochondrion; P – peroxisome; Pl – plastid. Each bar = 0.2 μm.

Another main reason for disfavouring the ER-vesiculation model were the widespread results with various organisms showing that peroxisomal proteins were added post-translationally, rather than co-translationally, to different types of peroxisomes (reviewed in Lazarow and Fujiki, 1985). Other reasons for loss of confidence in this model were disparate interpretations of perceived similarities in phospholipid and polypeptide composition between plant ER and glyoxysomes, and apparent glycosylation of some peroxisomal proteins. Cumulative evidence pointed away from a biogenetic relationship between ER and peroxisomes (Kindl and Lazarow, 1982; Huang *et al*, 1983; Trelease, 1984).

What, then, was considered the source of new peroxisomes? Small so-called 'pre-existing' (progenitor or rudimentary) peroxisomes became the focal point of a 'growth and division' model presented by Lazarow and Fujiki (1985). According to this model, pre-existing peroxisomes autonomously grow and replicate. The model was based chiefly on biogenetic studies of peroxisomal matrix proteins. However, Baker (1996) and Mullen and Trelease (1996) presented generalized models that illustrated similarities, as well as variations, in patterns of peroxisomal biogenesis among the most-studied organisms. Figure 2 is a modified version of the model presented by Mullen and Trelease (1996); updated references are given below.

In yeasts, peroxisomal synthesis (proliferation) is induced when certain carbon and/or organic nitrogen sources (e.g. (un)saturated fatty acids, ethanol, methanol, primary amines, D-amino acids and purines) are added to the growth medium. For example, a change in substrates in the medium of glucose-grown *Candida boidinii* and *Pichia pastoris* induces small pre-existing organelles to increase in number via fission/budding, then to enlarge to form 'mature' peroxisomes (Veenhuis and Goodman, 1990; Subramani, 1996) (Figure 2). However, the process is reversed in *Hansenula polymorpha* and *Yarrowia lipolytica* cells where a shift from glucose to methanol induces peroxisomal growth followed by division via fission of mature peroxisomes (Veenhuis *et al*, 2000; Titorenko and Rachubinski, 2001b). The means by which pre-existing mammalian peroxisomes proliferate varies considerably with cell culture conditions, organism diet, partial hepatectomy, etc. (Fahimi *et al*, 1993; Yamamoto and Fahimi, 1987). Figure 2 shows that proliferation of enlarged pre-existing organelles generally entails a budding/fission event, sometimes accompanied by 'segmentation' of progeny peroxisomes (Grabenbaurer *et al*, 2000). The peroxisomal reticulum formed in regenerating rat liver is a transient, dynamic structure that when present gives rise to spherical peroxisomes that bud from the reticulum. For plant peroxisomes, three studies (discussed below) describe proliferation in number of peroxisomes via a budding/fission of pre-existing peroxisomes, and abundant evidence exists for growth (differentiation) of pre-existing peroxisomes (see below) unaccompanied by a proliferative event. Overall, our knowledge of the means of plant peroxisomal proliferation is meager at best.

In summary, the schemes presented in Figure 2 illustrate division and growth, growth and division, division without growth, and growth without division. These patterns clearly show that the biogenesis of peroxisomes under varied conditions in different organisms (cells/tissues) does not specifically conform to the generalized model describing 'autonomous growth and division of pre-existing peroxisomes' (Lazarow and Fujiki, 1985) as assumed for many years by scientists in the field. Another important point is that the schemes shown in Figure 2 are incomplete, i.e. they do not indicate the source (site or means of origination) of the pre-existing peroxisomes in any of the situations. This is considered in subsequent paragraphs and sections.

As mentioned above and illustrated in Figure 2, enlargement/growth (differentiation) of pre-existing plant peroxisomes is a common event in certain organs and at certain stages of plant growth and development. Three well-documented examples of peroxisomal enlargement include: (a) glyoxysomal differentiation in cotyledons (not endosperm) during post-germinative growth of oilseed seedlings; (b) leaf peroxisomal differentiation in greening leaves (and cotyledons); and (c) root nodule peroxisomal differentiation in bacterial-uninfected nodule cells. In microscopic morphometric analyses of maturing and germinated cotton seeds, Kunce *et al* (1984) found that small glyoxysomes (0.2 to 0.5 μm in diameter) existed in mature cotyledon cells (just prior to seed desiccation), persisted without change in diameter through the desiccation period, and then increased seven-fold in volume over a thirty seven hour post-germinative period as the cotyledons expanded by cell enlargement, not by cell division. During the latter time period, the glyoxysomes differentiated via acquisition of acquired membrane and matrix enzymes needed for mobilization of the storage triacylglycerols (see Chapter 3 for details). Similar post-germinative enlargements of glyoxysomes have been observed microscopically in numerous other oilseed seedlings (see Olsen and Harada, 1995, and Olsen 1998, for references). In achlorophyllous bean leaves, relatively small (approximately 0.3 μm) undifferentiated leaf peroxisomes were observed often associated with ER (e.g. Figure 1D) dispersed in the cytoplasm, not specifically situated near the etioplasts (Gruber *et al*, 1973). After the leaves turned green and expanded fully, these leaf peroxisomes enlarged about five fold in diameter (to 1.5 μm) as they differentiated via acquisition of membrane components and glycolate oxidase, glyoxylate reductase, and catalase (and likely other photorespiratory enzymes), and became closely appressed to the green differentiated chloroplasts. In normal cortical root cells, the existing peroxisomes are relatively small (about 0.2 μm diameter) with little electron density in their matrix (e.g. Figure 1A, B). Development of soybean root nodules resulted in a dramatic five- to seven-fold increase in the diameter of peroxisomes in the bacterially-uninfected cells as the amount of uricase and catalase increased (Kaneko and Newcomb, 1987). As a corollary, Lee *et al* (1993) reported retardation of peroxisomal growth in mothbean root nodules expressing antisense uricase VN-35 RNA (see also Chapter 6). Other examples of peroxisomal enlargement in uninfected root nodule cells are found in studies of soybean (Newcomb and Tandon, 1981; Newcomb *et al*, 1985; Vaughn, 1985; Van den Bosch and Newcomb, 1986), black locust (Kaneko and Newcomb, 1990), and cowpea (Webb and Newcomb, 1987).

Pais and Carrapico (1982) used serial thick sections of germinating *Bryum capillare* spores to examine peroxisomal morphology during changeover in enzyme content (differentiation). They found that the peroxisomes formed an interconnected tubular compartment that assumed different functions with time. This is another manifestation of the enlargement (differentiation) phenomenon without proliferation in number of separate peroxisomes.

Documented examples of peroxisomal origination (replication/ division/proliferation) in plant cells are comparatively rare. Miyagishima *et al* (1998, 1999) convincingly showed that 'binary fission' was the means by which the single peroxisome (microbody) in the red alga *Cyanidioschyzon merolae* replicated without added proliferation-inducing substrates (represented in Figure 2, plant cells). Ferreria *et al* (1989) reported that peroxisomes in *Lemna* fronds grown in dim light underwent fission to form daughter peroxisomes in response to increased irradiance. Pais and Feijo (1987) described budding of existing peroxisomes in maturing orchid pollen grains. Others reported on proliferation of peroxisomes in other plants, but the mechanism by which the replication/division process ostensibly occurred was not examined (e.g. Vigil, 1970; Morre *et al*, 1990; Palma *et al*, 1991).

Figure 2: Schemes illustrating varied patterns of peroxisomal biogenesis relative to proliferation and enlargement, beginning with pre-existing peroxisomes, in the three most-studied groups of organisms

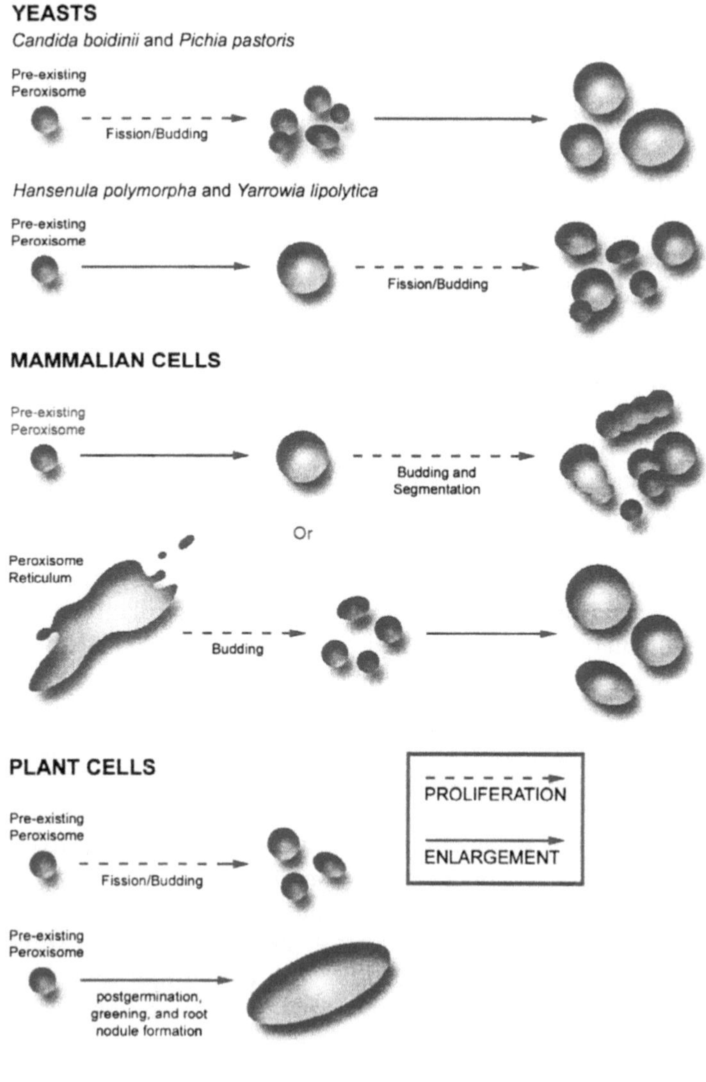

Figure 2. All schemes shown, except the proliferation of plant cells, reflect biogenesis in response to some induction stimulus (see text for details).

In summary, there is ample evidence that pre-existing, relatively undifferentiated plant peroxisomes undergo growth (enlargement) as they differentiate, but do not divide thereafter. Thus, the biogenesis of these plant peroxisomes seems to fit a 'division and growth' model, most similar to the biogenesis of *C. boidinii* and *P. pastoris* peroxisomes following a change to methanol in the culture medium (Figure 2). However, no published accounts describe the event or means for peroxisomal *origination* prior to plant peroxisomal growth in cotyledons, leaves, or nodules. Of the three studies that describe the means of peroxisomal division, only Ferreira *et al* (1989) suggested a "growth and division" event. However, they assumed, rather than showed, peroxisomal growth prior to division. Pais and Feijo (1987) concluded that the pollen grain peroxisomes budded without any notable peroxisomal growth before or after division. Miyagishima *et al* (1999) described the details of peroxisomal division as it occurred during division of the cells, not relative to any growth before or after division. Thus, when considering all of the available data, a consensus model that includes both origination and differentiation (enlargement) has not been constructed for the biogenesis of plant peroxisomes.

At this point, it seems reasonable to ask why are there so few data related to the means of peroxisomal origination (proliferation), especially when one considers the multitude of electron microscopic and biochemical studies that have been done with plant peroxisomes over the past thirty years. Baker (1996) also asked a similar question and speculated that perhaps there is no need for peroxisomal division in non-dividing plant cells because peroxisomal enlargement coupled with turnover of contents (differentiation) are sufficient to serve the non-dividing, differentiated cells. I think this is a reasonable assessment. Thus, plant biologists probably have not been studying peroxisomes in cells at stages when nascent peroxisomes are being formed, but when they are at non-dividing steady state, functional level or when they change their function and differentiate via enlargement, etc. The mechanisms involved in maintaining constant numbers of peroxisomes per cell during cell division under basal conditions are thought to be distinct from those that induce peroxisomal proliferation. Pex11p seems to be directly involved in regulating normal division of peroxisomes. Over-production of ScPex11p in *Saccharomyces cerevisiae* cells leads to the production of numerous smaller-than-normal peroxisomes, while cells deficient in ScPex11p have fewer 'giant' peroxisomes (Erdman and Blobel, 1995; Marshall *et al*, 1995, 1996). Pex11pα and Pex11pβ are postulated to control peroxisomal proliferation under induced and basal conditions respectively (Schrader *et al* 1998). Perhaps peroxisomal origination in higher plants could be elucidated in suspension cultured cells whose cell division can be synchronized experimentally (e.g. Samuels *et al*, 1998).

Thus, a major question is what is the likely origin (sources) of pre-existing peroxisomes in plant or other cells? Waterham *et al* (1993) concluded that yeasts such as *H. polymorpha* possess the alternative capability to produce peroxisomes *de novo*, although their observations that new peroxisomes invariably are in close association with "strands of ER" suggested to them that this compartment may be the source of new peroxisomes. South and Gould (1999) found that restoration of peroxisomal synthesis via addition of HsPex16p in cells from a Zellweger syndrome patient occurs in the absence of pre-existing peroxisomes, also suggesting to them a *de novo* synthesis of peroxisomes. However, they thought it unlikely that Pex16p expression 'catalyzed' virgin synthesis from cytosolic phospholipids and PMPs because membrane bound organelles typically arise from some already-existing membrane site, which most often is a "like" organelle. As mentioned earlier, pre-existing peroxisomes always exist in some cells (e.g. non-proliferative induced, Figure 2, plant cells) and are partitioned to daughter cells during cell division. Thus, the "like" organelle in these cases is "parent" pre-existing peroxisomes. Another postulated "like" source for peroxisomes are "earlier" peroxisomal vesicles such as P1 and P2 that were described by Titorenko and Rachubinski (2001a,b) based on data obtained with the yeasts *Y. lipolytica* and *P. pastoris*. In human cells, similar early vesicles are designated as "pre-peroxisomal vesicles" (South and Gould, 1999; Grabenbauer *et al*, 2000; Sacksteder and Gould, 2000). In the model for *Y. lipolytica,* the small P1 and P2 vesicles fuse to form larger vesicles that become capable of importing PMPs and matrix proteins, and these peroxisomes differentiate to form mature, functional peroxisomes. Formation of mature peroxisomes from pre-peroxisomal vesicles in human fibroblasts also seems to occur via a multi-step process, but apparently not via vesicle fusion. Peroxisomal vesicle fusion has been revealed thus far only in *Y. lipolytica*. Is there evidence for small, pre-peroxisomal vesicles in plant cells? In three different situations, catalase-containing peroxisomes (vesicles?) isolated in sucrose gradients exhibited progressively increased buoyant densities and enzyme content at different developmental stages (Choinski and Trelease, 1978; Feierabend and Beevers, 1972; Kudielka *et al*, 1981). Similar results were found for mammalian peroxisomes (Fahimi and Baumgart, 1999; Volkl *et al*, 1999, and references therein). Thus, working models suggest that the likely origin of pre-existing peroxisomes are 'pre-peroxisomal' or 'early peroxisomal' vesicles (pre-peroxisomes). Acceptance of this premise leads to the obvious question: what is the origin of pre-peroxisomes?

An obvious, dynamic source of membrane vesicles is the ER. Evidence has accumulated over the past five years to indicate that ER is implicated in one or more aspects of peroxisomal biogenesis (Erdmann *et al*, 1997; Kunau and Erdmann, 1998; Titorenko and Rachubinski, 1998; Mullen *et al*, 2001). Figure 3 illustrates a generalized, working model that attempts to incorporate and encompass notions and data for peroxisomal biogenesis that combine components of the 'growth and division', 'division and growth', and 'ER vesiculation' models into a more dynamic, multistep assembly scheme for all organisms. It depicts biogenetic variations that likely occur within different cells of the same organism, rather than variations among kingdoms as emphasized in Figure 2. The variously named pre-peroxisomes (immature/ early peroxisomal vesicles) are shown to originate by vesiculation (budding) from a specialized subdomain of the ER, named peroxisomal ER (pER) by Mullen *et al* (1999). These vesicles (pre-peroxisomes) possibly fuse with other ER-derived vesicles with varying cargo (import machinery?), or acquire this machinery and other proteins post-translationally from the cytosol, and form pre-existing peroxisomes. Thereafter, multiple options such as those shown in Figure 2 are possible. Information given in the following sections is presented within the context of this working model.

Figure 3: Overview model of peroxisomal biogenesis

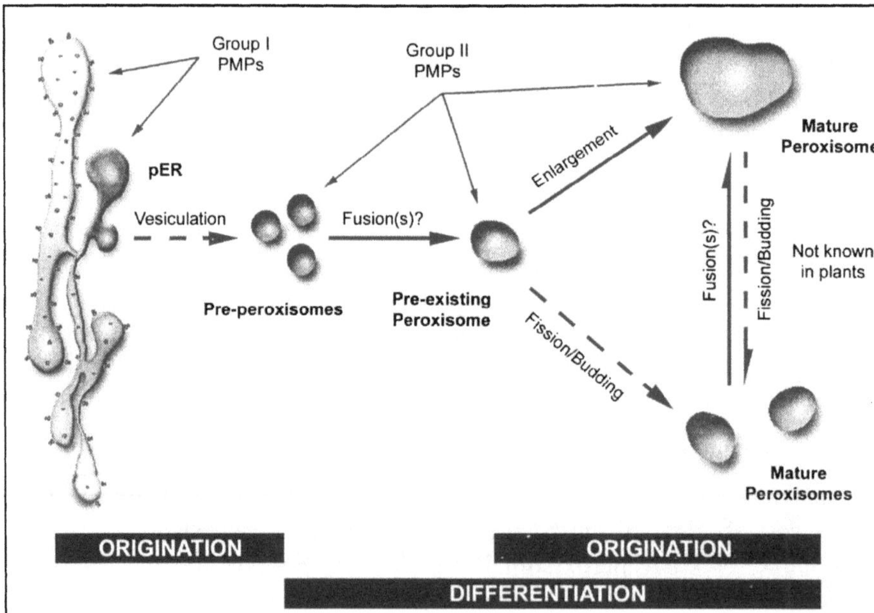

Figure 3. Overview model of peroxisomal biogenesis highlighting varied schemes of origination and differentiation of peroxisomes as they might occur in different cells/tissues of organisms.

MOLECULAR COMPOSITION OF PEROXISOMAL BOUNDARY MEMBRANES

Membrane lipids

Plant peroxisomal membranes possess several different phospholipids, e.g. phosphatidylcholine (PC), phosphatidylethanolamine (PE), phosphatidylinositol, and phosphatidylglycerol; cotyledon glyoxysomes possess in addition significant amounts of non-polar triacylglycerols (Donaldson and Beevers, 1977; Huang *et al*, 1983; Chapman and Trelease, 1991a). Peroxisomes in all organisms, however, are incapable of synthesizing their own membrane lipids (Donaldson, 1976; Beevers, 1979; Moore, 1982; Huang *et al*, 1983; Lord and Roberts, 1983; Lazarow and Fujiki, 1985; Chapman and Trelease, 1991a, b). ER, on the other hand, is a well-documented site of membrane lipid synthesis, thus early biogenetic models proposed ER as the primary source of lipids that existed in peroxisomal boundary membranes. Direct evidence for the transfer of phospholipids from ER to peroxisomes is limited to one study with glyoxysomes in the endosperm of castor bean seedlings (Kagawa *et al*, 1973). In another study of oilseed glyoxysomes, namely in cotyledons of cotton seedlings, oil bodies, not ER, were the source of peroxisomal lipids (Chapman and Trelease (1991a). Radiolabelled PC and triolein transferred directly from oil bodies to the glyoxysomes *in vitro*, and radiolabelled PC and PE were chased from ER to mitochondria, but not to glyoxysomes *in vivo*. In cotyledons of watermelon seedlings, Wanner *et al* (1982) suggested that oil body extensions called "lipid body appendices" gave rise to nascent glyoxysomes. Do these data reflect a significantly different source of peroxisomal membrane lipids? Perhaps not, because oil bodies (the apparent source of lipids in cotyledons) are formed on the rough ER by budding (Huang, 1992). Thus, ER may be the primary source of membrane lipids for oilseed glyoxysomes in all three cases, but with varied mechanisms for lipid transfer.

Membrane contact is regarded as a reasonable prerequisite for lipid movements between organelles such as between ER and mitochondria (Gaigg *et al*, 1995), plasma membranes (Pichler *et al* (2001), and peroxisomes (illustrated in Figure 1). Phospholipid carrier proteins have been considered as components of lipid transfer to peroxisomes, but have not been established (Olsen and Harada, 1995). The model illustrated in Figure 3 implies that ER-derived vesicles are a primary source and vector for peroxisomal membrane lipids. Taken together, substantial provocative data suggest ER is the primary source of peroxisomal membrane lipids, but definitive studies obviously are needed.

Identification and characterization of membrane proteins

Table 1 lists the plant PMPs that have been identified and at least partially characterized biochemically, immunologically, and/or sequentially (DNA). Early studies produced lists (polypeptide bands in SDS gels) of putative PMPs, e.g. for peroxisomes in leaves, oilseeds, and potato tubers (references in Table). These results spawned follow-up studies of polypeptides that were of interest for one or more reason (see most of the PMPs listed in the Table). Because contamination by non-peroxisomal membranes is inevitable for even the most highly purified peroxisomal fractions (e.g., Bunkelmann *et al*, 1995), and because PMP orthologues may not be localized to the same subcellular membranes, e.g. orthologues AtPex16 and YlPex16 (Lin *et al*, 1999 and Eitzen *et al*, 1997), it is important to verify the actual subcellular localization of each putative PMP within each tissues/species. One of the best means for accomplishing this is via reliable immunogold electron microscopic localization of specific PMP antigens to the boundary membrane of isolated, or *in situ*, peroxisomes. Those proteins that have been verified as authentic PMPs by this means are marked 'IEM' in Table 1. Of course, reliability of immunogold microscopy (and immunoblots) is dependent upon application of a specific antibody and employment of adequate controls.

PMPs generally are recovered in 'microsomal' pellets (100,000-200,000 g, one to two hour) following osmotic bursting of isolated peroxisomes. PMPs characterized as integral PMPs typically are insoluble in salt (e.g. 0.2-0.5 M KCl) or in alkaline (e.g. sodium carbonate, pH 11.5) solutions. Peripheral membrane proteins are recovered in supernatants following incubation in these solutions. Alternatively, or in addition, DNA sequences for a particular PMP are used to assess the presence/absence of putative transmembrane domain(s) (integral proteins) and/or prenylation signals (some peripheral proteins). Based on these criteria, PMPs are designated as integral (I) or peripheral PMPs (P) in Table 1. More detailed information is presented for several of these PMPs in other chapters of this book, e.g. antioxidants, superoxide-generating polypeptides, and NADH cyt c/ferricyanide reductases (Chapters 7 and 8); molecular chaperones (Chapter 11); porins (Chapter 5); peroxins (Chapters 9 and 12).

Two striking features of the listings in Table 1 are the few number of PMPs identified in plants, and the paucity of data for each of these PMPs. More PMPs are expected to exist within the varied types of peroxisomes as evidenced by their plasticity in biogenesis, function, and morphology that collectively led to writing of this book. In regard to the paucity of data, information on peroxins is a good example. Fifteen peroxin orthologues have been identified in Arabidopsis databases (Mullen *et al*, 2001; Charlton and Lopez-Huertas, Chapter 12). Twelve of these orthologues have been characterized in mammals and/or yeasts as integral or peripheral PMPs, yet only one orthologue, Pex14p from sunflower and Arabidopsis, has been characterized as an authentic PMP in plants. Genes coding for several putative peroxin PMPs have been cloned, e.g. peroxins 1, 6, 10, 16 (Lin *et al*, 1999; Schumann *et al*, 1999; Baker *et al*, 2000; Lopez-Huertas *et al*, 2000; Kaplan *et al*, 2001), but none of these gene products have been characterized as a PMP. It is expected that other PMPs/peroxins such as docking proteins (see Chapter 11, Mullen), translocon formers and ATPases (Pool *et al*, 1998), peptidases, etc. will be identified soon.

Table 1: Peroxisomal membrane proteins (polypeptides) (PMPs) identified in various types of plant peroxisomes.

Protein	Plant Source	Type; characterization	Refs.
Leaf peroxisomes			
Leaf PMPs (numerous)	Pea	I; SDS, W	1, 2
Oilseed glyoxysomes			
Cotyledon PMPs (numerous)	Cotton, cucumber, castor bean, pumpkin	I; SDS	3-5, 23
Tuber peroxisomes	Potato	I; SDS, W	6
Anti-oxidants			
Ascorbate peroxidase	Arabidopsis, cotton, cucumber, pea, pumpkin, spinach	I; As, C, IEM	2, 7-11
Mn – superoxide dismutase	Castor bean, watermelon	P; As, SDS	12, 13
Monodehydroascorbate reductase	Castor bean, cotton, pea, spinach	I; As, SDS, W	2, 8, 9, 14
Thioredoxin peroxidase (PMP 20)	Arabidopsis	P?; C	15
Superoxide-generating polypeptides			
PMPs 18, 29, 32	Pea	I; As, SDS	2, 16
NADH cyt c and ferricyanide reductases			
PMPs 29 – 36; PMP 53	Potato, castor bean	I; As, SDS	3, 6

Table 1 (continued)

Protein	Plant Source	Type; characterization	Refs.
Peroxin Orthologues			
Pex14p	Arabidopsis	I; C, W	17,18
Molecular chaperones			
PMP 73	Cucumber	I; IEM, SDS, W	19
DnaJ homolog (pmp61, pmp53)	Cucumber	P; C, SDS W	19, 20
Porins			
PMP 34-36	Castor bean, cucumber, spinach	I; As, SDS, W	21, 22
Specific PMPs (function?)			
PMP 28	Pumpkin	I; IEM, W	23
PMP 22	Arabidopsis	I; C, W	24

I = Integral protein; P = Peripheral protein; As = enzyme or component assay identification; C = cDNA clone available; IEM = localized via Immunogold Electron Microscopy; SDS = identified as polypeptide in SDS gel; W = identified on Western blot

1,2. López-Huertas *et al* (1995; 1999a).
3. Luster *et al* (1988).
4. Chapman and Trelease (1992).
5. Corpas *et al* (1994).
6. Struglics *et al* (1993).
7. Yamaguchi *et al* (1995a).
8. Bunkelmann and Trelease (1996).
9. Jiménez *et al* (1997).
10. Zhang *et al* (1997).
11. Ishikawa *et al* (1998).
12. del Río *et al* (1992).
13. del Río and Donaldson (1995).
14. Bowditch and Donaldson (1990).
15. Verdoucq *et al* (1999).
16. Corpas *et al* (2001).
17. Hayashi *et al* (2000).
18. López-Huertas *et al* (1999b).
19. Corpas and Trelease (1997).
20 Preisig-Müller and Kindl (1993).
21. Reumann (2000).
22. Corpas *et al* (2000).
23. Yamaguchi *et al* (1995b).
24. Tugal *et al* (1999).

SORTING PATHWAYS FOR PMPs

When considering the biogenesis of peroxisomes, several major questions related to these PMPs arise. For example: (a) what is the temporal and spatial regulation of PMP synthesis? (b) are the intracellular targeting and sorting mechanisms the same for all PMPs and for each target membrane (such as ER, pre-existing peroxisomes, or mature peroxisomes)? and, more specifically, (c) what are the requirements for PMP import? Titorenko and Rachubinski (2001a,b) and Figure 3 herein show that a 'limited subset' of yeast and plant PMPs (designated arbitrarily as Group I) are targeted to rough ER post-translationally by their amino-terminal or carboxy-terminal ER-targeting signals before ending up in mature peroxisomes via an '*indirect*' pathway. Another subset of PMPs (Group II) is shown targeted '*directly*' to peroxisomes (e.g. pre-peroxisomes, pre-existing peroxisomes, or mature peroxisomes) at varied stages of differentiation. Titorenko and Rachubinski (2001a, b) refer to this model as the "ER pre-peroxisomal endomembrane compartment" model. A second model presented by Titorenko and Rachubinski (2001a, b), referred to as "autonomous pre-peroxisomal endomembrane compartment", differs from the first model in that none of the PMPs are targeted first to ER; instead; two or more subsets of PMPs are shown sorting *directly* to pre-peroxisomal endomembrane 'vesicular structures' in the cytoplasm. This latter model is most consistent with available data for PMP sorting in mammalian cells (South and Gould, 1999; Sacksteder and Gould, 2000). A separate model similar to the latter 'autonomous' model is not presented in this chapter; however, the reader can visualize the autonomous model concept in Figure 3 simply by ignoring ER as the origination of pre-peroxisomes and the target for Group I PMPs. This subset of PMPs would be directed to the pre-peroxisomes that presumably would be derived by budding/fission from pre-existing peroxisomes.

Indirect sorting of PMPs to peroxisomes via ER/ER-derived vesicles

Indirect PMP sorting has been described for only one plant PMP, namely PMP31, cottonseed peroxisomal ascorbate peroxidase (GhAPX) (Mullen *et al*, 1999). Peroxisomal APX has been identified in membranes of glyoxysomes, leaf, and leaf-type (green cotyledon) peroxisomes (Table 1), and shown to persist in the boundary membranes as cotyledon peroxisomes change function from glyoxysomal to photorespiratory metabolism (Yamaguchi *et al*, 1995b; Bunkelmann and Trelease, 1997; Corpas and Trelease, 1998). The enzyme functions in the regeneration of NAD^+ and protection of the cell (particularly the peroxisomal membrane?) from toxic reactive oxygen species (Mullen and Trelease, 1996; del Rio and Donaldson, Chapters 8 and 9, respectively).

When peroxisomal GhAPX was expressed transiently in tobacco BY-2 cells, it sorted *in vivo* to a reticular/circular compartment and to peroxisomes (Figure 4 A-C). Solid arrows indicate endogenous catalase co-localized with expressed HA-epitope-tagged APX in peroxisomes. Identification of the reticular/circular compartment was provided in part by its co-localization with 3,3'dihexyloxacarbocyanine ($DiOC_6$) (Figure 4 C, D), a common stain for ER (Staehelin, 1997). Outlined arrows depict clear images of the circular reticular structures in Figures 4 A, C, D. These and other data were interpreted to indicate that APX sorted first to a distinct subdomain of ER, designated as peroxisomal ER (pER). The pER subdomain was perceived to be a smooth membrane portion of rough ER as illustrated in Figure 3. Nito *et al* (2001) recently confirmed peroxisomal APX sorting to ER in a cell fractionation study with pumpkin peroxisomal APX. Their magnesium-shift experiments revealed that the pumpkin APX was not associated with rough ER, but with smooth ER vesicles that they suggested may represent the pER subdomain previously described and illustrated in Figure 3. Mullen *et al* (1999) showed in addition that the *in vitro* insertion of peroxisomal GhAPX specifically into highly-purified ER membranes depended on the presence of ATP and one or two molecular chaperones, namely Hsp70 and AtJ2 (orthologue of *E. coli* DnaJ), on the nucleotide exchange factor AtE1 (orthologue of *E. coli* GrpE), and on a protease-sensitive factor.

It is not known whether peroxisomal APX (or other Group I proteins) initially are targeted and inserted directly into the pER subdomain, or are targeted first to 'general' ER and then transported to pER (Figure 3). Mullen *et al* (1999) discussed the "privileged site budding model" (Kuehn and Schekman, 1997) as a possible means for selective movement of PMPs to specific subdomains or "privileged sites" within ER. Putative gating proteins may regulate selectively cargo PMP entry into the proper ER subdomain from which the PMP is exported selectively by COPII vesicles. Titorenko and Rachubinski (2001a) incorporated gating proteins into their working model. Experiments with brefeldin A (BFA), a toxin that blocks vesicle-mediated protein exports from ER (and retrograde transport from Golgi to ER), helped elucidate cargo protein movement within sorting pathways. For example, in the presence of BFA, chimaeric peroxisomal APX accumulated in the ER of BY-2 cells (Mullen *et al*, 1999). Subsequent removal of BFA resulted in the redistribution of the APX chimaera into pre-existing peroxisomes, presumably via mobile pre-peroxisomal vesicles (Figure 3). In *H. polymorpha*, application of BFA resulted in the accumulation within ER of PMPs such as the PTS1 docking complex, that normally end up in pre-existing/mature peroxisomes (Salomons *et al*, 1997). While these results are informative, the dynamics of peroxisomal APX sorting needs to be examined further using varied experimental approaches.

Figure 4: Immunofluorescence micrographs of individual tobacco BY-2 cells

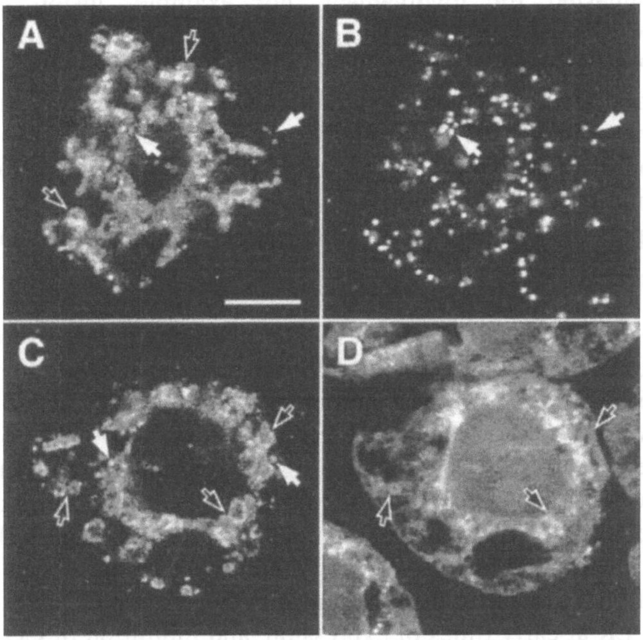

Figure 4. Immunofluorescence micrographs of individual tobacco BY-2 cells illustrating the subcellular localization of transiently-expressed cottonseed peroxisomal HA-epitope tagged APX (HA-APX) within a subdomain of ER (peroxisomal ER) and peroxisomes. A. Punctate Cyanine 2 (Cy2) fluorescence (solid arrows) and reticular/circular Cy2 fluorescence (outlined arrows) attributable to binding of anti-HA IgGs to overexpressed HA-APX; B. In the same cells as in panel A, punctate Cy5 fluorescence attributable to binding of anti-catalase IgGs to endogenous catalase is co-localized with punctate APX (solid arrows). C. Punctate and reticular/circular Cy5 fluorescence attributable to expressed HA-APX; D. In the same cell as in panel C, ER membranes stained with $DiOC_6$ partially co-localize (solid and outlined arrows) with HA-APX. Mullen et al, 1999 Plant Cell 11:2167 ©1999 American Society of Plant Biology. Reprinted with permission.

Other evidence that fuelled and supported the postulated indirect sorting of PMPs to ER is given in the following studies. The rat liver PMP50 was synthesized *in vitro* predominantly on membrane-bound polysomes, indicative of co-translational insertion into ER (Bodnar and Rachubinski (1991). This is the only reported exception to post-translational uptake of a PMP into any membrane. Mutants of *Y. lipolytica* defective in protein exit from the ER (*sec* mutations) also were defective in peroxisomal biogenesis, i.e. the cells exhibited a significant reduction in size and number of peroxisomes (Titorenko and Rachubinski, 1998). These authors also found that mutations in *PEX1* and *PEX6* genes resulted in an accumulation of

PMPs Pex2p and Pex16p in the ER. Several studies with Pex3p from several sources suggested ER involvement. For example, fusion proteins of HpPex3p were sorted first to ER en route to peroxisomes (Baerends *et al*, 1996), HsPex3p accumulated in ER of cells treated with BFA (Salomons *et al*, 1997), and HsPex3p overexpressed in COS7 cells led to proliferation of ER (Kammerer *et al*, 1998). Similarly, overexpressed ScPex15p led to the proliferation of ER that formed 'karmellae' (Elgersma *et al*, 1997). This was not surprising because ScPex15p is O-glycosylated under certain conditions and the topogenic signal for this yeast peroxin appears to include an ER targeting signal.

One major criticism of some of these interpretations is that overexpression, or truncation, of peroxins may yield misleading results. For example, overexpressed ScPat1p, a yeast homolog of rat PMP70, ended up in the nuclear envelope and the lateral ER (Hettema *et al*, 1999). The possibilities raised were that cryptic ER targeting signals exist in certain peroxins, and that ER may be a depository of excess protein, since it normally participates a quality control of aberrant (overproduced) proteins (e.g. Yokota *et al*, 2000). As mentioned before, virtually no evidence has been found for targeting of PMPs to ER in mammalian cells. For example, HsPex3p was not detected in ER at early times of expression, and inhibition of COPI function by BFA did not affect trafficking of HsPex3p to peroxisomes (South *et al*, 2000). Moreover, targeting of human fibroblast Pex2p, Pex3p, and Pex16p were not affected by applied inhibitors of COPI and COPII, and the peroxins were found only in peroxisomes (Voorn-Brouwer *et al*, 2001). Obviously, results such as these often are cited as negative support for sorting of PMPs to ER. Two counter arguments are: (a) that ER simply is not involved in PMP sorting within the narrow range of mammalian cells types examined thus far, but might be in other mammalian cell types; and (b) that the so-called 'existing endomembrane vesicles (pre-peroxisomes)' to which Pex3p and Pex16p are sorted to facilitate subsequent PMP import may be produced by vesicules from ER. The latter argument has been considered in the discussions of (e.g. South and Gould, 1999), but seemingly not taken seriously.

Direct sorting of PMPs from the cytosol to peroxisomes

PMPs of varied functions in varied organisms sort directly to peroxisomes. For example, Arabidopsis PMP22 (possible pore former) and sunflower Pex14p orthologue (docking protein in plants) inserted into isolated sunflower glyoxysomes (Tugal *et al*, 1999); *C. boidinii* PMP47 (a solute carrier protein) targeted and inserted into peroxisomes isolated from this yeast (McNew and Goodman, 1994; Dyer *et al*, 1996; Wang *et al*, 2001) and into peroxisomal membranes purified from potato tuber peroxisomes (Mullen *et al*, 1999). Rat liver PMP70 (ABC transporter; Imanaka *et al*, 1996), PAF-1 (RnPex2p) and PMP22 (Diestelkotter and Just, 1993; Pause *et al*, 1997, 2000), and PMP26 (RnPex11p, involved in peroxisomal proliferation; Passreiter *et al*, 1998) inserted specifically into purified rat liver peroxisomes. It is interesting to note that all of the direct-sorting studies have been done with 'mature' peroxisomes. It remains to be shown whether PMPs are sorted directly to 'pre-peroxisomes' as depicted in current models (e.g. Figure 3 this chapter, South and Gould, 1999; Titorenko and Rachubinski, 2001a, b).

IMPORT AND TARGETING OF PMPs

The available data seem sufficient to accept that certain PMPs in plant, yeast, and mammalian cells are sorted from the cytosol directly to peroxisomes, whereas certain other PMPs are sorted indirectly (through ER or pre-peroxisomal vesicles) from the cytosol to peroxisomes. However, more studies are needed to decide whether there is some categorical discrimination for the type(s) of PMPs that are sorted directly to pre-existing (or mature) peroxisomes versus those sorted otherwise (Figure 3). Salomons *et al* (1997) and Baerends *et al* (2000a) suggested that PMPs essential for biogenesis (i.e. peroxins) were sorted initially to ER, and those PMPs necessary for function (e.g. transporters, enzymes, etc.) were targeted directly to peroxisomes. Current data support this trend, but exceptions already have been reported, e.g. plant peroxisomal APX sorts first to ER (Mullen *et al*, 1999) and RnPex11p sorts directly to peroxisomes (Passreiter *et al*, 1998).

For any of the cases described above, questions arise as to: (a) how PMPs are added to membranes; and (b) what are the molecular targeting signals for membranes, and are certain signals specific for the various target membranes? As mentioned above, rat PMP50 appears to be the only PMP added co-translationally to a target membrane (ER) (Bodnar and Rachubinski, 1991), and *de novo* synthesis of membranes from precursor molecules in the cytosol is not a favored mechanism (South and Gould, 1999). A general scheme for PMP acquisition invokes a two-step process. First, positive sequence(s) within PMPs (membrane peroxisomal targeting signals, mPTS) are recognized by soluble cytosolic or membrane associated mPTS receptors. The second step is PMP insertion into the membrane and thermodynamic stabilization (folding) within the membrane. Studies related to this generalized scheme are discussed below.

Mechanisms and factors involved in the import of PMPs

Four of the twenty three known peroxins, namely Pex3p, Pex16p, Pex17p, and Pex19p, appear to be involved either in the assembly of peroxisomal membranes or in the import of PMPs into pre-peroxisomes or pre-existing peroxisomes (Sacksteder and Gould, 2000; Subramani *et al*, 2000; Terlecky and Fransen, 2000). With the exception of Pex17p, these conclusions are based on formation of functional peroxisomes in mutant yeast or mammalian cells (devoid of peroxisomes or vestiges thereof) following reintroduction of their wild type genes. Little is known about the specific function of Pex16p in any system; the only study published on this orthologue in plants (AtPex16p) did not suggest that AtPex16p participated in peroxisomal biogenesis (Lin *et al*, 1999). Overexpression of Pex3p resulted in multiplication of peroxisomes (Baerends *et al* 1996; Wiemer *et al*, 1996), suggesting a specific role in membrane synthesis, but no information has been reported for this orthologue in plants. Pex19p, a cytosolic protein in mammals (Sacksteder *et al*, 2000) and yeasts (Snyder *et al*, 1999a; Hettema *et al*, 2000), interacts with a diverse array of PMPs including yeast and mammalian peroxins and transporters. Pex19p currently is regarded as a cytosolic PMP receptor and/or chaperone, although new data on compatible binding sites raises questions about its receptor function (Snyder *et al*, 2000). Unfortunately, a receptor/chaperone function cannot be attributed to this orthologue in plants because no data are available on Pex19-protein interactions in plants. Most information on the involvement of Pex17p in membrane assembly is from studies with *P. pastoris* (Snyder *et al*, 1999b). A putative Arabidopsis orthologue exists, but there are no data published for any plant Pex17p.

Required cytosolic and membrane factors have been identified for one plant and several mammalian PMPs in *in vitro* import studies. Import of all PMPs examined thus far has been shown to be time and temperature dependent, involving two steps whereby PMPs bind to the membrane and subsequently insert into the bilayer. Binding of AtPMP22 to isolated sunflower peroxisomes was not dependent on the presence of ATP, whereas insertion of this PMP was stimulated in the presence of 2.4 mM ATP (Tugal *et al*, 1999). Insertion of rat PMP22, Pex2p (PAF-1) and PMP70 into liver peroxisomes does not require the hydrolysis of ATP (Imanaka *et al*, 1996; Pause *et al*, 1997). The rat PMP22 apparently forms two complexes, one associated with the chaperonin TriC, and the second with a 40 kDa polypeptide (P40). Pause *et al* (1997) suggest that nascent PMP22 binds first to TriC before transfer to P40, which may function as a cytosolic receptor (Pex19p?). Evidence suggests that ATP hydrolysis may be needed for final folding. Overall, it is apparent that acquisition of matrix and membrane proteins involves distinct receptors and transport machineries (Subramani *et al*, 2000; Holroyd and Erdmann, 2001; Mullen, Chapter 11).

Molecular targeting signals for PMPs

The targeting signals for PMPs (mPTS) are distinct from PTS1 and PTS2 for matrix proteins. Until recently, all mPTSs were thought to have in common a stretch of six to eight basic amino acids and often be oriented topologically on the matrix side of the boundary membrane (Subramani *et al*, 2000). Some of these mPTSs included an adjacent transmembrane domain (TMD) (ScPex15p – Elgersma *et al*, 1997; PpPex22p – Koller *et al*, 1999; GhAPX – Mullen and Trelease, 2000), whereas other mPTSs did not include a TMD (CbPMP47 – McCammon *et al*, 1994, Dyer *et al* 1996; PpPex3p – Wiemer *et al*, 1996; HsPex3p – Kammerer *et al*, 1998; Soukupova *et al*, 1999). A cytosol-oriented region of twenty two amino acid residues in RnPMP22 can target a different polypeptide directly to peroxisomes (Pause *et al*, 2000). The mPTS reported for peroxisomal GhAPX is the only plant mPTS described thus far. Mullen and Trelease (2000) found that a patch of five basic residues (RKRMK) within the hydrophilic C-terminal-most amino acid residues was necessary for sorting of peroxisomal APX to peroxisomes via ER. However, the C-terminal tail was not sufficient for sorting a passenger protein, whereas the peptide plus most of the immediately adjacent TMD was sufficient for sorting. The TMD did not to possess targeting information *per se*, but conferred the proper context for the C-terminal sorting signal to function. Thus, the mPTS for plant peroxisomal GhAPX seems similar to those cited above that are composed of a short basic stretch of basic amino acids plus an adjacent TMD, although this GhAPX is oriented topologically on the cytosolic side as are other 'tail-anchored' membrane proteins found within membranes of various organelles.

Thus, the signals described above appear to function in targeting both 'types' of PMPs, i.e. peroxins and functional PMPs, the latter represented by APX and the PMP47 transporter. The Pex3ps, ScPex15p, and APX appear to reach their target peroxisomes indirectly via the ER. The positively charged motif has been implicated in sorting PMPs directly to peroxisomes and indirectly to ER. How could this be? One possibility is involvement of two signals such as those described by differential loss-of-functions in ScPex15 (Elgersma *et al*, 1997). One signal guides the PMP to the ER (mPTS1), and a second one mediates subsequent routing to peroxisomes (mPTS2). Targeting results with GhAPX suggest that the C terminus of this PMP contains an overlapping ER (pER) targeting signal and a mPTS2. Evidence for this was that GhAPX inserted into ER membranes, but not into peroxisomal membranes, whereas in the same experiment, CbPMP47 with a mPTS1 inserted exclusively into peroxisomal membranes (Mullen *et al*, 1999).

Two recent studies indicate that reassessment of the mPTSs described in studies cited above is needed. Wang *et al* (2001) re-examined the sufficiency of the putative mPTS1 in CbPMP47 using a more sensitive quantitative localization assay. High fidelity targeting to peroxisomes *in vivo* required several elements, i.e. a cytosolic-oriented charged domain within cytoplasmic loop 1 and the adjacent TMD2 (the second of six TMDs), a short matrix loop containing a basic cluster, and a membrane-anchoring TMD. Thus, it is now believed that effective targeting of this PMP sorted directly to peroxisomes requires significantly more of the protein, i.e. specific regions of the protein that are located on both sides, and within the membrane. Wang *et al* (2001) proposed in a model that the composition of membrane targeting signal might depend on the protein orientation within the membrane. Single-span Type I proteins with the basic amino acid in the N-terminal region may have less complex signals than either single-span Type 2 proteins with a C-terminal basic cluster, or multi-spanning proteins. This may reflect more complex sorting pathways and different kinetics of membrane insertion compared with those of Type I proteins.

Jones *et al* (2001) showed that HsPMP34, the orthologue of CbPMP47, possesses at least two non-overlapping targeting signals, either of which is sufficient for targeting directly *in vivo* to fibroblast peroxisomes. They also obtained evidence for multiple independent targeting regions within HsPex13p. The two mPTSs identified in PMP34 corresponded to residues 1-147 and 244-307, which did not share any sequences (motifs) with mPTSs reported for any other PMP. Common features claimed by Jones *et al* are that mPTSs are relatively long (the shortest one they cite is twenty five amino acids long for Pex17p), and that they contain at least one membrane-spanning domain. Their data suggested that the basic residues might not be an essential feature of all mPTSs. An independent study of the same PMP by Honsho and Fujuki (2001) led to the conclusion that HsPMP34 had a single signal, and that the loop between TMDs 4 and 5 was essential for targeting. Similar targeting assays were employed; it seems that the major difference between the two reports is in the interpretation of the data. It is quite apparent that elucidation of targeting signals is quite complex, even for those signals that are involved in direct sorting to peroxisomes.

Unanswered questions related to understanding PMP sorting and assembly

Knowledge in this realm of plant peroxisomal biogenesis is severely lacking, and it is fragmented and incomplete for peroxisomes in other organisms. A major question is how cytosolic-synthesized PMPs are maintained in a non-aggregated state prior to and during import into a target membrane? Do cytosolic receptors participate in this process, and if so, how do they interact with the mPTS(s) on cargo PMPs? What actually constitutes a necessary and/or sufficient mPTS? Do PMPs bind to the target membrane in a folded or unfolded state, and how do they eventually fold and assemble into functional entities or complexes? Do 'early' peroxins involved in PMP acquisition function in the cytosol, or within the target membrane, or both? How much conservation is there among the mechanisms involved in the import of PMPs into the various target membranes? Do peroxisomal membranes assemble *de novo* in the cytosol? Is ER the primary source of membrane for all types of progenitor vesicles or peroxisomes?

ACKNOWLEDGEMENTS

Special thanks are extended to Cayle S. Lisenbee for his skill, efforts, and creativity in preparing all of the Figures (1-4), and for his critical reading of the manuscript. Dr. Robb Flynn created the original drawings that were modified to generate Figures 2 and 3. I also thank Drs. Eugene Vigil, Sue Ellen Frederick and Peter J. Gruber for their permissions to re-publish electron micrographs presented in Figure 1 A, B and D, respectively. Figure 1A reproduced with permission from Vigil *et al*, 1973. Figure 1B reproduced with permission from Frederick *et al*, 1968 ©Springer-Verlag, Berlin. Figures 1C and 1D reproduced with permission from Huang *et al*, 1983 ©Academic Press, Florida, USA. The word-processing help of Ms. Andra Williams is sincerely appreciated. NSF Grant MCB-0091826 and the William N. and Myriam Pennington Foundation provided financial support for writing and for some of the research presented in this chapter.

REFERENCES

Baerends, R.J.S, Faber, K.N., Kiel, J.A.K.W., van der Klei, I.J., Harder, W. and Veenhuis, M. (2000) Sorting and function of peroxisomal membrane proteins. *FEMS Micro. Rev.* **24**: 291-301.

Baerends, R.J.S., Rasmussen, S.W., Hilbrands, R.E., van der Heide, M., Klaas, N.F., Reuvekamp, P.T.W., Kiel, J.A.K.W., Cregg, J.M., van der Klei, I.J. and Veenhuis, M. (1996) The *Hansenula polymorpha PER9* gene encodes a peroxisomal membrane protein essential for peroxisome assembly and integrity. *J. Biol. Chem.* **271**: 8887-8894.

Baker, A. (1996) Biogenesis of plant peroxisomes. In *Membranes: Specialized Functions in Plants.* (Smallwood, M., Knox, J.P., Bowles, D.J. eds.). pp. 421-40. Bios Scientific Publishers Oxford, UK.

Baker, A., Charlton, W., Johnson, B., Lopez-Huertas, E., Oh, J., Sparkes, I. and Thomas, J. (2000) Biochemical and molecular approaches to understanding protein import into peroxisomes. *Biochem. Soc. Trans.* **28**: 499-504.

Beevers, H. (1979) Microbodies in higher plants. *Ann. Rev. Plant Physiol.* **30**: 159-193.

Bodnar, A.G. and Rachubinski, A. (1991) Characterization of the integral membrane polypeptides of rat liver peroxisomes isolated from untreated and clofibrate-treated rats. *Biochem. Cell Biol.* **69**: 499-508.

Bowditch, M.I. and Donaldson, R.P. (1990) Ascorbate free-radical reduction by glyoxysomal membranes. *Plant Physiol.* **94**: 531-537.

Bunkelmann, J. and Trelease, R.N. (1997) Expression of glyoxysomal ascorbate peroxidase in cotton seedlings during post-germinative growth. *Plant Sci.* **122**: 209-216.

Bunkelmann, J. and Trelease, R.N. (1996) Ascorbate peroxidase: A prominent membrane protein in oilseed glyoxysomes. *Plant Physiol.* **110**: 589-598.

Bunkelmann, J., Corpas, F.J. and Trelease, R.N. (1995) Four putative, glyoxysome membrane proteins are instead immunologically-related protein body membrane proteins. *Plant Sci.* **106**: 215-226.

Chapman, K.D. and Trelease, R.N. (1992) Characterization of membrane proteins in enlarging cottonseed glyoxysomes. *Plant Physiol. Biochem.* **30**: 1-10.

Chapman, K.D. and Trelease, R.N. (1991a) Intracellular localization of phosphatidylcholine and phosphatidylethanolamine synthesis in cotyledons of cotton seedlings. *Plant Physiol.* **95**: 69-76.

Chapman, K.D. and Trelease, R.N. (1991b) Acquisition of membrane lipids by differentiating glyoxysomes: role of lipid bodies. *J. Cell Biol.* **115**: 995-1007.

Choinski, J.S. and Trelease, R.N. (1978) Control of enzyme activities in cotton cotyledons during maturation and germination. II. Glyoxysomal enzyme development in embryos. *Plant Physiol.* **62**: 141-145.

Corpas, F.J. and Trelease, R.N. (1998) Differential expression of ascorbate peroxidase and a putative molecular chaperone in the boundary membrane of differentiating cucumber seedling peroxisomes. *J. Plant Physiol.* **153**: 332-338.

Corpas, F.J. and Trelease, R.N. (1997) The plant 73 kDa peroxisomal membrane protein (PMP73) is immunorelated to molecular chaperones. *Eur. J. Cell Biol.* **73**: 49-57.

Corpas, F.J., Barroso, J.B. and del Rio, L.A. (2001) Peroxisomes as a source of reactive oxygen species and nitric oxide signal molecules in plant cells. *Trends Plant Sci.* **4**: 145-150.

Corpas, F.J., Bunkelmann, J. and Trelease, R.N. (1994) Identification and immunochemical characterization of a family of peroxisome membrane proteins (PMPs) in oilseed glyoxysomes. *Eur. J. Cell Biol.* **65**: 280-290.

Corpas, F.J., Sandalio, L.M., Brown, M.J., del Rio, L.A. and Trelease, R.N. (2000) Identification of porin-like polypeptide(s) in the boundary membrane of oilseed glyoxysomes. *Plant Cell Physiol.* **41**: 1218-1228.

del Rio, L.A. and Donaldson, R.P. (1995) Production of superoxide radicals in glyoxysomal membranes from castor bean endosperm. *J. Plant Physiol.* **146**: 283-287.

del Rio, L.A., Sandalio, L.M., Palma, J.M., Bueno, P. and Corpas, F.J. (1992) Metabolism of oxygen radicals in peroxisomes and cellular implications. *Free Rad. Biol. Med.* **13**: 557-580.

Diestelkotter, P. and Just, W.W. (1993) *In vitro* insertion of the 22-kD peroxisomal membrane protein into isolated rat liver peroxisomes. *J. Cell Biol.* **123**: 1717-1725.

Donaldson, R.P. (1976) Membrane lipid metabolism in germinating castor bean endosperm. *Plant Physiol.* **57**: 510-515.

Donaldson, R.P. and Beevers, H. (1977) Lipid composition of organelles from germinating castor bean endosperm. *Plant Physiol.* **59**: 259-263.

Dyer, J.M., McNew, J.A. and Goodman, J.M. (1996) The sorting sequence of the peroxisomal integral membrane protein PMP47 is contained within a short hydrophilic loop. *J. Cell Biol.* **133**: 269-280.

Eitzen, G.A., Szilard, R.K. and Rachubinski, R.A. (1997) Enlarged peroxisomes are present in oleic acid-grown *Yarrowia lipolytica* overexpressing the *PEX16* gene encoding an intraperoxisomal peripheral membrane peroxin. *J. Cell Biol.* **137**: 1265-1278.

Elgersma, Y., Kwast, L., van den Berg, M., Snyder, W.B., Distel, B., Subramani, S. and Tabak, H.F. (1997) Overexpression of Pex15p, a phosphorylated peroxisomal integral membrane protein required for peroxisome assembly in *S. cerevisiae*, causes proliferation of the endoplasmic reticulum membrane. *EMBO J.* **16**: 7326-7341.

Erdmann, R. and Blobel, G. (1995) Giant peroxisomes in oleic acid-induced *Sacharomyces cerevisiae* lacking the peroxisomal membrane protein Pmp27. *J. Cell Biol.* **128**: 509-523.

Erdmann, R., Veenhuis, M. and Kunau, W-H. (1997) Peroxisomes: organelles at the crossroads. *Trends Cell Biol.* **7**: 400-407.

Fahimi, H.D. and Baumgart, E. (1999) Current cytochemical techniques for the investigation of peroxisomes: a review. *J. Histochem. Cytochem.* **47**: 1219-1232.

Fahimi, H.D., Baumgart, E. and Volkl, A. (1993) Ultrastructural aspects of the biogenesis of peroxisomes in rat liver. *Biochimie.* **75**: 201-208.

Feierabend, J. and Beevers, H. (1972) Developmental studies on microbodies in wheat leaves. II Ontogeny of particulate enzyme associations. *Plant Physiol.* **49**: 33-39.

Ferreira, R.M.B., Bird, B. and Davies, D.D. (1989) The effect of light on the structure and organization of *Lemna* peroxisomes. *J. Exp. Bot.* **40**: 1029-1035.

Frederick, S.E., Newcomb, E.H., Vigil, E.L. and Wergin, W.P. (1968) Fine-structure characterization of plant microbodies. *Planta.* **81**: 229-252.

Gaigg, B., Simbeni, R., Hrastnik, C., Paltauf, F. and Daum, G. (1995) Characterization of a microsomal subfraction associated with mitochondria of the yeast, *Saccharomyces cerevisiae.* Involvement in synthesis and import of phospholipids into mitochondria. *Biochim. Biophys. Acta.* **1234**: 214-220.

Grabenbauer, M., Satzler, K., Baumgart, E. and Fahimi, H.D. (2000) Three-dimensional ultrastructural analysis of peroxisomes in HepG2 cells. *Cell Biochem. Biophys.* **32**: 37-49.

Gruber, P.J., Becker, W.M. and Newcomb, E.H. (1973) The development of microbodies and peroxisomal enzymes in greening bean leaves. *J. Cell Biol.* **56**: 500-518.

Hayashi, M., Nito, K., Toriyama-Kato, K., Kondo, M., Yamaya, T. and Nishimura, M. (2000) *At*Pex14p maintains peroxisomal functions by determining protein targeting to three kinds of plant peroxisomes. *EMBO J.* **19**: 5701-5710.

Hettema, E.H., Distel, B. and Tabak, H.F. (1999) Import of proteins into peroxisomes. *Biochim. Biophys. Acta.* **1451**: 17-34.

Hettema, E.H., Girzalsky, W., van den Berg, M., Erdmann, R. and Distel, B. (2000) *Saccharomyces cerevisiae* Pex3p and Pex19p are required for proper localization and stability of peroxisomal membrane proteins. *EMBO J.* **19**: 223-233.

Holroyd, C. and Erdmann, R. (2001) Protein translocation machineries of peroxisomes. *FEBS Lett.* **501**: 6-10.

Honsho, M. and Fujiki, Y. (2001) Topogenesis of peroxisomal membrane protein requires a short, positively charged intervening-loop sequence and flanking hydrophobic segments. Study using human membrane protein pmp34. *J. Biol. Chem.* **276**: 9375-9382.

Huang, A.H.C. (1992) Oil bodies and oleosins in seeds. *Ann. Rev. Plant Physiol. Plant. Mol. Biol.* **43**: 177-200.

Huang, A.H.C., Trelease, R.N. and Moore, T.S. (1983) Plant Peroxisomes. *Amer. Soc. Pl. Physiol. Monograph Series.* Acad Press, NY.

Imanaka, T., Shiina, Y., Takano, T., Hashimoto, T. and Osumi, T. (1996) Insertion of the 70-kDa peroxisomal membrane protein into peroxisomal membranes *in vivo* and *in vitro*. *J. Biol. Chem.* **271**: 3706-3713.

Ishikawa, T., Yoshimure, K., Sakai, K., Tamoi, M., Takeda, T. and Shigeoka, S. (1998) Molecular characterization and physiological role of a glyoxysome-bound ascorbate peroxidase from spinach. *Plant Cell Physiol.* **39**: 23-34.

Jimenez, A. Hernandez, J.A., del Rio, L.A. and Sevilla, F. (1997) Evidence for the presence of the ascorbate-glutathione cycle in mitochondria and peroxisomes of pea leaves. *Plant Physiol.* **114**: 275-284.

Jones, J.M., Morrell, J.C. and Gould, S.J. (2001) Multiple distinct targeting signals in integral peroxisomal membrane proteins. *J. Cell Biol.* **153**: 1141-1149.

Kagawa, T., Lord, J.M. and Beevers, H. (1973) The origin and turnover of organelle membranes in castor bean endosperm. *Plant Physiol.* **51**: 61-65.

Kammerer, S., Holzinger, A., Welsch, U. and Roscher, A. (1998) Cloning and characterization of the gene encoding the human peroxisomal assembly protein Pex3p. *FEBS Lett.* **429**: 53-60.

Kaneko, Y. and Newcomb, E.H. (1990) Specialization for ureide biogenesis in the root nodules of black locust (*Robinia pseudoacacia* L), an amide exporter. *Protoplasma.* **157**: 102-111.

Kaneko, Y. and Newcomb, E.H. (1987) Cytochemical localization of uricase and catalase in developing root nodules of soybean. *Protoplasma.* **140**: 1-12.

Kaplan, C.P., Thomas, J.E., Charlton, W.L. and Baker, A. (2001) Identification and characterization of PEX6 orthologues from plants. *Biochim. Biophys. Acta.* **10414**: 1-8.

Kindl, H. and Laazarow, P.B. (1982) Peroxisomes and glyoxysomes. *Ann. NY Acad. Sci.* **386**: 550.

Koller, A., Snyder, W.B., Faber, K.N., Wenzel,T.J., Rangell, L., Keller, G.A. and Subramani, S. (1999) Pex22p of *Pichia pastoris*, essential for peroxisomal matrix protein import, anchors the ubiquitin-conjugating enzyme, Pex4p, on the peroxisomal membrane. *J. Cell Biol.* **146**: 99-112.

Kudielka, R.A., Kock, H. and Theimer, R.R. (1981) Substrate dependent formation of glyoxysomes in cell suspension cultures of anise (*Pimpinella anisum* L.). *FEBS Lett.* **136**: 8-12.

Kuehn, M.J. and Schekman, R. (1997) COPII and secretory cargo capture into transport vesicles. *Curr. Opin. Cell Biol.* **9**: 477-483.

Kunau, W-H. and Erdmann, R. (1998) Peroxisome biogenesis: back to the endoplasmic reticulum? *Curr. Biol.* **8**: R299-R302.

Kunce, C.M., Trelease, R.N. and Doman, D.C. (1984) Ontogeny of glyoxysomes in maturing and germinated cotton seeds – a morphometric analysis. *Planta.* **161**: 156-164.

Lazarow, P.B. and Fujiki, Y. (1985) Biogenesis of peroxisomes. *Ann. Rev. Cell Biol.* **1**: 489-530.

Lee, N-G., Stein, B., Suzuki, H. and Verma, D.P.S. (1993) Expression of antisense nodulin-35 RNA in *Vigna aconitifolia* transgenic root nodules retards peroxisome development and affects nitrogen availability to the plant. *Plant J.* **3**: 599-606.

Lin, Y., Nguyen, L.V., Rachubinski, R.A. and Goodman, H.M. (1999) The Pex16p homolog SSE1 and storage organelle formation in Arabidopsis seeds. *Science.* **284**: 328-330.

Lopez-Huertas, E. and Baker, A. (1999) Peroxisome Biogenesis. In *Transport of Molecules Across Microbial Membranes* (Broome-Smith, J.K., Baumberg, S., Stirling, C.J. and Ward, F.B. eds.). pp. 205-238. Proc. Symp. Soc. Gen. Microbiol., Cambridge University Press.

Lopez-Huertas, E., Charlton, W.L., Johnson, B., Graham, J.A. and Baker, A. (2000) Stress induces peroxisome biogenesis genes. *EMBO J.* **19**: 6770-6777.

Lopez-Huertas, E., Corpas, F.J., Sandalio, L.M. and del Rio, L.A. (1999a) Characterization of membrane polypeptides from pea leaf peroxisomes involved in superoxide radical generation. *Biochem J.* **337**: 531-536.

Lopez-Huertas, E., Oh, J. and Baker, A. (1999b) Antibodies against Pex14p block ATP-independent binding of matrix proteins to peroxisomes *in vitro*. *FEBS Lett.* **459**: 227-229.

Lopez-Huertas, E., Sandalio, L.M. and del Rio, L.A. (1995) Integral membrane polypeptides of pea leaf peroxisomes: characterization and response to plant stress. *Plant Physiol. Biochem.* **33**: 295-302.

Lord, J.M. and Roberts, L.M. (1983) Formation of glyoxysomes. *Int. Rev. Cytol.* **15**: 115-156.

Luster, D.G., Bowditch, M.I., Eldridge, K.M. and Donaldson, R.P. (1988) Characterization of membrane-bound electron transport enzymes from castor bean glyoxysomes and endoplasmic reticulum. *Arch. Biochem. Biophys.* **265**: 50-61.

Marshall, P.A., Dyer, J.M., Quick, M.E. and Goodman, J.M. (1996) Redox-sensitive homodimerization of Pex11p: a proposed mechanism to regulate peroxisomal division. *J. Cell Biol.* **135**: 123-137.

Marshall, P.A., Krimkevich, Y.I., Lark, R.H., Dyer, J.M., Veenhuis, M. and Goodman, J.M. (1995) Pmp27 promotes peroxisomal proliferation. *J. Cell Biol.* **129**: 345-355.

McCammon, M.T., McNew, J.A., Willy, P.J. and Goodman, J.M. (1994) An internal region of the peroxisomal membrane protein PMP47 is essential for sorting to peroxisomes. *J. Cell Biol.* **124**: 915-925.

McNew, J.A. and Goodman, J.M. (1994) An oligomeric protein is imported into peroxisomes *in vivo*. *J. Cell Biol.* **127**: 1245-1257.

Miyagishima, S., Itoh, R., Toda, K., Kuroiwa, H., Nishimura, M. and Kuroiwa, T. (1999) Microbody proliferation and segregation cycle in the single-microbody alga *Cyanidioschyzon merolae*. *Planta.* **208**: 326-336.

Miyagishima, S., Itoh, R., Toda, K., Takahashi, H., Kuroiwa, H. and Kuroiwa, T. (1998) Visualization of the microbody division in *Cyanidioschyzon merolae* with the fluorochrome brilliant sulphoflavin. *Protoplasma* **201**: 115-119.

Moore, T.S. Jr. (1982) Phospholipid biosynthesis. *Ann. Rev. Plant Physiol.* **33**: 235-259.

Morre, D.J., Sellden, G., Ojanpera, K., Sandelius, A.S., Egger, A., Morre, D.M., Chalko, C.M. and Chalko, R.A. (1990) Peroxisome proliferation in Norway spruce induced by ozone. *Protoplasma.* **155**: 58-65.

Mullen, R.T. and Trelease, R.N. (2000) The sorting signals for peroxisomal membrane-bound ascorbate peroxidase are within its C-terminal tail. *J. Biol. Chem.* **275**: 1-8.

Mullen, R.T. and Trelease, R.N. (1996) Biogenesis and membrane properties of peroxisomes: does the boundary membrane serve and protect? *Trends Plant Sci.* **1**: 389-394.

Mullen, R.T., Flynn, C.R. and Trelease, R.N. (2001) How are peroxisomes formed? The role of the endoplasmic reticulum and peroxins. *Trends Plant Sci.* **6**: 256-261.

Mullen, R.T., Lisenbee, C.S., Miernyk, J.A. and Trelease, R.N. (1999) Peroxisomal membrane ascorbate peroxidase is sorted to a membranous network that resembles a subdomain of the endoplasmic reticulum. *Plant Cell.* **11**: 2167-2185.

Newcomb, E.H. and Tandon, S.R. (1981) Uninfected cells of soybean root nodules: ultrastrucutre suggests key role in ureide production. *Science.* **212**: 1394-1396.

Newcomb, E.H., Tandon, S.R. and Kowal, R.R. (1985) Ultrastructural specialization for ureide production in uninfected cells of soybean root nodules. *Protoplasma.* **125**: 1-12.

Nito, K., Yamaguchi, K., Konda, M., Hayashi, M. and Nishimura, M. (2001) Pumpkin peroxisomal ascorbate peroxidase is localized on peroxisomal membranes and unknown membranous structures. *Plant Cell Physiol.* **42**: 20-27.

Olsen, L.J. (1998) The surprising complexity of peroxisome biogenesis. *Plant Mol. Biol.* **38**: 163-189.

Olsen, L.J. and Harada, J.J. (1995) Peroxisomes and their assembly in higher plants. *Ann. Rev. Plant Physiol.* **46**: 123-146.

Pais, M.S. and Feijo, J.A. (1987) Microbody proliferation during the microsporogenesis of *Ophrys lutea* Cav. (Orchidaceae). *Protoplasma.* **138**: 149-155.

Pais, M.S. and Carrapico, F. (1982) Microbodies – a membrane compartment. *Ann. NY Acad. Sci.* **386**: 510-513.

Palma, J.M., Garrido, M., Rodriguez-Garcia, M.I. and del Rio, L.A. (1991) *Arch. Biochem. Biophys.* **287**: 68-74.

Passreiter, M., Anton, M., Lay, D., Frank, R., Harter, C., Wieland, F.T., Gorgas, K. and Just, W.W. (1998) Peroxisome biogenesis: involvement of ARF and coatomer. *J. Cell Biol.* **141**: 373-383.

Pause, B., Diestelkotter, P., Heid, H. and Just, W.W. (1997) Cytosolic factors mediate protein insertion into the peroxisomal membrane. *FEBS Lett.* **414**: 95-98.

Pause, B., Saffrich, R., Hunziker, A., Ansorge, W. and Just, W.W. (2000) Targeting of the 22 kDa integral peroxisomal membrane protein. *FEBS Lett.* **00**: 1-6.

Pichler, H., Gaigg, B., Hrastnik, C., Achleitner, G., Kohlwein, S.D., Zellnig, G., Perktold, A. and Daum, G. (2001) A subfraction of the yeast endoplasmic reticulum associates with the plasma membrane and has a high capacity to synthesize lipids. *Eur. J. Biochem.* **268**: 2351-2361.

Pool, M.R., Lopez-Huertas, E. and Baker, A. (1998) Characterization of intermediates in the process of plant peroxisomal protein import. *EMBO J.* **17**: 6854-6862.

Preisig-Müller, R. and Kindl, H. (1993) Plant DnaJ homologue: molecular cloning, bacterial expression, and expression analysis in tissues of cucumber seedlings. *Arch. Biochem. Biophys.* **305**: 30-37.

Reumann, S. (2000) The structural properties of plant peroxisomes and their metabolic significance. *Biol. Chem.* **381**: 639-648.

Sacksteder, K.A. and Gould, S.J. (2000) The genetics of peroxisome biogenesis. *Ann. Rev. Genet.* **34**: 623-652.

Sacksteder, K.A., Jones, J.M., South, S.T., Li, X., Liu, Y. and Gould, S.J. (2000) PEX19 binds multiple peroxisomal membrane proteins, is predominantly cytoplasmic, and is required for peroxisome membrane synthesis. *J. Cell Biol.* **148**: 931-944.

Salomons, F.A., van der Klei, I.J., Kram, A.M., Harder, W. and Veenhuis, M. (1997) Brefeldin A interferes with peroxisomal protein sorting in the yeast *Hansenula polymorpha*. *FEBS Lett.* **411**: 133-139.

Samuels, A.L., Meehl, J., Lipe, M. and Staehelin, L.A. (1998) Optimizing conditions for tobacco BY-2 cell cycle synchronization. *Protoplasma.* **202**: 232-236.

Schrader, M., Reuber, B.E., Morrell, J.C., Jimenez-Sanchez, G., Obie, C., Stroh, T.A., Valle, D., Schroer, T.A. and Gould, S.J. (1998) Expression of PEX11β mediates peroxisome proliferation in the absence of extracellular stimuli. *J. Biol. Chem.* **273**: 29607-29614.

Schumann, U., Gietl, C. and Schmid, M. (1999) Sequence analysis of a cDNA encoding Pex10p, a zinc-binding peroxisomal integral membrane protein from *Arabidopsis thaliana* (Accession No. AF119572). (PGR99-025). *Plant Physiol.* **119**: 1147.

Snyder, W.B., Faber, K.N., Wenzel, T.J., Koller, A., Luers, G.H., Rangell, L., Keller, G.A. and Subramani, S. (1999) Pex19p interacts with Pex3p and Pex10p and is essential for peroxisome biogenesis in *Pichia pastoris*. *Mol. Biol. Cell.* **10**: 1745-1761.

Snyder, W.B., Koller, A., Choy, A.J. and Subramani, S. (2000) The peroxin Pex19p interacts with multiple, integral membrane proteins at the peroxisomal membrane. *J. Cell Biol.* **149**: 1171-1178.

Snyder, W.B., Koller, A., Choy, A.J., Johnson, M.A., Creg, J.M., Rangell, L., Keller, G.A. and Subramani, S. (1999b) Pex17p is required for import of both peroxisome membrane and lumenal proteins and interacts with Pex19p and the peroxisome targeting signal-receptor docking complex in *Pichia pastoris*. *Mol. Biol. Cell.* **10**, 4005-4019.

Soukupova, M., Sprenger, C., Gorgas, K., Kunau, W-H. and Dodt, G. (1999) Identification and characterization of the human peroxin PEX3. *Eur. J. Cell Biol.* **78**: 357-374.

South, S.T. and Gould, S.J. (1999) Peroxisome synthesis in the absence of pre-existing peroxisomes. *J. Cell Biol.* **144**: 255-266.

South, S.T., Sacksteder, K.A,. Li, X., Liu, Y. and Gould, S.J. (2000) Inhibitors of COPI and COPII do not block *PEX3*-mediated peroxisome synthesis. *J. Cell Biol.* **149**: 1345-1359.

Staehelin, L.A. (1997) The plant ER: a dynamic organelle composed of a large number of discrete functional domains. *Plant J.* **11**: 1151-1165.

Struglics, A., Fredlund, K.M., Rasmusson, A.G. and Moller, I.M. (1993) The presence of a short redox chain in the membrane of intact potato tuber peroxisomes and the assocation of malate dehydrogenase with the peroxisomal membrane. *Physiol. Plant.* **88**, 19-28.

Subramani, S. (1996) Protein translocation into peroxisomes. *J. Biol. Chem.* **271**, 32483-32486.

Subramani, S., Koller, A. and Snyder, W.B. (2000) Import of peroxisomal matrix and membrane proteins. *Ann. Rev. Biochem.* **69**, 399-418.

Terlecky, S.R. and Fransen, M. (2000) How peroxisomes arise. *Traffic.* **1**: 465-473.

Titorenko, V.I. and Rachubinski, R.A. (2001a) Dynamics of peroxisome assembly and function. *Trends Cell Biol.* **11**: 22-29.

Titorenko, V.I. and Rachubinski, R.A. (2001b) The life cycle of the peroxisome. *Nat. Rev.* **2**: 357-368.

Titorenko, V.I. and Rachubinski, R.A. (1998) Mutants of the yeast *Yarrowia lipolytica* defective in protein exit from the endoplasmic reticulum are also defective in peroxisome biogenesis. *Mol. Cell Biol.* **18**: 2789-2803.

Trelease, R.N. (1984) Biogenesis of glyoxysomes. *Ann. Rev. Plant Physiol.* **35**: 321-347.

Tugal, H.B., Pool, M. and Baker, A. (1999) Arabidopsis 22-kDa peroxisomal membrane protein. Nucleotide sequence analysis and biochemical characterization. *Plant Physiol.* **120**: 309-320.

Van den Bosch, K.A. and Newcomb, E.H. (1986) Immunogold localization of nodule-specific uricase in developing soybean root nodules. *Planta.* **167**: 425-436.

Vaughn, K.C. (1985) Structural and cytochemical characterization of three specialized peroxisome types in soybean. *Physiol Plantarum.* **64**: 1-12.

Veenhuis, M. and Goodman, J.M. (1990) Peroxisomal assembly: membrane proliferation precedes the induction of the abundant matrix proteins in the methylotrophic yeast *Candida boidinii*. *J. Cell Sci.* **96**: 583-590.

Veenhuis, M., Salomons, F.A. and Van der Klei, I.J. (2000) Peroxisome biogenesis and degradation in yeast: a structure/function analysis. *Micro. Res. Tech.* **51**: 584-600.

Verdoucq, L., Bignols, F., Jacquot, J-P., Chartier, Y. and Meyer, Y. (1999) *In vivo* characterization of a thioredoxin h target protein defines a new peroxiredoxin family. *J. Biol. Chem.* **274**: 19714-19722.

Vigil, E.L. (1973) Plant microbodies. *J. Histochem. Cytochem.* **11**: 958-962.

Vigil, E.L. (1970) Cytochemical and developmental changes in microbodies (glyoxysomes) and related organelles of castor bean endosperm. *J. Cell Biol.* **46**: 435-454.

Volkl, A., Mohr, H. and Fahimi, D. (1999) Peroxisome subpopulations of the rat liver: isolation by immune free flow electrophoresis. *J. Histochem. Cytochem.* **47**: 1111-1117.

Voorn-Brouwer, T., Kragt, A., Tabak, H.F. and Distil, B. (2001) Peroxisomal membrane proteins are properly targeted to peroxisomes in the absence of COPI- and COPII-mediated vesicular transport. *J. Cell Sci.* **114**: 2199-2204.

Wang, X., Unruh, M.J. and Goodman, J.M. (2001) Discrete targeting signals direct Pmp47 to oleate-induced peroxisomes in *Saccharomyces cerevisiae*. *J. Biol. Chem.* **276**: 10897-10905.

Wanner, G., Vigil, E.L. and Theimer, R.R. (1982) Ontogeny of microbodies (glyoxysomes) in cotyledons of dark-grown watermelon (*Citrullus vulgaris* Schrad.) seedlings. *Planta.* **156**: 314-325.

Waterham, H.R., Titorenko, V.I., Swaving, G.J., Harder, W. and Veenhuis, M. (1993) Peroxisomes in the methylotrophic yeast *Hansenula polymorpha* do not necessarily derive from pre-existing organelles. *EMBO J.* **12**: 4785-4794.

Webb, M.A. and Newcomb, E.H. (1987) Cellular compartmentation of ureide biogenesis in root nodules of cowpea (*Vigna unguiculata* (L.) Walp). *Planta.* **172**: 162-175.

Wiemer, E.A.C., Luers, G.H., Faber, K.N., Wenzel, T., Veenhuis, M. and Subramani, S. (1996) Isolation and characterization of Pas2p, a peroxisomal membrane protein essential for peroxisome biogenesis in the methylotrophic yeast *Pichia pastoris*. *J. Biol. Chem.* **271**: 18973-18980.

Yamaguchi, K., Mori, H. and Nishimura, M. (1995a) A novel isoenzyme of ascorbate peroxidase localized on glyoxysomal and leaf peroxisomal membranes in pumpkin. *Plant Cell Physiol.* **36**: 1157-1162.

Yamaguchi, K., Takeuchi, Y., Mori, H. and Nishimura, M. (1995b) Development of microbody membrane proteins during the transformation of glyoxysomes to leaf peroxisomes in pumpkin cotyledons. *Plant Cell Physiol.* **36**: 455-464.

Yamamoto, K. and Fahimi, H.D. (1987) Three-dimensional reconstruction of a peroxisomal reticulum in regenerating rat liver: evidence of interconnections between heterogeneous segments. *J. Cell Biol.* **105**: 713-722.

Yokota, S., Kamijo, K. and Oda, T. (2000) Aggregate formation and degradation of overexpressed wild-type and mutant urate oxidase proteins. Quality control of organelle-destined proteins by the endoplasmic reticulum. *Histochem. Cell Biol.* **114**: 433-446.

Zhang, H., Wang, J., Nickel, U., Allen, R.D. and Goodman, H.M. (1997) Cloning and expression of an Arabidopsis gene encoding a putative peroxisomal ascorbate peroxidase. *Plant Mol. Biol.* **34**: 967-971.

11

TARGETING AND IMPORT OF MATRIX PROTEINS INTO PEROXISOMES

Robert T. Mullen

University of Guelph, Guelph, Ontario, Canada

KEYWORDS

Accessory residues, catalase, chaperones, docking complex, glycolate oxidase, glyoxysome, import, *in vitro*, *in vivo*, isocitrate lyase, matrix, oligomer, peroxin, peroxisome, piggy-back, PTS, targeting signal, translocation, thiolase.

INTRODUCTION

Peroxisomes were recognized as cellular inclusions almost fifty years ago, but for most of the time since then their biogenesis, i.e. their assembly, differentiation, proliferation, and inheritance, has been poorly understood. For instance, compared with the impressive advances that were made in elucidating how protein constituents are sorted to the ER, mitochondria, chloroplasts, and nuclei, relatively little was known about these processes with respect to peroxisomes until the late 1980s. Then, three significant breakthroughs in our understanding of the biogenesis of peroxisomes led to intensified research that has since shifted the organelle from relative obscurity to the forefront of cell biology.

The first breakthrough came in 1987 when one of the two targeting signals now known to be responsible for sorting newly-synthesized proteins from the cytosol to the peroxisomal matrix was identified (Gould *et al*, 1987). Termed the type 1 peroxisomal targeting signal (PTS1), this topogenic

A. Baker and I.A. Graham (eds.), Plant Peroxisomes, 339–383.

determinant has been shown since to function in sorting proteins to peroxisomes in diverse organisms including yeasts, mammals, plants and trypanosomes (Gould *et al*, 1990a; Fung and Clayton, 1991; Keller *et al*, 1991; Motley *et al*, 2000). The second landmark was the development of genetic screens for yeast (Erdman *et al*, 1989; Cregg *et al*, 1990; Gould *et al*, 1992; Nuttley *et al*, 1992) and mammalian cultured cells (Zoeller *et al*, 1989; Morand *et al*, 1990) that had defects in the assembly of functional peroxisomes, particularly the import of proteins into the matrix. Extensive analysis of these mutants has since provided tremendous insight into the mechanisms involved in peroxisome biogenesis, as well as the identification of at least twenty three different protein factors, termed peroxins (pex), that are essential for peroxisome assembly (Distel *et al*, 1996). The third significant breakthrough during the late 1980s and early 1990s was the realization that peroxisomes have medicinal relevance, since genetic defects that impair the proper functioning of this organelle in humans, including protein targeting, were recognized as being responsible for a mostly lethal group of inborn disorders (reviewed in Fujiki, 2000; Gould and Valle, 2000).

The confluence of these three sets of discoveries has led to a significant amount of research on the biogenesis of peroxisomes in evolutionary diverse organisms, the majority of which has focused on the specific targeting signals and protein factors (peroxins) that are involved in protein import into the organelle. For instance, two distinct sorting pathways to the peroxisomal matrix have been identified, each one using a different PTS (PTS1 and PTS2) and cognate receptor protein (known as peroxins 5 and 7; Pex5p and Pex7p). Several other peroxins have also been discovered which function either as docking proteins that interact with both sets of receptor-cargo complexes at the peroxisome boundary membrane, or as part of a unique translocon that is also shared between the two matrix-protein sorting pathways. In addition, the energetics of import are beginning to be unravelled, as well as the roles that molecular chaperones play in peroxisome protein folding and assembly, sorting and translocation.

This burgeoning research on peroxisome biogenesis has led also to a number of unexpected discoveries related to the import of proteins into the organelle. The most notable of these – the observation that peroxisomes have the ability to import completely folded and oligomeric proteins – has challenged a classical paradigm which holds that a protein must be maintained in an unfolded conformation during translocation into an organelle (for review see McNew and Goodman, 1996; Smith and Schnell, 2001). The flexibility of the peroxisome membrane translocation machinery to import macromolecules also allows for an unusual shuttling pathway for the PTS1 receptor Pex5p that is not found in most other organelles, with the exception of the nucleus. This unique shuttling mechanism involves the translocation of Pex5p along with its PTS1-bearing protein cargo into the peroxisomal matrix and then recycling of the receptor back to the cytosol for another round of import (Dammai and Subramani, 2001). Other surprising discoveries related to peroxisome protein import include the existence, in at least some plants and yeasts, of preperoxisomal vesicles that originate from specialized segments of the ER. These vesicles while enroute to peroxisomes appear to serve as an import site for some matrix- and membrane-destined proteins and as the membrane source of nascent and enlarging peroxisomes (reviewed in Mullen *et al*, 2001a; Titorenko and Rachubinski, 2001a; 2001b). A consensus on the operation and maintenance of this ER-to-peroxisome pathway in different organisms and cells types, however, has yet to be reached (see Chapter 10).

Research carried out since the late 1980s on peroxisomes has opened up a wealth of information on the organelle's biogenesis. In this chapter, I review those studies that have focused on the targeting and import of proteins into the peroxisomal matrix, with particular emphasis on advances that have been made in understanding these processes in higher plants. It is worth stressing that when compared to other organisms such as yeasts and mammals, our knowledge of plant peroxisomal protein targeting and import (and peroxisome biogenesis in plants in general) is meagre. Therefore, where appropriate, I will briefly review data from mammals and yeast to provide information on how the current working models for peroxisome protein targeting and import in plant cells have evolved.

For more in-depth discussions of peroxisome biogenesis in yeast and mammalian cells the reader should refer to several recent comprehensive reviews by Titorenko and Rachubinski (2001b), Subramani *et al* (2000), Sacksteder and Gould (2000), and Terlecky and Fransen (2000). Other reviews by Olsen (1998), Crookes and Olsen (1999), Hayashi (2000), Lopez-Huertas and Baker (1999), Baker *et al* (2000) and Mullen *et al* (2001a) provide a good source of reference on plant peroxisomes.

PEROXISOMAL MATRIX TARGETING SIGNALS

Unlike chloroplasts and mitochondria, peroxisomes do not possess DNA or protein synthesizing machinery. All of the protein constituents of the peroxisomal matrix and boundary membrane are nuclear encoded, synthesized on free polyribosomes in the cytosol and sorted to organelle in a post-translational manner (Lazarow and Fujiki, 1985).

A number of targeting signals and pathways are involved in sorting newly-synthesized peroxisomal proteins (Table 1). For instance, several distinct targeting signals participate in the sorting of peroxisomal membrane proteins (PMPs). Probably the best-studied membrane peroxisomal targeting signal (mPTS) is the so-called type 1 mPTS. This signal is responsible for sorting several PMPs including the *Candidia boinidinii* 47-kDa PMP (PMP47) directly from the cytosol to peroxisomes and consists of a stretch of positively-charged amino acid residues located on the matrix-side (internal) of the boundary membrane and an immediately-adjacent transmembrane domain (Dyer *et al*, 1996; Wang *et al*, 2001). The type 2 mPTS (mPTS2) also consists of a positively-charged region and a nearby transmembrane domain. However, unlike the mPTS1, this targeting signal requires that the PMP be sorted first to the ER (via an overlapping or adjacent ER sorting signal) before it can direct the protein to the peroxisome boundary membrane (Baerends *et al*, 1996; Elgersma *et al*, 1997; Mullen *et al*, 1999). Cottonseed ascorbate peroxidase (APX) is one of a subset of PMPs that is sorted to peroxisomes by a mPTS2 and is the only plant PMP for which its targeting information has been delineated (Mullen *et al*, 1999; Mullen and Trelease, 2000). Recently, several other mPTSs, distinct from the mPTS1 and mPTS2, have been identified in various mammalian PMPs (Pause *et al*, 2000; Honsho and Fujiki, 2001; Jones *et al*, 2001). However, the precise nature of these targeting signals or the sorting pathway that these PMPs follow, i.e. directly from the cytosol to peroxisomes or indirectly via the ER, is unknown. For further discussion of the targeting signals and sorting pathways of PMPs refer to Chapter 10.

Table 1: Targeting signals for membrane- and matrix-destined peroxisomal proteins

Membrane PTSs (mPTSs)		
mPTS1	basic stretch of amino acids adjacent to TMD(s) and orientated on matrix side of peroxisomal of peroxisomal membrane (e.g., *C. boidinii* PMP47)[1,2]	
mPTS2	basic stretch of amino acids adjacent to TMD(s) and orientated on matrix side of peroxisomal membrane (e.g., Cottonseed APX)[3,4]	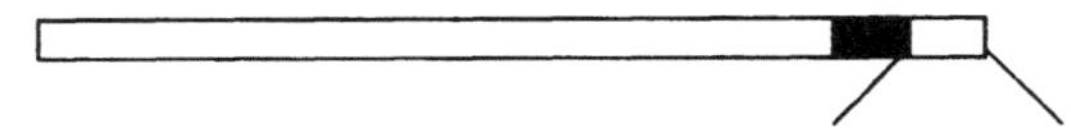
Others	not well defined - signal sequence varies depending upon the particular PMP (e.g., Rat liver PMP22)[5]	
Matrix PTSs		
PTS1	uncleaved, carboxy-terminal tripeptide (e.g., Cottonseed isocitrate lyase)[6]	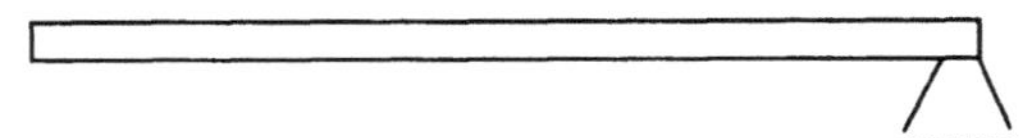
PTS2	cleaved, amino-terminal nonapeptide (e.g., Rat thiolase)[7,8]	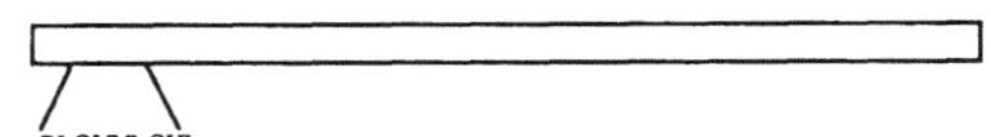
Internal PTS	precise signal sequence not defined (e.g., several yeast proteins without a recognizable PTS1 or PTS2)	

For examples of integral PMPs, putative transmembrane spanning domains (TMDs) are indicated as black boxes and basic clusters of amino acids residues within the mPTS1 or mPTS2 are underlined. 1. Dyer et al, 1996; 2. Wang et al, 2001; 3. Mullen et al, 1999; 4. Mullen and Trelease, 2000; 5. Pause et al, 2000; 6. Lee et al, 1997; 7. Swinkels et al, 1991; 8. Flynn et al, 1998.

At least three other distinct targeting signals are involved in sorting newly-synthesized proteins directly from the cytosol to the peroxisomal matrix (Table 1). It is important to point out that unlike some PMPs there is no direct evidence for a matrix-destined protein being sorted to peroxisomes via the ER. The most prominent matrix-targeting signal is the PTS1, an uncleaved carboxy-terminal SKL tripeptide motif (i.e. small-basic-hydrophobic residues) found in the majority of peroxisomal matrix-destined proteins. Another subset of peroxisomal matrix proteins possess a type 2 PTS (PTS2), an amino terminal nonapeptide (-R-X_6-H/Q-A/L/F-) which, in some organisms, is proteolytically removed following translocation into the peroxisome. Finally, a few peroxisomal proteins do not appear to possess a PTS1 or PTS2 and may be sorted by virtue of an internal PTS that remains to be characterized. Alternatively, these proteins might be imported into peroxisomes by associating with other PTS1- or PTS2-bearing proteins. In this section, I discuss each of these matrix PTSs, as well as the different peroxisome import assays that have been used to characterize PTSs and the parameters of peroxisomal protein import in plant cells.

The type 1 peroxisome targeting signal (PTS1)

Extensive experimental studies in evolutionarily diverse organisms including plants have provided a substantial amount of information on the nature of the PTS1. This targeting signal possesses four main characteristics. First, the PTS1 is a carboxy-terminal tripeptide motif (the so-called SKL motif) consisting usually of a small amino acid residue at the -3 position (e.g. serine or alanine), a basic residue at the -2 position (e.g. lysine or arginine) and a hydrophobic residue at the -1 position (e.g. leucine). However, a large (and growing) number of peroxisomal-destined proteins possess PTS1s with carboxy-terminal tripeptides that do not conform to the SKL motif. Second, proper functioning of the PTS1 is context dependent. That is, where a PTS1 does not conform to the SKL motif, residues upstream of the PTS1 called 'accessory residues' convey the proper context for the tripeptide signal to function (Elgersma *et al*, 1996; Mullen *et al*, 1997a). Third, the PTS1 functions only when at the carboxy terminus of proteins; the targeting signal does not function when either at the amino terminus or at internal locations (Gould *et al*, 1989; Miyazawa *et al*, 1989; Miura *et al*, 1992; Subramani, 1992). Fourth, unlike most other organelle-specific targeting signals (including the PTS2), the PTS1 is not proteolytically processed following import into the peroxisomal matrix.

Discovery of the PTS1 and its conservation among eukaryotes

The PTS1 was initially identified when the carboxy-terminal twelve amino acids residues of firefly luciferase (-LIKAKKGGKSKL-COOH) were demonstrated to be both necessary and sufficient for targeting of this protein to peroxisomes in cultured mammalian cells (Gould *et al*, 1987). Subsequent mutagenesis studies revealed that the PTS1 consists solely of the carboxy-terminal tripeptide -SKL (Gould *et al*, 1989) and that a number of conserved amino acid substitutions within this tripeptide preserved its targeting function (Gould *et al*, 1989; Swinkels *et al*, 1992; Miura *et al*, 1992). Based on these data, the PTS1 was defined as an SKL motif (i.e. small-basic-hydrophobic residues) with a minimal consensus sequence for mammalian cells of S/A/C-K/R/H-L. However, as discussed below, a large number of PTS1-containing proteins from various organisms including mammals have since been shown not to adhere strictly to this consensus sequence.

The realization that an insect PTS1 could function efficiently in cultured mammalian cells prompted researchers to test whether this targeting signal, and the mechanism for its recognition, is evolutionarily conserved throughout the eukaryotic kingdom. Strong support for this possibility came from additional heterologous expression studies with firefly luciferase in which the protein was found to be sorted to peroxisomes in the yeast *Saccharomyces cerevisiae* and in transgenic tobacco plants (Gould *et al*, 1990a). A carboxy-terminal SKL tripeptide was also sufficient for sorting other reporter proteins to glycosomes, a form of peroxisome in the parasitic trypanosomatid protozoa (Fung and Clayton, 1991), and to peroxisomes in the nematode *Caenorhabditis elegans* (Motley *et al*, 2000). Additional evidence for the evolutionary conservation of the PTS1 came from studies that employed antibodies raised against a peptide containing the carboxy-terminal SKL sequence. These antibodies recognized a number of peroxisomal proteins on either Western blots of protein extracts, cryo-fixed thin sections, and/or in permeabilized cultured cells from plants (castor bean seedling endosperm and suspension cultured tobacco and cotton cells), mammals, yeast, fungi, and trypanosomes (Gould *et al*, 1988; Gould *et al*, 1990b; Keller *et al*, 1991; Fransen *et al*, 1996; Trelease *et al*, 1996a; Usuda *et al*, 1999).

Since demonstrations that the import of PTS1-containing proteins is well conserved among eukaryotic organisms, plant scientists in particular have exploited this feature to test different aspects of peroxisomal protein import. For instance, the PTSs of a handful of mammalian and yeast proteins have been studied in tobacco BY-2 cells or transgenic Arabidopsis plants (Trelease *et al*, 1996; Flynn *et al*, 1998; Bongcam *et al*, 2000). Conversely, several targeting studies involving the sorting of plant proteins to mammalian and yeast peroxisomes have been reported (Trelease *et al*, 1994; Gietl *et al*, 1994; Taylor *et al*, 1996).

The PTS1 in plant cells – targeting of glycolate oxidase and isocitrate lyase

The first report on a functional role for the PTS1 in plant cells came from studies of glycolate oxidase, a key peroxisomal enzyme in the photorespiratory pathway (Volokita, 1991). Spinach glycolate oxidase possesses a carboxy-terminal tripeptide (ARL-COOH) that conforms to the mammalian PTS1 consensus sequence. To test whether this putative PTS1 could target a passenger protein to leaf peroxisomes *in vivo*, a chimaeric gene consisting of the bacterial enzyme β-glucuronidase (GUS) fused to the carboxy-terminal six amino acids of spinach glycolate oxidase was constructed and introduced stably into tobacco. Analysis of subcellular fractions from transgenic plants for GUS activity indicated that the glycolate oxidase carboxy terminus was weakly sufficient for targeting to leaf peroxisomes (Volokita, 1991). Interestingly, in another study that employed an *in vitro* peroxisome import assay, the carboxy-terminus of spinach glycolate oxidase was not necessary nor sufficient for import into isolated sunflower glyoxysomes (Horng *et al*, 1995). This apparent discrepancy has not yet been resolved.

The majority of studies on the plant PTS1 have focused on the glyoxylate cycle enzyme isocitrate lyase. In particular, whether the enzyme's carboxy-terminal tripeptide (either ARM or SRM depending upon the higher plant species) can function as a PTS1. Evidence against a targeting role for this tripeptide came from a number of *in vitro* and *in vivo* studies with castor bean isocitrate lyase. Baker and co-workers demonstrated that the carboxy terminus of the castor bean form of the enzyme, including its ARM-COOH, was dispensable for both *in vitro* import into isolated sunflower peroxisomes (Behari and Baker, 1993) and *in vivo* import into peroxisomes in both transgenic tobacco plants (Gao *et al*, 1996) and *S. cerevisiae* (Taylor *et al*, 1996). However, a similar carboxy-terminal peptide KARM was shown to be essential for targeting transiently-expressed cottonseed isocitrate lyase to peroxisomes in cultured Chinese hamster ovary cells (Trelease *et al*, 1994). The carboxy-terminal portion of *Brassica* isocitrate lyase (ending in -SRM-COOH) was also necessary, as well as sufficient, for *in vivo* import into peroxisomes in transgenic Arabidopsis plants (Olsen *et al*, 1993). In an attempt to resolve these discrepancies between the targeting of the three different versions of isocitrate lyase, Lee *et al* (1997) examined the import competency of each enzyme in the same *in vivo* import system, namely, suspension-cultured tobacco BY-2 cells. They found that carboxy-terminal residues of *Brassica*, castor bean, and cottonseed isocitrate lyase were all necessary (and their tripeptides sufficient) for targeting and import into BY-2 glyoxysomes.

Several possible reasons have been presented for the conflicting results on the functional role of the carboxy-terminal tripeptide in targeting isocitrate lyase to peroxisomes (reviewed by Olsen, 1998). Perhaps the most feasible is that, *in vitro*, modified versions of castor bean isocitrate lyase lacking their carboxy-terminal regions are imported into peroxisomes by a cryptic or 'hidden' targeting signal that is not exposed in the full-length wild-type protein, nor is efficient for sorting to peroxisomes *in vivo*. The identity of this so-called cryptic PTS sequence, however, remains to be determined.

Functional divergency of the plant PTS1

As we learn more about the nature of the PTS1 pathway, and as new genes encoding peroxisomal matrix proteins with PTS1-resembling sequences are cloned, it has become apparent that the PTS1 possesses a greater diversity of functional amino acid residues than those defined by the SKL motif. For example, a number of divergent residues, including non-basic ones at the -2 position of the carboxy-terminal tripeptide of phosphoglycerate kinase, function for glycosomal targeting in *Trypanosoma brucei* (Blattner *et al*, 1992). In *S. cerevisiae*, a wide range of SKL variants, including large residues at the -3 position and non-basic ones at the -2 position, are functional for import of malate dehydrogenase into peroxisomes (Elgersma *et al*, 1996). Several examples of divergence within the SKL motif also exist in a number of mammalian peroxisomal proteins (Motley *et al*, 1995; Purdue and Lazarow, 1996; Amery *et al*, 1998; Olivier *et al*, 2000), indicating that PTS1 variation can not be explained simply on the basis of species.

Perhaps the greatest range of functional residues within the PTS1 occurs in plant peroxisomal proteins (Table 2). This conclusion is based primarily on comprehensive mutational analyses of the PTS1 whereby selected carboxy-terminal peptides were appended to different non-peroxisomal reporter proteins, including chloramphenicol acetyltransferase (CAT) and GUS, and then tested for their ability (sufficiency) to redirect the resulting fusion protein to peroxisomes *in vivo* (Hayashi *et al*, 1996; Mullen *et al*, 1997a; Hayashi *et al*, 1997). Using a different experimental approach, Kragler *et al* (1998) demonstrated that a number of peptides ending in many of the same divergent carboxy-terminal tripeptides could interact specifically with the tobacco PTS1 receptor (NtPex5p) in a yeast two-hybrid system.

The range of carboxy-terminal tripeptides capable of targeting proteins to plant peroxisomes was further extended when the subcellular sorting of random GFP fusion proteins in transgenic Arabidopsis plants was examined (Cutler *et al*, 2000; see Table 2, tripeptides marked with an asterisk). Cutler and co-workers demonstrated that over a third of the transgenic lines studied by confocal microscopy (43 of 120 lines) displayed GFP fluorescence that co-localized with the immunostaining attributable to peroxisomal matrix catalase. However, they cautioned that GFP fusion proteins possessing highly divergent SKL carboxy-terminal tripeptides (e.g. -CMM, -TNL, -SHR, -EGP-COOH) could be sorted to peroxisomes by other targeting signals such as the PTS2.

Several pertinent conclusions on the nature of the plant PTS1 can been made based on results of *in vivo* targeting studies that are summarized in Table 2.

Table 2: Peroxisomal targeting efficiencies of various carboxy-terminal tripeptides in plant cells or transgenic plants†

Small-basic hydrophobic		**x-basic hydrophobic**		**Small x-hydrophobic**		**Small-basic-x**		**Others**	
arl	+	erl	–	anl	+	skd	–	egp*	+
arm	+	fkl	+/–	cmm*	+	sre	–	psi	–
ckl	+	frl	–	scl*	+	srk	–	psl*	+
crl	+	gkm*	+	sgl	–	srs	–	shr*	+
crm	+	krl	–	shl	+/–			slw*	+
gkl	+	lrl*	+/–	sil	–			tiy*	+
grl	–	pkl	–	sll	+			tnl	+
ski	+	prl	+	sml*	+				
skf	+	qkl*	+	snl*	+				
skl	+	tri*	+	spl*	+				
srf	+	wrl*	+	sql*	+				
sri	+	yrl	–	ssi*	+/–				
srl	+			ssl*	+/–				
srm	+			stl	–				
srv	–			syl*	+				
tkl	+								

Results shown were obtained from microscopic analyses of the subcellular localization of various reporter fusion proteins with carboxy-terminal tripeptides that either conformed or did not conform to the SKL motif, i.e. a small residues at position -3, a basic residue at postion -2, and a hydrophobic residue at position -1. Targeting efficiency of carboxy-terminal tripeptides for sorting fusion proteins are shown as either: + (peroxisomal), – (cytosolic or undetectable in peroxisomes), or +/– (peroxisomal or cytosolic depending upon the study).

† Based on data of Hayashi et al, 1996, 1997; Trelease et al, 1996; Lee et al, 1997; Mullen et al, 1997a, 1997b; Cutler et al, 2000.

** Based on the data of Cutler et al (2000) in which GFP-fusion proteins containing varied carboxy-terminal tripeptides were shown to be localized in leaf peroxisomes in transgenic Arabidopsis plants. Note that it is possible that other PTSs were responsible for sorting some of these GFP-fusion proteins to peroxisomes and therefore some tripeptides, particularly those highly divergent from the SKL motif, may not be true PTS1s.*

First, the PTS1 mostly conforms to the SKL motif. In fact, this motif likely constitutes the most favorable PTS1 signal because the majority of peroxisomal proteins in plants (and in other eukaryotes) possess the carboxy-terminal consensus sequence. Second, divergence within the PTS1 occurs primarily at the -3 and -2 positions; a tripeptide containing a non-hydrophobic residue at the -1 position has not been shown to be functional for targeting to peroxisomes. Finally, for those proteins that do not possess an 'ideal' (SKL) PTS1, amino acid residues directly preceding the carboxy-terminal tripeptide are necessary for sorting to peroxisomes. Termed 'accessory sequences' (Elgersma *et al*, 1996; Mullen *et al*, 1997a), these residues appear to convey the proper context for the divergent carboxy-terminal tripeptide to function efficiently as a PTS1.

Probably the best example for the context role of accessory residues in PTS1 targeting comes for studies of cottonseed catalase (Mullen *et al*, 1997b). This protein's carboxy-terminal tripeptide -PSI is conspicuously divergent from the SKL motif. Targeting studies revealed that the -PSI is necessary, but interestingly not sufficient, for sorting to peroxisomes. However, when an R (a residue commonly found at the -4 position in plant catalases) was added to the carboxy-terminal PSI, the tetrapeptide was sufficient for sorting a CAT fusion protein (CAT-RPSI) to peroxisomes. The role of a basic residue at the -4 position in conveying the proper context for a PTS1 to function has also been demonstrated for human catalase. Purdue and Lazarow (1996) showed that the divergent carboxy-terminal tripeptide -ANL of human catalase (non-basic at the -2 position) was not sufficient for sorting to *S. cerevisiae* peroxisomes, whereas -KANL was.

Accessory sequences may also constitute part of the PTS1 itself, directly interacting along with the carboxy-terminal tripeptide with the PTS1 receptor. Wolins and Donaldson (1997) showed that residues immediately adjacent to the acyl-CoA oxidase PTS1 bound with high-affinity to a putative PTS1 receptor integrated into castor bean glyoxysome membranes. In a yeast two-hybrid system, tobacco, human, and *S. cerevisiae*, PTS1 (Pex5p) receptors recognize a variety of SKL and non-SKL carboxy-terminal tripeptides, and the strength of these interactions is modulated by the position and type of upstream accessory residues (Kragler *et al*, 1998; Lametschwandtner *et al*, 1998). These data suggest that the observed species-specific variability in the PTS1 can be attributed to differences in the ability of an organism's PTS1 receptors to recognize the carboxy-terminal tripeptide and its accessory sequences (Hettema *et al*, 1999). They also imply that once the precise role of the accessory residues is characterized, the PTS1 may need to be recognized as consisting of more than just a carboxy-terminal tripeptide.

The type 2 peroxisome targeting signal (PTS2)

The PTS2 occurs in a small number of peroxisomal matrix-destined proteins and consists of two motifs: (1) a sequence-specific nonapeptide motif, i.e. -R-X_6-H/Q-A/L/F-; and (2) a structural motif that is dependent upon the physiochemical properties of residues within the nonapeptide. Unlike the PTS1, the PTS2 can function at internal positions; although the targeting signal is located most frequently near the amino-terminus of proteins. In plants and mammals, but not in yeast or trypanosomes, PTS2-containing proteins are synthesized as high molecular weight precursors and following import into the peroxisomal matrix the presequence containing the PTS2 is removed by a peptidase.

Evolutionary selection against the PTS2 pathway

In plants, just six PTS2-containing proteins have been identified, including glyoxysomal isoforms of thiolase, malate dehydrogenase, citrate synthase, aspartate aminotransferase, acyl-CoA oxidase, and a 70-kDa heat shock protein (Hsp70) (Gietl *et al*, 1994; Kato *et al*, 1996; Wimmer *et al*, 1997; Flynn *et al*, 1998; Gerbhardt *et al*, 1998; Hayashi *et al*, 1998a; Kato *et al*, 1998) (see Table 3). However, when compared to other organisms, the number of plant proteins that are sorted to peroxisomes by virtue of a PTS2 is relatively large. For instance, in *Yarrowia lipoytica* only the β-oxidation enzyme 3-ketoacyl-CoA thiolase contains a PTS2. Remarkably, the entire PTS2-targeting pathway is absent from *C. elegans* (Motley *et al*, 2000). Orthologues of typical PTS2-containing proteins such as thiolase and acyl-CoA oxidase instead have acquired a PTS1. Several peroxins required specifically for targeting and import of PTS2-containing proteins including the PTS2-receptor (Pex7p), Pex18p and Pex21p, are absent from the *C. elegans* genome sequence database. These observations have led to speculation that during evolution the selective pressure necessary for preserving the PTS2-targeting pathway has been relaxed (Motley *et al*,

2000). As a consequence, proteins once sorted by a PTS2, evolved carboxy-terminal tripeptides that, given the highly divergent nature of the amino acid residues within the targeting signal, function as a PTS1. The acquisition of the PTS1 by a PTS2 protein subsequently facilitated the loss of the PTS2 without affecting its intracellular localization. In the phylogenetic lineage leading to *C. elegans* the switch from PTS2 to PTS1 was apparently completed and, thereby, led to a loss of peroxin proteins specifically required for the PTS2-targeting pathway. In other organisms such as mammals, yeast and possibly less so in plants, the switch from PTS2 to PTS1 appears not to be complete, since some PTS2-containing proteins still exist, as well as some 'intermediate' proteins that possess both a PTS1 and PTS2 e.g. *Hansenula polymorpha* Pex8p (Waterham *et al*, 1994) and *S. cerevisiae* Dci1p (Karpichev and Small, 2000).

It is tempting to speculate also that in mammals, yeast and plants other more inconspicuous effects of the evolutionary switch from PTS2 to PTS1 occurred. For instance, the convergence at the peroxisome boundary membrane of the PTS1 and PTS2 sorting pathways might represent the remnants of what was once two separate translocation pathways.

Table 3: Plant peroxisomal proteins containing a known or putative PTS2

Acyl-CoA oxidase			
Pumpkin	MASPGEPNRTAEDESQAAAR	**RIERLSLHL**	TPIPLDDSQGVEMETCAAGKAKAKI-
Aspartate aminotransferase			
Arabidopsis	MKTTHFSSSSSSDR	**RIGALLRHL**	NSGSDSDNLSSLYASPTSGGTGGSV-
Soybean	MRPPVILKTTTSLLDSSSSSPPCDR	**RLNTLARHF**	LPQMASHDSISASPTSASDSVFNHL-
Thiolase			
Arabidopsis	MERAME	**RQKILLRHL**	NPVSSSNSSLKHEPSLLSPVNCVSE-
Arabidopsis	MEKAIE	**RQRVLLEHL**	RPSSSSSHNYEASLSASACLAGDSA-
Brassica	MEKAME	**RQRVLLEHL**	RPSSSSSHSFEGSLSASACLAGDSA-
Cucumber	MEKAIN	**RQSILLHHL**	RPSSSAYTNESSLSASVCAAGDSAS-
Mango	MEKAIN	**RQQVLLQHL**	RPSNSSSHNYESALAASVCAAGDSA-
Pumpkin	MEKAIN	**RQSILLHHL**	RPSSSAYSHESSLSASVCAAGDSAS-
Rice	MEKAIN	**RQRVLLAHL**	EPAASPAAAAPAITASACAAGDSAA-
Malate dehydrogenase			
Alfalfa	MEPNSYANS	**RITRIASHL**	NPPNLKMNEHGGSSLTNVHCRAKGG-
Brassica	MPHK	**RIAMISAHL**	QPSFTPQMEAKNSVMGLESCRAKGG-
Cucumber	MQPIPDVNQ	**RIARISAHL**	HPPKYQMEESSVLRRANCRAKGGAP-
Pumpkin	MKPIPDVNE	**RIARISAHL**	QPPKSQMEEGSVLRRANCRAKG-
Rice	MEDAAAAAR	**RMERLASHL**	RPPASQMEESPLLRGSNCRAKGAAP-
Soybean	NSGASD	**RISRIAGHL**	RPQREDDVCLKRSDCRAKGGVSGFK-
Watermelon	MQPIPDVNQ	**RIARISAHL**	HPPKSQMEESSALRRANCRAKGGAP-
Citrate synthase			
Pumpkin	MPTDMELSPSNVARH	**RLAVLAAHL**	SAASLEPPVMASSLEAHCVSAQTMV-
Hsp70			
Watermelon	MRKSNHVSS	**RTVFFGQKL**	GNSSAFPTATFLKLRSNISRRNSSV-
		-R-X_6-H/K-L/F-	

PTS2 nonapeptide sequences are shown in bold. Cysteine residues located at known or putative processing sites are underlined. The plant PTS2 consensus sequence is -R-X_6-H/K-L/F-. Adapted from Flynn et al, 1998 and Kato et al, 2000.

Functionality of residues within the PTS2

Several mutational analysis studies have focused on the functionality of amino acid residues within the PTS2 nonapeptide in mammals (Osumi *et al*, 1991; Swinkels *et al*, 1991), yeasts (Erdmann, 1994; Glover *et al*, 1994a), trypansomes (Blattner *et al*, 1995), and plants (Gietl *et al*, 1994; Kato *et al*, 1996, 1998). Based on collective data, the PTS2 has been defined as an -R-L/I/V-X_5-H/Q-L/A-consensus sequence, and most substitutions of the conserved amino acid residues at positions 1, 2, 8 or 9 abolish peroxisomal targeting. However, results obtained from more recent targeting experiments in tobacco BY-2 cells suggest that this PTS2 is oversimplified. Flynn *et al* (1998) presented a comprehensive alignment of over two dozen PTS2 nonapeptides from various proteins in different organisms showing that with the exception of R at position 1 and H or Q at position 8, variations in residues exist at all other positions. Based on these alignment data, they constructed (via site-directed mutagenesis) nonapeptide variants of the rat thiolase PTS2 and then tested whether they were sufficient for targeting a passenger protein (CAT) to BY-2 glyoxysomes. Results revealed that a variety of residues at divergent positions with the PTS2 nonapeptide, particularly within the so-called arbitrary -X_5- domain, either diminished or abolished targeting of thiolase-CAT fusion proteins to glyoxysomes.

Taken together, the data obtained from studies with rat thiolase reveal a number of new characteristics of PTS2 targeting in plant cells (Flynn *et al*, 1998). First, a revised PTS2 consensus sequence -R/K-X_6-H/Q-A/L/F- (specifically -R-X_6-H/K-L/F- for plants) can be formulated, with the second position of the nonapeptide changed to an X because more than three different amino acid residues can exist at this position. Second, unlike the PTS1, the targeting action of the PTS2 nonapeptide is not enhanced by a context-specific environment conferred by adjacent (accessory) amino acid residues. For example, similar glyoxysomal targeting occurs for CAT thiolase fusion proteins with a presequence consisting of a mango thiolase nonapeptide within the context of either adjacent rat thiolase or mango thiolase amino-terminal residues. Context within the nonapeptide itself, however, is critical for efficient targeting. Several CAT thiolase fusion proteins possessing single amino-acid substitutions within the nonapeptide are not sorted to peroxisomes, unlike other fusion proteins possessing more than one of the same substitutions, which are. Apparently, residues within the nonapeptide, including the -X_6- domain, are capable of operating in a synergistic manner to convey maximum PTS2 function. This implies also that PTS2 should not be defined solely by a sequence-specific motif, but also by the physiochemical properties of its residues within the nonapeptide. This model is similar to the specific structural characteristics that define other targeting signals located in amino-terminal presequences (e.g. mitochondria, chloroplasts and ER).

Proteolytic processing of PTS2-containing proteins

In plants and mammals, but not in yeast or trypansosomes, the amino-terminal presequence of proteins containing a PTS2 is proteolytically processed following import. This contrasts with the maturation of mitochondrial, chloroplast, and secretory proteins, where the import of the preprotein and cleavage of the presequence are coupled.

Studies on peroxisomal isoforms of pumpkin citrate synthase (Kato *et al*, 1996) and malate dehydrogenase (Kato *et al*, 1998) have demonstrated that a conserved cysteine residue located downstream of the PTS2 at the carboxy-terminal end of the presequence is essential for proteolytic processing following import of both enzymes (see Table 3). Deletion or substitution of this cysteine residue with another amino acid does not affect targeting but does prevent cleavage of the amino-terminal presequence of both enzymes in transgenic Arabidopsis peroxisomes. Interestingly, a similar cysteine residue exists in at least two other plant PTS2-containing enzymes, acyl-CoA and thiolase (Kato *et al*, 2000; Table 3). In addition, the conserved cysteine exists in mammalian PTS2-containing proteins, but not in enzymes from yeast or trypansomes, suggesting that the cysteine-dependent proteolytic mechanism is evolutionarily conserved at least among plants and mammals (Kato *et al*, 2000). It remains to be determined for those plant PTS2-containing proteins that do not possess the conserved cysteine residue (i.e. aspartate aminotransferase and Hsp70; Table 3) whether they are cleaved following import into peroxisomes.

Only two proteins have been implicated in the proteolytic processing of PTS2-containing proteins, i.e. a rat 110-kDa metalloendoprotease (Authier *et al*, 1995) and a castor bean endosperm cysteine endopeptidase (Gietl *et al*, 1997). However, neither peptidase can proteolytically process a PTS2-containing preprotein *in vivo*. Moreover, the castor bean enzyme was recently demonstrated not to be localized in peroxisomes, but rather in ricinosomes; small spherical electron-dense organelles (>0.5 um in diameter) involved in protein mobilization during post-germinative growth (Schmid *et al*, 1998).

Internal peroxisome targeting signals

A small number of peroxisomal matrix-destined proteins in yeast (but not in plants or mammals) do not contain a recognizable PTS1 or PTS2 sequence; they may be sorted by an alternative targeting signal(s) located internally within the protein – possibly a PTS3? However, the precise nature of each of these internal targeting signals or the existence of distinct receptors for them has not been defined. In fact, it is now generally held that most proteins once proposed to be sorted to peroxisomes by internal PTSs, are in fact sorted by virtue of their association with *bona fide* PTS1- or PTS2-containing proteins. Evidence in favor of this is mounting. For example, in *S. cerevisiae* cells lacking a PTS1 receptor a number of proteins containing putative internal targeting signal(s) are mis-localized to the cytosol, along with PTS-1 containing proteins (e.g. Elgersma *et al*, 1995; Yang *et al*, 2001). Indeed, one of these mislocalized proteins (delta3,delta2-enoyl-CoA isomerase) that lacks a recognizable PTS1 or PTS2 has been shown to be targeted to peroxisomes as hetero-oligomers with the PTS1-containing protein Dci1p (Yang *et al*, 2001).

In plants, no targeting signal other than the PTS1 and PTS2 has been shown to be both necessary and sufficient for targeting a protein to the peroxisomal matrix.

Experimental systems used to elucidate PTSs and the mechanisms of protein import into peroxisomes

Assays that have been employed to study protein targeting and import into peroxisomes in plant cells fall into two broad categories: (1) *in vitro* assays using purified peroxisomes, and (2) *in vivo* assays with either transgenic plants or transiently-transformed suspension-cultured cells. In general, *in vivo* assays have focused on the targeting signals for peroxisomal proteins, whereas *in vitro* assays have allowed researchers to mimic the process of peroxisome protein import.

In vitro peroxisomal import assays

In vitro import systems using purified peroxisomes have provided a powerful means to dissect the molecular mechanisms involved in the import of proteins containing a PTS1 (Imanaka *et al*, 1987; Small *et al*, 1987; Heinemann and Just, 1992; Behari and Baker, 1993; Brickner *et al*, 1997) or a PTS2 (Miura *et al*, 1994). As illustrated in Figure 1, in a typical peroxisomal *in vitro* import assay *in vitro*-translated or recombinant protein substrate (test protein) are initially mixed with peroxisome-rich fractions recovered by high-speed density gradient centrifugation. Following an appropriate incubation period, the peroxisomes are treated with protease to degrade non-imported proteins, re-isolated and then protein constituents analyzed by SDS-PAGE and fluorography or immunoblotting. Several important control reactions, including disruption of peroxisomes prior to protease treatment and demonstration of dependence upon ATP for protein import are included in the assay to confirm that aggregation or inherent protein folding are not responsible for the observed protease protection (Baker, 1996).

Figure 1: Schematic representation of an in vitro import assay. Isolated peroxisomes are mixed with a test protein under the appropriate conditions.

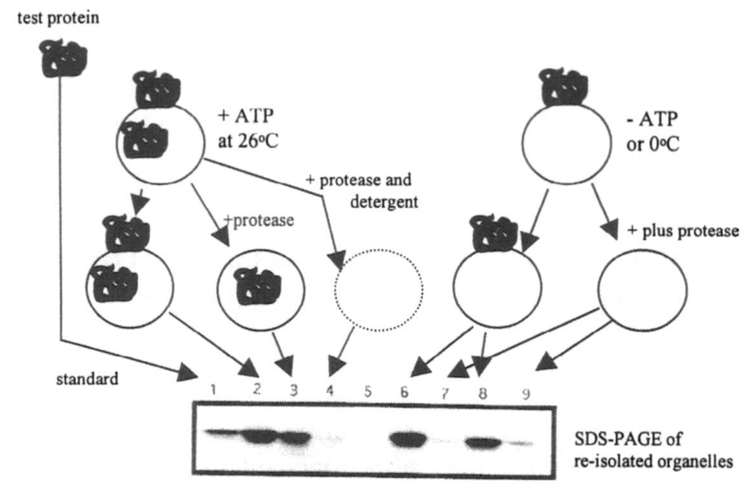

Figure 1. In the presence of ATP and at 26°C, protein binds to the surface and is imported. Bound and imported protein can be distinguished by protease treatment before the organelles are re-isolated and subjected to SDS-PAGE and an appropriate detection method (lanes 2 and 3). As controls to ensure that the appearance of protease protected protein is genuinely due to import a sample is re-isolated, and treated with protease in the presence of detergent (lane 4). Lane 5 is a control where test protein is subjected to import conditions in the absence of organelles. This tests for the protein's solubility under import conditions, as aggregation and precipitation could give rise to protease protection and therefore the artefactual appearance of import. In the absence of ATP or in the presence of ATP at low temperature the import of most proteins is inhibited, but they are still able to bind to the organelle surface. This is shown in lane 6 (incubation in the absence of ATP), lane 7 (incubation in the absence of ATP followed by protease treatment), lane 8 (incubation at 0°C) and lane 9 (incubation at 0°C followed by protease treatment). Typically, a known fraction of the amount of test protein added to each assay is loaded on the gel (lane 1) as a standard to allow estimation of the efficiency of the import reaction. Based on Baker (1996).

Peroxisomal PTS1 import has been achieved using glyoxysomes from sunflower (Behari and Baker, 1993; Onyeocha *et al*, 1993) and pumpkin (Brickner *et al*, 1997; 1998) as well as peroxisomes from rat liver (Imanaka *et al*, 1987) and *C. tropicalis* (Small *et al*, 1987). In each assay system acquisition of PTS1-containing proteins is energy-, temperature-, time- and targeting-signal-dependent. Experiments with pumpkin glyoxysomes also have demonstrated that the molecular chaperones Hsp70 and Hsp90 are necessary for efficient protein import (Crookes and Olsen, 1998). Other studies with sunflower glyoxysomes have revealed that protein import is regulated by NADPH (Pool *et al*, 1998a) and directly dependent upon the docking protein Pex14p located on the cytosolic face of the peroxisome boundary membrane (Lopez-Huertas *et al*, 1999). Sunflower glyoxysomes have also been used to identify import intermediates during the process of protein translocation through the boundary membrane (Pool *et al*, 1998b). Characterization of these intermediates should allow for future chemical cross-linking experiments to be carried out so that protein factors constituting part of the translocation complex can be identified (reviewed in Baker, 1996).

Unfortunately, *in vitro* peroxisomal import assays are not without their shortcomings. The principal problem encountered is reproducibility, primarily due to the instability (fragility) of purified peroxisomes and the fact that matrix proteins can differentially leak out of organelles *in vitro*. Import assays with purified peroxisomes also suffer from a lack of an easily identifiable hallmark for protein acquisition, such as the preprotein cleavage used in the study of translocation into other organelles (e.g. ER, chloroplasts and mitochondria). The common criterion for *in vitro* import of proteins into peroxisomes is based on protease-protection (reviewed in Subramani, 1993). However, certain peroxisomal proteins (e.g. catalase) and reporter proteins (e.g. CAT) are inherently resistant to protease digestion. Another important and often overlooked consideration of *in vitro* import assays with purified peroxisomes is the heterogeneity in the population of (pre)peroxisomes in a cell with respect to density, protein composition and capacity to import nascent proteins (reviewed by Mullen *et al*, 2000b; Titorenko and Rachubinski, 2001a). Finally, in the few studies aimed at characterizing PTSs *in vitro*, targeting determinants have been identified strictly by a sequence's necessity for peroxisomal protein import, i.e. whether import is abolished when a particular polypeptide sequence is deleted from the protein. No *in vitro* study using purified peroxisomes has demonstrated the sufficiency, or perhaps more importantly the lack of sufficiency, for any peroxisomal protein sequence to import a passenger (protein) into peroxisomes.

In mammalian cells, two additional *in vitro* peroxisomal import assays have been developed: microinjection (Walton *et al*, 1992a, 1994) and semi-permeabilized cells (Rapp *et al*, 1993; Wendland and Subramani, 1993; Terlecky *et al*, 2001). Both assays have proven to be effective in providing mechanistic insights into peroxisomal import and in circumventing the shortcomings associated with *in vitro* assays that utilize purified peroxisomes. Results obtained with microinjection and semi-permeabilized cells have been consistent with those using purified peroxisomes and also have demonstrated several novel properties of the import process for PTS1- and PTS2-containg proteins (Wendland and Subramani, 1993; Legakis and

Terlecky, 2001). The feasibility of either of these two *in vitro* peroxisomal import assays in plants remains to be tested.

In vivo peroxisomal protein targeting assays

One *in vivo* approach for examining the targeting signals responsible for sorting proteins to peroxisomes employes transgenic tobacco and Arabidopsis plants. This targeting assay involves agrobacterium-mediated plant transformation followed by an examination, using a variety of microscopic and biochemical analyses, of the subcellular localization of ectopically-expressed proteins. In most studies, the expressed protein has been a chimera consisting of a polypeptide sequence that contains a putative PTS1 or PTS2 fused to a readily detectable reporter protein, such as GUS or CAT. For instance, Nishimura and co-workers have analyzed the targeting signals of several glyoxysomal enzymes by fusing sequences coding for the protein's putative PTS1 or PTS2 to GUS and then ectopically expressing the fusion protein in transgenic Arabidopsis plants (for review see Kato *et al*, 2000). Other *in vivo* peroxisomal targeting studies with transgenic plants have focused on wild-type (full-length) proteins, i.e. without an appended reporter protein. Onyeocha *et al* (1993) and Marrison *et al* (1993) both showed that castor bean isocitrate lyase is sorted into leaf peroxisomes in transgenic tobacco plants, and Olsen *et al* (1993) reported that *Brassica* isocitrate lyase is targeted into leaf and root peroxisomes in transgenic Arabidopsis plants. In each case, the localization (via cell fractionation and microscopy) of the introduced isocitrate lyase protein was possible because the enzyme is not expressed endogenously in leaves or roots.

The targeting studies with ectopically-expressed glyoxysomal isocitrate lyase in transgenic plants also demonstrate an important general feature of plant peroxisomal protein import: that the different classes of peroxisomes in higher plants have common protein import machineries and that transitions in peroxisomal function (e.g. the switch from glyoxysomes to leaf peroxisomes in cotyledons during greening or from leaf peroxisomes back to glyoxysomes during senescence) are not regulated primarily at the level of protein import. Rather, the different classes of plant peroxisomes appear to change their metabolic function simply as a consequence of the changing pool of distinct sets of peroxisomal-destined proteins that are synthesized *de novo* in the cytosol. Of course, this point underlies the conservation of peroxisomal protein import that has been reported to exist among almost all eukaryotic cells.

A second *in vivo* peroxisomal import assay employs tobacco BY-2 suspension-cultured cells and has proved to be an effective alternative to transgenic plants because of the short amount of time required to carry out targeting studies, i.e. days with BY-2 cells versus months with transgenic plants. The assay, developed by Banjoko and Trelease (1995) and based on a comparable *in vivo* import assay with mammalian cells (Trelease *et al*, 1994), relies on transient transformation (via biolistic particle bombardment) with DNAs coding for either wild-type or modified (e.g. truncated) peroxisomal proteins, or various reporter fusion proteins (Figure 2). To distinguish between endogenous BY-2 and transiently-expressed peroxisomal proteins, the introduced protein usually contains an in-frame epitope tag (e.g. HA epitope) located at either the carboxy or amino terminus. Following bombardment, cells are incubated for approximately 24 hours to allow gene expression and protein sorting and then chemically fixed and permeabilized with pectolyase and Triton X-100. Targeting to the peroxisome (or any other subcellular site) is assessed by indirect immunofluorescence microscopy; specifically, comparing immunostaining of introduced expressed proteins with an endogenous peroxisomal protein such as matrix catalase. Import of introduced proteins into peroxisomes is demonstrated using differential permeabilziation with digitonin rather than triton X-100, a detergent that permeabilizes plasmalemma, but not organellar, membranes (Lee *et al*, 1997).

Figure 2: Tobacco BY-2 cells as in vivo peroxisomal import assay

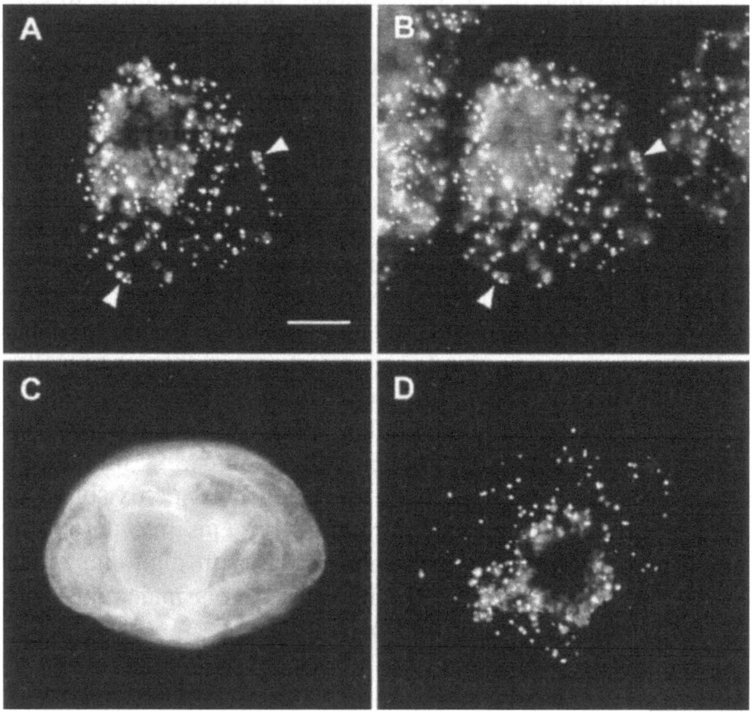

Figure 2. Immunofluorescence patterns attributable to either (A) transiently-expressed myc-epitope-tagged rice multifunctional (MFP) protein in individual peroxisomes in a single transformed BY-2 cell; (B) endogenous peroxisomal catalase in the same cell as A as well as surrounding non-transformed cells (arrows in A and B show obvious colocalizations); (C) transiently-expressed chloramphenicol acyltransferase (CAT) in the cytosol of an single transformed cell; and (D) transiently-expressed CAT-SRM (CAT appended to the carboxy-terminal tripeptide of the rice MFP protein; SRM-COOH) in individual peroxisomes. Transiently-expressed proteins were introduced into BY-2 cells via biolistic bombardment. All cells were fixed and processed for immunofluorescence microscopy as described by Dyer and Mullen (2001). Bar in A is 10 µm. (Murphy and Mullen, unpublished).

BY-2 peroxisomes (redefined as glyoxysomes because they posses the glyoxylate cycle enzymes isocitrate lyase and malate synthase) are abundant and easily distinguishable from other cytoplasmic organelles by electron and fluorescence microscopy (Banjoko and Trelease, 1995; Trelease *et al*, 1996a; Lee *et al*, 1997; Mullen *et al*, 2001b). BY-2 cell cultures are also readily amenable to stable-transformation (using Agrobacterium) of peroxisomal-destined proteins and to biochemical studies of peroxisome biogenesis

(Mullen *et al*, 2001b). Other advantages offered by BY-2 cell cultures are that the environment and growth conditions are completely controllable and that the population of cells can be synchronized. These attributes are difficult to meet with intact transgenic plants. However, BY-2 cells have not been as effective as *in vitro* import assays in deciphering the specific mechanisms of peroxisomal protein import. Thus, a combination of both *in vivo* and *in vitro* approaches will most likely provide the best understanding of peroxisomal protein targeting and import.

OLIGOMERIC PROTEIN IMPORT

For organelles such as mitochondria and the ER, newly-synthesized proteins in the cytosol must first be unfolded before they can be threaded vectorially through a gated translocon on the organelle's boundary membrane. For peroxisomes, however, incoming matrix proteins can transverse the boundary membrane fully folded and as oligomeric protein complexes (for reviews see McNew and Goodman, 1996; Crookes and Olsen, 1999; Smith and Schnell, 2001). Evidence in support of this unique feature comes from a diverse set of experiments. For example, kinetic studies of peroxisomal protein import and assembly have shown that a number of enzymes including human catalase (Middelkoop *et al*, 1991; *S. cerecisiae* malate dehydrogenase (Elgersma *et al*, 1996) and *Y. lipoytica* thiolase, isocitrate lyase and alcohol oxidase all oligomerize in the cytosol prior to translocation into peroxisomes (Titorenko *et al*, 1998). Another study with aminopterin, a folate-analogue which binds to and stabilizes dihydrofolate reductase (DHFR) in a folded conformation, revealed that the drug does not inhibit the import of a DHFR-PTS1 fusion protein into glycosomes (Hausler *et al*, 1996). In *in vitro* assays based on microinjection (Walton *et al*, 1992a) and cell permeabilization (Wendland and Subramani, 1993), pre-folded proteins such as luciferase or human serum albumin (HSA) conjugated to multiple PTS1-containing peptides were all imported into mammalian peroxisomes. Perhaps even more remarkable was the demonstration that microinjected 9 nm colloidal gold particles coated with HSA-PTS1 peptides can also be imported into the peroxisomal matrix in mammalian cells (Walton *et al*, 1995) (Figure 3).

Figure 3: Peroxisomal import of gold particles in mammalian cells

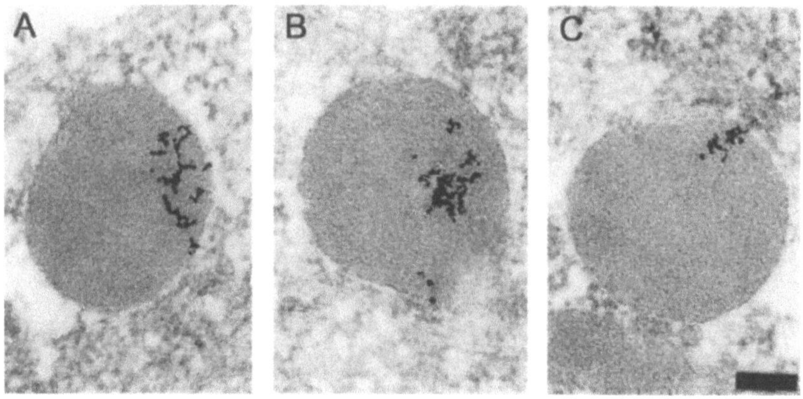

Figure 3. Colloidal gold particles (4-9 nm in diameter) conjugated to human serum albumin (HSA) containing a carboxy-terminal SKL were microinjected into the cytosol of Hs68 cells (A-C).

Following an incubation of 16 h at 37°C, cells were fixed and processed for electron microscopy as described elsewhere (Walton et al, 1995). Bar in C is 0.2 μm.

Other convincing evidence that peroxisomes have the capacity to acquire oligomeric protein complexes has come from so-called 'piggy-back-import' experiments in various organisms, including *S. cerevisiae* (McNew and Goodman, 1994; Glover *et al*, 1994b), mammalian cultured cells (McNew and Goodman, 1994: Leiper *et al*, 1996), tobacco BY-2 cells (Lee *et al*, 1997; Flynn *et al*, 1998), and transgenic Arabidopsis (Kato *et al*, 1999). In these experiments, researchers exploited the fact that peroxisomal proteins exist as oligomers. Protein subunits lacking a PTS, but containing an epitope tag were co-expressed in the same cell with subunits possessing a PTS (but lacking an epitope tag). The subsequent immunodetection in peroxisomes of epitope-tagged subunits indicated that they had associated in the cytosol with PTS-containing subunits and that the resulting oligomeric complex was imported into the organelle. When the ability of the subunits to associate in the cytosol was disrupted, however, (by deleting the protein's oligomerization domain) those subunits lacking a PTS could not enter peroxisomes (Flynn *et al*, 1998). Glover *et al* (1994b) showed in *S. cerevisiae* that the ratio of thiolase dimers consisting of one PTS2-containing subunit and one mutated thiolase subunit (lacking a PTS2) is the same inside and outside of peroxisomes. Thus, oligomers containing piggy-backed subunits do not dissociate on the surface of the peroxisomes to allow monomers to be imported and reassembled. Recently, a modified version of the 'piggy-backed-import' assay was used to demonstrate that the ability of the peroxisomes to acquire fully folded and oligmeric substrates can be extended to the PTS1 receptor, Pex5p. Dammi and Subramani, (2001) showed that a human Pex5p reporter protein is able to bind its PTS1-containing cargo in the cytosol and then deliver it to the peroxisomal boundary membrane. From there Pex5p is imported along with its cargo into the peroxisomal matrix and then exported back into the cytosol for another round of import. These intriguing results have reinforced the widely held belief that the peroxisomal import machinery is more similar to that of the nucleus than that of other organelles such as ER and mitochondria. For example, both the peroxisome and nucleus allow transport of folded and oligomeric proteins and both possess receptors (nuclear importins and peroxisomal Pex5p) that can shuttle between the cytosol and the interior of the organelle (reviewed by Smith and Schnell, 2001). It remains to be determined whether PTS1 receptors in other organisms such as plants also shuttle between the cytosol and peroxisomal matrix during import, although preliminary evidence suggests that such as mechanism does indeed exist for Pex5p in watermelon (Wimmer *et al*, 1998) and in several species of yeast (reviewed in Titorenko and Rachubinski, 2001b).

Despite growing evidence in support of oligomeric protein import into peroxisomes, some proteins are imported into these organelles only as monomers. For instance, alcohol oxidase in various yeast (other than *Y. lipolytica*; Titorenko *et al*, 1998) is initially targeted to peroxisomes as monomers and then assembled into enzymatically-active homo-octomers in the peroxisomal matrix (Distel *et al*, 1987; Evers *et al*, 1995; Waterham *et al*, 1997). *In vitro* studies with sunflower glyoxysomes revealed that *Brassica* isocitrate lyase can be imported either as an oligomer or as a monomer (Crookes and Olsen, 1998). However, the import efficiency of the monomeric form of the enzyme is greater than that of the fully-assembled tetrameric isocitrate lyase molecule. Although folded and oligomeric proteins are import-competent, loosely folded proteins or protein subunits may be the preferred substrates for translocation (reviewed in Crookes and Olsen, 1998).

Several models have been proposed to describe the possible mechanisms for importing folded and monomeric proteins into peroxisomes (reviewed in McNew and Goodman, 1996; Subramani, 1996; Smith and Schnell, 2001; Titorenko and Rachubinski, 2001b). In two early models, the transport of proteins into the peroxisomal matrix was proposed to occur either through static pores in the boundary membrane or by a vesicle transport system that resembles modified endocytosis. However, there is no direct evidence to support either of these two mechanisms, although electron microscopic images of 'micropore channel-like structures' in membranes of rat peroxisomes (Makita, 1995) and peroxisomal membrane invaginations (McNew and Goodman, 1994; 1996) are intriguing. In another model, the import of peroxisomal proteins occurs through a dynamic translocation apparatus that consists of a pore that can be quickly assembled and disassembled. Formation of this translocon is purportedly induced by the binding of a protein substrate (either oligomeric or monomeric) to a docking site on the peroxisome boundary membrane. This involves the assembly of a specific set of peroxin proteins that are readily adaptable in forming a pore of a size necessary to accommodate the transported substrate. Following import, the translocon disassembles rapidly, thereby maintaining the permeability properties of the peroxisome. The dynamic nature of this process could also help explain why large or small pores have not been detected microscopically. It is anticipated that with the recent development of *in vitro* systems that simulate protein import and allow for the isolation of translocation intermediates (Pool *et al*, 1998b; Terlecky *et al*, 2001; Legakis and Terlecky, 2001), the tools necessary to dissect the mechanism of protein translocation across the peroxisomal membrane are now available.

It is important to note that peroxisomes (and the nucleus) are not the only organelles capable of importing folded and oligomeric proteins in plant cells. Recently, the translocation machineries of the chloroplast, specifically the envelope and the ΔpH-dependent thylakoid pathways, were shown to import *in vitro*-folded proteins that had assembled with cofactors in the cytosol (reviewed in Keegstra and Cline, 1999). These results are particularly surprising given that the mechanisms of protein import into the chloroplast must occur in a manner that does not compromise the ability of thylakoids to maintain a transmembrane ion gradient. The nature of these chloroplast translocation processes is not well understood, but further research should 'shed light' on them and perhaps provide some additional insight into the mechanisms involved in peroxisomal protein translocation.

ROLE OF MOLECULAR CHAPERONES IN PEROXISOMAL PROTEIN IMPORT

Molecular chaperones are a diverse group of proteins that function in concert in a variety of processes that include: protein folding, unfolding, degradation, assembly, and the transport of proteins into specific intracellular compartments (Hartl, 1996; Miernyk, 1999; Frydman, 2001). For peroxisomes, several soluble and membrane-bound molecular chaperones have been proposed to be involved at various stages in the import of matrix proteins (Table 4) (reviewed in Crookes and Olsen, 1999; Hettema *et al*, 1999). It is noteworthy that, unlike other areas of peroxisome biogenesis, the identification of molecular chaperones and an understanding of their roles in peroxisomal protein import has come primarily from studies with plants.

Table 4: Molecular chaperones proposed to be involved in various aspects of peroxisomal protein import

Stage and proposed function(s)	(Co)-chaperone	Organism	Subcellular location	Reference
Translation				
Proper folding	Hsp70	Pumpkin Human	Cytosol	1-3
	TRiC	Human	Cytosol	2, 3
	Hsp40	Human	Cytosol	2, 3
Post-translation				
Assembly and peroxisomal targeting	Hsp70	Pumpkin Human, Rat	Cytosol	1, 4-7
	Hsp90	Pumpkin	Cytosol	1
	Djp1p (J-domain protein)	*S. cerevesiae*	Cytosol	5
Import and intraperoxisomal assembly	J-domain protein	Cucumber	Peroxisomal boundary membrane	8
	PMP61 (J-domain protein)	Cucumber	Peroxisomal boundary membrane	9
	PMP73 (Hsp70-like)	Cucumber	Peroxisomal boundary membrane	9
	HsP70	Watermelon Cucumber	Peroxisomal matrix	6, 10
	J-domain protein	Cucumber	Peroxisomal matrix	6

1. Crookes and Olsen, 1998; 2. Frydman et al, 1994; 3. Frydman and Hartl, 1996; 4. Walton et al, 1994; 5. Hettema et al, 1998; 6. Preisig-Muller et al, 1994; 7. Harano et al, 2001; 8. Diefenbach and Kindl, 2000; 9. Corpas and Trelease, 1997; 10. Wimmer et al, 1997.

Cytosolic molecular chaperones

At least three cytosolic molecular chaperones have been implicated in the folding of peroxisomal proteins during translation. For instance, one member of the 70-kDa family of heat shock or stress related proteins (Hsp70) is necessary for the folding of glyoxysomal isocitrate lyase in heat-stressed pumpkin seedlings (Crookes and Olsen, 1998). A cytosolic Hsp70 and co-chaperone Hsp40 (a DnaJ orthologue), as well as the chaperonin TRiC (t-complex polypeptide 1 ring complex) are involved in the folding of newly-synthesized peroxisomal firefly luciferase in mammalian cells (Frydman *et al*, 1994; Frydman and Hartl, 1996).

Assembly and targeting of nascent peroxisomal matrix proteins also appears to be mediated by molecular chaperones. Crookes and Olsen (1998) demonstrated that antibodies directed against either cytosolic wheat germ Hsp70 or *Escherichia coli* Hsp90 inhibit import of proteins into isolated pumpkin glyoxysomes *in vitro*. Interestingly, Hsp70 directly interacts with pumpkin peroxisomal proteins, whereas Hsp90 does not, suggesting that the mode of action of these two chaperones is different (see below). Hsp70 antibodies also inhibit peroxisomal protein import in cultured mammalian cells (Walton *et al*, 1994). In *S. cerevisiase,* the import of peroxisomal proteins requires the cytosolic J-domain protein Djp1p (Hettema *et al*, 1998). This co-chaperone functions in concert with Hsp70, either maintaining the import-competent conformation of newly-synthesized matrix proteins in the cytosol, or stimulating the recognition of peroxisomal proteins by their cognate receptors by exposing their PTSs and/or by acting on the receptors directly (reviewed in Hettema *et al*, 1999).

Multiple roles for Hsp90 in peroxisomal protein import in plant cells also have been proposed (Crookes and Olsen, 1998; Pratt *et al*, 2001). Crookes and Olsen (1998) speculated that Hsp90 can function in a cytoplasmic chaperone heterocomplex with the Hsp70 and maintains the import competence of peroxisomal matrix-destined proteins. Alternatively, Hsp90 and Hsp70 may stimulate or 'prime' PTS receptors for interaction with their target proteins. Indeed, Hsp70 does interact directly with, and regulates Pex5p in an ATP-dependent manner (Harano *et al*, 2001). Hsp90 might also target the PTS1-receptor (Pex5p) and its protein cargo to peroxisomes. Pratt *et al* (2001) noted that Pex5p resembles an immunophilinin FKBP52 because both proteins contain tetratricopeptide repeat (TPR) domains that are known to bind Hsp90. In mammalian cells, FKBP52 binds a Hsp90-glucocorticoid receptor complex and then mediates its targeting to the nucleus along microtubular pathways. Pratt and co-workers proposed that Pex5p functions in a similar manner to FKBP52. That is, Hsp90 binds Pex5p and PTS1-containing proteins and then targets, by some unknown mechanism, the entire complex to the peroxisome along microtubules. Evidence in support of this intriguing hypothesis has not been reported.

Peroxisomal membrane-bound and intraperoxisomal molecular chaperones

Several studies with plant peroxisomes have revealed that molecular chaperones exist on the organelle boundary membrane. For instance, a prenylated J-domain protein is localized to the cytosolic face of cucumber glyoxysomal membranes (Preisig-Muller *et al*, 1994) and binds to a specific species of cytosolic Hsp70 (Diefenbach and Kindl, 2000). Thus, the membrane-bound Hsp40 could function to release incoming PTS1-receptor-cargo from accompanying cytosolic Hsp70, a process that would also account for the *in vitro* ATP hydrolysis in peroxisomal protein import (see Olsen, 1998 for review). A similar role in protein import could exist for two other presumptive chaperones that are localized in the glyoxysomal boundary membrane. Corpas and Trelease (1997) demonstrated that a 61-kDa PMP and a 73-kDa PMP are immunorelated to a J-domain protein, and a member of the Hsp70 family of chaperones, respectively. However, neither the chaperone activity of these two PMPs, nor their topological orientation in glyoxysomal membranes has yet been demonstrated.

Several Hsp70 proteins have been localized to the peroxisomal matrix in plant cells. In watermelon, one Hsp70 gene contains two in-frame translation start sites that can generate two polypeptides with presequences of different lengths; synthesis from the first start site yields a longer presequence that targets the protein to a plastid *in vitro* (Wimmer *et al*, 1997). Translation from the second start site yields a shorter presequence that contains a PTS2 capable of directing a reporter protein to *H. polymorpha* peroxisomes *in vivo*. Apparently, when translation begins at the first start site, the PTS2 is not functional *in vitro* within the context of the longer presequence. Other studies with cucumber have revealed two Hsp70 proteins and a J-domain protein localized to the glyoxysomal matrix (Diefenbach and Kindl, 2000). Each of these putative intraperoxisomal molecular chaperones likely has the same function as their cytosolic counterparts, i.e. a role in facilitating folding and in the proper assembly of proteins that are imported as monomers. In addition, the Hsp70s might have a more direct function in translocation on any unfolded substrates; for instance, serving as 'ratchet-like' proteins analogous to an Hsp70 protein involved in mitochondrial protein import.

In consideration of the large (and growing) number of reports that molecular chaperones are involved with plant peroxisomal protein import, it is apparent that their roles in this process are complex. This is particularly intriguing in light of the fact that protein unfolding is not a prerequisite for matrix protein import. A precise role for any chaperone in peroxisome protein import needs to be clarified and research with plants will likely continue to lead the way.

MECHANISM OF PEROXISOMAL PROTEIN IMPORT

A number of the components (peroxins) of the molecular machinery underlying the import of proteins into the peroxisomal matrix have been identified in yeast and mammalian cells (reviewed in Sacksteder and Gould, 2000; Subramani *et al*, 2000; Terlecky and Fransen, 2000; and Titorenko and Rachubinski, 2001b). A detailed description of known PEX genes is presented in Chapter 12. As summarized in Figure 4a and b, matrix protein import in both yeasts and mammals involves a complex sequence of events that begin in the cytosol and culminate with the translocation of the cargo protein into the peroxisomal matrix. Although several mechanistic differences and unique peroxins exist between yeast and mammals, the overall process of peroxisomal matrix protein import is quite similar. The first step is the recognition in the cytosol of PTS1- and PTS2-containing proteins by their cognate receptors, Pex5p and Pex7p, respectively. As discussed earlier, several molecular chaperones have been implicated in the folding of peroxisomal proteins in the cytosol and in the PTS-receptor recognition. Following formation of the receptor-cargo-protein (-chaperone?) complexes they are targeted to the surface of the peroxisomal boundary membrane where they bind initially to the PMP Pex14p. Interestingly, evidence for this membrane-bound peroxin as the point of convergence of the two matrix protein sorting pathways, comes from studies whereby the loss of Pex14p in all organisms examined to date (including plants; see below) results in defects in the import of both PTS1- and PTS2-containing proteins. Other membrane-bound components of the docking complex include Pex13p in mammalian cells and Pex13p and Pex17p in yeast. After docking, the fate of the PTS-receptor-cargo complex and the role of other membrane-bound peroxins in the translocation process is not well established. Nevertheless, in most working models of peroxisomal protein import in yeasts and mammals, the PTS-receptor-cargo complex is transferred to a translocation machine made up of several RING finger membrane proteins including Pex2p, Pex10p, and Pex12p. At this point, the cargo protein along with its receptor is translocated across the peroxisomal boundary membrane into the interior or the organelle. Several other peroxins including Pex2p, Pex4p, Pex8p, Pex22p, and two members of the AAA ATPase family of proteins that are localized in the cytosol (Pex1p and Pex6p) have been implicated in the recycling of the PTS receptors back to the cytosol.

Figure 4: Models for peroxisomal matrix protein import

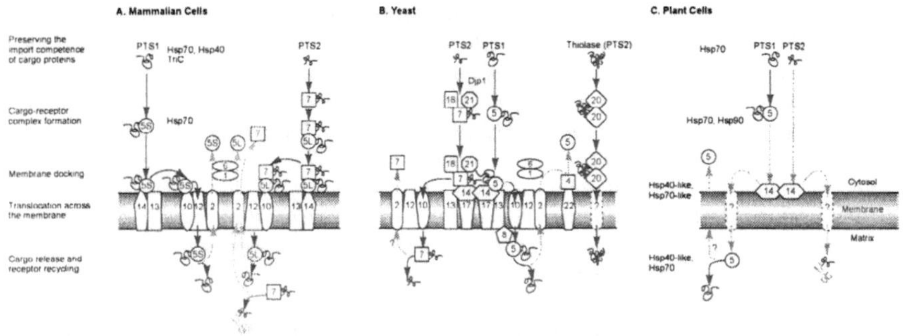

Figure 4. The fundamental steps of the import process are described on the left in A. For references and further information on specific steps described below for mammalian cells (A) and yeast (B) see reviews by Sacksteder and Gould (2000), Subramani et al (2000) and Titorenko and Rachubinski (2001b).

A. Mammalian cells. PTS1- and PTS2-containing proteins are recognized in the cytosol by their specific cognate receptors, Pex5p and Pex7p, respectively. Two isoforms of Pex5 exist in mammalian cells; the short isoform (5S) that interacts with PTS1-containing proteins and the long isoform (5L) that interacts with PTS2-containing proteins and Pex7p carrying its cargo protein. The initial docking site for both the sets of receptor-cargo complexes is the integral membrane protein Pex14p. After docking, the receptor-cargo complexes are transferred to other membrane-bound components of the import machinery including Pex13, Pex2p, Pex10 and Pex12. These latter three peroxins have been implicated in the translocation of the receptors and cargo (monomers and oligomers) across the boundary membrane. Following import, receptor-cargo dissociation occurs, PTS2-containing proteins are proteolytically processed, and receptors are recycled back to the cytosol (Dammai and Subramani, 2001). Receptor recycling appears to involve some of the same membrane-bound peroxins that are involved in protein translocation into the organelle (Pex2, Pex10, Pex12p), as well as two cytosolic peroxins, Pex1 and Pex6.

B. Yeast. PTS1- and PTS2-containing proteins are recognized in the cytosol by their specific cognate receptors, Pex5p and Pex7p, respectively. In addition to Pex7p, two other cytosolic peroxins, Pex18p and Pex21p, play a role in the delivery of PTS2 proteins to yeast peroxisomes. In Y. lipolytica, the cytosolic peroxin, Pex20p, and not Pex7p, is necessary for import of thiolase into peroxisomes. Both cargo-loaded Pex5p and Pex7p (along with Pex18p-Pex21p) are transported to the boundary membrane, where the peripheral membrane protein, Pex14p, functions as a convergent component of a docking apparatus. Membrane-bound Pex13p and Pex17p are also components of the docking apparatus. Translocation across the boundary membrane of the receptor and PTS-cargo involves the integral membrane proteins Pex2p, Pex10p and Pex12p. The membrane-associated protein, Pex8p, has been implicated in the dissociation during import of Pex5p from its cargo. Unlike in mammalian and plant cells, in yeast PTS2-containing proteins are not proteolytically processed following import. Several peroxins have been implicated in the recycling of Pex5p back to the cytosol, including Pex1p, Pex2p, Pex4p, Pex6p and Pex22p. C. Plant cells. PTS1- and PTS2-containing proteins are synthesized on cytosolic ribosomes where the former subset of proteins are recognized by their cognate receptor, Pex5p (Kragler et al, 1998). *...continued*

Figure 4: Models for peroxisomal matrix protein import (continued)

Recognition of PTS2-containing proteins by Pex7p has not yet been demonstrated, although the receptor protein from Arabidopsis has been cloned (Schumann et al, 1999a). Import of both monomeric and oligomeric PTS1- and PTS2-containing proteins into peroxisomes requires the docking protein Pex14p (Hayashi et al, 2000). Notably, Pex14p is a peripheral PMP in Arabidopsis (shown) (Hayashi et al, 2000), but in sunflower glyoxysomes Pex14p is an integral PMP (Lopez-Huertas et al 1999). Following import, PTS2-containing proteins are proteolytically processed (Kato et al, 2000). Several cytosolic, membrane-bound and intraperoxisomal molecular chaperones have been identified in plant cells. Pex5p appears to shuttle between the cytosol and the interior of the peroxisome (Wimmer et al, 1998).

Panels A and B reprinted by permission from Nature Reviews Molecular Cell Biology (Titorenko and Rachubinski, 2001b) copyright (2001) Macmillan Magazines Ltd.

In plants, our understanding of the events involved in the targeting, docking and translocation of matrix proteins is significantly less than that for mammals and yeast (Figure 4c). This is due in large part to the fact that only two peroxins, Pex5p and Pex14p, have been demonstrated experimentally to participate in peroxisomal matrix protein import.

Plant Pex5 orthologues have been identified in several species including tobacco (Kragler *et al*, 1998), watermelon (Wimmer *et al*, 1998), and Arabidopsis (Brickner *et al*, 1998: Zolman *et al*, 2000). Kragler *et al* (1998) employed a two-hybrid assay to isolate the cDNA encoding the tobacco Pex5p (NtPex5p). Interestingly, NtPex5p shares a greater amino acid sequence similarity with its homologue from humans than from yeast; an observation that has been noted also for most other plant peroxins (Mullen *et al*, 2001a). The function of NtPex5p as a PTS1 receptor was confirmed by demonstrating that expression of the protein in a pex5 deletion strain of *S. cerevisiae* restored the import of a PTS1-containing reporter protein (Kragler *et al*, 1998). These heterologous expression experiments also indicated that other components of the protein import machinery, specifically those that directly interact with Pex5p such as the docking proteins Pex13p and Pex14p, are functionally conserved among plants and yeast. Two-hybrid assays revealed that the same region of NtPex5p that was demonstrated in other Pex5 proteins to be both necessary and sufficient for interaction with the PTS1, i.e. seven tetratricopeptide repeat motifs, was capable of interacting with peptides containing carboxy-terminal tripeptides that mostly conformed to the SKL motif.

Studies with watermelon Pex5 (CvPex5) also showed that the expression of a plant PTS1 receptor in a yeast pex5 mutant was able to restore, albeit in a inefficient manner, the PTS1 import defect (Wimmer *et al*, 1998). Wimmer *et al* (1998) noted that since the restoration of protein import was partial, the

PTS1-binding domain of CvPex5p might contain specific conformations for carboxy-terminal tripeptides that are found only in plant peroxisomal matrix proteins. Interestingly, localization studies with CvPex5p revealed that the receptor was localized mainly in the cytosol, but also inside peroxisomes. These data suggest that, similar to the PTS1 receptors in yeast and mammalian cells, the plant PTS1 receptor shuttles its cargo from the cytosol to the peroxisome interior and then exits for another round of import. However, this possibility remains to be proven experimentally.

Pex14p was isolated by genetic analyses, i.e. positional cloning, of an Arabidopsis mutant called *ped2* with defects in glyoxysomal fatty acid β-oxidation (Hayashi *et al*, 2000). Consistent with the role of Pex14p in mammals and yeast, AtPex14p appears to function as a docking protein; serving as the point of convergence at the peroxisome boundary membrane for the PTS1- and PTS2-receptor-cargo complexes. Support for this conclusion comes from observations that in the Arabidopsis *ped2* mutant all of the PTS1- and PTS2-containing proteins examined were partially mislocalized to the cytosol (Hayashi *et al*, 2000). The role of Pex14p as a docking protein has also been demonstrated *in vitro*. Lopez-Huertas *et al* (1999) showed that antibodies against human Pex14p inhibit the ATP-independent binding to sunflower glyoxysomal membranes of matrix proteins, but not integral PMPs. Based on these observations, the authors suggested that the docking of matrix cargo proteins via Pex14p at the boundary membrane occurs before the ATP requiring step of protein import. What remains to be determined is whether Pex14p directly interacts with the PTS receptors or what other membrane-bound peroxins function along with Pex14p in the docking complex.

It is important to note that thirteen other plant *PEX* orthologues genes have been identified in Arabidopsis (Mullen *et al*, 2001a), many of which have been implicated in matrix protein import in other organisms. However, only a handful of these Arabidopsis genes have been cloned including Pex1p (Lopez-Huertas *et al*, 2000), Pex6p (Kaplan *et al*, 2001), Pex7p (Schumann *et al*, 1999a), Pex10p (Baker *et al*, 2000; Schumann *et al*, 1999b) and Pex16p (Lin *et al*, 1999), none of which have yet to be proven experimentally to participate in plant peroxisome biogenesis.

PERSPECTIVES

Over the past decade, research on the biogenesis of plant peroxisomes has revealed that different topogenic sequences are responsible for directing nascent proteins from the cytosol to the peroxisomal matrix. What now remains to be understood is the molecular machinery that decodes these targeting signals and thereby permits the correct sorting, docking and translocation of incoming peroxisomal proteins. Toward this end, investigators will undoubtedly focus on isolating and characterizing plant orthologues of yeast and mammalian peroxins that have been shown to participate in various aspects of protein import. Of course, it is also possible that other peroxins novel to plants will be identified by using selection screens for mutants with peroxisomal defects (e.g. Hayashi *et al*, 1998b; Zolman *et al*, 2000) (see Chapter 9).

The involvement of specific molecular chaperones at various stages of the peroxisomal import process will need to be determined. In particular, what role do chaperones play in protein folding and unfolding at the peroxisomal boundary membrane and within the organelle? Answers to these questions should provide insight into how monomeric and oligomeric proteins enter the peroxisomal matrix. Also, if a similar scenario exists in plants as it does in humans, we will learn how receptor proteins (folded?) can exit the matrix so that they can recycle back to the cytosol for another round of import. The development of biochemical assays for examining specific steps in peroxisomal protein translocation (Pool *et al*, 1998b) should help elucidate many of these processes.

The challenge for plant scientists in the next few years will be to keep an open mind and not to assume that plants are simply 'green yeasts'. Given the unique presence of diverse peroxisomal pathways and the different classes of peroxisomes in plants it is not unreasonable to predict that novel principles will be discovered.

ACKNOWLEDGMENTS

I gratefully acknowledge Dr. Derek Bewley (University of Guelph) for critically reviewing the manuscript, Dr. Jan Mierynk (University of Missouri-Columbia) for advice on molecular chaperones, Dr. John Dyer (USDA-ARS, New Orleans) for his helpful comments throughout the writing of this chapter, and Ms. Mary Murphy (University of Guelph) for acquiring microscopic images shown in Figure 1. I am also grateful to Ian Smith (University of Guelph) for constructing Figure 3 and to Ms. Stephanie Mullen (Guelph) for her assistance with references. I thank Drs. Alison Baker (Leeds University), Paul Walton (University of Western Ontario) and Richard Rachubinski (University of Alberta) for contributing materials or giving permission to use their published findings. I apologize to those researchers whose work I was unable to cite because of space limitations.

This work was supported by a grant (No. 217291) from the Natural Sciences and Engineering Council of Canada (NSERC).

REFERENCES

Amery, L., Brees, C., Baes, M., Setoyama, C., Miura, R., Mannaerts, G. and Van Veldhoven P. (1998) C-terminal tripeptide Ser-Asn-Leu (SNL) of human D-asparate oxidase is a functional peroxisome-targeting signal. *Biochem.* **336**: 367-371.

Authier, F., Bergeron, J.J., Ou, W.J., Rachubinski, R.A., Posner, B.I. and Walton, P.A. (1995). Degradation of the cleaved leader peptide of thiolase by a peroxisomal proteinase. *Proc. Natl. Acad. Sci. USA*. **92**: 3859-3863.

Baerends, R.J., Rasmussen, S.W., Hilbrands, R.E., van der Heide, M., Faber, K.N., Reuvekamp, P.T., Kiel, J.A., Cregg, J.M., van der Klei, I.J. and Veenhuis M. (1996) The *Hansenula polymorpha* PER9 gene encodes a peroxisomal membrane protein essential for peroxisome assembly and integrity. *J. Biol. Chem.* **271**: 8887-88894.

Baker, A. (1996) *In vitro* systems in the study of peroxisomal protein import. *Experientia*. **52**: 1055-1062.

Baker, A., Charlon, W., Johnson, B., Lopez-Huertas, E., Oh, J., Sparkes, I. and Thomas, J. (2000) Biochemical and molecular approaches to understanding protein import into peroxisomes. *Biochem. Soc. Trans.* **28**: 499-504.

Banjoko, A. and Trelease, R. (1995) Development and application of an *in vivo* plant peroxisome import system. *Plant Physiol.* **107**: 1201-1208.

Behari, R. and Baker, A. (1993) The carboxyl terminus of isocitrate lyase is not essential for import into glyoxysomes in an *in vitro* system. *J. Biol. Chem.* **268**: 7318-7322.

Blattner, J., Dorsam, H. and Clayton, C. (1995) Function of N-terminal import signals in trypanosome microbodies. *FEBS Lett.* **360**: 310-314.

Blattner, J., Swinkels, B., Dorsam, H., Prospero, T., Subramani, S. and Clayton, C. (1992) Glycosome assembly in trypanosomes: variations in the acceptable degeneracy of a COOH-terminal microbody targeting signal. *J. Cell Biol.* **119**: 1129-1136.

Bongcam, V., MacDonald-Comber Petetot, J., Mittendorf, V., Robertson, E.J., Leech, R.M., Qin, Y.M., Hiltunen, J.K., and Poirier Y. (2000) Importance of sequences adjacent to the terminal tripeptide in the import of a peroxisomal *Candida tropicalis* protein in plant peroxisomes. *Planta.* **211**: 150-157.

Brickner, D. G. and Olsen, L. (1998) Nucleotide triphosphates are required for the transport of glycolate oxidase into peroxisomes. *Plant Physiol.* **116**: 309-317.

Brickner, D.G, Brickner, J.H. and Olsen, L.J. (1998) Sequence analysis of a cDNA encoding Pex5p, a peroxisomal targeting signal type 1 receptor from Arabidopsis. *Plant Physiol.* **118**: 330.

Brickner, D.G., Harada, J. and Olsen L. (1997) Protein transport into higher plant peroxisomes. *Plant Physiol.* **113**: 1213-1221.

Corpas, F. and Trelease, R. (1997) The plant 73 kDa peroxisomal membrane protein (PMP73) is immunorelated to molecular chaperones. *Eur. J. Cell Biol.* **73**: 49-57.

Cregg, J.M., van der Klei, I.J., Sulter, G.J., Veenhuis, M. and Harder, W. (1990). Peroxisome-deficient mutants of *Hansenula polymorpha. Yeast.* **6**: 87-97.

Crookes, W. and Olsen, L. (1999) Peroxin puzzles and folded freight: peroxisomal protein import in review. *Naturwissenschaften.* **86**: 51-61.

Crookes, W. and Olsen, L. (1998) The effects of chaperones and the influence of protein assembly on peroxisomal protein import. *J. Biol. Chem.* **273**: 17236-17242.

Cutler, S., Ehrhardt, D., Griffits, J. and Somerville, C. (2000) Random GFP:cDNA fusions enable visualization of subcellular structures in cells of Arabidopsis at a high frequency. *Proc. Natl. Acad. Sci. USA.* **97**: 3718-3723.

Dammai, V. and Subramani, S. (2001) The human peroxisomal targeting signal receptor, Pex5p, is translocated into the peroxisomal matrix and recycled to the cytosol. *Cell.* **105**: 187-196.

Diefenbach J. and Kindl, H. (2000) The membrane-bound DnaJ protein located at the cytosolic site of glyoxysomes specifically binds the cytosolic isoform 1 of Hsp70, but not other Hsp70 species. *Eur. J. Biochem.* **267**: 746-754.

Distel, B., Erdmann, R., Gould, S., Blobel, G., Crane, D., Cregg, J., Dodt, G., Fujiki, Y., Goodman, J., Just, W., Kiel, J., Kunau, W., Lazarow, P., Mannaerts, G., Moser, H., Osumi, T., Rachubinski, R., Roscher, A., Subramani, S., Tabak, H., Tsukamoto, T., Valle, D., van der Klei, I., van Veldhoven, P. and Veenhuis, M. (1996) A unified nomenclature for peroxisome biogenesis factors. *J. Cell Biol.* **135**: 1-3.

Distel, B., Veenhuis, M. and Tabak, H.F. (1987) Import of alcohol oxidase into peroxisomes of *Saccharomyces cerevisiae. EMBO J.* **10**: 3111-3116.

Dyer, J.M. and Mullen, R.T. (2001) Immunocytological localization of two plant fatty acid desaturases in the endoplasmic reticulum. *FEBS Lett.* **494**: 44-47.

Dyer, J.M., McNew, J.A. and Goodman, J.M. (1996) The sorting sequence of the peroxisomal integral membrane protein PMP47 is contained within a short hydrophilic loop. *J. Cell Biol.* **133**: 269-280.

Elgersma, Y., Kwast, L., van den Berg, M., Snyder, W.B., Distel, B., Subramani, S. and Tabak, H.F. (1997) Overexpression of Pex15p, a phosphorylated peroxisomal integral membrane protein required for peroxisome assembly in *S.cerevisiae*, causes proliferation of the endoplasmic reticulum membrane. *EMBO J.* **16**: 7326-7341.

Elgersma, Y., van Roermund, C.W., Wanders, R.J. and Tabak, H.F. (1995) Peroxisomal and mitochondrial carnitine acetyltransferases of *Saccharomyces cerevisiae* are encoded by a single gene. *EMBO J.* **14**: 3472-3479.

Elgersma, Y., Vos, A., van den Berg, M., van Roermund, C., van der Sluijs, P., Distel, B. and Tabak, H. (1996) Analysis of the carboxyl-terminal peroxisomal targeting signal 1 in a homologous context in *Saccharomyces cerevisiae*. *J. Biol. Chem.* **271**: 26375-26382.

Erdmann, R. (1994) The peroxisomal targeting signal of 3-oxoacyl-CoA thiolase from *Saccharomyces cerevisiae*. *Yeast*. **10**: 935-944.

Erdmann, R., Veenhuis, M., Mertens, D. and Kunau, W. (1989) Isolation of peroxisome-deficient mutants of *Saccharomyces cerevisiae*. *Proc. Natl. Acad. Sci. USA*. **86**: 5419-5423.

Evers, M., Harder,W. and Veenhuis, M. (1995) In vitro dissociation and re-assembly of peroxisomal alcohol oxidases of *Hansenula polymorpha* and *Pichia pastoris*. *FEBS Lett*. **368**: 293-296.

Flynn, C., Mullen, R. and Trelease, R. (1998) Mutational analyses of a type 2 peroxisomal targeting signal that is capable of directing oligomeric protein import into tobacco BY-2 glyoxysomes. *Plant J*. **16**: 709-720.

Fransen, M., Brees, C., Van Veldhoven, P. and Mannaerts, G. (1996) The visualization of preoxisomal proteins containing a C-terminal targeting sequence on western blot by using the biotinylated PTS1-receptor. *Anal. Biochem*. **242**: 26-30.

Frydman, J. (2001) Folding of newly translated proteins *in vivo*: the role of molecular chaperons. *Ann. Rev. Biochem*. **70**: 603-647.

Frydman, J. and Hartl, F.U. (1996) Principles of chaperone-assisted protein folding: differences between *in vitro* and *in vivo* mechanisms. Science. **272**: 1497-1502.

Frydman, J., Nimmesgern, E., Ohtsuka, K. and Hartl, F. (1994) Folding of nascent polypeptide chains in a high molecular mass assembly with molecular chaperones. *Nature*. **14**: 96-97.

Fujiki, Y. (2000) Peroxisome biogenesis and peroxisome biogenesis disorders. *FEBS Lett*. **476**: 42-46.

Fujiki, Y. and Lazarow, P. (1985) Post-translational import of fatty acyl-coA oxidase and catalase into peroxisomes of rat liver *in vitro*. *J. Biol. Chem*. **260**: 5603-5609.

Fung, K. and Clayton, C. (1991) Recognition of a peroxisomal tripeptide entry signal by the glycosomes of *Trypanosoma brucei*. *Mol Biochem. Parasitol*. **45**: 261-264.

Gao, X., Marrison, J.L., Pool, M.R., Leech, R.M. and Baker, A. (1996) Castor bean isocitrate lyase lacking the putative peroxisomal targeting signal is imported into plant peroxisomes both *in vivo* and *in vitro*. *Plant Physiol*. **112**: 1457-1464.

Gerbhardt, J.S., Wadsworth, G.J. and Matthews, B. (1998) Characterization of a single soybean cDNA encoding cytosolic and glyoxysomal isozymes of aspartate aminotransferase. *Plant Mol. Biol*. **37**: 99-108.

Gietl, C., Faber, K., Van Der Klei, I. and Veenhuis, M. (1994) Mutational analysis of the N-terminal topogenic signal of watermelon glyoxsomal malate dehydrogenase using the heterologous host *Hansenula polymorpha*. *Proc. Natl. Acad. Sci. USA*. 91: 3151-3155.

Gietl, C., Wimmer, B., Adamec, J. and Kalousek, F. (1997) A cysteine endopeptidase isolated from castor bean endosperm microbodies processes the glyoxysomal malate dehydrogenase precursor protein. *Plant Physiol.* **113**: 863-871.

Glover, J., Andrews, D. and Rachubinski, A. (1994b) *Saccharomyces cerevisiae* peroxisomal thiolase is imported as a dimer. *Proc. Natl. Acad. Sci. USA.* **91**: 10541-10545.

Glover, J., Andrews, D., Subramani, S. and Rachubinski, R. (1994a) Mutagenesis of the amino targeting signal of *Saccharomyces cerevisiae* 3-ketoacyl-CoA thiolase reveals conserved amino acids required for import into peroxisomes *in vivo*. *J. Biol. Chem.* **269**: 7558-7563.

Gould, S.J. and Valle, D. (2000) Peroxisome biogenesis disorders: genetics and cell biology. *Trends Genet.* **16**: 340-345.

Gould, S.J., Keller, G.A. and Subramani, S. (1988) Identification of peroxisomal targeting signals located at the carboxy terminus of four peroxisomal proteins. *J. Cell Biol.* **107**: 897-905.

Gould, S.J., Keller, G.A. and Subramani, S. (1987) Identification of a peroxisomal targeting signal at the carboxy terminus of firefly luciferase. *J. Cell Biol.* **105**: 2923-2931.

Gould, S.J., Keller, G.A., Hosken, N., Wilkinson, J. and Subramani, S. (1989) A conserved tripeptide sorts protein to peroxisomes. *J. Cell Biol.* **108**: 1657-1664.

Gould, S.J., Keller, S., Schneider, M., Howell, S., Garrard, L., Goodman, J., Distel, B., Tabak, H. and Subramani, S. (1990a) Peroxisomal protein import is conserved between yeast, plants, insects and mammals. *EMBO J.* **9**: 85-90.

Gould, S.J., Krisans, S., Keller, G. and Subramani, S. (1990b) Antibodies directed against the peroxisomal targeting signal of firefly luciferase recognize multiple mammalian peroxisomal proteins. *J. Cell Biol.* **110**: 27-34.

Gould, S.J., McCollum, D., Spong, A.P., Heyman, J.A. and Subramani, S. (1992) Development of the yeast *Pichia pastoris* as a model organism for a genetic and molecular analysis of peroxisome assembly. *Yeast.* **8**: 613-628.

Harano, T., Nose, S., Uezu, R., Shimizu, N. and Fujiki, Y. (2001) Hsp70 regulates the interaction between the peroxisome targeting signal type 1 (PTS1)-receptor Pex5p and PTS1. *Biochem.* **357**: 157-165.

Hartl, F.U. (1996) Molecular chaperones in cellular protein folding. *Nature.* **381**: 571-579.

Hausler, T., Stierhof, Y., Wirtz, E. and Clayton, C. (1996) Import of a DHFR hybrid protein into glycosomes *in vivo* is not inhibited by the folate-analogue aminopterin. *J. Cell Biol.* **132**: 311-324.

Hayashi, H. (2000) Plant peroxisomes: molecular basis of the regulation of their functions. *J. Plant Res.* **113**: 103-109.

Hayashi, H., De Bellis, L., Yamaguchi, K., Kato, A., Hayashi, M. and Nishimura, M. (1998) Molecular characterization of a glyoxysomal long chain acyl-CoA oxidase that is synthesized as a precursor of higher molecular mass in pumpkin. *J. Biol. Chem.* **273**: 8301-8307.

Hayashi, M., Aoki, M., Kato, A., Kondo, M. and Nishimura, M. (1996) Transport of chimaeric proteins that contain a carboxy-terminal targeting signal into microbodies. *Plant J.* **10**: 225-234.

Hayashi, M., Aoki, M., Kondo, M. and Nishimura, M. (1997) Changes in targeting efficiencies of proteins to plant microbodies caused by amino acid substitutions in the carboxy-terminal tripeptide. *Plant Cell Physiol.* **38**: 759-768.

Hayashi, M., Toriyama-Kato, K., Kondoa, M. and Nishimura, M. (1998) 2,4-dichlorophenoxybutyric acid-resistant mutants of Arabidopsis have defects in glyoxysomal fatty acid β-oxidation. *Plant Cell.* **10**: 183-196.

Hayashi, M., Nito, K., Toriyama-Kato, K., Kondo, M., Yamaya, T. and Nishimura, M. (2000) *At*Pex14p maintains peroxisomal functions by determining protein targeting to have three kinds of plant peroxisomes. *EMBO J.* **19**: 5701-5710.

Heinemann, P. and Just, W. (1992) Peroxisomal protein import *in vivo* evidence for a novel translocation competent compartment. *FEBS Lett.* **300**: 179-182.

Hettema, E.H., Distel, B. and Tabak, H.F. (1999) Import of proteins into peroxisomes. *Biochim. Biophys. Acta.* **1451**: 17-34.

Hettema, E., Ruigrok, C., Koerkamp, M., van de Berg, M., Tabak, H. and Braakman, I. (1998) The cytosolic DnaJ-like protein Djp1p is involved specifically in peroxisomal protein import. *J. Cell Biol.* **142**: 421-434.

Honsho, M. and Fujiki, Y. (2001) Topogenesis of peroxisomal membrane protein requires a short, positively charged intervening-loop sequence and flanking hydrophobic segments. Study using human membrane protein PMP34. *J. Biol. Chem.* **276**: 9375-9382.

Horng, J., Behari, R., Burke, C. and Baker, A. (1995) Investigation of the energy requirement and targeting signal for the import of glyocate oxidase into glyoxysomes. *Eur. J. Biochem.* **230**: 157-163.

Imanaka, T., Small, G. and Lazarow, P. (1987) Translocation of acyl-CoA oxidase into peroxisomes requires ATP hydrolysis, but not a membrane potential. *J. Cell Biol.* **105**: 2915-2922.

Jones, J.M., Morrell, J.C. and Gould, S.J. (2001) Multiple distinct targeting signals in integral peroxisomal membrane proteins. *J. Cell Biol.* **153**: 1141-1150.

Kaplan, C.P., Thomas, J.E., Charlton, W.L. and Baker, A. (2001) Identification and characterisation of PEX6 orthologues from plants. *Biochim. Biophys. Acta.* **1539**: 173-180.

Karpichev, I. and Small, G. (2000) Evidence for a novel pathway for the targeting of a *Saccharomyces cerevisiae* peroxisomal protein belonging to the isomerase/hydratase family. *J. Cell Sci.* **113**: 533-544.

Kato, A., Hayashi, M. and Nishimura, M. (1999) Oligomeric proteins containing N-terminal targeting signals are imported into peroxisomes in transgenic Arabidopsis. *Plant Cell Physiol.* **40**: 586-591.

Kato, A., Hayashi, M., Kondo, M. and Nishimura, M. (2000) Transport of peroxisomal proteins synthesized as large precursors in plants. *Cell Biochem. Biophys.* **32**: 269-275.

Kato, A., Hayashi, M., Kondo, M. and Nishimura, M. (1996) Targeting and processing of a chimaeric protein with the N-terminal presequence of the precursor to glyoxysomal citrate synthase. *Plant Cell.* **8**: 1601-1611.

Kato, A., Takedda-Yoshikawa, Y., Hayashi, M., Kondo, M., Hara-Nishimura, I. and Nishimura, M. (1998) Glyoxysomal malate dehydrogenase in pumpkin: cloning of a cDNA and functional analysis of its presequence. *Plant Cell Physiol.* **39**: 186-195.

Keegstra, K. and Cline, K. (1999) Protein import and routing systems of chloroplasts. *Plant Cell.* **11**: 557-570.

Keller, G., Krisans, S., Gould, S., Sommer, J., Wang, C., Schliebs, W., Kunau, W., Brody, S. and Subramani, S. (1991) Evolutionary conservation of a microbody targeting signal that targets proteins to peroxisomes, glyoxysomes and glycosomes. *J. Cell Biol.* **114**: 893-904.

Kragler, F., Lametschwandtner, G., Christmann, J., Hartig, A. and Harada, J. (1998) Identification and analysis of the plant peroxisomal targeting signal 1 receptor NtPEX5. *Proc. Natl. Acad. Sci. USA.* **95**: 13336-13341.

Lametschwandtner, G., Brocard, C., Fransen, M., Van Veldhoven, P., Berger, J. and Hartig, A. (1998) The difference in recognition of terminal tripeptides as peroxisomal targeting signal 1 between yeast and human is due to different affinities of their receptor Pex5p to the cognate signal and to the residues adjacent to it. *J. Biol. Chem.* **273**: 33635-33643.

Lazarow, P. and Fujiki, Y. (1985) Biogenesis of peroxisomes. *Ann. Rev. Cell Biol.* **1**: 489-530.

Lee, M., Mullen, R. and Trelease, R. (1997) Oilseed isocitrate lyases lacking their essential type 1 peroxisomal targeting signal are piggybacked to glyoxysomes. *Plant Cell.* **9**: 185-197.

Legakis J. and Terlecky, S. (2001) PTS2 protein import into mammalian cells. *Traffic.* **2**: 252-260.

Leiper, J., Oatey, P. and Danpure, C. (1996) Inhibition of alanine: glyoxyate aminotransferase 1 dimerization is a prerequisite for its peroxisome-to-mitochondrion mistargeting in primary hyperoxaluria type 1. *J. Cell Biol.* **135**: 939-951.

Lin, Y., Sun, L., Nguyen, L.V., Rachubinski, R.A. and Goodman, H.M. (1999) The Pex16p homologue SSE1 and storage organelle formation in Arabidopsis seeds. *Science.* **284**: 328-330.

Lopez-Huertas, E. and Baker, A. (1999) Peroxisome Biogenesis. In: *Transport of Molecules Across Microbial Membranes.* (Broome-Smith, J.K., Baumberg, S., Stirling, C.J., Ward, F.B., eds.). pp. 205-238. Proc. Symp. Soc. Gen. Microbiol. Cambridge University Press.

Lopez-Huertas, E., Charlton, W.L., Johnson, B., Graham, I.A. and Baker, A. (2000) Stress induces peroxisome biogenesis genes. *EMBO J.* **19**: 6770-6777.

Lopez-Huertas, E., Oh, J. and Baker, A. (1999) Antibodies against Pex14p block ATP-independent binding of matrix proteins to peroxisomes *in vitro*. *FEBS Lett.* **459**: 227-229.

Makita, T. (1995) Molecular organization of hepatocyte peroxisomes. *Int. Rev. Cytol.* **160**: 303-352.

Marrison, J., Onyeocha, I., Baker, A. and Leech, R. (1993) Recognition of peroxisomes by immunofluorescence in transformed and untransformed tobacco cells. *Plant Physiol.* **103**: 1055-1059.

McNew, J. and Goodman, J. (1996) The targeting and assembly of peroxisomal proteins: some old rules do not apply. *Trends Biochem. Sci.* **21**: 54-58.

McNew, J. and Goodman, J. (1994) An oligomeric protein is imported into peroxisomes *in vivo*. *J. Cell Biol.* **127**: 1245-1257.

Middelkoop, E., Strijland, A. and Tager, J. (1991) Does aminotriazeole inhibit import of catalase into peroxisomes by retarding unfolding? *FEBS Lett.* **279**: 79-82.

Miernyk, J.A. (1999) Protein folding in the plant cell. *Plant Physiol.* **121**: 695-703.

Miura, S., Kasuya-Arai, I., Mori, H., Miyazawa, S., Osumi, T., Hashimoto, T. and Fujiki, Y. (1992) Carboxyl-terminal consenus Ser-Lys-Leu-related tripeptide of peroxisomal proteins functions *in vitro* as a minimal peroxisome-targeting dignal. *J. Biol. Chem.* **267**: 14405-14411.

Miura, S., Miyazawa, S., Osumi, T., Hashimoto, T. and Fujiki, Y. (1994) Post-translational import of 3-ketoacyl-CoA thiolase into rat liver peroxisomes i*n vitro*. *J. Biochem.* **115**: 1064-1068.

Miyazawa, S., Osumi, T., Hashimoto, T., Ohno, K. and Miura, S. (1989) Peroxisome targeting signal of rat liver acyl-coenzyme A oxidase resides at the carboxy terminus. *Mol. Cell. Biol.* **9**: 83-91.

Morand, O.H., Allen, L.A., Zoeller, R.A. and Raetz, C.R. (1990) A rapid selection for animal cell mutants with defective peroxisomes. *Biochim. Biophys. Acta.* **1034**: 132-141.

Motley, A., Hettma, E., Ketting, R., Plasterk, R. and Tabak, H. (2000) *Caenorhabditis elegans* has a single pathway to target matrix proteins to peroxisomes. *EMBO Reports.* **1**: 40-46.

Motley, A., Lumb, M., Oatey, P., Jennings, P., De Zoysa, P., Wanders, R., Tabak, H. and Danpure, C. (1995) Mammalian alanine/glyoxyate aminotransferase 1 is imported into peroxisomes via the PTS1 translocation pathway. Increased degeneracy and context specificity of the mammalian PTS1 motif and implications for the peroxisomes-to-mitochondrion mistargeting of AGT in primary hyperoxauria type 1. *J. Cell Biol.* **131**: 95-109.

Mullen, R.T. and Trelease, R.N. (2000) The sorting signals for peroxisomal membrane-bound ascorbate peroxidase are within its C-terminal tail. *J. Biol. Chem.* **275**: 16337-16344.

Mullen, R.T., Flynn, C.R. and Trelease, R.N. (2001a) How are peroxisomes formed? The role of the endoplasmic reticulum and peroxins. *Trends Plant Sci.* **6**: 256-261.

Mullen, R.T., Lee, M. and Trelease, R.N. (1997b) Identification of the peroxisomal targeting signal for cottonseed catalase. *Plant J.* **12**: 313-322.

Mullen, R.T., Lee, M., Flynn, C.R. and Trelease, R.N. (1997a) Diverse amino acid residues functions within the type 1 peroxisomal targeting signal. *Plant Physiol.* **115**: 881-889.

Mullen, R.T., Lisenbee, C.S., Flynn, C.R. and Trelease, R.N. (2001b) Stable and transient expression of chimaeric peroxisomal membrane proteins induces an independent 'zippering' of peroxisomes and an endoplasmic reticulum subdomain. *Planta*. In press.

Mullen, R.T., Lisenbee, C.S., Miernyk, J.A. and Trelease, R.N. (1999) Peroxisomal membrane ascorbate peroxidase is sorted to a membranous network that resembles a subdomain of the endoplasmic reticulum. *Plant Cell.* **11**: 2167-2185.

Nuttley, W.M., Brade, A.M., Gaillardin, C., Eitzen, G.A. and Glover, J.R. (1992) Rapid identification and characterization of peroxisomal assembly mutants in *Yarrowia lipolytica*. *Yeast.* **9**: 507-517.

Olivier, L., Kovacs, W., Masuda, K., Keller, G. and Krisans, S. (2000) Identification of peroxisomal targeting signals in cholesterol biosynthetic enzymes: AA-CoA thiolase HMG-CoA synthase, MPPD, and FPP synthase. *J. Lipid Res.* **41**: 1921-1935.

Olsen, L.J. (1998) The surprising complexity of peroxisome biogenesis. *Plant Mol. Biol.* **38**: 163-189.

Olsen, L.J., Ettinger, W.F., Damsz, B., Matsudaira, K., Webb, M.A. and Harada, J.J. (1993) Targeting of glyoxysomal proteins to peroxisomes in leaves and roots of a higher plant. *Plant Cell.* **5**: 941-952.

Onyeocha, I., Beehari, R. and Baker, A. (1993) Targeting of castor bean glyoxysomal isocitrate lyase to tobacco leaf peroxisomes. *Plant Mol. Biol.* **22**: 385-396.

Osumi, T., Tsukamoto, T., Hata, S., Yokota, S., Miura, S., Fujiki, Y., Hijikata, M., Miyazawa, S. and Hashimoto T. (1991) Amino-terminal presequence of the precursor of peroxisomal 3-ketoacyl-CoA thiolase is a cleavable signal peptide for peroxisomal targeting. *Biochem. Biophys. Res. Commun.* **181**: 947-954.

Pause, B., Saffrich, R., Hunziker, A., Ansorge, W. and Just, W. (2000) Targeting of the 22 kDa integral peroxisomal membrane protein. *FEBS Lett.* **471**: 23-28.

Pool, M., Lopez-Huertas, E. and Baker, A. (1998b) Characterization of intermediates in the process of plant peroxisomal protein import. *EMBO J.* **17**: 6854-6862.

Pool, M., Lopez-Huertas, E., Horng, J. and Baker, A. (1998a) NADPH is a specific inhibitor of protein import into glyoxysomes. *Plant J.* **15**: 1-14.

Pratt, W., Krishna, P. and Olsen, L. (2001) Hsp90-binding immunophilins in plants: the protein movers. *Trends Plant Sci.* **6**: 54-58.

Preisig-Müller, R., Muster, G. and Kindl, H. (1994) Heat shock enchances the amount of prenylated DnaJ protein at membranes of glyoxysomes. *Eur. J. Biochem.* **219**: 57-63.

Purdue, P. and Lazarow, P. (1996) Targeting of human catalase to peroxisomes is dependent upon a novel COOH-terminal peroxisomal targeting sequence. *J. Cell Biol.* **134**: 849-862.

Rapp, S., Soto, U., Just, W. (1993) Import of firefly luciferase into peroxisomes of permeabilized Chinese hamster ovary cells: A model system to study peroxisomal protein import *in vitro*. *Exp. Cell Res.* **205**: 59-65.

Sacksteder, K.A. and Gould, S.J. (2000) The genetics of peroxisome biogenesis. *Ann. Rev. Genet.* **34**: 623-652.

Schmid, M., Simpson, D., Kalousek, F. and Gietl, C. (1998) A cysteine endopeptidase with a C-terminal KDEL motif isolated from castor bean endosperm is a marker enzyme for the ricinosome, a putative lytic compartment. *Planta.* **206**: 466-475.

Schumann, U., Gietl, C. and Schmid, M. (1999a) Sequence analysis of a cDNA encoding Pex7p, a peroxisomal targeting signal 2 receptor from *Arabidopsis thaliana. Plant Physiol.* **120**: 339.

Schumann, U., Gietl, C. and Schmid, M. (1999b) Sequence analysis of a cDNA encoding Pex10p, a zinc-binding peroxisomal integral membrane protein from *Arabidopsis thaliana. Plant Physiol.* **119**: 1147.

Small, G.M., Imanaka, T., Shio, H. and Lazarow, P.B. (1987) Efficient association of in vitro translation products with purified stable *Candida tropicalis* peroxisomes. *Mol. Cell. Biol.* **7**: 1848-1855.

Smith, M. and Schnell, D. (2001) Peroxisomal protein import: The paradigm shifts. *Cell.* **105**: 293-296.

Subramani, S. (1996) Convergence of model systems for peroxisome biogenesis. *Curr. Opin. Cell Biol.* **8**: 513-518.

Subramani, S. (1993) Protein import into peroxisomes and biogenesis of the organelle. *Ann. Rev. Cell Biol.* **9**: 445-478.

Subramani, S. (1992) Targeting of proteins into the peroxisomal matrix. *J. Mem. Biol.* **125**: 99-106.

Subramani, S., Koller, A. and Snyder, W.B. (2000) Import of peroxisomal matrix and membrane proteins. *Ann. Rev. Biochem.* **69**: 399-418.

Swinkels, B., Gould, S. and Subramani, S. (1992) Targeting efficiencies of various permutations of the consensus C-terminal peroxisomal targeting signal. *FEBS Lett.* **305**: 133-136.

Swinkels, B., Gould, S., Bodnar, A., Rachubinski, A. and Subramani, S. (1991) A novel, cleavable peroxisomal targeting signal at the amino-terminus of the rat 3-ketoacyl-CoA thiolase. *EMBO J.* **10**: 3255-3262.

Taylor, K., Kaplan, C., Gao, X. and Baker, A. (1996) Localization and targeting of isocitrate lyases in *Saccharomyces cerevisiae. Biochem J.* **319**: 255-262.

Terlecky, S.R. and Fransen, M. (2000) How peroxisomes arise. *Traffic.* **1**: 465-473.

Terlecky, S., Legakis, J., Hueni, S. and Subramani, S. (2001) Quantitative analysis of peroxisomal protein import *in vitro*. *Exp. Cell Res.* **263**: 98-106.

Titorenko, V.I. and Rachubinski, R.A. (2001a) Dynamics of peroxisome assembly and function. *Trends Cell Biol.* **11**: 22-29.

Titorenko, V.I. and Rachubinski, R.A. (2001b) The life cycle of the peroxisome. *Nat. Rev. Mol. Cell Biol.* **2**: 357-368.

Titorenko, V.I., Smith, J., Szilard, R. and Rachubinski, R. (1998) Pex20p of the yeast *Yarrowia lipolytica* is required for the oligomerization of thiolase in the cytosol and for its targeting to the peroxisome. *J. Cell Biol.* **142**: 403-420.

Trelease, R., Choe, S. and Jacobs, B. (1994) Conservative amino acid substitutions of the C-terminal tripeptide (Ala-Arg-Met) on cottonseed isocitrate lyase preserve import *in vivo* into mammalian cell peroxisomes. *Eur. J. Cell Biol.* **65**: 269-279.

Trelease, R., Lee, M., Banjoko, A. and Bunkelmann J. (1996a) C-Terminal polypeptides are necessary and sufficient for *in vivo* targeting of transiently-expressed proteins to peroxisomes in suspension-cultured plant cells. *Protoplasma.* **195**: 156-167.

Trelease, R.N., Xie, W., Lee, M.S. and Mullen, R.T. (1996b) Rat liver catalase is sorted to peroxisomes by its C-terminal tripeptide Ala-Asn-Leu, not by the internal Ser-Lys-Leu motif. *Eur. J. Cell Biol.* **71**: 248-258.

Usada, N., Johkura, K., Hachiya, T. and Nakazawa, A. (1999) Immunoelectron microscopy of peroxisomes employing the antibody for the SKL sequence PTS1 c-terminus common to peroxisomal enzymes. *J. Histochem. Cytochem.* **47**: 1119-1126.

Volokita, M. (1991) The carboxy-terminal end of glycolate oxidase directs a foreign protein into tobacco leaf peroxisomes. *Plant J.* **1**: 361-366.

Walton, P., Gould, S., Feramisco, J. and Subramani, S. (1992a) Transport of microinjected proteins into peroxisomes of mammalian cells: inability of Zellweger cell lines to import proteins with the SKL tripeptide peroxisomal targeting signal. *Mol. Cell Biol.* **12**: 531-541.

Walton, P., Hill, P. and Subramani, S. (1995) Import of stably folded proteins into peroxisomes. *Mol. Biol. Cell.* **6**: 675-683.

Walton, P., Wendland, M., Subramani, S., Rachubinski, R. and Welch, W. (1994) Involvement of 70-kD heat-shock proteins in peroxisomal import. *J. Cell Biol.* **125**: 1037-1046.

Wang, X., Unruh, M.J. and Goodman, J.M. (2001) Discrete targeting signals direct Pmp47 to oleate-induced peroxisomes in *Saccharomyces cerevisiae*. *J. Biol. Chem.* **276**: 10897-10905.

Waterham, H.R., Russell, K., de Vries, Y. and Cregg, J.M. (1997) Peroxisomal targeting, import, and assembly of alcohol oxidase in *Pichia pastoris*. *J. Cell Biol.* **139**: 1419-1431.

Waterham, H.R., Titorenko, V.I., Haima, P., Cregg, J.M., Harder, W. and Veenhuis, M. (1994) The *Hansenula polymorpha* PER1 gene is essential for peroxisome biogenesis and encodes a peroxisomal matrix protein with both carboxy- and amino-terminal targeting signals. *J. Cell Biol.* **127**: 737-749.

Wendland, M. and Subramani, S. (1993) Cytosol-dependent peroxisomal protein import in a permeabilized cell system. *J. Cell Biol.* **120**: 675-685.

Wimmer, C., Lottspeich, F., Van Der Klei, I., Veenhuis, M. and Gietl, C. (1997) The glyoxysomal and plastid molecular chaperones (70-kDa Heat shock protein) of watermelon cotyledons are encoded by a single gene. *Proc. Natl. Acad. Sci. USA*. **94**: 13624-13629.

Wimmer, C., Schmid, M., Veenhuis, M. and Gietl C. (1998) The plant PTS1 receptor: similarities and differences to its human and yeast counterparts. *Plant J.* **16**: 453-464.

Wolins, N. and Donaldson, R. (1997) Binding of the peroxisomal targeting sequence SKL is specified by a low-affinity site in castor bean glyoxysomal membranes. *Plant Physiol.* **113**: 943-949.

Yang, X., Purdue, P. and Lazarow, P. (2001) Eci1p uses a PTS1 to enter peroxisomes: either its own or that of a partner, Dci1p. *Eur. J. Cell Biol.* **80**: 126-138.

Zoeller, R.A., Allen, L.A., Santos, M.J., Lazarow, P.B., Hashimoto, T., Tartakoff, A.M. and Raetz, C.R. (1989) Chinese hamster ovary cell mutants defective in peroxisome biogenesis. Comparison to Zellweger syndrome. *J. Biol. Chem.* **264**: 21872-21878.

Zolman, B.K., Yoder, A. and Bartel, B. (2000) Genetic analysis of indole-3-butyric acid responses in *Arabidopsis thaliana* reveals four mutant classes. *Genet.* **156**: 1323-1337.

12

PEX GENES IN PLANTS AND OTHER ORGANISMS

Wayne Charlton[1] and Eduardo Lopez-Huertas[2]

[1]*University of Leeds, Leeds, UK.* [2]*Puleva Biotech, Granada, Spain*

KEY WORDS

Peroxins; homologues; Arabidopsis; peroxisome biogenesis; protein import.

INTRODUCTION

Peroxisomes were the last of the major cell organelles to be discovered. They are small (approximately 0.1-1.5μM in diameter) structures surrounded by a single membrane embedded with numerous integral and peripheral membrane proteins. Together, the proteins and lipids provide a selective barrier (Chapter 8), thus allowing compartmentalization of the anabolic and catabolic reactions taking place within the organelle. Such reactions include lipid mobilization (Chapters 2 and 3), nitrogen metabolism (Chapter 6) and photorespiration (Chapter 5).

Within an organism, specialized peroxisomes take various forms to perform different developmental and tissue-specific functions. In plants, at least, rather than specific peroxisomes being formed *de novo*, existing peroxisomes are inter-converted from one peroxisome type to another through the import of the necessary proteins. For example, glyoxysomes (specialized peroxisomes in which the glyoxylate cycle takes place) carry out lipid mobilization to provide energy for development prior to the onset of photosynthesis during seed germination. However, once the photosynthetic machinery is up and running and the seed's lipid reserves are exhausted, the

A. Baker and I.A. Graham (eds.), Plant Peroxisomes, 385–426.

enzymes specific to the now redundant glyoxysomes are degraded. New proteins and enzymes are then imported to equip the organelle as a leaf-type peroxisome for functions such as photorespiration and stress management.

Despite the inter-conversion between peroxisome types in plants, new peroxisomes have to be formed as an organism grows and develops. The mechanism of peroxisome biogenesis, however, remains obscure. As membranes are not believed to be formed *de novo*, they have to arise via one of two pathways. They are formed either by the division of pre-existing peroxisomes or the membranes are taken from other vesicular structures such as the endoplasmic reticulum (ER) and the necessary proteins imported to mature the peroxisome. This topic is addressed in detail in Chapter 10.

Recent years have seen a remarkable increase in research attributed to the field of peroxisome biogenesis and protein import. Initially, through genetic approaches and subsequently through the approaches of bioinformatics and homology based cloning, a total of twenty three peroxisome biogenesis (*PEX*) genes have been cloned from yeast. The protein products of *PEX* genes are termed peroxins.

In mammals, thirteen *PEX* genes are known while the number reported for plants is only seven, although research in plant peroxisome biogenesis predates that in yeast, (see Table 1). However, with the recent completion of the *Arabidopsis thaliana* genome sequencing project, a possible eight more *PEX* genes have been identified, bringing the total in plants to fifteen; see Table 2.

Table 1: PEX genes currently cloned from plant species

Gene	Species	Accession Number	Reference
PEX1	*Arabidopsis thaliana*	AF275382	Lopez-Huertas *et al*, 2000
PEX5	*Arabidopsis thaliana*	AF074843	Brickner *et al*, 1998
	Citrullus vulgaris	AF068690	Wimmer *et al*, 1998
	Nicotiana Tabacum	AF053104	Kragler *et al*, 1998
PEX6	*Helianthus annuus*	AJ305171	Kaplan *et al*, 2001
PEX7	*Arabidopsis thaliana*	AF130973	Schumann *et al*, 1999
PEX10	*Arabidopsis thaliana*	AF119572	Schumann *et al*, 1999
		AJ276134	Baker *et al*, 2000
PEX14	*Arabidopsis thaliana*	AJ251524	Lopez-Huertas *et al*, 1999
		AB037538	Hayashi *et al*, 2000
PEX16	*Arabidopsis thaliana*	AF085354	Lin *et al*, 1999

Table 2: PEX homologues of Arabidopsis identified through a search of the Arabidopsis genome database

Name, subcellular localisation and function	Possible matches[a]	Map position and markers
PEX2 PMP RING-finger motif Matrix protein import	Acc # AC011717.4 At1g79810 (BAC = F19K16.23)	Ch 1, 127-132cM Mi157 SGCSNP142
PEX3 PMP Early peroxisome assembly	No 1: Acc # AB020749.1 Protein id = BAB02028.1 At3g18160 (BAC = MRC8.15)	Ch 3, 25-30cM Mi289 M228 Gln1-3
	No 2: Acc # AC073555 At1g48640 BAC = F11I4.17	Ch 1, 75-80cM SGCSNP209 Mi208 Mi441
PEX4 PMP Ubiquitin-conjugated enzyme. May function in receptor recycling	Acc AC005405* At5g25760 BAC = F18A17.10	Ch 5 (45-50cM) TNY SGCSNP193
PEX6 Mainly cytosolic AAA protein. Early peroxisome biogenesis and receptor recycling	Acc # = AC006550.2 At1g03000 (BAC = F10O3.17)	Ch 1 (2-7cM) RS10 T7I23
PEX11 PMP Peroxisome proliferation	No 1[b]: Acc # AC012463 At1g47750 BAC = T2E6.18	Ch 1 (73-78cM) SGCSNP322 Mi 291
	No 2[b]: Acc # AL096860 At3g47430 BAC = T21L8.180	Ch 3 (73-78cM) SGCSNP216 ASN1
	No 3[b]: Acc # AF332441 Protein id = AAG48804.1 BAC = T1N6.24	Ch 1 (0-5cM) SGCSNP5 SGCSNP131
	No 4[c]: Acc # AC004665 At2g45740 BAC = F4I18.28	Ch 2 (90-95cM) UBIQUE VeO19
	No 5[c]: Acc # ATT27I15	Ch 3 (93-98cM)

Name, subcellular localisation and function	Possible matches[a]	Map position and markers
	At3g61070 BAC = T27I15.160	SGCSNP74 SGCSNP87
PEX12 PMP RING-finger motif Matrix protein import	Acc # = AC022287 At3g04460 (BAC = T27C4.11)	Ch 3, 4-9cM Mi74b
PEX13 PMP Matrix protein import	No 1: Acc # AL021890.1 (Protein id = CAA17157.1) BAC = T8o5-110)	Ch 4, 72-77cM SGCSNP121 ATML1 SGCSNP211
	No 2*: Acc # AC009853 At3g07560 BAC = F21O3.27	Ch 4 (7-12cM) Mi 357 SGCSNP115
PEX17 PMP Docking protein	Acc # AL021713* At4g18200 BAC = T9A21.50	Ch 4 (61-66cM) SGCSNP349 Mi32
PEX19 Mainly cytosolic. Chaperone for PMPs	Acc # AL391151 Protein id = CAC01899.1 Clone = K10A8	Ch 5, 30-35cM Mi438
	Acc # AC009895-7 At3g03490 BAC = T21P5.9	Ch 3, 2-7cM Mi74b Mi199 Nga172

To obtain the Arabidopsis PEX homologues, protein sequences from known human and yeast PEX genes were used to search the Arabidopsis genome database (www.arabidopsis.org/blast/). Putative Arabidopsis matches were then used to re-screen the Swissprot protein database (analysis.molbiol.ox.ac.uk/blastall.cgi) to re-confirm similarity to known PEX genes

a. *Accession numbers are as designated by the National Centre for Biotechnology Information (NCBI). Where possible, protein sequence identities are given as designated by the Arabidopsis Genome. Otherwise the protein is given the protein id number (e.g. CAC01899.1) indicated in the annotation data supplied as part of the accession number data. The genomic locus of each gene is given as a BAC (Bacterial Artificial Chromosome) number; e.g. T21P5.9 is the ninth gene on BAC T21P5.*

b. c. *Protein sequence alignments of the five putative PEX11 genes revealed two separate classes indicated by sequence similarity and similarity to the human Pex11pβ (b) and Pex11pα (c) peroxins.*

**Sequence identified by Mullen et al, (2001) in addition to those discovered from our searches.*

Despite the identification of a number of *PEX* genes from the plant, animal and yeast kingdoms, surprizingly little is known concerning the mechanism of biogenesis and protein import into the organelle.

In this chapter, we plan to take a comprehensive look at all twenty three *PEX* genes currently known. We will reveal when and how (see complementation analysis below) the first example of each *PEX* gene was discovered, along with what is known about the function and subcellular localization of each peroxin. In addition, we will expand on the few *PEX* genes cloned and characterized from plants. Finally, we present the results of a systematic search of the Arabidopsis genome for orthologues of the remaining *PEX* genes and attempt to address the question "what is the minimum number of *PEX* genes required to form a fully functional peroxisome?".

Complementation analysis

From the information presented below, it will become evident that most *PEX* genes were initially identified in yeasts by what is termed 'complementation analysis'. This is a technique whereby a DNA sequence is identified and confirmed as being involved in a particular metabolic pathway if its gene product can complement a cell line defective in that pathway. In the field of peroxisome research, cell lines (yeasts, plant or animal) are usually chemically mutated (for example, using the chemical mutagen ethyl methanesulphonate) and a selection for defective peroxisome biogenesis performed.

Under certain growth conditions, (for example, when supplemented with simple sugars such as glucose), many yeast species are not reliant upon peroxisomes for growth. Therefore, selection for defective peroxisome biogenesis is usually via reduced, (or the complete lack of), growth on a medium (for example oleic acid) for which peroxisomes become essential for growth. Replica plating on a glucose medium thus ensures maintenance of the mutants. Having obtained a cell line defective in peroxisome biogenesis, complementation analysis is carried out to identify the defective gene. This is achieved by transforming the mutant cell line with a genomic or cDNA library and screening the transformants for growth, once again, on media requiring peroxisomes for growth. For a cell to survive, it will have been transformed with a DNA sequence expressing a number of proteins; one of which complements the defective endogenous gene. Thus, complementation analysis identifies the gene responsible for the mutant phenotype when defective and also confirms its function in one step.

PEX GENES

PEX genes can be classed as either cytoplasmic, membrane associated (integral or peripheral) or a combination of the two. During the growth and maturation of new peroxisomes, the recruitment of lipids and membrane proteins is required in order to build up the selectively permeable membrane. Although the properties of most of these *PEX* gene proteins (peroxins) are still being investigated, together, they perform the functions of early peroxisome biogenesis, peroxisome membrane protein (PMP) import, targeting receptors, chaperones, matrix protein import, receptor recycling and proliferation, (see Figure 1).

For clarity, the following *PEX* gene review has been arranged in an order reflecting their function.

Figure 1: PEX genes and peroxisomal protein import

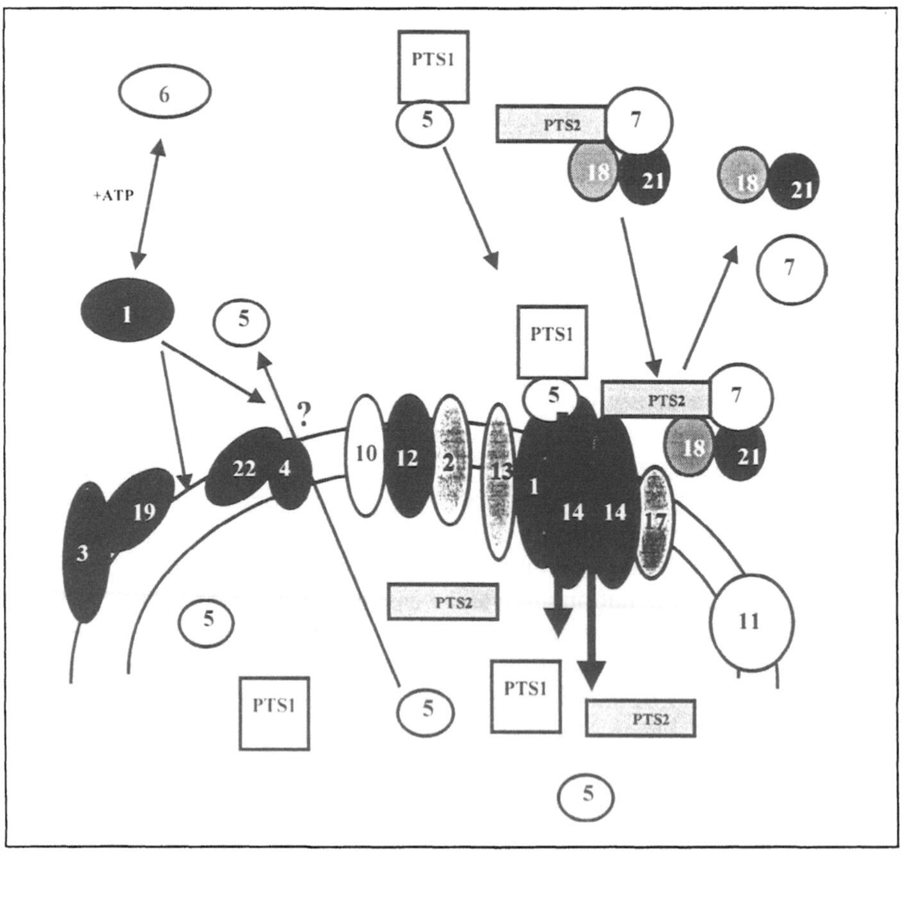

Figure 1. Schematic of peroxisome biogenesis. Shown are a compilation of processes and interactions from many species and do not represent the interactions from any one species and do not necessarily represent the views of other authors. The peroxisome membrane contains a number of peroxisome membrane proteins (PMP), including Pex2p, Pex10p, Pex12p, Pex13p, Pex14p and Pex17p which, along with the inherent properties of the lipid bilayer, imparts selectivity on the membrane. These PMPs are inserted into the immature peroxisome membrane during peroxisome maturation with Pex3p and Pex19p, believed to play a role in this process. Once a full complement of PMPs is in place, the peroxisome is competent to import matrix proteins for the final stages of maturation and for interconversion between peroxisome types.

Matrix proteins present in the cytosol possess one of two targeting signals (PTS1 or PTS2) and are targeted to the peroxisome by the target receptor proteins Pex5p (PTS1) and Pex7p (PTS2). In S. cerevisiae, transfer of the Pex7p-PTS2 protein to the peroxisome membrane is reportedly mediated by Pex18p and Pex21p in the form of accessory proteins. In humans, Pex5p coupled to its PTS1 protein has recently been shown to be translocated into the peroxisomal matrix prior to dissociation and recycling of Pex5p back to the cytosol. Recycling of Pex5p is believed to involve the peroxins Pex1p, Pex4p, Pex6p and Pex22p.

Docking and transfer of the matrix proteins across the membrane is still largely a mystery, although it is clear that many of the PMPs, including Pex2p, Pex10p, Pex12p, Pex13p, Pex14p and Pex17p play a role in matrix protein import. Of these peroxins, Pex14p is believed to be the initial point of docking and recent reports suggest the protein forms a pentameric complex in the form of a membrane-spanning pore through which it may be possible for proteins to translocate. Following import, the target receptors are recycled to the cytosol. Four peroxins, Pex1p, Pex6p, Pex22p and Pex4p have been implicated as playing a role in recycling of the PTS1 target receptor protein, Pex5p, although the specific roles of these peroxins remains unknown. No peroxins have been implicated in the recycling of the PTS2 target receptor protein Pex7p.

Early peroxisome biogenesis genes

PEX1

PEX1 has been cloned from representatives of the plant, animal and yeast kingdoms, including *Arabidopsis thaliana (*Lopez-Huertas *et al*, 2000), *Saccharamyces cerevisiae* (Erdmann *et al*, 1991) and humans (Reuber *et al*, 1997; Tamura *et al*, 1998a). It was first cloned from *S. cerevisiae* via complementation analysis (Erdmann *et al*, 1991). Subcellular localization of the *PEX1* protein (Pex1p) is still subject to debate. Reports vary between being cytoplasmic in humans (Yahraus *et al*, 1996; Tamura *et al*, 1998b) to loosely associated with the peroxisome membrane in *Pichia augusta* (=*H. polymorpha*), (Kiel *et al*, 1999), while reports on *P. pastoris* indicate it to be associated with small non-peroxisome vesicles (Faber *et al*, 1998). Pex1p proteins are members of the *A*TPase *A*ssociated with diverse cellular *A*ctivities (AAA) proteins; a family of proteins involved in a diverse array of functions, including cell cycle regulation and vesicle-mediated protein transport. *PEX1* expression is induced during various stress responses such as pathogen attack and wounding (Lopez-Huertas *et al*, 2000). The Pex1p protein has been demonstrated to interact with another AAA protein, Pex6p in *P. pastoris* (Faber *et al*, 1998), *Hansenula polymorpha,* (Kiel *et al*, 1999) and humans (Tamura *et al*, 1998b; Geisbrecht *et al*, 1998), with the interaction being shown to be ATP-dependent in *P. pastoris* (Faber *et al*, 1998).

The function of Pex1p is interesting as reports indicate two widely differing roles. Pex1p and Pex6p are apparently involved in early peroxisome biogenesis with deletion of either of these two genes in *P. pastoris* or *H. polymorpha* resulting in small vesicular peroxisome remnants, (Faber *et al*, 1998, Kiel *et al*, 1999). In addition, Titorenko and co-workers (Titorenko *et al*, 2000; Titorenko and Rachubinski, 2000), have identified and purified six subforms of peroxisomes at different stages of maturity from *Y. lipolytica* and shown fusion of the earliest two forms to be ATP-dependent and inhibited by antibodies to Pex1p and Pex6p.

Surprisingly, cells deficient for Pex1p or Pex6p have revealed the protein to be required for Pex5p stability in *P. pastoris* (Collins *et al*, 2000) and humans (Yahraus *et al*, 1996). Pex1p and Pex6p are also believed to act in the later stages of peroxisomal matrix protein import, i.e. after matrix protein translocation (Collins *et al*, 2000) and it is postulated that both Pex1p and Pex6p could play a role in Pex5p recycling.

Pex1p and Pex6p are likely, therefore, to be performing dual roles at opposing ends of peroxisome biogenesis; namely early peroxisome biogenesis and receptor recycling. The fact that both functions are observed in *P. pastoris* rules out the possibility that different species have recruited these proteins for widely different functions. As these proteins are ATPases, one of their two roles may be merely as a source of energy. Further work will be required to confirm or reject these very different models of Pex1p and Pex6p action. Perhaps these proteins perform multiple functions at different stages of peroxisome biogenesis. We currently have no information on the function of *PEX1* and *PEX6* in plants although both have recently been cloned (see below).

PEX6

PEX6 has been cloned from representatives of the plant, animal and yeast kingdoms, including sunflower (*Helianthus annuus)* (Kaplan *et al*, 2001), *S. cerevisiae (*Voorn-Brouwer *et al*, 1993), and humans (Yahraus *et al*, 1996). First cloned from *S. cerevisiae* by Voorn-Brouwer *et al*, (1993) via complementation analysis, PEX6, as with PEX1, is member of the AAA protein family. Pex6p interacts with Pex1p in *P. pastoris* (Faber *et al*, 1998), *H. polymorpha* (Kiel *et al*, 1999) and humans (Tamura *et al*, 1998a; Geisbrecht *et al*, 1998). The interaction has been demonstrated to be ATP-dependent in *P. pastoris* (Faber *et al*, 1998). Subcellular localization of Pex6p is unclear with reports from *H. polymorpha* suggesting Pex6p to be associated with the outer peroxisome membrane (Kiel *et al*, 1999), while studies on *P. pastoris* tend to suggest that it is associated with non-peroxisome vesicles (Faber *et al*, 1998). As with Pex1p, Pex6p has been reported to be involved both in the early and late stages of peroxisome biogenesis; see *PEX1* section for details.

Early peroxisome biogenesis genes in plants

PEX1

A *PEX1* cDNA from Arabidopsis (At*PEX1*) was isolated by Lopez-Huertas and co-workers (2000) through a database homology search with the *S. cerevisiae PEX1* sequence (see Table 1). Following identification of a putative *PEX1* genomic sequence, the corresponding cDNA was amplified via the polymerase chain reaction (PCR) and used to screen a three day seedling hypocotyl cDNA library. The deduced amino acid sequence of AtPex1p comprises 1117 amino acids with a predicted molecular weight of 122 kDa. Pex1p possesses two ATP binding motifs predicted in the sequence as also reported for its human and yeast counterparts. The protein shares identity with known Pex1p proteins from humans, *S. cerevisiae, P. pastoris* and *Y. lipolytica* and can be immunoprecipitated by antibodies raised against other Pex1p proteins (Lopez-Huertas *et al*, 2000) which confirms the identity of the isolated clone.

An important feature of peroxisomal metabolism is that these organelles are very sensitive to external signals. Hypolipidaemic drugs and plasticizers induce proliferation of the peroxisomal population in the liver of rodents (Van den Bosch *et al*, 1992). In plants, peroxisomes are induced in response to different stress situations (for example, herbicides, xenobiotics, senescence) and a role in the cell response to stress has been proposed for these organelles (Palma *et al*, 1991; del Río *et al*, 1998; see also Chapter 7). In order to study the expression of peroxisome biogenesis genes *in vivo* during conditions of stress, a 1Kb promoter fragment of At*PEX1* was linked to luciferase and used to transform Arabidopsis plants (*PEX1*-LUC plants). As discussed earlier, studies have linked *PEX1* with a role in early peroxisome biogenesis (Faber *et al*, 1998, Kiel *et al*, 1999; Titorenko *et al*, 2000; Titorenko and Rachubinski, 2000). It was, therefore, interesting to establish if the *PEX1* gene is regulated by conditions that influence peroxisome biogenesis.

Analysis of young healthy *PEX1*-LUC plants revealed luciferase activity mainly in the meristems and young developing leaves, with little expression in mature, healthy, leaves. These results suggest that *PEX1* expression (and, thus, possibly peroxisome biogenesis) is greatest in actively dividing tissue and relatively low in tissues where organelle maintenance (rather than biogenesis) is required.

Different stress situations yield reactive oxygen species, such as hydrogen peroxide (H_2O_2), which act as signal molecules that trigger gene expression of various transcription factors and cellular protectants, (Levine *et al*, 1994; Karpinski *et al*, 1999; Guan *et al*, 2000). The effect of the stress signal molecule H_2O_2 on the expression of *PEX* genes was investigated. Luciferase expression increased rapidly and dramatically throughout the plant when *PEX1*-LUC plants were sprayed with 1 mM H_2O_2. This effect was also seen when only one leaf was incubated with 1mM or 5-10μM H_2O_2; the latter concentration mimicking the level of H_2O_2 generated by the oxidative burst during pathogen attack. In these experiments, luciferase expression began at the point of contact with the H_2O_2 solution and extended systemically throughout the plant within ninety minutes. Analysis of *PEX1, PEX5, PEX10, PEX14* and *GST* genes by reverse transcription polymerase chain reaction (RT-PCR) and/or Northern analysis in wild type plants and a Chinese hamster ovary (CHO) cell line (Lopez-Huertas *et al*, 2000), revealed these also to be induced by H_2O_2. The CHO cell results suggest that stress signalling molecules such as H_2O_2 are shared between plants and animals.

Two situations in which peroxisomes have been reported to proliferate are clofibrate treatment (Palma *et al*, 1991) and natural senescence (Pastori and del Rio, 1994), both resulting in a significant increase in *PEX1*-luciferase expression in whole plants. Wounding and pathogen attack were also studied. *PEX1*-LUC plants were inoculated in one leaf with *Pseudomonas syringae* pv. *Tomato* which induces an oxidative burst and hypersensitive response. Expression in the inoculated leaf was maximal at about two hours after inoculation and, although the pathogen was inoculated at the tip of the leaf, gene expression was mainly localized around the vascular tissue. However, the response to the pathogen was localized rather than systemic. In wounded leaves, *PEX1*-luciferase expression peaked between 45 and 120 minutes, then declined. Again it was mainly observed localized around the vascular tissues in the veins and the periphery of the leaf. Together, these results suggest that stress and the stress signalling molecule H_2O_2 induces peroxisome biogenesis not only in plants, but also in animal cells (Lopez-Huertas *et al*, 2000).

PEX6

A *PEX6* cDNA was recently isolated from sunflower (*Helianthus annuus*). RT-PCR was used to generate a probe which was subsequently used to screen a cDNA library, (Kaplan *et al*, 2000). The isolated cDNA encodes a 909 amino acid protein with a calculated molecular mass of 99.5 kDa which is 5% less than that determined empirically, which may indicate an incomplete cDNA. Northern and Western blot analysis of *PEX6* during germination revealed peak expression of the gene and protein to be around day two to three post-seed imbibition. This deviates slightly from that shown for other *PEX* genes of Arabidopsis, but most probably reflects the difference in developmental timescale between the two species. *PEX6* mRNA levels were also shown to increase in senescent tissue. Direct functional evidence for the involvement of Ha*PEX6* in peroxisome biogenesis has yet to be obtained.

Peroxisome membrane protein (PMP) import genes

Only Pex19p and Pex16p have, to date, been shown to have a function directly related to membrane protein import, however, as Pex3p has been reported to interact with Pex19p, it has been included in this section.

PEX3

PEX3 has been cloned from yeasts and animals, including *S. cerevisiae* (Hohfeld *et al*, 1991), and humans (Kammerer *et al*, 1998; Soukupova *et al*, 1999). It was first cloned in *S. cerevisiae* by Hohfeld and co-workers (1991) via complementation analysis. All Pex3p proteins studied reveal a transmembrane domain at the amino-terminus of the protein which anchors the protein to the membrane, leaving the majority of the protein exposed to the cytoplasm. An interaction between the C-terminus of Pex3p and Pex19p has been reported, (Gotte *et al*, 1998, Snyder *et al*, 1999b; Sacksteder *et al*, 2000), although the role of this interaction remains unclear.

PEX16

First cloned via complementation analysis of a *pex16-1* mutant in *Y. lipolytica* by Eitzen *et al*, (1997), *PEX16* has since been cloned from *A. thaliana* (Lin *et al*, 1999) and humans (Honsho *et al*, 1998; South and Gould, 1999). In *Y. lipolytica*, Pex16p has been demonstrated to be sorted via the ER, (Titorenko and Rachubinski, 1998). An interaction between Pex16p and Pex19p has been shown, however, the function of this interaction is unclear (Sacksteder *et al*, 2000). The subcellular localization of Pex16p appears to vary with species. In *Y. lipolytica* it is believed to be a peripheral membrane protein associated with the matrix side of the peroxisome, (Eitzen *et al*, 1997), while in humans it behaves as an integral membrane protein with both its termini exposed to the cytoplasm (South and Gould, 1999).

PEX16 is somewhat of an enigma. Although in humans it appears to be required for the early stages of peroxisome biogenesis, in plants its claim to being a peroxin is dubious (see later) while in *S. cerevisiae* the gene is absent from its genome database.

Human cells defective for *PEX16* give rise to one of a class of human genetic diseases, known as Zellweger syndrome, characterized by the lack of matrix protein import. Recently, South and Gould (1999) have characterized a cell line from a Zellweger patient with defective membrane protein import. The mutation in this cell line was *PEX16*. Having cloned the human *PEX16* gene, they showed that, following *PEX16* injection into cells from this patient, synthesis and peroxisomal membrane protein import was detected within two to three hours and was followed by matrix protein import. These findings thus support a role for Pex16p in the import of peroxisome membrane proteins and, more importantly, support the idea that peroxisomes may be formed in the absence of pre-existing peroxisomes.

PEX19

PEX19 was first cloned following functional complementation of an EMS mutated *S. cerevisiae* cell line (Gotte *et al*, 1998). However, a putative orthologue (*PxF*) was identified four years earlier by James and co-workers (1994) in a screen for novel farnesylated proteins in Chinese hamster ovary cells. Since then, *PEX19* has also been cloned from *P. pastoris* (Snyder *et al*, 1999b) and humans (Matsuzono *et al*, 1999). The protein contains a C-terminal prenyl group binding site (Gotte *et al*, 1998, James *et al*, 1994). Prenylation is believed to mediate protein-membrane interactions, yet it is unknown whether the Pex19p interaction is a prenyl-protein or prenyl-lipid interaction. Early studies, using indirect immunofluorescence, revealed punctate staining of the PxF antibody indicative of peroxisome structures, suggesting Pex19p to be peroxisomal (James *et al*, 1994). However, recent studies using a variety of techniques including immunocytochemistry and sedimentation analysis, have shown that it is largely cytosolic with a small amount associated with the peroxisome (Gotte *et al*, 1998; Snyder *et al*, 1999a; Sacksteder *et al*, 2000). Studies of Snyder *et al* (1999a) suggest the peroxisomal association to possibly be via the binding of Pex3p. Matsuzono and co-workers (1999) have revealed farnesylated Pex19p to be partly, if not totally, anchored in the peroxisomal membrane, with its N-terminal region exposed to the cytosol.

Human Pex19p has been shown to interact with fourteen peroxisome proteins, including, Pex3p, Pex10p, Pex11pα and β, Pex12p, Pex13p and Pex14p (Sacksteder *et al*, 2000). A number of these proteins have also been shown to interact with the yeast Pex19p. Combined research data would, thus, suggest that the function of Pex19p is to bind peroxisome membrane proteins and facilitate their insertion into the peroxisome membrane. However, work of Snyder *et al* (2000) suggests that Pex19p is not the peroxisome membrane protein receptor protein. Their evidence for such a conclusion is three-fold. Firstly, although they showed that Pex19p interacted with six integral peroxisome membrane proteins, these interactions were not reduced following the inhibition of new protein synthesis as would be expected for interactions of the peroxisome membrane protein receptor. Secondly, the binding domains of these six integral membrane proteins do not overlap. Thirdly, co-immunoprecipitation of Pex19p from fractions containing cytosol or peroxisomes revealed that the interactions were predominantly between cytosolic integral membrane proteins and Pex19p. From these data, the authors believe Pex19p to play a chaperone role.

Membrane protein import genes in plants

PEX16

As mentioned earlier, the peroxin label for a *PEX16* homologue in plants is debatable. In Arabidopsis plants, its claim lies with a protein SSE1 (see Table 1) that is required for protein and oil body biogenesis, both of which are endoplasmic reticulum-dependent. Mutants of SSE1 (*sse1*), isolated from a transferred DNA (T-DNA) transgenic line, give rise to a shrunken seed phenotype in which starch predominates as the major seed storage compound in place of the usual lipids and proteins. SSE1 possesses 26% identity with Pex16p of *Y. lipolytica* and partially complements *Y. lipolytica pex16* mutants defective in peroxisome biogenesis and transportation of proteins associated with plasma membranes and cell walls (Lin *et al*, 1999).

However, the work of Lin and co-workers failed to report any clear phenotype to suggest that the gene was, in fact, a peroxin. Simple experiments such as seed germination and growth on media supplemented with sucrose or 2,4-dichlorophenoxybutyric acid (Hayashi *et al*, 1998) would have indicated whether the peroxisomes were functional. The authors state that SSE1 expression is likely to be required during germination yet fail to show an expression profile during germination to test whether expression of SSE1 follows that of peroxisome biogenesis. Obviously, more work is needed in order to address the involvement of SSE1 in peroxisome biogenesis.

Peroxisome targeting signals and related proteins

Two peroxisome targeting receptors have been identified to date: Pex5p and Pex7p. Both proteins bind peptides possessing the appropriate targeting sequence motif and transport them to the peroxisome surface for import into the peroxisome lumen. Also included in this section are Pex18p, Pex20p and Pex21p as they are believed to have a role in PTS2 targeted import.

PEX5

PEX5 is the most studied of the *PEX* genes which is reflected in the number (around fourteen) of species from which it has been cloned. These include *A. thaliana* (Brickner *et al*, 1998), *Nicotiana tobacum* (Kragler *et al*, 1998),

S. cerevisiae (Van der Leij *et al*, 1993) and humans (Dodt *et al*, 1995; Wiemer *et al*, 1995; Fransen *et al*, 1995). *PEX5* was first identified in *S. cerevisiae* by Van der Leij and co-workers (1993), via functional complementation of the *pas10-1* mutant isolated from a genetic screen using hydrogen peroxide lethality as selection (H_2O_2 is produced during β-oxidation of fatty acids). The protein, Pex5p, is a member of the tetratricopeptide repeat (TPR) family of proteins and functions as the Peroxisome Targeting Signal 1 (PTS1) receptor protein. However, in humans, Pex5p is also required for PTS2 protein import. To accomplish this, *PEX5* transcripts undergo alternative splicing to produce two mature *PEX5* mRNAs and proteins of differing lengths; termed *PEX5L* and *PEX5S*. The longer of the two (*PEX5L*), but not the shorter, is required for the import of PTS2 targeted proteins (Otera *et al*, 1998).

The crystal structure of a fragment of Pex5p (in a complex with a PTS1 peptide) that contains all the seven predicted TPRs (the region that interacts with the PTS1 protein) has been published (Gatto *et al*, 2000). The structure reveals three of the TPRs to almost completely surround the PTS1 peptide with TPR4 forming a hinge region, thus allowing the two sets of TPRs (1-3 and 5-7) to form a single binding site.

Once bound to its PTS1 protein, the Pex5p-PTS1 protein complex is believed to dock with Pex14p at the peroxisome membrane surface. The mechanism following docking is still a hotly debated topic, with much speculation surrounding the role of Pex5p during the translocation event. The two schools of thought are that Pex5p either shuttles the PTS1 protein to the docking site, dissociates and then returns to the cytosol (Simple Shuttle Model) or that Pex5p enters and exits the peroxisome during each import cycle (Extended Shuttle Model). Until recently, evidence for the presence of intraperoxisomal Pex5p has been limited to the yeasts *Y. lipolytica* (Szilard *et al*, 1995) and *H. polymorpha* (van der Klei *et al*, 1998). However, recent work by Dammai and Subramani (2001), have shown that human Pex5p, bound to its PTS1 protein, enters the peroxisome matrix and recycles back to the cytosol in a manner analogous to receptors involved in nuclear import. For a comprehensive account of matrix protein import, see Chapter 11.

PEX7

PEX7 was initially isolated from *S. cerevisiae* by Marzioch *et al*, (1994) by complementation analysis. It has since been widely studied and cloned from a number of species, including *A. thaliana* (Schumann *et al*, 1999b) and humans (Bravermann *et al*, 1997; Purdue *et al*, 1997; Motley *et al*, 1997). Pex7p is a member of the WD40 repeat family of proteins and functions as the Peroxisome Targeting Signal 2 (PTS2) receptor. Subcellular localization of Pex7p is unclear. In *P. pastoris* it is reported to be both cytoplasmic and intraperoxisomal (Elgersma *et al*, 1998). In *S. cerevisiae* some reports suggest it to be both cytoplasmic and peroxisomal associated with the latter being thiolase-dependent, (Marzioch *et al*, 1994). Others believe it is totally intraperoxisomal (Zhang and Lazarow, 1996). See Chapter 11 for a more detailed account of *PEX7*.

PEX18 and *PEX21*

PEX18 and *PEX21* have only been cloned from the yeast *S. cerevisiae* (Purdue *et al*, 1998). Both proteins were isolated from a two hybrid screen using, as bait, the Pex7p protein. They are largely cytosolic, although a small amount (<10%) of Pex18p has been found to be associated with peroxisomes where it is believed to be associated with the outer face of the membrane, but an intraperoxisomal localization cannot be ruled out. Pex18p is weakly homologous to Pex21p and both display partial functional redundancy. Cells lacking Pex18p and Pex21p abolished PTS2 targeting of thiolase, indicating the Pex18p and Pex21p are key components of the PTS2-targeting pathway (Purdue *et al*, 1998).

PEX20

PEX20 has been cloned only from the yeast *Y. lipolytica* (Titorenko *et al*, 1998), being identified via complementation analysis. The protein is predominantly cytoplasmic with a small amount believed to be associated with the periphery of the peroxisome membrane. It is required for the import of thiolase, a PTS2 protein, however, it does not exhibit homology to the PTS2 receptor Pex7p. Titorenko and co-workers (1998) suggest that, in the cytosol, monomeric Pex20p binds monomeric thiolase and promotes the formation of a heterotetrameric complex. This complex then binds to the peroxisomal membrane with translocation of the thiolase homodimer, resulting in the release of Pex20p monomers back to the cytosol.

Peroxisome targeting signal receptors in plants

Although *PEX5* and *PEX7* have been cloned from plant species, characterization has only been carried out on *PEX5*; see Table 1.

PEX5

PEX5 has been cloned from three plant species; tobacco, watermelon and Arabidopsis, independently, by a variety of techniques. Tobacco *PEX5* (Nt*PEX5*; Kragler *et al*, 1998), was isolated via the two-hybrid screen, using, as bait, an artificially created PTS1 protein (the Gal4 binding domain fused to subunit IV of cytochrome c-oxidase with a C-terminal SKL). Watermelon *PEX5* (Cv*PEX5*; Wimmer *et al*, 1998), was isolated via PCR using primers designed to the highly conserved regions of *PEX5* from human and four yeasts. Arabidopsis *PEX5* (Brickner *et al*, 1998), was identified through a homology search of the Arabidopsis genome database using the mouse *PEX5* sequence.

In plants, the *PEX5* gene transcribes a single transcript, giving rise to a single Pex5p protein. Transformation of a *H. polymorpha* strain deficient for *PEX5* with Cv*PEX5* resulted in partial restoration of peroxisome formation. In fractionation experiments, CvPex5p was found mainly in the cytosol with a small amount associated with peroxisomes and could, therefore, function as a cycling receptor between the cytosol and peroxisome, as has been proposed for the human Pex5p and the Pex5ps of some yeast peroxisomes (Wimmer *et al*, 1998).

NtPex5p shares greater sequence identity with human than with yeast Pex5p. The function of NtPex5p was also established by showing that the plant protein restored protein translocation into peroxisomes when expressed in a yeast *pex5* deletion mutant. Two hybrid assays showed that NtPex5p interacts with PTS1 variants that are also recognised by its homologue from humans, suggesting both sequence and functional similarities between the plant and human Pex5ps (Kragler *et al*, 1998).

Matrix protein import

By far the greatest number of *PEX* genes identified to date play roles in the import of matrix proteins into the peroxisome lumen. For a detailed account of matrix protein import, see Chapter 11.

PEX2

PEX2 is unusual in that it was first cloned from rats rather than one of the yeast species. A rat cDNA library was used to complement a strain (Z65) of Chinese Hamster Ovary (CHO) cells defective in peroxisome assembly, (Tsukamoto *et al*, 1991). Since this initial report, it has also been cloned from yeasts, including *S. cerevisiae* (Liu *et al*, 1996) and *Y. lioplytica* (Eitzen *et al*, 1996), and humans (Schimozawa *et al*, 1992).

As with Pex16p, Pex2p from *Y. lipolytica* has been shown to be sorted via the ER, (Titorenko and Rachubinski, 1998). Pex2p proteins are integral membrane proteins that have both termini projecting into the cytosol (Harano *et al*, 1999). All Pex2p proteins possess a C-terminus zinc RING finger (a motif that is also present in Pex10p and Pex12p) which is believed to be involved in protein-protein interactions. Although Pex2p is required for peroxisomal localization of PTS1- and PTS2-containing proteins, work by Huang *et al*, (2000), have shown that a cysteine -258 to tyrosine substitution in the RING finger motif results in a complete defect only in the PTS1 pathway but not the PTS2 pathway.

PEX8

PEX8 has only been cloned from yeasts; being initially cloned from *H. polymorpha* via complementation analysis (Waterham *et al*, 1994). Since this time it has also been cloned from *S. cerevisiae* (Rehling *et al*, 2000) and *P. pastoris* (Liu *et al*, 1995). The sequence identity between Pex8p proteins from different species is quite large. Although all Pex8p proteins so far studied possess a PTS1 sequence, the Pex8p protein of *H. polymorpha* also possesses a PTS2 sequence. In *P. pastoris* the PTS2 sequence only poorly conforms to the PTS2 consensus while in *S. cerevisiae* and *Y. lipolytica* it is absent. The subcellular localization is reflected in this sequence variation with *H. polymorpha* Pex8p being proposed as a matrix protein (Waterham *et al*, 1994) while *P. pastoris* and *S. cerevisiae* Pex8ps are reported to be membrane proteins facing the peroxisomal matrix (Liu *et al*, 1995, Rehling *et al*, 2000). Pex8p interacts with Pex5p and is required for PTS1 and PTS2 import (Rehling *et al*, 2000).

PEX9

To date, *PEX9* has only been cloned from the yeast *Y. lipolytica,* (Eitzen *et al*, 1995), where it was identified via complementation analysis. It is an integral membrane protein. As yet Pex9p has not been shown to interact with any proteins, although reports suggest that the protein is essential for peroxisomal matrix protein import, but not the initial steps of peroxisomal membrane proliferation (Eitzen *et al*, 1995).

PEX10

PEX10 was first cloned from *P. pastoris* (Kalish *et al*, 1995) through complementation analysis. Since this initial report, it has been cloned from a number of species including *A. thaliana* (Schumann *et al*, 1999; Baker *et al*, 2000) and humans (Warren *et al*, 1998). Although unpublished, a putative *S. cerevisiae PEX10* is present in the *S. cerevisiae* database; accession number Q05568. All PEX10 proteins contain a C-terminal zinc RING finger, as do Pex2p and Pex12p, and a totally conserved motif (TLGEEYV) at the amino-terminus of unknown function (Kalish *et al*, 1995; Tan *et al*, 1995; Warren *et al*, 1998). The discovery of a 2 bp deletion immediately upstream of the RING motif in a patient suffering from a peroxisomal biogenesis dissorder enabled Okumoto and co-workers (1998) to shown that the C-terminal portion of *PEX10*, including the RING finger, is required for biological function.

Pex10p appears to be an integral membrane protein with its C and N termini projecting into the peroxisomal lumen (Kalish *et al*, 1995, Okumoto *et al*, 1998a). The function of Pex10p is unclear; human cell lines deficient in Pex10p are defective in peroxisome matrix protein import, but not peroxisome membrane protein import. In addition, these mutant cell lines show no reduction in the association of Pex5p with peroxisomes, suggesting that Pex10p is involved after the docking of the Pex5p-PTS1 protein complex with Pex14p, (Chang *et al*, 1999).

PEX12

PEX12 has been cloned from animals and yeasts, including *P. pastoris* (Kalish *et al*, 1996) and humans (Chang *et al,* 1997; Okumoto *et al*, 1998b). It was initially identified in *P. pastoris* using cells engineered to express a PTS1-GFP (Green Fluorescent Protein) construct. From this cell line, cells deficient in the import of the GFP were identified by fluorescence microscopy (Kalish *et al*, 1996), leading to the identification of the *pas10* mutant. This, in turn, lead to the cloning of the *PAS10* (*PEX12*) gene through functional complementation with a genomic library. As with *PEX10*, a *S. cerevisiae PEX12* sequence is present in the *S. cerevisiae* genome database (accession number Q04370), but nothing has been published on this peroxin for *S. cerevisiae*. All Pex12p proteins possess a zinc RING finger (as do Pex2p and Pex10p), believed to be involved in protein-protein interactions (Borden, 1998). Similar to Pex2p and Pex10p, Pex12p is an integral membrane protein with its C-terminus (containing the zinc RING finger) predicted to project into the cytosol (Chang *et al*, 1997; Okumoto and Fujiki, 1997). Studies have revealed that Pex12p interacts with Pex10p and Pex5p and, as with Pex10p, human cell lines deficient in functional Pex12p proteins are defective in peroxisome matrix, but not membrane, protein import. No decrease in the association of Pex5p with peroxisomes is observed in these mutant cell lines, suggesting that Pex12p is involved after the docking of the Pex5p-PTS1 protein complex with Pex14p, (Chang *et al*, 1999).

PEX13

PEX13 has been cloned from yeasts and animals including *S. cerevisiae* (Elgersma *et al*, 1996; Erdmann and Blobel, 1996) and humans (Fransen *et al*, 1998; Bjorkman *et al*, 1998). In 1996, *PEX13* was cloned independently by three groups from *S. cerevisiae* and *P. pastoris*. The *S. cerevisiae PEX13* gene was cloned by Elgersma *et al* (1996) and Erdmann and Blobel (1996). The method adopted by Elgersma and co-workers was that of complementation analysis, whereas the approach taken by Erdmann and Blobel was to purify Pex13p from isolated peroxisome membranes, previously induced by growth on oleate (a medium used to induce peroxisome proliferation in yeasts). The *PEX13* gene of *P. pastoris* was isolated by Gould *et al* (1996). They used complementation analysis to isolate the gene, which was then used to search genomic databases. From these searches, they also identified *PEX13* homologues from *S. cerevisiae*, the nematode *Caenorhabditis elegans* and humans. Pex13p is an integral membrane protein with its termini exposed to the cytosol (Girzalsky *et al*, 1999). Pex13p proteins possess a Src homology 3 (SH3) domain that is thought to be involved in protein-protein interactions (Gould *et al*, 1996). Proteins with this motif vary greatly in their function (e.g. actin-binding proteins, tyrosine kinases, phospholipases, etc.). Pex13p has been demonstrated to bind Pex5p, Pex7p and Pex14p (Gould *et al*, 1996; Erdmann and Blobel, 1996; Elgersma *et al*, 1996; Albertini *et al*, 1997; Brocard *et al*,

1997; Girzalsky *et al*, 1999). The Pex13p-Pex5p interaction has been shown to be weaker when Pex5p was bound to a PTS1-containing peptide; this being opposite to that observed for the Pex14p-Pex5p interaction (Urquhart *et al*, 2000).

PEX14

PEX14 was first cloned from the yeast *H. polymorpha* by complementation analysis (Komori *et al*, 1997). Since then it has been cloned from a number of species, covering the the animal, plant and yeast kingdoms, including: *A. thaliana* (Hayashi *et al*, 2000; Lopez-Huertas *et al*, 1999), *S. cerevisiae* (Albertini *et al*, 1997) and humans (Fransen *et al*, 1998; Will *et al*, 1999). The protein is a key component of the protein import machinery of peroxisomes and is believed to be the initial docking site of Pex5p laden with its PTS1-containing peptide (Otera *et al*, 2000). In *S. cerevisiae*, Pex14p interacts with Pex5p, Pex7p, Pex13p and Pex17p (Albertini *et al*, 1997; Brocard *et al*, 1997; Will *et al*, 1999). In rats, it has been shown to bind peroxisome-associated Pex5p in the ratio of 5:1 Pex14p:Pex5p. Although speculation, it is thus proposed to form a homopentamer complex possessing multiple membrane-spanning regions in the form of a pore-like structure for protein translocation, with the carboxy and amino termini exposed to the cytoplasm and peroxisome lumen, respectively, (Gouveia *et al*, 2000). Despite this research, its topology within the peroxisome membrane appears to vary among different species. In humans it is reported to be an integral membrane protein with it C-terminus and N-terminus exposed to the cytosol and peroxisomal lumen, respectively (Will *et al*, 1999). In *S. cerevisiae,* Brocard *et al* (1997) report Pex14p as a peroxisome membrane protein lacking a membrane spanning domain. Albertini *et al* (1997), report the protein as being attached to the outer surface of the peroxisome membrane. In Arabidopsis, Pex14p has been shown to be membrane-associated, with at least a part of the protein exposed to the cytosol (Hayashi *et al*, 2000). It will be interesting to see if future research provides a consensus to the topology of Pex14p or if the variation currently reported between organisms holds. See Chapters 9 and 11 for a more comprehensive account of *PEX14* function.

PEX17

PEX17 has been cloned only from the yeasts *S. cerevisiae* (Huhse *et al*, 1998) and *P. pastoris* (Snyder *et al*, 1999a). It was initially cloned by Huhse *et al* (1998) via complementation analysis of an EMS derived *pex17-1* mutant. The protein contains one predicted membrane spanning domain and two coiled-coil structures. In *P. pastoris*, Pex17p is believed to be an integral membrane protein with the C-terminus exposed to the cytosol (Snyder, *et al*, 1999a). However, localization studies in *S. cerevisiae* show it to be located at the surface of the peroxisomes (Huhse *et al*, 1998), while in *Y. lipolytica* it is suggested to be peripherally located on the lumen side of the peroxisome membrane (Smith *et al*, 1997). Pex17p interacts with Pex14p and with Pex5p; the latter binding being Pex14p-dependent (Huhse *et al*, 1998). The studies of Huhse and co-workers in *S. cerevisiae* revealed that mutants defective in *PEX17* showed defects in PTS1 and PTS2 import, yet localization of two PMPs, Pex11p and Pex3p, was unaffected. In contrast, Snyder and co-workers (1999a) have shown that *PEX17* mutants in *P. pastoris*, in addition to displaying defects in matrix protein import, also showed defects in the localization of PMPs such as Pex3p and Pex2p into the membrane. However, the cells were not totally incompetent in importing Pex3p. Therefore, Snyder *et al* suggest that either Pex17p's role in PMP import was missed in the work of Hushe *et al* (1998) or that Pex17p has an additional role in PMP import that *S. cerevisiae* lacks.

PEX23

PEX23 has only been cloned from the yeast *Y. lipolytica* (Brown *et al*, 2000). As with most of the other *PEX* genes, this again was by complementation analysis. It is an integral membrane protein largely protected from the cytosol. *PEX23* knockout mutants appear to mis-localize matrix proteins to the cytoplasm. However, this mis-localization is not absolute, as a small amount of matrix proteins appear to co-localize with small vesicular structures that contain some peroxisomal matrix and membrane proteins and cannot, therefore, be classed as true 'ghost' peroxisomes, (Brown *et al*, 2000).

Matrix protein import in plants

Two membrane proteins involved in matrix protein import have been cloned from plants, namely Pex10p and Pex14p; see Table 1. However, only Pex14p has been characterized to any extent.

PEX10

In plants, *PEX10* has only been cloned from Arabidopsis. It was first cloned by Schumann *et al* (1999a), using human and yeast *PEX10* sequences to search the Arabidopsis genome database. Since then, a second *PEX10* sequence has been identified by Baker *et al* (2000) differing in two amino acids; however, no further studies have so far been published.

PEX14

Similarly with *PEX10*, *PEX14* has been cloned only from one member of the plant kingdom; unsurprisingly this being Arabidopsis (Hayashi *et al*, 2000; J. Oh, Ph.D. thesis, 2001, University of Leeds). Different cloning approaches were adopted by the two groups. Oh combined database searches with PCR while Hayashi *et al* used a combination of positional cloning and complementation analysis of a peroxisome deficient (*ped*) mutant known as *ped2*. Arabidopsis Pex14p is a peroxisome membrane protein of about 75 kDa, which makes it considerably longer than the mammalian and yeast orthologues. AtPex14p shares all the major motifs with yeast and mammalian Pex14p. These include a region showing strong homology to a region in human Pex14p that forms the Pex5p binding site, two predicted transmembrane domains and a coiled-coil region (Hayashi *et al*, 2000). The carboxy-terminal region of the protein is heavily charged and rich in proline. Two independent studies have shown the involvement of AtPex14p in the import of PTS1 and PTS2 type proteins. In the first, mutant Arabidopsis plants lacking *PEX14* (*ped2*) resulted in both PST1 and PTS2 proteins being localised to the cytosol and peroxisomal matrix (Hayashi *et al*, 2000). Interestingly, this mutant showed a reduced but not a total loss of import of matrix proteins. Peroxisomes from this mutant also showed abnormal morphology that resembled peroxisomes of patients suffering peroxisome biogenesis disorders. Mutation of At*PEX14* (*ped2*) was not lethal, however sucrose was required for post-germinative growth since reduced activity of glyoxysomal β-oxidation prevented sucrose production from the seed's lipid reserves. A normal phenotype was recovered when the *ped2* plants were grown under high CO_2 conditions (Hayashi *et al*, 2000).

In a second study using an *in vitro* import system from sunflower, antibodies raised against HsPex14p inhibited binding of both PTS1 and PTS2 type proteins, but not the integral membrane protein PMP22 to peroxisomes. These results suggest Pex14p to be involved in the import of matrix, but not membrane proteins (Lopez-Huertas *et al*, 1999). Pex14p behaves as an integral membrane protein as it is resistant to carbonate extraction, and protease digestion experiments indicated that the amino terminus is oriented towards the cytosol (J. Oh, Ph.D. thesis, 2001, University of Leeds). Interestingly, research on human Pex14p indicates the amino-terminal portion to be exposed to the peroxisomal lumen (Will *et al*, 1999). Recently, it has been reported by Gouveia and co-workers (2000) that 15% of Pex5p is associated with the peroxisome in rat and that each peroxisome-associated Pex5p protein is bound by five Pex14p molecules. From this, the authors speculate that five Pex14p molecules form a transmembrane pore-like structure that shields Pex5p from the membrane's hydrophobic environment.

Receptor recycling

Four peroxins have been linked with a role in recycling of the peroxisome targeting receptor Pex5p; Pex1p, Pex6p, Pex4p and Pex22p. Pex1p and Pex6p are also involved in early peroxisome biogenesis and have been described above.

PEX4

PEX4 has only been cloned from yeasts, including *S. cerevisiae* (Wiebel and Kanau, 1992), *P. pastoris* (Crane *et al*, 1994) and *H. polymorpha* (van der Klei *et al*, 1998). First cloned from *S. cerevisiae* via complementation analysis (Wiebel and Kanau, 1992), it is a member of the ubiquitin-conjugating protein family and is localized to the outer surface of the peroxisome membrane (Crane *et al*, 1994). In *P. pastoris*, Pex4p has been shown to interact with Pex22p which is believed to anchor Pex4p at the peroxisomal membrane (Koller *et al*, 1999). Pex4p (and Pex22p) functions in the latter stages of matrix protein import, i.e. after matrix protein translocation. In *P. pastoris* and *H. polymorpha pex4* (and *pex22*) mutants, the steady-state abundance of Pex5p is severely reduced, (van der Klei *et al*, 1998; Collins *et al*, 2000). Interestingly, in *H. polymorpha*, this PTS1 import defect could be suppressed by over-expression of Pex5p in a dosage-dependent manner. However, unlike in wild-type cells in which Pex5p is localized throughout the matrix, in the *pex4* mutant, Pex5p accumulated at the inner surface of the peroxisome membrane. From this evidence, the authors propose that Pex4p plays a role the normal functioning of Pex5p, possibly via a role in recycling of the receptor protein (van der Klei *et al*, 1998).

PEX22

PEX22 has been cloned only from the yeast *P. pastoris*; the method once again being complementation analysis (Koller *et al*, 1999). However, through database searches, the authors have also identified a possible *S. cerevisiae PEX22* gene. Pex22p is an integral membrane protein with its C-terminus exposed to the cytoplasm and its N-terminus in the peroxisomal lumen. Pex22p interacts with Pex4p (a protein reported to be involved in receptor recycling), and is believed to be involved in Pex4p localization (Koller *et al*, 1999). Collins and co-workers, (2000) have shown that Pex22p acts after matrix protein translocation.

Proliferation

PEX11

PEX11 has been cloned from yeasts and animals, including *S. cerevisiae* (Erdmann and Blobel, 1995; Marshall *et al*, 1995) and humans (Abe *et al*, 1998; Abe and Fujiki, 1998; Schrader *et al*, 1998). *PEX11* was first cloned from *S. cerevisiae* independently by two groups; Erdmann and Blobel, (1995) and Marshall *et al*, (1995). Interestingly, both groups opted to clone the gene by first purifying the protein from isolated peroxisome membranes. Thus, Pex11p is a peroxisome membrane protein. It is believed to be inserted via two transmembrane domains with its termini exposed to the cytosol (Passreiter *et al*, 1998; Abe *et al*, 1998). In higher eukaryotes, two *PEX11* genes exist; *PEX11*α and *PEX11*β (Schrader *et al*, 1998). Pex11pα contains a C-terminal dilysine motif which has been linked to coatomer binding (membrane coats comprizing a complex of proteins that surround vesicles involved in intracellular trafficking) that mediate membrane budding and vesiculation) (Passreiter *et al*, 1998). It is, therefore, speculated to play a role in peroxisome proliferation through its involvement with coatamer-dependent vesiculation of the endoplasmic reticulum to form immature peroxisome vesicles. This motif is not present in yeasts and it remains unproven as to whether Pex11p binds coatamer *in vivo*.

Despite this, a role for Pex11p in peroxisome proliferation does not appear to be in question. Sakai *et al* (1995) have shown that Pex11p from *S. cerevisiae* complements a mutation in the Pmp30 gene of *Candida boidinii*, a gene shown to be induced by peroxisome proliferators. Marshall and co-workers (1995) have shown that disruption of the *PEX11* gene leads to a phenotype showing a few normal sized peroxisomes plus one or two very large peroxisomes in *S. cerevisiae* cells grown on oleate, (a substrate known to induce peroxisome biogenesis, Erdmann and Blobel, 1995). Over-expression of the *PEX11* gene resulted in an increased number of normal sized peroxisomes. Finally, it has been shown by Schrader *et al* (1998) that human Pex11pβ induces peroxisome proliferation when over-expressed; over-expression of Pex11pα showed a less dramatic increase. Interestingly, mRNA levels of *PEX11β* was shown to be insensitive to peroxisome proliferating agents while *PEX11α* mRNA levels revealed a >10-fold increase. From these data, the authors suggest that *PEX11α* plays a role in peroxisome proliferation in response to extracellular stimuli while *PEX11β* is involved more in the constitutive control of peroxisome abundance.

Other PEX genes

Included in this section are *PEX* genes of unknown function; of which there is currently only one.

PEX15

To date *PEX15* has only been cloned from the yeast *S. cerevisiae*. (Elgersma *et al*, 1997); the method of identification, unsurprizingly, being that of complementation analysis. Pex15p is an integral membrane protein with its C-terminus projecting into the peroxisome matrix and the N-terminus exposed to the cytosol. No proteins have yet been identified that interact with Pex15p.

THE SEARCH FOR FURTHER PLANT PEX GENES

The recent completion of the Arabidopsis genome sequence (The Arabidopsis Genome Initiative, 2000) presents an opportunity for a comprehensive search for *PEX* genes in a higher plant genome. Table 2 shows the results of a search of the Arabidopsis genome database, conducted at the protein level, using sequences from human and yeast *PEX* genes. The chromosomal location of these newly identified genes, and the seven cloned genes, are shown in Figure 2.

Figure 2: Schematic showing approximate relative positions of fifteen Arabidopsis thaliana PEX genes

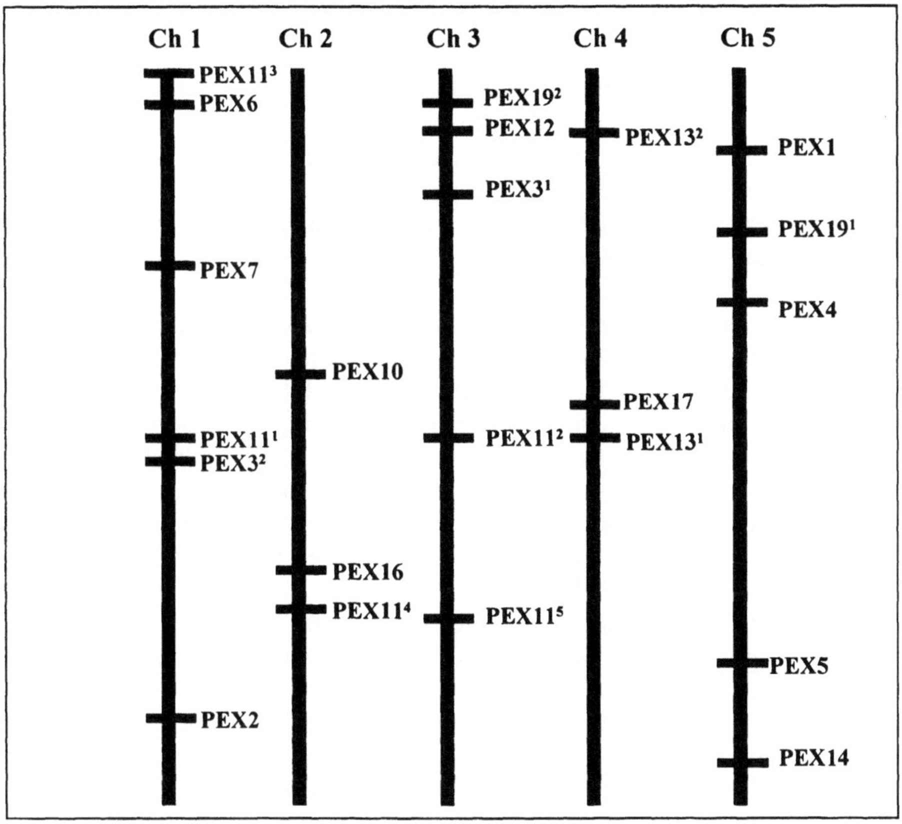

The functions of all these genes in plants remains to be established. However, assuming for the purposes of this chapter that they are *bona fide* orthologues, analysis of the 23 *PEX* genes identified so far highlights a number of interesting points. The first is the greater number of multicopy genes present in the Arabidopsis genome compared to those found in animals and yeasts. To date, only *PEX11* appears to be present as a multicopy gene across the plant, animal and yeast kingdoms (Moreno *et al*, 1994; Schrader *et al*, 1998). However, as can be seen in Figure 2, four Arabidopsis PEX genes display this phenomenon; *PEX3*, *PEX11*, *PEX13* and *PEX19*. Considering that 58% of the Arabidopsis genome is duplicated (The Arabidopsis Genome Initiative, 2000), it is surprising that more *PEX* genes are not present in multiple copies.

PEX11 is an interesting gene family as expression of the two human forms are reported to be different. Pex11pα is responsive to extracellular stimuli while Pex11pβ is constitutively expressed. Of the five Arabidopsis *PEX11* genes, two are similar in protein sequence to Pex11pα and three are similar to Pex11pβ. It is, therefore, plausible to speculate that those showing similarity to Pex11pα initiate peroxisome proliferation in response to extracellular stimuli while the three showing similarity to Pex11pβ play a role in the constitutive abundance of peroxisomes. The fact that more than one gene exists for each Pex11p type could be due to tissue- or developmental-specific expression.

The second point of interest is that eight of the 23 *PEX* genes are absent from the Arabidopsis genome with ten *PEX* genes being absent from humans. Interestingly, all eight absent in Arabidopsis are also absent in humans, suggesting peroxisome biogenesis to be highly conserved between mammals and plants. *Saccharomyces cerevisiae* lacks only four of the twenty three known peroxins. This would tend to suggest that either yeasts display a more elaborate and complex mechanism of peroxisome biogenesis, or that a number of *PEX* genes still remain hidden in the Arabidopsis and human genome databases due to the lack of sensitivity existing in search algorithms in detecting highly divergent sequences. It could also mean that non-homologous proteins have evolved to perform similar tasks.

The concept that plants show greater conservation with humans than with yeasts is intriguing. A similar search of the Arabidopsis database for *PEX* homologues by Mullen *et al*, (2001), also revealed a high degree of similarity in primary amino acid sequence with human counterparts.

Despite this sequence conservation between plant and human *PEX* genes, functionally it could be said that plants display greater similarity to yeasts. For example, *PEX4*, which is believed to be involved in recycling of the PTS1 receptor protein Pex5p, is absent in humans yet present in plants and yeasts. If the absence of *PEX4* from the human genome database is not merely due to being camouflaged by genetic variation and can be extrapolated to include all animals, it would imply that a completely different recycling mechanism for Pex5p occurs in yeasts and plants compared to that in animals.

In addition, a number of peroxins such as Pex2p and Pex16p in *Y. lipolytica* (Titorenko and Rachubinski, 1998) and Pex15p in *S. cerevisiae* (Engelsma *et al*, 1997) have been suggested to be sorted indirectly to peroxisomes via the ER and ER-derived vesicles. In plants, the peroxisome matrix protein ascorbate peroxidase has been shown to be post-translationally sorted to the peroxisome from the cytosol via a distinct subdomain of the ER (Mullen *et al*, 1999; Mullen and Trelease, 2000). In contrast, no evidence exists to suggest an involvement of the ER in peroxisome biogenesis in humans, (South and Gould, 1999; South *et al*, 2000). Much conflicting evidence exists in the literature arguing for and against ER involvement in peroxisome biogenesis in yeasts. As these arguments cannot be done justice in this review, we refer the reader to a recent review by Mullen *et al* (2001) and to Chapter 10.

Until recently, the mechanism of targeting proteins to the peroxisome via the PTS1 and PTS2 import systems appeared to be uniform across all three kingdoms (see above). However, recent work by Motley *et al* (2000) has shown that the nematode *Caenorhabditis elegans* lacks genes encoding proteins of the PTS2-import pathway. The authors of the research propose that initially, *C. elegans* likely possessed both import pathways but abandoned the PTS2 pathway after the PTS2 proteins had acquired a PTS1 signal peptide. Considering the PTS1 signal peptide is a rather degenerate SKL motif, such a proposal is not unreasonable. To further corroborate their theory, the authors reveal that although *Drosophila melanogaster* possesses a putative PTS2 receptor, a number of proteins that are targeted by the PTS2 signal in other organisms, including three thiolases and alkyldihydroxyacetonephosphate synthase, all contain PTS1 peptides instead of PTS2 motifs in *D. melanogaster*.

If *D. melanogaster* is indeed in the process of converting wholly to the PTS1-targeting mechanism, then it begs the question why have more complex organisms such as the mammals and plants stayed with the two-receptor approach? What advantage does the PTS1 method of protein targeting confer on *C. elegans* and *D. melanogaster* that has not been conferred on humans and plants. Obviously, answers cannot be gained from our current knowledge of *PEX* genes.

ESSENTIAL GENES FOR PEROXISOME BIOGENESIS

From the above analysis of genetic similarities and differences of *PEX* genes from the plant, animal and yeast kingdoms, one can speculate as to which *PEX* genes are necessary and sufficient for a fully functional peroxisome. To put it another way, which genes are the essential components of the organelle and which are merely 'optional extras' designed to improve and adapt the organelle to particular conditions. As a starting point we can look to those peroxins that are common to the plant, animal and yeast kingdoms. Obviously such conservation must read as 'essential component' in the peroxisome blueprint. This would amount to twelve of the 23 peroxins: Pex1p, Pex2p, Pex3p, Pex5p, Pex6p, Pex7p Pex10p, Pex11p, Pex12p, Pex13p, Pex14p, and Pex19p if comparing human, *S. cerevisiae* and Arabidopsis *PEX* genes. However, with the recent research on *C. elegans* revealing the absence of Pex7p, this brings the total down to eleven as illustrated in Figure 3.

Figure 3: Essential peroxisomal biogenesis genes

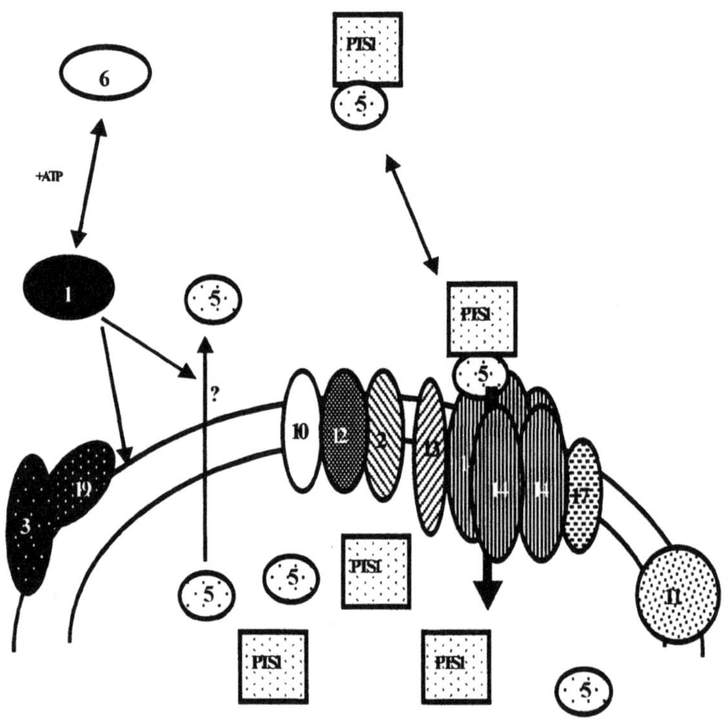

Figure 3. Schematic of essential peroxisome biogenesis genes. Shown is the minimum number of peroxins hypothesised by the authors (see main text) to form a functional peroxisome. Briefly, Pex1p and Pex6p function in early peroxisome biogenesis. Pex3p and Pex19p function to import peroxisome membrane proteins (PMP) into the peroxisome membrane to mature the peroxisome. Pex5p provides a targeting receptor and Pex2p, Pex10p, Pex12p, Pex13p and Pex14p form the docking site and translocation complex for transfer of the receptor protein plus cargo across the peroxisome membrane. Pex1p and Pex6p, in addition to being involved in early peroxisome biogenesis have also been implicated with a role in receptor recycling and thus functions to recycle Pex5p from the peroxisomal lumen to the cytosol. Finally, Pex11p, in its various forms, functions as the peroxisome proliferating protein.

If one was to make an educated guess as to those peroxins necessary for a basic 'no frills' peroxisome (as explained below), it would match very closely those peroxins identified above based on common occurrence across kingdoms, with the addition of a peroxisome membrane protein targeting receptor. As yet, the gene for this peroxin has still to be identified from any species, although the receptor versus chaperone argument is still not resolved for Pex19p.

We can compare the genes we have identified as common to the three kingdoms elements to those we feel would be essential. New peroxisome formation is the first step in peroxisome biogenesis. Genes identified as being involved are *PEX1* and *PEX6*, both of which are present in the three kingdoms. Peroxisome membrane protein import would then be required to mature the peroxisome and develop the matrix protein import complex and to these ends *PEX3* and *PEX19* are also ubiquitous. These two proteins have been linked with a role in binding peroxisome membrane proteins and facilitating their insertion into the peroxisome membrane. At this stage we lack the PMP targeting protein, although it may be that Pex19p proves to perform this function. The next stage is the PMPs themselves that make up the matrix protein import complexes. For this we have *PEX13* and *PEX14* to establish a docking site for the PTS1 protein laden with its cargo. The matrix protein targeting signal receptor is supplied by *PEX5* to bring matrix proteins to the peroxisome membrane surface. Translocation through the peroxisome membrane could be via a Pex14p homopentamer or the Pex2p/Pex10p/ Pex12p complex; both of which have been proposed to form membrane spanning structures. It is also possible that translocation is mediated by both complexes. Whatever the mechanism, all four peroxins have been identified and are present in plants, animals and yeasts.

In addition to being involved in early peroxisome biogenesis, Pex1p and Pex6p have also been demonstrated to be required for Pex5p stability and thus possibly play a role in Pex5p recycling. Thus, the individual components required to form a mature but basic peroxisome are complete. Once mature, proliferation is accomplished by *PEX11*.

Having identified a basic set of peroxisome biogenesis genes, this leaves the remaining twelve *PEX* genes as 'optional extras'. Of these twelve, eight appear to be involved in matrix protein import. Those not involved in protein import are involved in early peroxisome biogenesis, receptor recycling or are of unknown function.

The idea of a set of essential peroxisome biogenesis genes common to the three kingdoms is interesting. It predicts that those *PEX* genes not part of this set will not, by definition, be common to all organisms. Many may be specific to sub-groups. These sub-groups may range from kingdoms to a single species. Even within an individual organism these non-essential *PEX* genes may not be universally recruited; it is feasible that different tissues tailor their requirements by employing different groups of these non-essential *PEX* genes for slightly different functions.

It follows, then, that database searches for *PEX* homologues of a particular species using known *PEX* genes are likely to be less productive as the searches move from the essential set of *PEX* genes to those accessory *PEX* genes peculiar to particular sub-groups. This may prove to be the case for *PEX9* and *PEX20* of *Y. lipolytica* or *PEX18* and *PEX21* of *S. cerevisiae.* Other approaches may need to be employed to identify species specific *PEX* genes performing accessory roles. One such approach is to use a two-hybrid screen and systematically use, as bait, each *PEX* gene in turn, until all possible interactions have been tested.

THE FUTURE

The completion of the Arabidopsis genome sequencing project has given rise to an almost instantaneous increase (over 100%) in the number of *PEX* genes identified in plants. This, in turn, has enabled us to consider variations that exist between species and kingdoms and to build up a theoretical universal set of peroxisome biogenesis genes. In the future, as more genomes are sequenced, more and more pieces of the peroxisome jigsaw will fall into place. One major piece is the identification of the PMP receptor protein.

However, sequencing only helps in the putative identification of genes. As the genome of each organism is sequenced it must be followed up with the appropriate research. This includes techniques such as import assays and crosslinking techniques with either radiolabelled or purified proteins that have been successfully used for the identification of proteins and factors involved in the import of proteins into other organelles, including mitochondia and chloroplasts. Fortunately, this is currently an area of active research in the field of peroxisome biogenesis.

REFERENCES

Abe, I. and Fujiki, Y. (1998) cDNA cloning and characterization of a constitutively expressed isoform of the human peroxin Pex11p. *Biochem. Biophys. Res. Comm.* **252**(2): 529-533.

Abe, I., Okumoto, K., Tamura, S., Fujiki, Y. (1998) Clofibrate-inducible 28-kDa peroxisomal integral membrane protein is encoded by PEX11. *FEBS Lett.* **431**: 468-472.

Albertini, M., Rehling, P., Erdmann, R., Girzalsky, W., Kiel, J.A., Veenhuis, M. and Kunau, W.H. (1997) Pex14p, a peroxisomal membrane protein binding both receptors of the two PTS-dependent import pathways. *Cell.* **89**(1): 83-92.

Baker, A., Charlton, W., Johnson, B., Lopez-Huertas, E., Oh, J., Sparkes, I. and Thomas, J. (2000) Biochemical and molecular approaches to understanding protein import into peroxisomes. *Biochem. Soc. Trans.* **28**(4): 499-504.

Bjorkman, J., Stetten, G., Moore, C.S., Gould, S.J. and Crane, D.I. (1998) Genomic structure of PEX13, a candidate peroxisome biogenesis disorder gene. *Genomics.* **54**(3): 521-528.

Borden, K.L. (1998) RING fingers and B-boxes: zinc-binding protein-protein interaction domains. *Biochem. Cell Biol.* **76**: 351-358.

Braverman, N., Steel, G., Obie, C., Moser, A., Moser, H., Gould, S.J. and Valle, D. (1997) Human PEX7 encodes the peroxisomal PTS2 receptor and is responsible for rhizomelic chondrodysplasia punctata. *Nat. Genet.* **15**(4): 369-376.

Brickner, D.G., Brickner, J.H. and Olsen, J.J. (1998) Sequence analysis of a cDNA encoding Pex5p, a peroxisomal targeting signal type 1 receptor from Arabidopsis. *Plant Physiol.* **118**: 330.

Brocard, C., Lametschwandtner, G., Koudelka, R. and Hartig, A. (1997) Pex14p is a member of the protein linkage map of Pex5p. *EMBO J.* **16**: 5491-5500.

Brown, T.W., Titorenko, V.I. and Rachubinski, R.A. (2000) Mutants of the *Yarrowia lipolytica* PEX23 gene encoding an integral peroxisomal membrane peroxin mis-localize matrix proteins and accumulate vesicles containing peroxisomal matrix and membrane proteins. *Mol. Biol. Cell.* **11**: 141-152.

Chang, C.C., Lee, W.H., Moser, H., Valle, D. and Gould, S.J. (1997) Isolation of the human PEX12 gene, mutated in group 3 of the peroxisome biogenesis disorders. *Nat Genet.* **15**(4): 385-388.

Chang, C.C., Warren, D.S., Sacksteder, K.A. and Gould, S.J. (1999) PEX12 interacts with PEX5 and PEX10 and acts downstream of receptor docking in peroxisomal matrix protein import. *J. Cell Biol.* **147**: 761-774.

Collins, C.S., Kalish, J.E., Morrell, J.C. and Gould, S.J. (2000) The peroxisome biogenesis factors Pex4p, Pex22p, Pex1p, and Pex6p act in the terminal steps of peroxisomal matrix protein import. *Mol. Cell. Biol.* **20**: 7516-7526.

Crane, D.I., Kalish, J.E. and Gould, S.J. (1994) The *Pichia pastoris* PAS4 gene encodes a ubiquitin-conjugating enzyme required for peroxisome assembly. *J. Biol. Chem.* **269**(34): 21835-21844.

Dammai, V. and Subramani, S. (2001) The human peroxisomal targeting signal receptor, Pex5p, is translocated into the peroxisomal matrix and recycled to the cytosol. *Cell.* **105**: 187-196.

del Rio, L.A., Pastori, G.M., Palma, J.M., Sandalio, L.M., Sevilla, F., Corpas, F.J., Jimenez, A., Lopez-Huertas, E. and Hernandez, J. (1998) The activated oxygen role of peroxisomes in senescence. *Plant Physiol.* **116**: 1195-1200.

Dodt, G., Braverman, N., Wong, C., Moser, A., Moser, H.W., Watkins, P., Valle, D. and Gould, S.J. (1995) Mutations in the PTS1 receptor gene, PXR1, define complementation group 2 of the peroxisome biogenesis disorders. *Nat. Genet.* **9**(2): 115-125.

Eitzen, G.A., Aitchison, J.D., Szilard, R.K., Veenhuis, M., Nuttley, W.M. and Rachubinski, R.A. (1995) The *Yarrowia lipolytica* gene PAY2 encodes a 42-kDa peroxisomal integral membrane protein essential for matrix protein import and peroxisome enlargement, but not for peroxisome membrane proliferation. *J. Biol. Chem.* **270**(3): 1429-1436.

Eitzen, G.A., Itiorenko, V.I., Smith, J.J., Veenhuis, M., Szilard, R.K. and Rachubinski, R.A. (1996) The *Yarrowia lipolytica* gene PAY5 encodes a peroxisomal integral membrane protein homologous to the mammalian peroxisome assembly factor PAF-1. *J. Biol. Chem.* **271**(34): 20300-20306.

Eitzen, G.A., Szilard, R.K. and Rachubinski RA. (1997). Enlarged peroxisomes are present in oleic acid-grown *Yarrowia lipolytica* over-expressing the PEX16 gene encoding an intraperoxisomal peripheral membrane peroxin. *J. Cell. Biol.* **137**(6):1265-1278.

Elgersma, Y., Elgersma-Hooisma, M., Wenzel, T., McCaffery, J.M., Farquhar, M.G. and Subramani, S. (1998) A mobile PTS2 receptor for peroxisomal protein import in *Pichia pastoris. J. Cell. Biol.* **140**(4): 807-820.

Elgersma, Y., Kwast, L., Klein, A., Voorn-Brouwer, T., van den Berg, M., Metzig, B., America, T., Tabak, H.F. and Distel, B. (1996) The SH3 domain of the *Saccharomyces cerevisiae* peroxisomal membrane protein Pex13p functions as a docking site for Pex5p, a mobile receptor for the import PTS1-containing proteins. *J. Cell. Biol.* **135**(1): 97-109.

Elgersma, Y., Kwast, L., van den Berg, M., Snyder, W.B., Distel, B., Subramani, S. and Tabak, H.F. (1997) Over-expression of Pex15p, a phosphorylated peroxisomal integral membrane protein required for peroxisome assembly in *S.cerevisiae*, causes proliferation of the endoplasmic reticulum membrane. *EMBO J.* **16**: 7326-7341.

Erdmann, R. and Blobel, G. (1996) Identification of Pex13p a peroxisomal membrane receptor for the PTS1 recognition factor. *J. Cell. Biol.* **135**(1): 111-121.

Erdmann, R. and Blobel, G. (1995) Giant peroxisomes in oleic acid-induced *Saccharomyces cerevisiae* lacking the peroxisomal membrane protein Pmp27p. *J. Cell Biol.* **128**(4): 509-523.

Erdmann, R., Wiebel, F.F., Flessau, A., Rytka, J,, Beyer, A., Frohlich, K.U. and Kunau, W.H. (1991) PAS1, a yeast gene required for peroxisome biogenesis, encodes a member of a novel family of putative ATPases. *Cell.* **64**(3): 499-510.

Faber, K.N., Heyman, J.A. and Subramani, S. (1998) Two AAA family peroxins, PpPex1p and PpPex6p, interact with each other in an ATP-dependent manner and are associated with different subcellular membranous structures distinct from peroxisomes. *Mol. Cell. Biol.* **18**(2): 936-943.

Fransen, M., Brees, C., Baumgart, E., Vanhooren, J.C., Baes, M., Mannaerts, G.P., Van Veldhoven, P.P. (1995) Identification and characterization of the putative human peroxisomal C-terminal targeting signal import receptor. *J. Biol. Chem.* **270**(13): 7731-7736.

Fransen, M., Terlecky, S.R. and Subramani, S. (1998) Identification of a human PTS1 receptor docking protein directly required for peroxisomal protein import. *Proc. Natl. Acad. Sci. USA.* **95**(14): 8087-8092.

Gatto, Jr., G.J., Geisbrecht, B.V., Gould, S.J. and Berg, J.M. (2000) Peroxisomal targeting signal-1 recognition by the TPR donmains of human PEX5. *Nature Struct. Biol.* **12**: 1091-1095.

Geisbrecht, B.V., Collins, C.S., Reuber, B.E., Gould, S.J. (1998) Disruption of a PEX1-PEX6 interaction is the most common cause of the neurologic disorders Zellweger syndrome, neonatal adrenoleukodystrophy, and infantile Refsum disease. *Proc. Natl. Acad. Sci. USA.* **95**: 8630-8635.

Girzalsky, W., Rehling, P., Stein, K., Kipper, J., Blank, L., Kunau, W.H. and Erdmann, R. (1999) Involvement of Pex13p in Pex14p localization and peroxisomal targeting signal 2-dependent protein import into peroxisomes. *J. Cell Biol.* **144**: 1151-1162.

Gotte, K., Girzalsky, W., Linkert, M., Baumgart, E., Kammerer, S., Kunau, W.H. and Erdmann, R. (1998) Pex19p, a farnesylated protein essential for peroxisome biogenesis. *Mol. Cell. Biol.* **18**(1): 616-628.

Gould, S.J., Kalish, J.E., Morrell, J.C., Bjorkman, J., Urquhart, A.J. and Crane, D.I. (1996) Pex13p is an SH3 protein of the peroxisome membrane and a docking factor for the predominantly cytoplasmic PTs1 receptor. *J. Cell Biol.* **135**(1): 85-95.

Gouveia, A.M.M., Reguenga, C., Oliveira, M.E.M., Sa-Miranda,C. and Azevedo, J.E. (2000) Characterisation of peroxisomal Pex5p from rat liver. *J. Biol. Chem.* **275**: 32444-32451.

Guan, L.M., Zhao, J. and Scandalios, J. (2000) *Cis* elements and trans-factors that regulate expression of the maize Cat1 antioxidant gene in response to ABA and osmotic stress:H_2O_2 is the likely intermediary signalling molecule for the response. *Plant J.* **22**: 87-95

Harano T, Shimizu N, Otera H, Fujiki Y. (1999) Transmembrane topology of the peroxin, Pex2p, an essential component for the peroxisome assembly. *J. Biochem.* (Tokyo) **125**: 1168-1174.

Hayashi, M., Nito, K., Toriyama-Kato, K., Kondo, M., Yamaya, T. and Nishimura, M. (2000) AtPex14p maintains peroxisomal functions by determining protein targeting to three kinds of plant peroxisomes. *EMBO J.* **19**(21): 5701-5710.

Hayashi, M., Nito, K., Toriyama-Kato, K., Kondo, M., Yamaya, T. and Nishimura, M. (1998) AtPex14p maintains peroxisomal functions by determining protein targeting to three kinds of plant peroxisomes. *EMBO J.* **19**: 5701-5710.

Hohfeld, J., Veenhuis, M. and Kunau, W.H. (1991) PAS3, a *Saccharomyces cerevisiae* gene encoding a peroxisomal integral membrane protein essential for peroxisome biogenesis. *J. Cell Biol.* **114**(6): 1167-1178.

Honsho, M., Tamura, S., Shimozawa, N., Suzuki, Y., Kondo, N. and Fujiki, Y. (1998) Mutation in PEX16 is causal in the peroxisome-deficient Zellweger syndrome of complementation group D. *Am. J. Hum. Genet.* **63**(6): 1622-1630.

Huang, Y., Ito, R., Miura, S., Hashimoto, T. and Ito, M. (2000) A mis-sense mutation in the RING finger motif of PEX2 protein disturbs the import of peroxisomal targeting signal 1 (PTS1)-containing protein but not the PTS2-containing protein. *Biochem. Biophys. Res. Commun.* **270**: 717-721.

Huhse, B., Rehling, P., Albertini, M., Blank, L., Meller, K., Kunau, W.-H. (1998) Pex17p of *Saccharomyce cerevisiae* is a novel peroxin and component of the peroxisomal protein translocation machinery. *J. Cell Biol.* **140**: 49-60.

James, G.L., Goldstein, J.L., Pathak, R.K., Anderson, R.G. and Brown, M.S. (1994) PxF, a prenylated protein of peroxisomes. *J. Biol. Chem.* **269**: 14182-14190.

Kalish, J.E., Keller, G.A., Morrell, J.C., Mihalik, S.J., Smith, B., Cregg, J.M. and Gould, S.J. (1996) Characterization of a novel component of the peroxisomal protein import apparatus using fluorescent peroxisomal proteins. *EMBO J.* **15**(13): 3275-3285.

Kalish, J.E., Theda, C., Morrell, J.C., Berg, J.M. and Gould, S.J. (1995) Formation of the peroxisome lumen is abolished by loss of *Pichia pastoris* Pas7p, a zinc-binding integral membrane protein of the peroxisome. *Mol. Cell Biol.* **15**(11): 6406-6419.

Kaplan, C.P., Thomas, J.E., Charlton, W.L. and Baker, A. (2001) Identification and characterisation of PEX6 orthologues from plants. *Biochim. Biophys. Acta.* **1539**: 173-180.

Karpinski, S., Reynolds, H., Karpinska, B., Wingsle, G., Creissen, G. and Mullineaux, P. (1999) Systemic signaling and acclimation in response to excess excitation energy in Arabidopsis. *Science.* **284**: 654-657.

Kiel, J.A., Hilbrands, R.E., van der Klei, I.J., Rasmussen, S.W., Salomons, F.A., van der Heide, M., Faber, K.N., Cregg, J.M. and Veenhuis, M. (1999) *Hansenula polymorpha* Pex1p and Pex6p are peroxisome-associated AAA proteins that functionally and physically interact. *Yeast.* **15**(11): 1059-1078.

Koller, A., Snyder, W.B., Faber, K.N., Wenzel, T.J., Rangell, L., Keller, G.A. and Subramani, S. (1999) Pex22p of *Pichia pastoris*, essential for peroxisomal matrix protein import, anchors the ubiquitin-conjugating enzyme, Pex4p, on the peroxisome membrane. *J. Cell Biol.* **146**: 99-112.

Komori, M., Rasmussen, S.W., Kiel, J.A.K.W., Baerends, R.J.S., Cregg, J.M., van der Klei, I.J. and Veenhuis, M. (1997) The *Hansenula polymorpha* PEX14 gene encodes a novel peroxisomal membrane protein involved in matrix protein import. *EMBO J.* **16**: 44-53.

Kragler, F., Lametschwandtner, G., Christmann, J., Hartig, A. and Harada, J.J. (1998) Identification and analysis of the plant peroxisomal targeting signal 1 receptor NtPEX5. *Proc. Natl. Acad. Sci. USA*, **95**: 13336-13341.

Kammerer, S., Holzinger, A., Welsch, U. and Roscher, A.A. (1998) Cloning and characterization of the gene encoding the human peroxisomal assembly protein Pex3p. *FEBS Lett.* **429**(1): 53-60.

Levine, A., Tenhaken, R., Dixon, R. and Lamb, C. (1994) H_2O_2 from the oxidative burst orchestrates the plant hypersensitive disease resistance response. *Cell.* **79**: 589-593.

Lin, Y., Sun, L., Nguyen, L.V., Rachubinski, R.A. and Goodman, H.M. (1999) The Pex16p homolog SSE1 and storage organelle formation in Arabidopsis seeds. *Science.* **284**: 328-330.

Liu, H., Tan, X., Russerll, K.A., Veenhuis, M. and Cregg, J.M. (1995) PER3, a gene required for peroxisomal biogenesis in *Pichia pastoris*, encodes a peroxisomeal membrane protein involved in protein import. *J. Biol. Chem.* **220**(18): 10940-10951.

Liu, Y., Gu, K.L. and Dieckmann, C.L. (1996) Independent regulation of full-length and 5'-truncated PAS5 mRNAs in *Saccharomyces cerevisiae. Yeast.* **12**(2): 135-143.

Lopez-Huertas, E., Charlton, W.L., Johnson, B., Graham, I.A. and Baker, A. (2000) Stress induces peroxisome biogenesis genes. *EMBO J.* **19**(24): 6770-6777.

Lopez-Huertas, E., Oh, J. and Baker, A. (1999) Antiboodies against Pex14p block ATP-independent binding of matrix proteins to peroxisomes *in vitro. FEBS Lett.* **459**: 227-229.

Marshall, P.A., Krimkevich, Y.I., Lark, R.H., Dyer, J.M., Veenhuis, M. and Goodman, J.M. (1995) Pmp27 promotes peroxisomal proliferation. *J. Cell. Biol.* **129**(2): 345-355.

Marzioch, M., Erdmann, R., Veenhuis, M. and Kunau, W.H. (1994) PAS7 encodes a novel yeast member of the WD-40 protein family essential for import of 3-oxoacyl-CoA thiolase, a PTS2-containing protein, into peroxisomes. *EMBO J.* **13**(20): 4908-4918.

Matsuzono, Y., Kinoshita, N., Tamura, S., Shimozawa, N., Hamasaki, M., Ghaedi, K., Wanders, R.J., Suzuki, Y., Kondo, N. and Fujiki, Y. (1999) Human PEX19: cDNA cloning by functional complementation, mutation analysis in a patient with Zellweger syndrome, and potential role in peroxisomal membrane assembly. *Proc. Natl. Acad. Sci. USA.* **96**(5): 2116-2121.

Moreno, M., Lark, R., Campbell, K.L. and Goodman, J.M. (1994) The peroxisomal membrane proteins of *Candida boidinii*: gene isoltion and expression. *Yeast.* **10**(11): 1447-1457.

Motley, A.M., Hettema, E.H., Hogenhout, E.M., Brites, P., ten Asbroek, A.L., Wijburg, F.A., Baas, F., Heijmans, H.S., Tabak, H.F., Wanders, R.J. and Distel, B. (1997) Rhizomelic chondrodysplasia punctata is a peroxisomal protein targeting disease caused by a non-functional PTS2 receptor. *Nat. Genet.* **15**(4): 377-380.

Motley, A.M., Hettema, E.H., Ketting, R., Plasterk, R. and Tabak, H.F. (2000) *Caenorhabditis elegans* has a single pathway to target matrix proteins to peroxisomes. *EMBO Rep.* **1**: 40-46.

Mullen, R.T. and Trelease, R.N. (2000) The sorting signals for peroxisomal membrane-bound ascorbate peroxidase are within its C-terminal tail. *J. Biol. Chem.* **275**: 16337-16344.

Mullen, R.T., Flynn, C.R. and Trelease, R.N. (2001) How are peroxisomes formed? The role of the endoplasmic reticulum and peroxins. *Trends Plant Sci.* **6**: 273-278.

Mullen, R.T., Lisenbeea, C.S., Miernyk, C.A. and Treleasea, R.N. (1999) Peroxisomal membrane ascorbate peroxidase is sorted to a membranous network that resembles a subdomain of the endoplasmic reticulum. *Plant Cell.* **11**: 2167-2185.

Oh, J. (2001). Identification and characterisation of PEX14, a peroxisomal assembly gene from Arabidopsis. Ph.D. thesis, University of Leeds.

Okumoto, K. and Fujiki, Y. (1997) PEX12 encodes an integral membrane protein of peroxisomes. *Nat. Genet.* **17**: 265-266.

Okumoto, K., Itoh, R., Shimozawa, N., Suzuki, Y., Tamura, S., Kondo, N. and Fujiki, Y. (1998) Mutations in PEX10 is the cause of Zellweger peroxisome deficiency syndrome of complementation group B. *Hum. Mol. Genet.* **7**: 1399-1405.

Okumoto, K., Shimozawa, N., Kawai, A., Tamura, S., Tsukamoto, T., Osumi, T., Moser, H., Wanders, R.J., Suzuki, Y., Kondo, N. and Fujiki, Y. (1998a) PEX12, the pathogenic gene of group III Zellweger syndrome: cDNA cloning by functional complementation on a CHO cell mutant, patient analysis, and characterization of Pex12p. *Mol. Cell. Biol.* **18**(7): 4324-4336.

Otera, H., Harano, T., Honsho, M., Ghaedi, K., Mukai, S., Tanaka, A., Kawai, A., Shimizu, N. and Fujiki, Y. (2000) The mammalian peroxin Pex5pL, the longer isoform of the mobile peroxisome targeting signal (PTS) type 1 transporter, translocates the Pex7p.PTS2 protein complex into peroxisomes via its initial docking site, Pex14p. *J. Biol. Chem.* **275**(28): 21703-21714.

Otera, H., Okumoto, K., Tateishi, K., Ikoma, Y., Matsuda, E., *et al.* (1998) Peroxisome targeting signal type-1 (PTS1) receptor is involved in import of both PTS1 and PTS2: studied with PEX5-defective CHO cell mutants. *Mol. Cell Biol.* **18**: 388-399.

Palma, J.M., Garrido, M., Rodriguez-Garcia, M.I. and del Rio, L.A. (1991) Peroxisome proliferation and oxidative stress mediated by activated oxygen species in plant peroxisomes. *Arch. Biochem. Biophys.* **287**: 68-74.

Passreiter, M., Anton, M., Lay, D., Frank, R., Harter, C., Wieland, F.T., Gorgas, K. and Just, W.W. (1998). Peroxisome biogenesis: involvement of ARF and coatomer. *J. Cell Biol.* **141**: 373-83.

Pastori, G.M. and del Rio, L.A. (1994) An activated oxygen-mediated role for peroxisomes in the mechanism of senescence of *Pisum sativum* L. *Planta.* **193**: 385-391.

Purdue, P.E., Yang, X. and Lazarow, P.B. (1998) Pex18p and Pex21p, a novel pair of related peroxins essential for peroxisomal targeting by the PTS2 pathway. *J. Cell Biol.* **143**: 1859-1869.

Purdue, P.E., Zhang, J.W., Skoneczny, M. and Lazarow, P.B. (1997) Rhizomelic chondrodysplasia punctata is caused by deficiency of human PEX7, a homologue of the yeast PTS2 receptor. *Nat. Genet.* **15**(4): 381-384.

Rehling, P., Skaletz-Rorowski, A., Girzalsky, W., Voorn-Brouwer, T., Franse, M.M., Distel, B., Veenhuis, M., Kunau, W.H. and Erdmann, R. (2000) Pex8p, an intraperoxisomal peroxin of *Saccharomyces cerevisiae* required for protein transport into peroxisomes binds the PTS1 receptor Pex5p. *J. Biol. Chem.* **275**(5): 3593-3602.

Reuber, B.E., Germain-Lee, E., Collins, C.S., Morrell, J.C., Ameritunga, R., Moser, H.W., Valle, D. and Gould, S.J. (1997) Mutations in PEX1 are the most common cause of peroxisome biogenesis disorders. *Nat. Genet.* **17**(4): 445-448.

Sacksteder, K.A., Jones, J.M., South, S.T., Li, X., Liu, Y. and Gould, S.J. (2000) PEX19 binds multiple peroxisomal membrane proteins, is predominantly cytoplasmic, and is required for peroxisome membrane synthesis. *J. Cell. Biol.* **148**: 931-944.

Sakai, Y., Marshall, P.A., Saiganji, A., Takabe, K., Saiki, H., Kato, N. and Goodman, J.M. (1995) The *Candida boidinii* peroxisomal membrane protein Pmp30 has a role in peroxisomal proliferation and is functionally homologous to Pmp27 from *Saccharomyces cerevisiae*. *J. Bacteriol.* **177**: 6773-6781.

Schrader, M., Reuber, B.E., Morrell, J.C., Jimenez-Sanchez, G., Obie, C., Stroh, T.A., Valle, D., Schroer, T.A. and Gould, S.J. (1998) Expression of PEX11β mediates peroxisome proliferation in the absence of extracellular stimuli. *J. Biol. Chem.* **273**(45): 29607-29614.

Schumann, U., Gietl, C. and Schmid, M. (1999a) Sequence analysis of a cDNA encoding Pex10p, a zinc-binding peroxisomal integral membrane protein from *Arabidopsis thaliana*. *Plant Physiol.* **119**(3): 1147.

Schumann, U., Gietl, C. and Schmid, M. (1999b) Sequence analysis of a cDNA encoding Pex7p, a peroxisomal targeting signal 2 receptor from Arabidopsis. *Plant Physiol.* **120**: 339.

Shimozawa, N., Tsukamoto, T., Suzuki, Y., Orii, T., Shirayoshi, Y., Mori, T. and Fujiki, Y. (1992) A human gene responsible for Zellweger syndrome that affects peroxisome assembly. *Science*. **255**: 1132-1134.

Smith, J.J., Szilard, R.K., Marelli, M. and Rachubinski, R.A. (1997) The peroxin Pex17p of the yeast *Yarrowia lipolytica* is associated peripherally with the peroxisomal membrane and is required for the import of a subset of matrix proteins. *Mol. Cell. Biol.* **17**(5): 2511-2520.

Snyder, W.B., Koller, A., Choy, A.J., Johnson, M.A., Cregg, J.M., Rangell, L., Keller, G.A. and Subramani, S. (1999a) Pex17p is required for import of both peroxisome membrane and lumenal proteins and interacts with Pex19p and the peroxisome targeting signal-receptor docking complex in *Pichia pastoris*. *Mol. Biol. Cell.* **10**: 4005-4019.

Snyder, W.B., Faber, K.N., Wenzel, T.J., Koller, A., Luers, G.H., Rangell, L., Keller, G.A. and Subramani, S. (1999b) Pex19p interacts with Pex3p and Pex10p and is essential for peroxisome biogenesis in *Pichia pastoris*. *Mol. Biol. Cell.* **10**: 1745-1761.

Snyder, W.B., Koller, A., Choy, A.J. and Subramani, S. (2000) The peroxin Pex19p interacts with multiple, integral membrane proteins at the peroxisomal membrane. *J. Cell Biol.* **149**: 1171-1178.

Soukupova, M., Sprenger, C., Gorgas, K., Kunau, W.H. and Dodt, G. (1999) Identification and characterization of the human peroxin PEX3. *Eur. J. Cell Biol.* **78**(6): 357-374.

South, S.T. and Gould, S.J. (1999) Peroxisome synthesis in the absence of pre-existing peroxisomes. *J. Cell. Biol.* **144**(2): 255-266.

South, S.T., Sackstedera, K.A., Lia, X., Liua, Y. and Gould, S.J. (2000) Inhibitors of COPI and COPII do not block PEX3-mediated peroxisome synthesis. *J. Cell Biol.* **149**: 1345-1359.

Szilard, R.K., Titorenko, V.I., Veenhuis, M. and Rachubinski, R.A. (1995) Pay32p of the yeast *Yarrowia lipolytica* is an intraperoxisomal component of the matrix protein translocation machinery. *J. Cell Biol.* **131**: 1453-1469.

Tamura, S., Okumoto, K., Toyama, R., Schimozawa, N., Tsukamoto, T., Susuki, Y., Osumi, T., Kondo, N. and Fujiki, Y. (1998) Human PEX1 cloned by functional complementation of a CHO cell mutant is responsible for peroxisome-deficient Zellweger syndrome of complementation group 1. *Proc. Natl. Acad. Sci. USA.* **95**: 4350-4355.

Tamura, S., Shimozawa, N., Suzuki, Y., Tsukamoto, T., Osumi, T. and Fujiki, Y. (1998a) A cytoplasmic AAA family peroxin, Pex1p, interacts with Pex6p. *Biochem. Biophys. Res. Commun.* **245**: 883-886.

Tan, X., Waterham, H.R., Veenhuis, M. and Cregg, J.M. (1995) The *Hansenula polymorpha* PER8 gene encodes a novel peroxisomal integral membrane protein involved in proliferation. *J. Cell Biol.* **128**: 307-319.

The Arabidopsis Genome Initiative (2000) Analysis of the genome sequence of the flowering plant *Arabidopsis thaliana*. *Nature.* **408**: 796-815.

Titorenko, V.I. and Rachubinski, R.A. (2000) Peroxisomal membrane fusion requires two AAA family ATPases, Pex1p and Pex6p. *J. Cell Biol.* **150**: 881-886.

Titorenko, V.I. and Rachubinski, R.S. (1998) Mutants of the yeast *Yarrowia lipolytica* defective in protein exit from the endoplasmic reticulum are also defective in peroxisome biogenesis. *Mol. Cell Biol.* **18**: 2789-2803.

Titorenko, V.I., Chan, H. and Rachubinski, R.A. (2000) Fusion of small peroxisomal vesicles in vitro reconstructs an early step in the *in vivo* multistep peroxisome assembly pathway of *Yarrowia lipolytica*. *J. Cell Biol.* **148**: 29-43.

Titorenko, V.I., Smith, J.J., Szilard, R.K. and Rachubinski, R.A. (1998) Pex20p of the yeast *Yarrowia lipolytica* is required for the oligomerization of thiolase in the cytosol and for its targeting to the peroxisome. *J. Cell Biol.* **142**: 403-420.

Tsukamoto, T., Miura, S. and Fujiki, Y. (1991) Restoration by a 35K membrane protein of peroxisome assembly in a peroxisome-deficient mammalian cell mutant. *Nature*. **350**: 77-81.

Urquhart, A.J., Kennedy, D., Gould, S.J. and Crane, D.I. (2000) Interaction of Pex5p, the type 1 peroxisome targeting signal receptor, with the peroxisomal membrane proteins Pex14p and Pex13p. *J. Biol. Chem*. **275**: 4127-4136.

Van den Bosch, H., Schutgens, R.B., Wanders, R.J. and Tager, J.M. (1992) Biochemistry of peroxisomes. *Ann. Rev. Biochem*. **61**: 157-197.

van der Klei, I.J., Hilbrands, R.E., Kiel, J.A.K.W., Rasmussen, S.W., Cregg, J.M. and Veenhuis, M. (1998) The ubiquitin-conjugating enzyme Pex4p of *Hansenula polymorpha* is required for efficient functioning of the PTS1 import machinery. *EMBO J*. **17**: 3608-3618.

Van der Leij, I., Franse, M.M., Elgersma, Y., Distel, B. and Tabak, H.F. (1993) PAS10 is a tetratricopeptide-repeat protein that is essential for the import of most matrix proteins into peroxisomes of *Saccharomyces cerevisiae*. *Proc. Natl. Acad. Sci. USA*. **90**(24): 11782-11786.

Voorn-Brouwer, T., van der Leij, I., Hemrika, W., Distel, B. and Tabak, H.F. (1993) Sequence of the PAS8 gene, the product of which is essential for biogenesis of peroxisomes in *Saccharomyces cerevisiae*. *Biochim. Biophys. Acta*. **1216**(2): 325-328.

Warren, D.S., Morrell, J.C., Moser, H.W., Valle, D. and Gould, S.J. (1998) Identification of PEX10, the gene defective in complementation group 7 of the peroxisome-biogenesis disorders. *Am. J. Hum. Genet*. **63**: 347-359.

Waterham, H.R., Titorenko, V.I., Haima, P., Cregg, J.M., Harder, W. and Veenhuis, M. (1994) The *Hansenula polymorpha* PER1 gene is essential for peroxisome biogenesis and encodes a peroxisomal matrix protein with both carboxy- and amino-terminal targeting signals. *J. Cell Biol*. **127**(3): 737-749.

Wiebel, F.F. and Kunau, W.H. (1992) The Pas2 protein essential for peroxisome biogenesis is related to ubiquitin-conjugating enzymes. *Nature*. **359**(6390): 73-76.

Wiemer, E.A., Nuttley, W.M., Bertolaet, B.L., Li, X., Francke, U., Wheelock, M.J., Anne, U.K., Johnson, K.R. and Subramani, S. (1995) Human peroxisomal targeting signal-1 receptor restores peroxisomal protein import in cells from patients with fatal peroxisomal disorders. *J. Cell Biol*. **130**(1): 51-65.

Will, G.K., Soukupova, M., Hong, X., Erdmann, K.S., Kiel, J.A., Dodt, G., Kunau, W.H. and Erdmann, R. (1999) Identification and characterization of the human orthologue of yeast Pex14p. *Mol. Cell. Biol*. **19**(3): 2265-2277.

Wimmer, C., Schmid, M., Veenhuis, M. and Gietl C. (1998) The plant PTS1 receptor: similarities and differences to its human and yeast counterparts. *Plant J*. **16**: 453-464.

Yahraus, T., Braverman, G., Dodt, G., Kalish, J.E., Morrell, J.C., Moser, H.W., Valle, D. and Gould, S.J. (1996) The peroxisome biogenesis disorder group 4 gene, PXAAA1, encodes a cytoplasmic ATPase required for stability of the PTS1 receptor. *EMBO J*. **15**: 2914-2923.

Zhang, W. and Lazarow, P.B. (1996). Peb1p (Pas7p) is an intraperoxisomal receptor for the NH2-terminal, type 2, peroxisomal targeting sequence of thiolase: Peb1p itself is targeted to peroxisomes by an NH2-terminal peptide. *J. Cell Biol.* **132**: 325-334.

13

PROSPECTS FOR INCREASING STRESS RESISTANCE OF PLANT PEROXISOMES

Ruth Grene Alscher

Virginia Tech, Blacksburg, USA

KEYWORDS

Reactive oxygen species (ROS), hydrogen peroxide, peroxisome, glutathione, protein methionine sulphoxide reductase, heat shock proteins, HSP27, HSP70, chaperones, protein repair; AN*GR1* or AN*GR2*, antisense glutathione reductase 1 or 2; est, expressed sequence tag; GR, glutathione reductase; nt, nucleotide; ROS.

INTRODUCTION

Any circumstance in which cellular redox homeostasis is disrupted can lead to oxidative stress, or the generation of reactive oxygen species or ROS (Asada, 1994). Orchestrated defense/antioxidant processes ensue in response to the imposition of stress (Figure 1). Functional roles of these responses include the protection of redox-sensitive enzymatic processes, the preservation of membrane integrity and the protection of DNA and proteins (Scandalios, 1997). Redox-sensitive regulatory enzymes such as fructose-1, 6-bisphosphatase (FbPase) can be protected from oxidation/inactivation by the action of antioxidants such as glutathione. Plant cells respond defensively to oxidative stress by removing the ROS and maintaining antioxidant defense compounds at levels that reflect ambient environmental conditions (Scandalios, 1997). The mechanisms that act to adjust antioxidant levels to afford protection include changes in antioxidant gene expression.

A. Baker and I.A. Graham (eds.), Plant Peroxisomes, 427–443.

Figure 1: Thiol redox control and stress defense

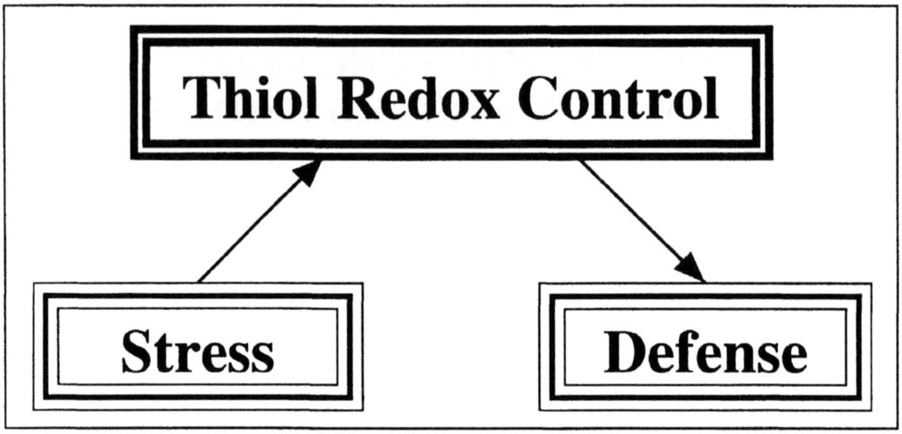

Responses to reactive oxygen species (ROS) in plant cells

Production of ROS during environmental stress is one of the main causes for decreases in productivity, injury, and death that accompany these stresses in plants. Diverse oxidative stresses of environmental origin cause increases in intracellular ROS, posing a toxic chemical threat to the integrity of DNA, proteins, and cell membranes. Organelles such as the peroxisome, where ROS are being produced at a relatively high rate, are especially at risk. Metabolic containment mechanisms for ROS involving antioxidant genes and associated processes are likely to have predated or co-evolved with the appearance of aerobiosis and represent fundamental adaptations of aerobic systems to an oxygen dependent metabolism (Scandalios, 1997). Developmental transitions such as seed maturation, in which peroxisomes play an important role, also involve oxidative stress (Leprince *et al*, 1990; Walters, 1998).

ROS themselves play a role in intracellular redox sensing, activating antioxidant resistance mechanisms, among other adaptive processes (Karpinski *et al*, 1997; May *et al*, 1998a; Toledano and Leonard, 1991).

In some instances, the plant successfully adapts to its changed environment. Our understanding of the cellular and repair mechanisms contributing to successful adaptation is fragmented and scanty at present. Limiting the production of ROS during stress is thought to be achieved by suppression of cellular pathways involved in ROS production, as well as induction of various ROS scavenging and repair mechanisms. A co-ordinated global cellular shift in gene expression is most likely involved in this metabolic and molecular adjustment to unfavorable conditions (Cushman and Bohnert, 2000).

ROS are produced at high levels in peroxisomes. Hydrogen peroxide is produced in the peroxisomal respiratory pathway by flavin oxidase. Fatty acid beta oxidation and glycolate oxidase action are other sources of hydrogen peroxide production in the peroxisome (see discussion by del Rio, Chapter 7) for a comprehensive account of ROS production in peroxisomes).

Metabolic defense mechanisms: limiting ROS-mediated damage

Antioxidant defence molecules have several roles. Ascorbic acid, glutathione and α-tocopherol have each been shown to act as antioxidants in the detoxification of ROS. These compounds have central and interrelated roles, acting both non-enzymatically and as substrates in enzyme-catalysed detoxification reactions (Foyer, 1993; Hausladen and Alscher, 1994; Hess, 1993; Winkler *et al*, 1994). An anti-ROS response includes the induction of genes that belong to ROS scavenging mechanisms, which are localized to several different subcellular compartments. The best known of these are the ascorbate-glutathione cycle, located in the chloroplast, the cytosol, the mitochondrion and the peroxisome, and catalase, localized only in peroxisomes and related microbodies (Willekens *et al*, 1997; Corpas *et al*, 2001; Dat *et al*, 2001). One function of catalase, which exists in higher plants as a multigene family (Willekens *et al*, 1994), is to remove the bulk of the hydrogen peroxide produced in peroxisomes during photorespiration and from other subcellular sources under stress conditions (Willekens *et al*, 1997; Dat *et al*, 2000). Plants which were antisense with respect to catalase were more sensitive to paraquat, hydrogen peroxide and high-light than untransformed controls (Dat *et al*, 2001; Willekens *et al*, 1997). Catalase is inactivated by light and by the superoxide anion (Corpas *et al*, 2001). Ascorbate peroxidase (APX), which functions to remove smaller amounts of hydrogen peroxide in many subcellular locations, is inhibited by high levels of its own substrate; levels which would arise when catalase is not functioning and/or is overwhelmed by ROS. Thus, APX and catalase act in tandem to scavenge hydrogen peroxide throughout the plant cell. A detailed treatment of these scavenging pathways appears in Chapter 7 and will not be discussed further here. Since these scavenging mechanisms are themselves susceptible to attack by ROS, it seems likely that alternative or additional resistance and repair systems exist, especially in environments such as the interior of the peroxisome where ROS levels are high. A discussion of additional resistance and repair processes follows below.

GLUTATHIONE AS ANTIOXIDANT AND REDOX SENSOR

A major role in the adaptation process is played by glutathione and glutathione reductase (GR) (Mullineaux and Creissen, 1997). GR has the important role of maintaining the glutathione pool in its useful reduced form. Integral parts of the response of glutathione metabolism to stress are net synthesis of glutathione and increases in GR activity to maintain the larger glutathione pool in the reduced state. Stress-mediated increases in GR activity may occur as a result of gene activation.

Glutathione acts as a redox sensor of environmental cues and forms part of the multiple regulatory circuitry coordinating defense gene expression. The redox status of intracellular glutathione, which is present at high (mM) levels, is regulated by the action of GR (Noctor and Foyer, 1998). There are two known genes encoding GR (Creissen *et al*, 1995a; Creissen *et al*, 1995b; Madamanchi *et al*, 1992). The redox state of the GSH/GSSG couple may act as a direct link between environmental cues and crucial molecular adaptive responses of plant cells (Hausladen, 1993; Broadbent, 1995; Roxas, 1997; May *et al*, 1998; Noctor *et al*, 1998; May *et al*, 1998a and b; Baginsky, 1999).

Reduced glutathione has a protective role in ROS containment and glutathione reductase (GR) functions to maintain cellular pools of glutathione in the reduced state. Previous biochemical data demonstrated the existence of multiple isoforms of GR, which have been assigned to various organelles, including the peroxisome (Edwards *et al*, 1990). However, only two Arabidopsis GR genes have been identified in the Arabidopsis genome (Donahue *et al*, submitted). A peroxisomal GR gene has not been characterized to date, although a peroxisomal GR protein has been described (Corpas *et al*, 2001).

Glutathione has been reported to regulate rates of cell division (Sanchez-Fernandez *et al*, 1997) and the induction of antioxidant defences, as exemplified by the induction of Cu/Zn superoxide dismutase (Herouart *et al*, 1993). Glutathione has also been suggested as an intermediary in a redox sensing signaling pathway in plants involving the ROS-mediated oxidation of membrane lipids to oxylipins as the initial step (Ball *et al*, 2001).

Metabolic cycles located within the peroxisome successively oxidize and re-reduce glutathione, using NAD(P)H as the ultimate electron donor. Upon the imposition of oxidative stress, the existing cellular pool of reduced glutathione (GSH) is converted to oxidized glutathione (GSSG); GR acts to reduce the newly formed GSSG, and glutathione biosynthesis is stimulated (Madamanchi *et al*, 1994; May and Leaver, 1993). The rate-limiting step for glutathione synthesis is thought to be gamma-glutamylcysteine synthetase, which is feedback regulated by GSH and is controlled primarily by the level of available L-cysteine (Griffith, 1999; Noctor and Foyer, 1998). Glutathione biosynthesis is thought to occur in the chloroplast and in the cytosol (Noctor and Foyer, 1998). The possibility of a glutathione biosynthetic pathway in the peroxisome may not have been explored. Thus, for effective protection of peroxisomal processes by reduced glutathione transport across the bounding peroxisomal membrane may be necessary.

GR levels affect glutathione metabolism and, by extension, ROS protection mechanisms. GR activities increase as the glutathione pool increases through a multi-level control mechanism, which includes co-ordinate activation of genes encoding glutathione biosynthetic enzymes and GR (Xiang and Oliver, 1998). Control by ROS of glutathione biosynthesis is exerted at the translational level (Xiang and Oliver, 1998).

Increasing glutathione biosynthetic capacity has been shown to enhance resistance to oxidative stress (Arisi *et al*, 1998; Zhu *et al*, 1999). The ability of the plant cell to maintain and increase reduced glutathione levels is an important factor in protecting photosynthesis against sulphur dioxide (Alscher *et al*, 1987), which depends at least in part on GR. Transgenic tobacco overexpressing (3x) an *E. coli* gene (*gor*) encoding GR targeted to the plastid was more resistant to paraquat and sulphur dioxide than were non-transformed plants (Aono *et al*, 1993). Over-expression of *gor* in the chloroplasts of poplar led to increases in both total glutathione pool sizes and in the ratio of GSH to GSSG. Transformed plants showed enhanced protection against chloroplast-localized oxidative stress (Foyer *et al*, 1995). Transgenic tobacco plants with increased levels of GSSG were found to grow better under salinity and chilling stress than their non-transformed counterparts, suggesting that resistance/adaptive pathways were stimulated by GSSG (Roxas *et al*, 1997). We have characterized transgenic antisense *Arabidopsis thaliana* (L) Heynh. plants that are depleted in the expression of one or other of the two known Arabidopsis glutathione reductase genes (*GR1*, organellar and *GR2*, cytosolic), (Donahue *et al*, submitted). Our working hypothesis is that activity levels of GR play a pivotal role in redox sensing and adjustment processes as well as a direct role in the maintenance of reduced glutathione. Abnormal seed morphologies and/or embryo abortions first appeared in antisense *GR2* plants (AN*GR2)* eight days after flowering in at least four transformant lines. An altered phenotype expressed as a change in seedling growth habit was observed in antisense *GR2*, but not in antisense *GR1* lines. AN*GR2* lines showed an increased time to flowering. GR genes may play important individual roles in seed development, and subsequent seedling development related to the levels of glutathione and effects of ROS. Since the peroxisome contains GR, its ROS sensing mechanism(s) of peroxisomes are likely to respond to alterations in glutathione metabolism.

MOLECULAR CHAPERONES

Four distinct functions have been assigned to molecular chaperones. They can act as repair proteins, they can remove proteins that are irretrievably damaged, and they can facilitate the import of newly synthesized proteins into the interior of organelles such as the peroxisome. The fourth function is as antioxidant molecules themselves in conjunction with protein methionine-sulphoxide reductase.

Molecular chaperones interact to protect against heat and water stress through the repair of denatured proteins

Evidence is accumulating to suggest an important role for heat shock proteins/molecular chaperones in stress resistance in plant and animal systems (Gustavsson *et al*, 1999; Harndahl *et al*, 2001; Wehmeyer and Vierling, 2000). Increased expression of HSPs of the 70, 101 and sHSP classes were observed in drought acclimated rooted cuttings of loblolly pine (Heath *et al*, 2001, unpublished data). HSP70 is known to occur in glyoxysomes; in fact, the glyoxysomal protein is encoded by the same gene as the chloroplast form of the protein (Wimmer *et al*, 1997). The small heat shock proteins that are localized to the cytosol appear to respond to specific developmental signals associated with the acquisition of desiccation tolerance that occur during seed development (Wehmeyer and Vierling, 2000). A small HSP (HSP18.1) has been shown to interact with HSP70 to reactivate heat-denatured luciferase (Lee and Vierling, 2000). HSP90 was ineffective in their reactivation system, and was also found to be unresponsive to drought stress in loblolly pine (Heath *et al*, 2001, unpublished data). The cytosolic HSPs prevent heat-mediated and water-stress-mediated aggregation of proteins. These HSPs may prevent loss of conformation in low-water conditions. Denatured substrate proteins are bound to the sHSP oligomers *in vitro* presumably by hydrophobic regions (Harndahl *et al*, 2001; Lee and Vierling, 1998). It is thought that the sHSPs act to bind denatured proteins and to maintain them in a state that allows for ATP-dependent refolding by larger HSPs/molecular chaperones (Lee and Vierling, 2000). No small heat shock proteins have been specifically associated with peroxisomes or glyoxysomes to date. However, since the HSP70 present in those organelles is so similar to the corresponding chloroplast HSP70, it is possible that just such an interaction between sHSPs and the larger molecules may in fact exist. A protein of the DnaK (J) class was found to be essential for the HSP70-mediated refolding/repair mechanism (Lee and Vierling, 2000). A DnaJ protein has been found in association with glyoxysomes (Diefenbach and Kindl, 2000), albeit at the membrane surface. This protein reacts specifically with a particular cytosolic isoform of HSP70, and not with other forms of HSP70.

Molecular chaperones and methionine sulphoxidation

Surface methionine residues are preferentially oxidized in proteins. Methionine residues act as an antioxidant protein reservoir (Figure 2). Amino acid residues in proteins are one of the major targets of ROS attack. The side chains of methionine and cysteine are more sensitive to oxidation than the side chains of other amino acids. This differential sensitivity has been exploited by nature to create a protective mechanism against ROS-mediated protein damage. Cysteine residues are maintained in a reduced state through the action of glutathione. Surface-exposed methionine residues are available for oxidation by molecules such as hydrogen peroxide, thus effectively lowering the degree of the threat. In the case of glutamine synthetase from *E. coli*, it was found that eight of the sixteen methionine residues present in the protein could be oxidized with little effect in overall enzyme activity (Levine *et al*, 1999). They are re-reduced through the action of protein methionine sulphoxide reductase (PMSR), using thioredoxin as a source of reductant (Hoshi and Heinemann, 2001; Lowther *et al*, 2000). PMSR is a highly conserved protein from *E.coli* to human (Brot and Weissbach, 2000). Combined with the reductive action of PMSR, which restores the methionines to their original state, a larger mechanism within the cell is now able to maintain a stable configuration of protein state. It is thought that the degree to which methionine oxidation occurs may be underestimated due to imprecision of assay methods (Squier and Bigelow, 2000). A proposed overall scheme for redox regulation of cellular defense processes involving thiol redox control is represented in Figure 2. sHSPs contain a unique methionine rich domain at its N terminus, which exists as an amphipathic alpha helix. Some of these methionine residues exist at the surface of the protein, and are oxidized very readily, with resultant loss of chaperone activity (Harndahl *et al*, 2001). The chloroplast HSP21 is easily and reversibly oxidized by the same concentration of hydrogen peroxide as that which brings about HSP21 oxidation *in vivo* in Arabidopsis leaves (Harndahl *et al*, 1999) and is a candidate for the PMSR-mediated series of reactions shown in Figure 2. The oxidation of surface methionine residues, which is mediated by the hydroxyl radical, ozone and peroxynitrile, as well as hydrogen peroxide, has been proposed to function in an antioxidant capacity in animal and yeast systems (Moskovitz *et al*, 1997; Preville *et al*, 1999).

Figure 2: Oxidation and reduction of methionine residues as an antioxidant reservoir

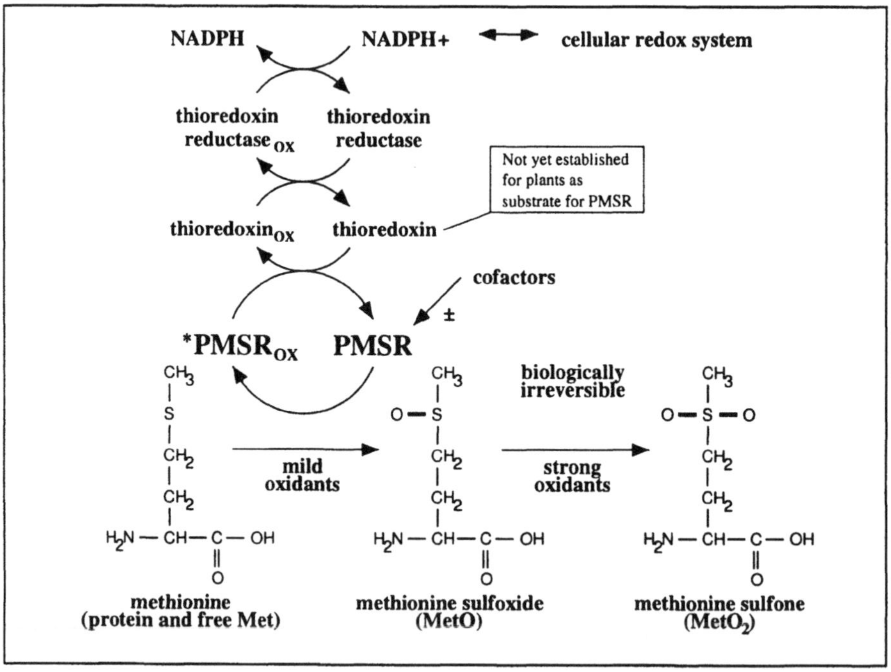

Figure 2. (Modified from Hoshi and Heinemann, 2001). Cytosolic and chloroplast forms of PMSR are known in plants. The thioredoxin activation system in the chloroplast is well known as a source of reductant for redox regulated enzymes. Candidate substrates for PMSR action include the heat shock proteins.

Molecular chaperones can interact with glutathione to protect against oxidative stress

Mehlen *et al*, 1996) and (Garrido *et al*, 1998) present evidence of an interaction between sHSPs and glutathione in mammalian cells that resulted in increased resistance to cell death induced by tumour necrosis factor or by hydrogen peroxide. Resistance was dependent both on increases in reduced glutathione and on increases in expression of sHSPs, and was shown to decrease the levels of cellular ROS. (Preville *et al*, 1999) demonstrated that sHSPS can protect against oxidative stress in L929 cells. This interaction has not yet been investigated in plant systems. However, much evidence points to the importance of glutathione biosynthesis in protection against ROS damage (Alscher, 1989; Noctor and Foyer, 1998; May *et al*, 1998a).

THE CENTRAL ROLE OF HYDROGEN PEROXIDE IN SIGNALLING OXIDATIVE STRESS RESISTANCE RESPONSES

Hydrogen peroxide is thought to act a signalling molecule in the induction of stress defense responses, within certain limits (Dat *et al*, 2001). Beyond that threshold level, hydrogen peroxide exerts its oxidative action, and cellular injury occurs. Its levels are modulated through the joint action of catalase and ascorbate peroxidase, as is illustrated by the finding that antisense catalase tobacco plants were more susceptible to oxidative stress than control plants (see above), as were antisense APX Arabidopsis plants (Wang *et al*, 1999).

Exposure to hydrogen peroxide has been reported to result in the induction of at least 100 genes in Arabidopsis (Neill *et al*, 2001). Hydrogen peroxide accumulation in barley that had been inoculated with powdery mildew was shown to induce glutathione biosynthesis during the hypersensitive response (Vanacker *et al*, 2000). Molecular chaperones, glutathione-S-transferases, various protein kinases (see below), and redox-sensitive transcription factors (Pastori and Foyer, 2001) are known ROS-responsive genes. Some of these genes, e.g. the molecular chaperones, are also induced by long term exposure to drought (Heath *et al*, 2001). Hydrogen peroxide is thought to exert its inductive effect through oxidation of cysteine and methionine residues, oxidation of particular membrane lipids which act as oxylipin receptors in a signaling cascade that involves glutathione (Ball *et al*, 2001; Hamberg, 1999) and direct influences on protein kinase cascades (Kovtun *et al*, 2000).

Kovtun *et al*, 2000 demonstrated the existence of a hydrogen peroxide mediated protein kinase cascade, which signals the activation of defense genes such as HSP18 (small heat shock protein) and glutathione S-transferase.

Hydrogen peroxide mediated peroxisomal biogenesis

Hydrogen peroxide is thought to be an important ROS that is involved in signalling stress mechanisms in plants. Lopez-Huertas *et al* (2000) reported a stimulatory effect of hydrogen peroxide on peroxisomal biogenesis. Hydrogen peroxide is both a means of induction of stress resistance, and a product of stress imposition in the plant cell. Consequently, it has been widely used as a tool to investigate the mounting of defense mechanisms and the response of defense pathways. In the case of peroxisomes, their response to the presence of hydrogen peroxide consists in an increase in peroxisomal biogenesis. A major component of peroxisomal biogenesis is the import of cytosolic-synthesized proteins into the interior of the organelle (Corpas *et al*, 2001). The HSP 70 interacts with a DnaJ-like protein at the peroxisomal surface and may be involved in protein import into the organelle.

The imposition of oxidative stress gives rise to organelle proliferation, thus adding another layer of complexity to stress responses (Lopez-Huertas *et al*, 2000). Since hydrogen peroxide was used as the only representative ROS in the peroxisomal proliferation experiments, there is no information as yet on effects of other ROS on peroxisomal biogenesis. Charged species such as the superoxide anion cannot cross membranes, and thus are not so likely as candidates for mediating signals across bounding membranes. The inhibition of catalase by the superoxide anion would result in increased levels of hydrogen peroxide, however, which could constitute the initiation of a peroxisome-specific signalling cascade. Peroxisomal proliferation also occurs during senescence (Corpas *et al*, 2001) and may result from the well-documented increases in ROS levels that occur with ageing.

PROSPECTS FOR IMPROVING STRESS RESISTANCE OF PEROXISOMAL PROCESSES

Engineering superior co-operation between catalase and APX in maintaining peroxisomal hydrogen peroxide levels within the threshold for a signal function is one possibility for improving stress resistance of peroxisomes. To become a practical goal, however, a great deal more would have to understood about the transcriptional and post-transcriptional regulation of individual members of both gene families.

Glutathione remains a prime candidate for engineering increased stress resistance in plant cells, especially in organelles such as the peroxisome where ROS are produced at high levels. Since the ascorbate/glutathione scavenging pathway is present in peroxisomes, it is likely that glutathione levels are high, and that a mechanism for providing GSH to the organelle exists. Transport of glutathione across the peroxisomal bounding membrane is one possible focus for engineering increased stress resistance. Engineering increased flux through the cytosolic glutathione biosynthetic pathway is another possibility; a strategy that has proved to provide increased protection to photosynthesis in poplar (Foyer *et al*, 1995). HSP70s are known to function in transport of proteins into the interior of the peroxisome and a member of the HSP70 gene family has been identified in the interior of the organelle. It is not yet known if the peroxisome contains sHSPs as well as HSP70s. Exposure to hydrogen peroxide results in the induction of peroxisomal biogenesis. Kovtun *et al*, 2000 demonstrated the existence of a hydrogen peroxide mediated protein kinase cascade, which signals the activation of defense genes such as HSP18 (small heat shock protein) and glutathione S-transferase. The ROS-sensing mechanism that elicits peroxisomal biogenesis is unknown, but it could also involve sHSPs and glutathione in a manner analogous to the mammalian system.

Although a cytosolic and a plastidic form of PMSR have been described in plants (Sadanandom *et al*, 2000), little information concerning their respective roles in antioxidant defense is yet available. A search for the *in vivo* substrates of PMSR in the peroxisome and the cytosol could yield valuable information. The PMSR mechanism may constitute an important additional antioxidant mechanism. Peroxisomal HSPs containing the methionine rich region present in HSP21 found in plastids are good candidates for PMSR substrates. This mechanism may also form part of the ROS sensing signalling process that gives rise to peroxisomal biogenesis. Taken together, the overall goal would be to improve the interactions shown in Figure 3, so as to increase the speed and efficiency of the signalling pathways that give rise to the mobilization of antioxidant defense mechanisms. Thiol redox control is proposed to play an essential and central role in mediating plant cell antioxidant responses to the imposition of oxidative stress.

Figure 3: Redox regulation of gene expression. (1), (2) and (3) can function as redox sensors

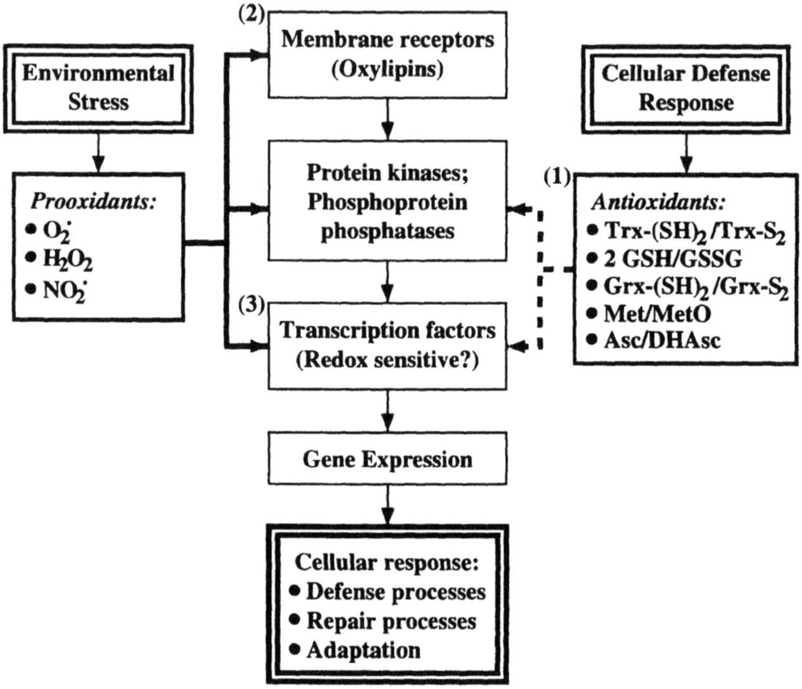

Figure 3 is modified from Figure 2 Arner and Holmgren (2000) Eur. J. Biochem. ***267****: 6102-6109.©Springer-Verlag. Suggested levels of regulation of cellular responses are indicated. The balance between levels of pro- and antioxidant levels at any time determines the degree of activation of cellular responses.*

$Trx(SH)_2/Trx\text{-}S_2$ = thioredoxin redox couple
$Grx\text{-}(SH)/Grx\text{-}S_2$ = glutaredoxin couple
Met/MetO = protein methionine residues
Asc = ascorbic/dehydroascorbic acid

REFERENCES

Alscher, R.G. (1989) Biosynthesis and antioxidant function of glutathione in plants. *Physiol.Plant.* **77**: 457-464.

Alscher, R., Bower, J. and Zipfel, W. (1987) The basis for different sensitivities of photosynthesis to SO2 in two cultivars of pea. *J.Exp.Bot.* **38**: 99-108.

Aono, M., Kubo, A., Saji, H., Tanaka, K. and Kondo, N. (1993). Enhanced tolerance to photooxidative stress of transgenic *Nicotiana tabacum* with high chloroplastic glutathione reductase activity. *Plant Cell Physiol.* **34,** 129-135.

Arisi, A.M., Cornic, G., Jouanin, L. and Foyer, C.H. (1998) Overexpression of iron superoxide dismutase in transformed poplar modifies the regulation of photosynthesis at low CO_2 partial pressures or following exposure to the prooxidant herbicide methyl viologen. *Plant Physiol.* **117**: 565-74.

Arner, E.S.J. and Holmgren (2000). Physiological functions of thioredoxin and thioredoxin reductase. *Eur J. Biochem.* **267**: 6102-6109.

Asada, K. (1994) Production and action of active oxygen species in photosynthetic tissue. In *Causes of Photooxidative Stress and Amelioration of Defense Systems in Plants* (Foyer, C.H. and Mullineaux, P.M., eds.). pp. 77-104. Boca Raton: CRC Press.

Baginsky, S., Tiller, K., Pfannschmidt, T. and Link, G. (1999) PTK, the chloroplast RNA polymerase-associated protein kinase from mustard (*Sinapis alba*), mediates redox control of plastid *in vitro* transcription. *Plant Mol. Biol.* **5**: 1013-23.

Ball, L., Richard, O., Bechtold, U., Penkett, C., Reynolds, H., Kular, B., Creissen, G., Karpinski, S., Schuch, W. and Mullineaux, P. (2001) Changes in global gene expression in response to excess excitation energy in *Arabidopsis thaliana*. *J. Exp. Bot.* **52**:

Broadbent, P., Creissen, G.P., Kular, B., Wellburn, A.R. and Mullineaux, P.M. (1995) Oxidative stress responses in transgenic tobacco containing altered levels of glutathione reductase activity. *Plant J.* **8**: 0.

Brot, N. and Weissbach, H. (2000) Peptide methionine sulphoxide reductase: biochemistry and physiological role. *Biopolymers.* **55**: 288-296.

Corpas, F.J., Barroso, J.B. and del Rio, L.A. (2001) Peroxisomes as a source of reactive oxygen species and nitric oxide signal molecules in plant cells. *Trends Plant Sci.* **6**: 145-150.

Creissen, G., Broadbent, P., Stevens, R., Wellburn, A.R. and Mullineaux, P. (1995a) Manipulation of glutathione metabolism in transgenic plants. *Biochem. Soc. Trans.* **24**: 465-469.

Creissen, G., Reynolds, H., Xue, Y. and Mullineaux, P. (1995b) Simultaneous targeting of pea glutathione reductase and of a bacterial fusion protein to chloroplasts and mitochondria in transgenic tobacco. *Plant J.* **8**: 167-175.

Cushman, J. C. and Bohnert, H. J. (2000) Genomic approaches to plant stress tolerance. *Curr. Opin. Plant Biol.* **3**: 117-124.

Dat, J.F., Inze, D. and Van Breusegem, F. (2001) Redox Report. **6:** 37-42.

Dat, J.F., Vandenabeele, E., Vranova, M., Van Montagu, M., Inze, D. and Van Breusegem, F. (2000) Dual action of the active oxygen species during plant stress responses. *Cell. Mol. Life Sci.* **57**: 779-795.

Diefenbach, J. and Kindl, H. (2000). The membrane-bound DnaJ protein located at the cytosolic site of glyoxysomes specifically binds the cytosolic isoform 1 of Hsp70 but not other Hsp70 species. *Eur. J. Biochem.* **267**: 746-754.

Edwards, E.A., Rawsthorne, S. and Mullineaux, P.M. (1990) Subcellular distribution of multiple forms of glutathione reductase in leaves of pea (*Pisum sativum* L.). *Planta.* **180**: 278-284.

Foyer, C.H. (1993) Ascorbic acid. In *Antioxidants in Higher Plants* (R.G. Alscher and J.L. Hess, eds), pp. 31-58. CRC Press, Boca Raton, Fl. *ISBN* **0,** 31-58.

Foyer, C. H., Sourian, N., Perret, S., Lelandais, M., Kunert, K.-J., Pruvost, C. and Jouanin, L. (1995) Overexpression of glutathione reductase, but not glutathione synthetase leads to increases in antioxidant capacity and resistance to photoinhibition in poplar trees. *Plant Physiol.* **109**: 1047-1057.

Garrido, C., Fromentin, A., Bonnotte, B., Favre, N., Moutet, M., Arrigo, A. P., Mehlen, P. and Solary, E. (1998). Heat shock protein 27 enhances the tumourigenicity of immunogenic rat colon carcinoma cell clones. *Cancer Res.* **58**: 5495-5499.

Griffith, O. (1999) Biological and pharmacological regulation of mammalian glutathione synthesis. *Free Rad. Biol. Med.* **27**: 922-935.

Gustavsson, N., Harndahl, U., Emanuelsson, A., Roepstorff, P. and Sundby, C. (1999) Methionine sulphoxidation of the chloroplast small heat shock protein and conformational changes in the oligomer. *Protein Sci.* **8**: 2506-2512.

Hamberg, M. (1999) An epoxy alcohol synthase pathway in higher plants: biosynthesis of antifungal trihydroxy oxylipins in leaves of potato. *Lipids.* **34**: 1131-1142.

Harndahl, U., Hall, R. B., Osteryoung, K. W., Vierling, E., Bornman, J. F. and Sundby, C. (1999) The chloroplast small heat shock protein undergoes oxidation-dependent conformational changes and may protect plants from oxidative stress. *Cell Stress Chap.* **4**: 129-138.

Harndahl, U., Kokke, B. P., Gustavsson, N., Linse, S., Berggren, K., Tjerneld, F., Boelens, W. C. and Sundby, C. (2001) The chaperone-like activity of a small heat shock protein is lost after sulphoxidation of conserved methionines in a surface-exposed amphipathic alpha-helix. *Biochim. Biophys. Acta.* **1545**: 227-237.

Hausladen, A. and Alscher, R. (1994) Purification and characterization of glutathione reductase isozymes specific for the state of cold hardiness of red spruce. *Plant Physiol.* **105**: 205-213.

Hausladen, A. and Alscher, R.G. (1993) Glutathione. In *Antioxidants in higher plants.* (Alscher, R.G. and Hess, J.L., eds.). pp. 1-30: CRC Press, Inc.

Herouart, D., Van Montagu, M. and Inze, D. (1993) Redox-activated expression of the cytosolic copper/zinc superoxide dismutase gene in *Nicotiana*. *Proc. Natl. Acad. Sci. USA*.. **90**: 3108-3112.

Hess, J.L. (1993) Vitamin E, alpha-tocopherol. In *Antioxidants in higher plants* (Alscher, R.G. and Hess, J.L., eds.). pp. 111-134. Boca Raton, FL: CRC.

Hoshi, T. and Heinemann, S. (2001) Regulation of cell function by methionine oxidation and reduction. *J. Physiol.* **531**: 1-11.

Karpinski, S., Escobar, C., Karpinska, B., Creissen, G. and Mullineaux, P.M. (1997) Photosynthetic electron transport regulates the expression of cytosolic ascorbate peroxidase genes in *Arabidopsis* during excess light stress. *Plant Cell.* **9**: 627-640.

Kovtun, Y., Chiu, W.L., Tena, G. and Sheen, J. (2000) Functional analysis of oxidative stress-activated mitogen-activated protein kinase cascade in plants. *Proc. Natl. Acad. Sci. USA.* **97**: 2940-2945.

Lee, G.J. and Vierling, E. (2000) A small heat shock protein cooperates with heat shock protein 70 systems to reactivate a heat-denatured protein. *Plant Physiol.* **122**: 189-198.

Lee, G.J. and Vierling, E. (1998) Expression, purification, and molecular chaperone activity of plant recombinant small heat shock proteins. *Methods Enzymol.* **290**: 350-365.

Leprince, O., Thorpe, P.C., Deltour, R., Atherton, N.M. and Hendry, G.A.F. (1990) The role of free radicals and radical processing systems in loss of desiccation tolerance in germinating maize. *New Phytol.* **116**: 573-580.

Levine, R.L., Berlett, B.S., Moskovitz, J., Mosoni, L. and Stadtman, E.R. (1999) Methionine residues may protect proteins from critical oxidative damage. *Mech. Ageing Dev.* **107**: 323-332.

López-Huertas, E., Charlton, W.L., Johnson, B., Graham, I.A. and Baker, A. (2000) Stress induces peroxisome biogenesis genes. *EMBO J.* **19**: 6770-6777.

Lowther, W.T., Brot, N., Weissbach, H. and Matthews, B.W. (2000) Structure and mechanism of peptide methionine sulphoxide reductase, an "anti-oxidation" enzyme. *Biochem.* **39**: 13307-13312.

Madamanchi, N., Anderson, J., Alscher, R., Cramer, C. and Hess, J. (1992) Purification of multiple forms of glutathione reductase from pea (*Pisum sativum* L.) seedlings and enzyme levels in ozone fumigated pea leaves. *Plant Physiol.* **100**: 138-145.

Madamanchi, N.R., Donahue, J.V., Cramer, C.L., Alscher, R.G. and Pedersen, K. (1994) Differential response of Cu, Zn superoxide dismutase in two pea cultivars during a short term exposure to sulphur dioxide. *Plant Mol. Biol.* **26**: 95-103.

May, M.J. and Leaver, C.J. (1993) Oxidative stimulation of glutathione synthesis in *Arabidopsis thaliana* suspension cultures. *Plant Physiol.* **103**: 621-627.

May, M.J., Vernoux, T., Leaver, C., Van Montagu, M. and Inze, D. (1998a) Glutathione homeostasis in plants: implications for environmental sensing and plant development. *J. Exp. Bot.* **49**: 649-667.

May, M.J., Vernoux, T., Sanchez-Fernandez, R., Van Montagu, M. and Inze, D. (1998b) Evidence for post-transcriptional activation of gamma-glutamylcysteine synthetase during plant stress responses. *Proc. Natl. Acad. Sci. USA.* **95**: 12049-12054.

Mehlen, P., Kretz-Remy, C., Preville, X. and Arrigo, A.P. (1996) Human HSP27, Drosophila HSP27 and human αB-crystallin expression-mediated increase in glutathione is essential for the protective activity of these proteins against TNFα-induced cell death. *EMBO J.* **15**: 2695-2706.

Moskovitz, J., Berlett, B.S., Poston, J.M. and Stadtman, E.R. (1997) The yeast peptide-methionine sulfoxide reductase functions as an antioxidant *in vivo*. *Proc. Natl. Acad. Sci. USA.* **94**: 9585-9589.

Mullineaux, P.M. and Creissen, G.P. (1997) Glutathione reductase: regulation and role in oxidative stress. In *Oxidative Stress and the Molecular Biology of Antioxidant Defenses.* (Scandalios. J.G., ed.). pp. 667-713. Plainview: Cold Spring Harbor.

Neill, S.J., Desikan, R., Clarke, A., Hurst, R. and Hancock, J.T. (2001) Hydrogen peroxide and nitric oxide as signalling molecules in plants. *J. Exp. Bot.* **52**: in press.

Noctor, G. and Foyer, C.H. (1998) Ascorbate and glutathione: keeping active oxygen under control. *Ann. Rev. Plant Physiol. Plant Mol. Biol.* **49**: 249-279.

Pastori, G.M. and Foyer, C.H. (2001) Identifying oxidative stress responsive genes by transposon tagging. *J. Exp. Bot.* **52**, in press.

Preville, X., Salvemini, F., Giraud, S., Chaufour, S., Paul, C., Stepien, G., Ursini, M.V. and Arrigo, A.P. (1999) Mammalian small stress proteins protect against oxidative stress through their ability to increase glucose-6-phosphate dehydrogenase activity and by maintaining optimal cellular detoxifying machinery. *Exp. Cell Res.* **247**: 61-78.

Roxas, V.P., Smith, J., Roger K., Allen, E.R. and Allen, R.D. (1997) Overexpression of glutathione S-transferase/glutathione peroxidase enhances the growth of transgenic tobacco seedlings during stress. *Nature Biotech.* **15**: 988-991.

Sadanandom, A., Poghosyan, Z., Fairbairn, D.J. and Murphy, D.J. (2000) Differential regulation of plastidial and cytosolic isoforms of peptide methionine sulfoxide reductase in Arabidopsis. *Plant Physiol.* **123**: 255-264.

Sanchez-Fernandez, R., Fricker, M., Corben, L.B., White, N.S., Sheard, N., Leaver, C.J. and Van Montagu, M. (1997) Cell proliferation and hair tip growth in the Arabidopsis root are under mechanistically different forms of redox control. *Proc. Nat. Acad. Sci. USA.* **94**: 2745-2750.

Scandalios, J.G. (1997). Molecular genetics of superoxide dismutases in plants. In *Oxidative Stress and the Molecular Biology of Antioxidative Defenses.* (Scandalios, J.G., ed.). pp. 527-568. Plainview: Cold Spring Harbor.

Squier, T.C. and Bigelow, D.J. (2000) Protein oxidation and age-dependent alterations in calcium homeostasis. *Front Biosci.* **5**: D504-526.

Toledano, M.B. and Leonard, W.J. (1991) Modulation of transcription factor NF-kappa B binding activity by oxidation-reduction *in vitro*. *Proc. Nat. Acad. Sci. USA*. **88**: 4328-4332.

Vanacker, H., Carver, T.L. and Foyer, C.H. (2000) Early H_2O_2 accumulation in mesophyll cells leads to induction of glutathione during the hyper-sensitive response in the barley-powdery mildew interaction. *Plant Physiol.* **123**: 1289-1300.

Walters, C. (1998). Understanding the mechanisms and kinetics of seed ageing. *Seed Sci.Res.* **8**: 223-244.

Wang, J., Zhang, H. and Allen, R.D. (1999) Overexpression of an Arabidopsis peroxisomal ascorbate peroxidase in tobacco increases protection against oxidative stress. *Plant Cell Physiol.* **40**: 725-732.

Wehmeyer, N. and Vierling, E. (2000) The expression of small heat shock proteins in seeds responds to discrete developmental signals and suggests a general protective role in desiccation tolerance. *Plant Physiol.* **122**: 1099-1108.

Willekens, H., Chamnongpol, S., Davey, M. Schraudner, M., Langebartels, C., Van Montagu. M., Inze, D., and Van Camp. W. (1997) Catalase is a sink for H_2O_2 and is indispensable for stress defence in plants. *EMBO J.* **16**: 4806-4816.

Willekens, H., Villarroel, R., Van Montagu, M., Inze, D. and Van Camp. W. (1994) Molecular identification of catalases from *Nicotiana plumbaginifolia* (L.) *FEBS Lett.* **352**: 79-83.

Wimmer, B., Lottspeich, F., van der Klei, I., Veenhuis, M. and Gietl, C. (1997) The glyoxysomal and plastid molecular chaperones (70-kDa heat shock protein) of watermelon cotyledons are encoded by a single gene. *Proc. Natl. Acad. Sci. USA*. **94**: 13624-13629.

Winkler, B.S., Orselli, S.M. and Rex, T.S. (1994) The redox couple between glutathione and ascorbic acid: a chemical and physiological perspective. *Free Rad. Biol.Med.* **17**: 333-349.

Xiang, C. and Oliver, D. (1998) Glutathione metabolic genes co-ordinately respond to heavy metals and jasmonic acid in Arabidopsis. *Plant Cell.* **10**: 1539-1550.

Zhu E, P.-S. E., Tarun, A.S., Weber, S.U., Jouanin, L. and Terry, N. (1999) Cadmium tolerance and accumulation in Indian mustard is enhanced by overexpressing γ-glutamylcysteine synthetase. *Plant Physiol.* **121**: 1169-1178.

14

FUTILE CYCLING THROUGH β-OXIDATION AS A BARRIER TO INCREASED YIELDS OF NOVEL OILS

Elizabeth Rylott and Tony Larson

University of York, York, U.K

KEYWORDS

Fatty acids; TAG; acyl CoA; lipid synthesis; transgenic oilseeds.

INTRODUCTION

Why should we engineer novel fatty acids into plants?

Oils have long been vital in the production of synthetic polymers, varnishes, detergents, lubricants, foodstuffs, and pharmaceuticals. The possibility of producing oil-based complex chemicals more cheaply in plants than can currently be made by chemical synthesis methods is an attractive incentive for manufacturers. For this reason, the use of plants to directly produce valuable novel oils is now a major target for the biotechnology industry. Furthermore, the over-production of conventional crops by farmers in the Developed World represents a large, untapped future market for the farming of these alternative crops. There is also a need to develop alternative sources in the face of the steady decline in the levels of currently used raw materials; crude oil, coal, and fossil-derived hydrocarbons. In addition, unlike conventional production methods, making novel oils in plants is environmentally favourable.

A. Baker and I.A. Graham (eds.), Plant Peroxisomes, 445–463.

Limitations of engineering novel fatty acids into plants

Plant species produce a vast array of useful, unusual oils (Kinney, 1998), and many of these oils have potential commercial applications (Ohlrogge, 1994; Alonso and Maroto, 2000). However, for the most part, the yields of novel oils from plants are not at economically viable levels and whilst some plants do already produce high levels of unusual oils naturally, many of these plants have poor agronomic traits. Traditional breeding programmes have had only limited success in improving such species for wide-scale agriculture. An attractive and viable alternative to overcome this is the use of genetic manipulation to engineer the genes required for novel fatty acids directly into crop species.

However, despite the efforts by several research groups to date, to engineer novel fatty acids into crop plants, there have been relatively few commercial successes. The only novel fatty acid produced from transgenic plants that is commercially available is lauric acid, produced in *Brassica napus* by Calgene (Monsanto Company). This is despite a decade of research that has seen the systematic introduction of a number of genes responsible for the synthesis of novel fatty acids into the model oilseed plants, *Arabidopsis thaliana* and *B. napus*.

There are three main problems in genetic engineering programs that need to be overcome for the successful commercialization of novel fatty acid production in plants.

First, the gene or genes necessary for the production of the novel oil need to be successfully transformed into a plant species. Initially, this can be into a model plant species such as Arabidopsis or directly into a crop species. This has been achieved for the synthesis of several novel fatty acids, including those containing shortened acyl chains (Voelker, 1992), useful functional groups, such as acetylenic and epoxy groups (Lee *et al*, 1998), or acyl chains with modified degrees of desaturation (Broun *et al*, 1998). It is only a matter of time before the remaining genes that have potential commercial application, for example those responsible for the synthesis of myristoleic, nervonic, and polyunsaturated fatty acids, are successfully transformed into plants.

Second, the transformed plants need to produce viable seed, with the novel lipid reserves being metabolized to support seedling establishment. Although this has, so far, not proved a major problem in genetically engineered plants, as discussed in this chapter, increased levels of novel oils may have adverse effects on membrane stability, affecting seedling establishment and subsequent plant development.

The third and most important problem to solve is that of yield. Plants genetically transformed to produce novel fatty acids have, for the most part, delivered poor yields. The levels of several novel fatty acids in transformed plants are reported to be only 5-20% of those achieved from the donor plant species (Ohlrogge, 1999). Thus, there is great scope to improve our understanding of the mechanisms preventing the substantial accumulation of novel fatty acids in genetically engineered plants.

EVIDENCE FOR FUTILE CYCLING OF NOVEL FATTY ACIDS DURING LIPID SYNTHESIS

β-oxidation in plant peroxisomes

In plant peroxisomes, β-oxidation catabolises fatty acids *via* the repeated cleavage of 2-carbon acetyl-CoA units from the thiol ends of activated fatty acids. Each turn of the β-oxidation cycle requires the action of three enzymes: (1) acyl CoA oxidase (ACX); (2) the multi-functional protein (MFP) which exhibits 2-*trans*-enoyl CoA hydratase, L-3-hydroxyacyl CoA dehydrogenase, D-3-hydroxyacyl CoA epimerase and Δ^3,Δ^2-enoyl CoA isomerase activities; and (3) L-3-ketoacyl CoA thiolase (KAT). Plants contain a family of ACX isozymes with different chain-length specificities (Kirsch *et al,* 1986; Hooks *et al,* 1996). Four Arabidopsis *ACX* isogenes have been characterised and encode enzymes with long (*ACX2),* medium-long (*ACX1),* medium (*ACX3*) and short (*ACX4*) chain substrate specificities (Hayashi *et al,* 1999; Hooks *et al,* 1999a; Eastmond *et al*, 2000; Froman *et al,* 2000). Whilst, multiple isozymes exist in plants for both MFP (Behrends *et al*, 1988; Preisig Müller and Kindl, 1993; Preisig-Müller *et al*, 1994; Gühnemann-Schäfer and Kindl, 1995) and KAT (Kato *et al*, 1996; Hayashi *et al*, 1998), in *Arabidopsis*, only one major *MFP* (*MFP2*) (Eastmond and Graham, 2000) and one major *KAT* (*KAT2*) (Hayashi *et al*, 1998; Germain *et al*, 2001) gene are strongly expressed during germination and post-germinative growth.

Expression during the plant life cycle

Expression during germination and early post-germinative growth

The expression levels of β-oxidation genes change dramatically during the life cycle of an oilseed plant. The first peak of β-oxidation expression occurs during germination and early post-germinative growth. This correlates with a massive breakdown of the triacylglycerol (TAG) lipid storage reserves. Free fatty acids are cleaved from the TAG backbone by the activity of lipases. At the peroxisomal membrane, the fatty acids are condensed with CoA to form acyl CoA esters by the action of acyl CoA synthetase (ACS). Inside the peroxisome, these acyl CoAs are fed into the β-oxidation spiral to produce acetyl-CoA, which then enters the glyoxylate cycle. Enzymes of the glyoxylate cycle convert the acetyl CoA to malate and succinate, which is subsequently exported to the gluconeogenesis pathway in the cytosol to be converted to sugars for seedling growth and establishment (Beevers, 1961). Once the seedling becomes photosynthetically competent, the expression of β-oxidation genes in the leaves declines to much lower levels.

Seed maturation

In developing Arabidopsis seeds lipid is rapidly synthesised. In the plastids, acetyl CoA carboxylase (ACCase) catalyses the first committed step, carboxylating acetyl CoA to form malonyl CoA. After conversion to malonyl ACP by transacylase, further acetyl CoA units are sequentially added by enzymes in the Fatty Acid Synthase (FAS) complex to produce acyl ACPs. In most plants, C16:0 and C18:0 acyl ACPs are the end-products of FAS and these may then be desaturated to produce palmitoyl ACP and oleoyl ACP, which make up the predominant acyl chains in plant lipids. Acyl ACPs may then be used for glycerolipid synthesis in the plastid or the ACP cleaved to release free fatty acids by the action of thioesterases. Free fatty acids are condensed with CoA by the action of membrane bound ACSs before being exported for use in membranes, storage lipid biosynthesis or other lipid-requiring processes. For TAG synthesis, acyl CoAs are transferred to the three *sn* positions on the glycerol backbone by the activity of acyltransferases, e.g. diacylglycerol acyltransferase (DAGAT) which adds acyl CoAs to the glycerol backbone of diacylglycerol (DAG) to form triacylglycerol (TAG) (for reviews see Harwood, 1988; Ohlrogge and Jaworski, 1997).

Surprizingly, β-oxidation is also active during the process of lipid deposition in maturing seeds. In the late stages of normal oilseed development, after the majority of TAG has accumulated, a second peak of β-oxidation activity is seen. This is thought to play a scavenging role, removing and recycling acyl CoAs that have not been incorporated into storage triacylglycerol (TAG) (Gerhardt, 1992).

Senescing tissues

Whilst the level of β-oxidation in mature leaves is relatively low and thought to be involved primarily in lipid turnover and recycling (Gerhardt, 1992), there is a third peak of β-oxidation in senescing leaves and other senescing tissues. This is involved in the recycling of membrane lipids and other cellular components. β-oxidation is also seen in carbohydrate-starved suspension cell cultures and in roots (Dieuaide *et al*, 1992; Pistelli *et al*, 1992; Lee *et al*, 1998). Plant tissues respond to these low carbohydrate conditions by increasing the expression of β-oxidation and glyoxylate cycle genes needed to catabolise lipid into carbohydrate.

Simultaneous operation of lipid synthesis and β-oxidation – a futile cycle

The production of commercially viable yields of novel fatty acids in genetically engineered plants requires an adequate accumulation of these fatty acids in storage TAGs during seed maturation. This process requires the co-ordination of plastidial fatty acid synthesis, transport of acyl intermediates to the cytosol, and acylation onto glycerol-3-phosphate in the Kennedy pathway. However, peroxisomal β-oxidation, which is normally a catabolic process associated with lipid breakdown during seed germination, may use as substrates the fatty acids and acyl CoA esters generated during lipid synthesis (Figure 1). This provides the opportunity for a futile cycle to function during lipid synthesis, where selected metabolic intermediates might be catabolized rather than be incorporated into triacylglycerol. The following sections provide evidence as to how this process may operate during the synthesis of novel oils in transgenic plants.

Figure 1: Futile cycle of fatty acid synthesis and catabolism in plants.

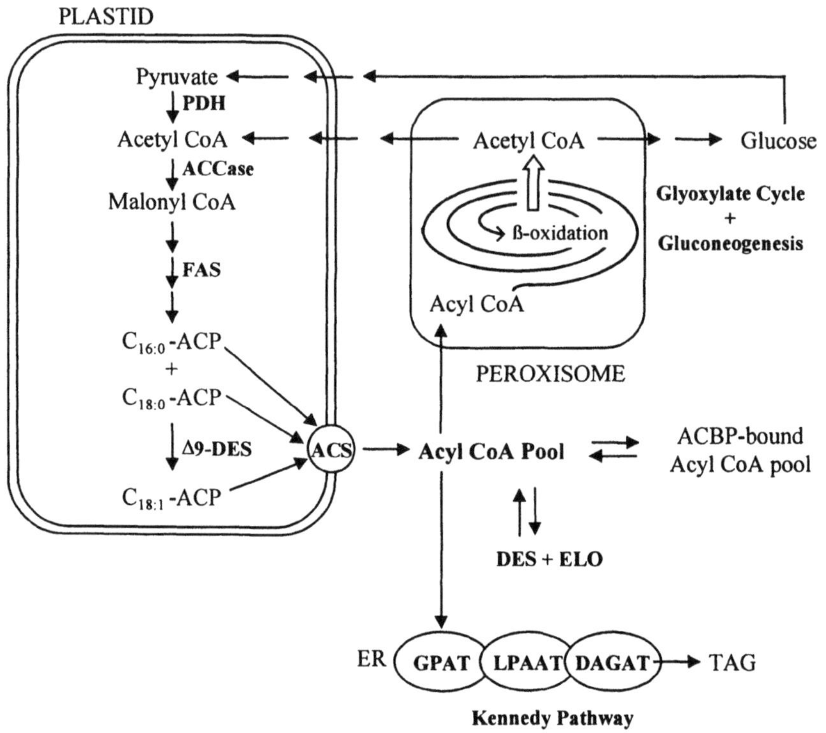

Figure 1: PDH = pyruvate dehydrogenase complex; ACCase = acyl CoA synthetase; FAS = fatty acid synthesis complex; DES = desaturase; ACS = acyl CoA synthetase; ELO = elongases; GPAT = glycerol-3-phosphate transferase; LPAAT = lysophoshatidic acid transferase; DAGAT = diacylglycerol acyltransferase; ACBP = acyl CoA binding protein.

Cases for the operation of β-oxidation during lipid synthesis

In theory, β-oxidation might operate during lipid synthesis to remove metabolic intermediates that, for whatever reason, cannot be or are only inefficiently incorporated into triacylglycerols. This would be necessary to prevent an increase of poorly utilized free fatty acids or acyl CoA esters to concentrations that would destabilise cell membranes. The following sections will present examples of this induction of lipid breakdown during novel lipid synthesis in transgenic plants.

Induction of β-oxidation in transgenic plants expressing a medium chain thioesterase

Whilst lipid storage reserves in *B. napus* and Arabidopsis seeds consist of greater than 90% C16- 18 chain length fatty acids, some species, including coconut, *Cuphea sps* and California bay have been shown to accumulate the commercially valuable medium chain fatty acids in their seed oils. Thioesterases from *Cuphea sps* and California bay exhibit medium chain length specificity whilst in *B. napus* they hydrolyse predominantly long chain fatty acid ACPs. This specificity was targeted to increase the levels of commercially valuable lauric acids in *B. napus* by transforming it with a lauroyl-acyl carrier protein (ACP) thioesterase (MCTE) from California bay (Voelker *et al*, 1992).

Independent MCTE transformants showed both high levels of transgene expression and a range of laurate levels in the mature seed up to 59 mol%. In MCTE transformants under the control of the 35S promoter, high levels of MCTE activity was also seen in the leaves, and isolated chloroplasts were shown to accumulate lauric acid up to 34%. Despite this, no accumulation of laurate was seen in the leaves of these transgenics (Eccleston *et al*, 1996). The authors proposed that lauric acid in the leaves of these lines was being turned over by induced β-oxidation activity the *B. napus* leaves. When they measured the level of β-oxidation activity in developing seeds of both the 35S and napin promoter lines, they found that medium chain ACX activity was induced. Furthermore, the ACX activity was approximately proportional to the levels of MCTE activity with the highest MCTE activity lines also having the highest ACX activity. Interestingly, the increased ACX activity was specifically medium chain (C12:0), whilst long chain (C16:0) ACX activity remained unaltered. Malate synthase and ICL activities were also shown to be upregulated, suggesting that accumulating lauric acid can stimulate glyoxylate cycle expression and thus the alteration of peroxisomes into glyoxysomes.

Despite the upregulation of lipid catabolizing genes, no decrease in overall lipid content in the seed was observed. Further investigations revealed an upregulation in the levels of both the biotin carboxylase and biotin carboxyl carrier protein (BCCP) subunits of ACCase, stearoyl ACP desaturase and 3-ketoacyl ACP synthase III at the mid-stage of seed development. Like the ACX activities, these were approximately proportional to the level of MCTE activity in the transgenic lines (Eccleston and Ohlrogge, 1998). In conclusion, the potential drain on lipid production created by the specific catabolism of lauric acid was compensated for by an increase in lipid synthesis.

In Arabidopsis MCTE over-expressing lines, lipid catabolizing genes were not induced in the leaves (Hooks *et al*, 1999b). This is probably due to the basal levels of β-oxidation in *Arabidopsis* being sufficient to catabolize the lauric acid produced in these transgenic lines. Indeed, a recent catalogue of Arabidopsis leaf EST databases showed a high number of transcripts for both thiolase and ACX β-oxidation genes (Mekhedov *et al*, 2000). These reports provided the first evidence that futile cycling through β-oxidation is a factor limiting the production of unusual fatty acids in plants.

Accumulation of novel β-oxidation intermediates in transgenic plants expressing polyhydroxyalkenone synthase

The most common explanations suggested for poor yields of novel oils in transgenic plants have been either inadequate rates of fatty acid synthesis, or inefficient incorporation of acyl CoA intermediates into TAG. However, based on the MCTE studies, a third mechanism, the selective catabolism of biosynthetic intermediates through peroxisomal β-oxidation now appears to be a likely contributing factor. This is effectively a futile cycle, because the intermediates are converted, via peroxisomal β-oxidation, back to acetyl CoA, the primary molecule in fatty acid synthesis.

The plastic PHA is synthesised by the polymerisation of 3-hydroxyacyl CoA monomers, intermediate compounds in the β-oxidation of lipids, by PHA synthase. Arabidopsis transformed with a peroxisomally targeted PHA synthase gene from *Psudomonas aeruginosa* accumulate PHA in their peroxisomes (Mittendorf *et al*, 1999). This accumulation of PHA mirrors the pattern of β-oxidation activity, with high levels of PHA synthesis during germination and senescence and lower levels of PHA accumulation during photosynthetic growth.

Double transgenic plants expressing the bacterial PHA synthase gene in combination with either a MCTE gene from the lauric acid-accumulating *Cuphea lanceolata* or mutant *dagat* lines, have been used to monitor the flux through β-oxidation in developing Arabidopsis seeds. PHA synthase/MCTE double transgenic lines were shown to store 18-fold more PHA than plants transformed with the PHA synthase alone, indicating a significant flow of medium chain acyl CoA through β-oxidation in these lines (Poirier *et al*, 1999). When the PHA synthase lines were crossed with *dagat* mutants, which accumulate C18:3 fatty acids, PHA levels were also shown to increase. Analysis of the PHA composition profiles revealed an increase in the proportion of catabolised products of C18:3 as the monomer substrates as excess C18:3-CoA was recycled through β-oxidation (Poirier *et al*, 1999).

STRATEGIES TO CONTROL FUTILE CYCLING

Reducing biosynthetic substrate supply to β-oxidation

One strategy that might be effective in curtailing the futile cycling of novel fatty acids in transgenic plants is to minimise the availability of biosynthetic intermediates, such as free fatty acids and acyl CoAs, to β-oxidation. This could be achieved by either increasing the rate of TAG synthesis in the Kennedy pathway, so that these intermediates are kept at low concentrations, or by making this pool of intermediates physiologically unavailable for catabolism. Both these possibilities will now be discussed.

Optimizing acyltransferase specificity and TAG synthesis

In the case of MCTE transformants, it has been known for some time that plastidial synthesis of medium chain acyl ACPs is quite effective when the appropriate thioesterases from species such as California bay are engineered into *B. napus* (Kinney, 1998). It is also known that species that naturally accumulate medium chain fatty acids in their seed oils have high, specific acyltransferase activities for medium chain acyl CoAs (Battey and Ohlrogge, 1989). However, this is not the case in transgenic plants, where it has been shown that DAGATs with low specificities for medium chain acyl CoAs, relative to lysophophatidylcholine acyltranferase (LPAAT), result in the accumulation of medium chain fatty acids in phosphatidylcholine rather than TAG (Wiberg *et al*, 1997, 2000). Whilst the species-specific substrate discrimination of thioesterases was the basis behind engineering the transgenic MCTE *B. napus* plants, the levels of lauric acid achieved in these lines were much lower than in the donor species, California bay. Comparisons of the MCTE activities with lauric acid accumulation showed that in transgenic plants producing up to 35 mol% lauric acid, the increase in MCTE activity was linear with laurate production. However, above the level of 35 mol%, further increases in MCTE activity resulted in proportionally much smaller changes in lauric acid accumulation, For example, an eight to

10-fold increase in MCTE activity was reported to give only a two-fold increase in mol% lauric acid. Analysis of the transgenic *B. napus* TAG composition revealed that lauric acid was mainly incorporated at the *sn*-1 and *sn*-3 positions, with only 5% at the *sn*-2 position (Voelker *et al*, 1996). This was presumed to be due to the negative substrate discrimination of the *B. napus sn*-2 acyl transferase and was partially overcome by transforming a coconut lysophosphatidic acid *sn*-2 acyltransferase gene into the MCTE overexpressing *B. napus* line (Knutzon *et al*, 1999).

The incorporation of novel fatty acids into membrane lipids in transgenic plants may result in membrane instability. This may then trigger protective mechanisms, which result in the recycling of the novel fatty acids by β-oxidation (Eccleston and Ohlrogge, 1998). Indeed, plants that naturally accumulate novel lipids into TAG do not readily incorporate them into their membrane lipids. Dahlqvist *et al*, (2000) studied both *Crepis palustris*, which incorporates vernolic acid into TAG, and castor bean, which incorporates both vernolic and ricinoleic acids. In these species they demonstrated a transient flux of novel lipids into the major membrane lipid phosphatidylcholine (PC) before subsequent, selective transfer from PC to TAG by an acyl CoA-independent phospholipid: diacylglycerol acyltransferase (PDAT) activity. Dahlqvist *et al* (2000) suggested that PDAT might play an important role in the specific channelling of bilayer-disturbing fatty acids, such as ricinoleic and vernolic from PC into TAG. Furthermore, whilst PDAT was present in developing seeds of both *C. palustris* and castor bean, sunflower, which produces only common fatty acids, lacked this PDAT activity during seed development. This evidence strongly suggests that the lack of specific PDAT activity, along with other substrate-specific genes are likely to be factors limiting the accumulation of novel oils in transgenic plants.

Minimizing loss of acyl CoAs to β-oxidation during lipid synthesis

Whilst optimizing acyltransferase specificity for novel fatty acyl CoAs has already been shown to improve the accumulation of the equivalent fatty acids in TAG (Knutzon *et al*, 1999), other mechanisms might also operate to prevent the loss of these biosynthetic intermediates to β-oxidation. In the *B. napus* MCTE lines, although lauric acid is catabolized by induced β-oxidation activity in the maturing seeds, acyl CoA concentrations have also been shown to accumulate (Larson and Graham, 2001). Acyl CoAs also accumulate during TAG catabolism in germinating seedlings of the Arabidopsis *kat2* ketoacyl CoA thiolase knockout mutant (Germain *et al*,

2001) (Figure 2). These data suggest that, if the novel fatty acid components within the acyl CoA pool are made unavailable to peroxisomal β-oxidation, futile cycling might be ameliorated.

Figure 2: Acyl CoA profiles in transgenic oilseeds

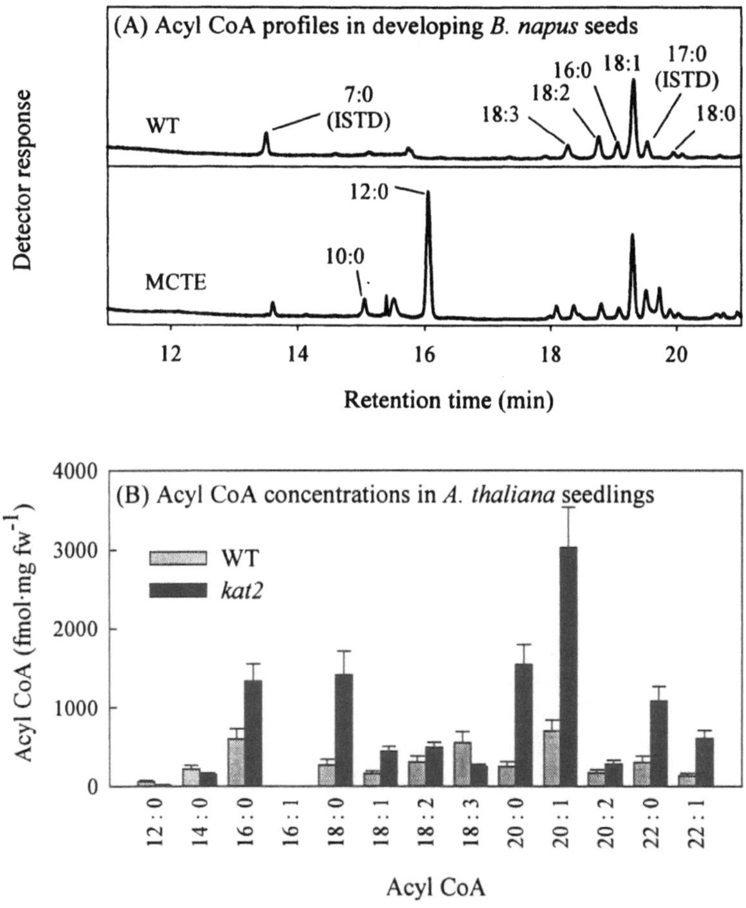

Figure 2. Panel A shows the HPLC elution profile of acyl CoA esters from wild type (WT) and transgenic (MCTE) B. napus seeds sampled 40 days after flowering. The MCTE plants contain a California bay medium chain thioesterase construct that leads to the accumulation of lauric acid in the seed oils. Panel B illustrates the retention of long-chain acyl CoAs (particularly 20:1) associated with incomplete TAG catabolism in transgenic kat2 Arabidopsis seedlings lacking ketoacyl CoA thiolase activity. Wild-type (WT) and kat2 seedlings were sampled five days after imbibition.

It has been proposed that only a very small proportion of free long chain acyl CoAs exist in the cytosol of oilseeds, the remainder being bound to acyl CoA binding proteins (ACBPs) (Figure 1) in an approximately equimolar ratio (Brown *et al*, 1998). It is essential that free long-chain acyl CoAs are kept below their critical micellar concentrations to prevent the detergent-effects mediating disruptions in cellular membranes, but also to control both the availability and regulatory effects of these molecules on other areas of metabolism. In yeast and mammals, long chain acyl CoAs are known to have a number of regulatory roles, including inhibition of acyltransferases, ACCase, and FAS (Faergeman and Knudsen, 1997). For example, sub-micromolar levels of long chain acyl CoAs inhibit ACCase in yeast (Ogiwara *et al*, 1978). Work, also in yeast demonstrated that exogenously supplied long chain free fatty acids need to be activated to acyl CoAs by ACS before lipid biosynthesis is inhibited (Kamiryo *et al*, 1976).

In yeast and mammals, the cytosolic concentration of acyl CoAs is controlled, in part, by binding to fatty acid binding proteins (FABP) and ACBP (Faergeman and Knudsen, 1997). In a yeast mutant deficient in ACBP, the long-chain acyl CoA pool was increased, although the rate of lipid synthesis did not increase (Scherling *et al*, 1996). There is also evidence that ACBP may have some part in regulating acyl CoA concentrations and lipid synthesis in oilseeds. In *B. napus*, low levels of free long chain acyl CoAs have been demonstrated to inhibit the glucose-6-phosphate transporter. This inhibition was reversed by the addition of ACBPs (Fox *et al*, 2000). A subsequent study has shown that this repression of sugar import also inhibits the rate of fatty acid synthesis (Fox *et al*, 2001). Brown *et al* (1998) found that ACBP was expressed during seed development and oil deposition in *B. napus*, but not during the germination and the mobilisation of lipid reserves. Using immunogold labelling to an ACBP identified in Arabidopsis (ACBP1), Chye *et al* (1999) found that ACBP1 was not associated with TAG synthesis, but with vesicles involved in membrane assembly, during seed development. Together, these data suggest that ACBPs may be responsible for mediating the long-chain acyl CoA pool size and chaperoning membrane lipid synthesis, but are not directly involved in TAG assembly. In the absence of ACBP or in the case of an increased acyl CoA: ACBP ratio, excess long chain acyl CoAs might be expected to be catabolized in a futile cycle, or be preferentially incorporated into TAG rather than membrane lipids, assuming that acyltransferase specificity is sufficient. This has yet to be demonstrated experimentally.

Regulating β-oxidation

Restricting β-oxidation gene expression during the period of novel lipid accumulation with the aim of reducing futile cycling may be a way to enhance the level of novel oil production.

Characterised β-oxidation mutants *ped1* and *acx3*

Arabidopsis mutants in two of the β-oxidation pathway genes expressed strongly during germination and post-germinative growth have been characterised. Two mutants in the *KAT2* gene, *ped1* (Hayashi *et al*, 1998) and *kat2* (Germain *et al,* 2001), which contains a T-DNA insertion in the *KAT2* gene have been identified. In both homozygous *ped1* and *kat2* mutants, seedling growth is dependent upon exogenous sugar, and lipid bodies persist, unmetabolized in green cotyledons. In *kat2*, storage triacylglycerol (TAG) has also been shown to accumulate and extracts from *kat2* seedlings have significantly decreased thiolase activity, using aceto-acetyl CoA (C4: 0) as substrate.

The second mutant is a promoter-trapped T-DNA insertion in the medium-chain length acyl CoA oxidase gene (*ACX3*) (Eastmond *et al*, 2000). The activity of medium chain ACX was reduced by 95% of wild type in this mutant, however, unlike *kat2*, lipid catabolism and seedling development were unaltered. This is presumed to be due to the overlapping substrate specificities of the remaining ACX genes. Of these two genes, the thiolase appears to be a suitable target for blocking novel oil breakdown.

Engineering plants to degrade novel fatty acids during seedling establishment

Genetically modified plants need to be capable of catabolising novel oil-containing TAG reserves to provide energy for germination and seedling establishment. Although plant β-oxidation activity appears able to catabolise novel fatty acids, the capacity of other enzymes involved in lipid breakdown may be limited. The vast majority of oil seed crop species produce mainly unsaturated storage lipids of C16-C20 chain lengths and exhibit broad lipase substrate specificities. However, species with unusual fatty acids stored in TAG, or species with greater than 80% saturated fatty acids in TAG have lipases with narrow substrate specificities (Hellyer *et al*, 1999). For example, the seeds of some *Cuphea* species have over 80% capric (C10:0) fatty acid incorporated into TAG, and have lipases with increased C10:0 specificity. Conversely *B. napus* lipases have been shown to discriminate against novel fatty acids in TAG (Hills *et al*, 1990). It will be necessary to identify such unique steps in the breakdown of unusual fatty acids in donor plant species, and to investigate their effects on catabolism and seedling establishment in novel oil producing crop species.

Microarray data: new insights into lipid metabolism gene expression/ EST libraries to find novel genes

Technical advances and the reduced costs in sequencing DNA mean that sequencing Expressed Sequence Tags (ESTs) from libraries of different species and at different developmental stages is now a routine method for identifying new genes. For example, genes involved in the conjugated double bond formation in α-eleostearic and α-parinaric fatty acids from *Momordica charantia* and *Impatiens balsamina* were identified by sequencing developing seed EST libraries (Cahoon *et al*, 1999). Libraries of ESTs have also been used to provide 'digital Northerns' based on the relative abundance of the ESTs for each gene (Mekhedov *et al*, 2000, White *et al*, 2000). In addition, previously unknown genes have been identified, for example candidate FAD5, 3-ketoacyl-acyl carrier protein (ACP) synthase, and BCCP genes were all identified during a cataloguing of Arabidopsis EST databases (Mekhedov *et al*, 2000).

Microarrays have emerged as a powerful new tool for analyzing the gene expression of complete biosynthetic pathways and even whole genomes. Girke *et al* (2000) have used microarrays to analyze the mRNA levels of over 2,600 genes expressed in developing Arabidopsis seeds relative to those in leaves, and roots. These included 113 genes involved in lipid metabolism, although only 10 of these genes were found to have a greater than ten-fold higher expression in the seeds when compared to leaves and roots. These included lipases, oleosins and a fatty acid elongase 1 (FAE1). The arrays were also hybridized to *B. napus* mRNA, however, the signal intensities were reduced by two-fold, resulting in only 50% of the genes being detected above the background signal intensity. Whilst this means that weakly expressed genes may not be detected in experiments with heterologous probes, microarrays still provide an extremely valuable instrument to analyze gene expression in related species. Of the 2,600 genes arrayed, over 40% were of unknown function. This included a number of uncharacterised seed-specific transcription factors, further analysis of which may provide insights into the transcriptional control of seed metabolism and possibly roles directly related to lipid metabolism. The microarray results have also provided a large number of seed-specific genes, promoter analysis of which may identify potential targets and promoters for use in biotechnology. With the power of microarrays, in a single step a very large initial gene pool has been reduced to relatively few, promising candidate genes for further analysis. This represents large savings in both cost and research development time.

CONCLUSIONS

The two main causes of the low yields of novel oils in transgenic plants compared to those in donor plants are firstly, recycling of novel oils through β-oxidation, and secondly, subsequent substrate discrimination of enzymes downstream in lipid biosynthesis. Engineering additional genes with novel substrate specificities absent in the recipient plant (for example thioesterases, *sn*-2 acyl transferases and PDATs) has already been shown to increase yields. Further increases may be achievable by restricting β-oxidation during seed development to prevent novel lipid turnover. This could be approached by the use of specific or inducible promoters to restrict β-oxidation activity during seed development whilst permitting β-oxidation during germination and seedling establishment. Transgenic plants with novel oil containing TAG may also require specific lipases to ensure TAG catabolism and subsequent seedling establishment.

However engineering plants with two or more different genes is more complex with increased developmental time, production costs and additionally, the resulting plants may exhibit transgene instability and silencing. Recently a 'triple construct' was used to transform all three bacterial genes necessary for the biosynthetic route to polyhydroxybutyrate (PHB) in a single transformation event. The resulting transformants accumulated a significant, four-fold increase on previously published levels (Bohmert *et al*, 2000). Such advances suggest that whole pathways will be routinely engineered into transgenic plants and more cost effectively than additional individual genes. With the advances in technology and subsequent knowledge gained from the analysis of transformed plants, the target of commercial production of novel oils in crop species is fast becoming a reality.

REFERENCES

Alonso, D.L. and Maroto, F.G. (2000) Plants as 'chemical factories' for the production of polyunsaturated fatty acids. *Biotechnol. Adv.* **18**: 481-497.

Battey, J.F. and Ohlrogge, J.B. (1989) A comparison of the metabolic fate of fatty acids if different chain lengths in developing oilseeds. *Plant. Physiol.* **90**: 835-840.

Beevers, H. (1961) Metabolic production of sucrose from fat. *Nature.* **191**: 433-436.

Behrends, W., Engeland, K. and Kindl, H. (1988) Characterisation of two forms of the multifunctional protein acting in fatty acid β-oxidation. *Arch. Biochem. Biophys.* **263**: 161-169.

Bohmert, K., Balbo, I., Kopka, J., Mittendorf, V., Naweath, C., Poirier, Y., Tischendorf, G., Trethewey, R.N. and Willmittzer, L. (2000) Transgenic Arabidopsis plants can accumulate polyhydroxybutyrate to up to 4% of their fresh weight. *Planta.* **211**: 841-845.

Broun, P., Boddupalli, S. and Somerville, C. (1998) A bifunctional oleate 12-hydroxylase: desaturase from *Lesquerella fendleri*. *Plant J.* **13**: 201-210.

Brown, A.P., Johnson, P., Rawsthorne, S. and Hills, M.D. (1998) Expression and properties of acyl-CoA binding protein from *Brassica napus*. *Plant Physiol. Biochem.* **36**: 629-635.

Cahoon, E.B., Carlson, T.J., Ripp, K.G., Schweiger, B.J., Cook, G.A., Hall, S.E. and Kinney, A.J. (1999) Biosynthetic origin of conjugated double bonds: production of fatty acid components of high value drying oils in transgenic soybean embryos. *Proc. Natl. Acad. Sci. USA.* **96**: 12935-12940.

Chye, M., Huang, B. and Zee, S.Y. (1999) Isolation of a gene encoding Arabidopsis membrane associated acyl-CoA binding protein and immunolocalization of its product. *Plant J.* **18**: 205-214.

Dahlqvist, A., Stahl, U., Lenman, M., Banas, A., Lee, M., Sandager, L., Ronne, H. and Stymne, S. (2000) Phospholipid:diacylglycerol acyltransferase: An enzyme that catalyzes the acyl-CoA-independent formation of triacylglycerol in yeast and plants. *Proc. Natl. Acad. Sci. USA.* **97**(12): 6487-6492.

Dieuaide, M., Couée, I., Pradet, A. and Raymond, P. (1992) Effects of glucose starvation on the oxidation of fatty acids by maize root tip mitochondria and peroxisomes: evidence for mitochondrial fatty acid β-oxidation and acyl-CoA dehydrogenase activity in a higher plant. *Biochem. J.* **296**: 199-207.

Eastmond, P.J. and Graham, I.A. (2000) The multifunctional protein AtMFP2 is coordinately expressed with other genes of fatty acid β-oxidation during seed germination in *Arabidopsis thaliana*. *Biochem. Soc. Trans.* **28**: 95-99.

Eastmond, P.J., Hooks, M.A., Williams, D., Lange, P., Bechtold, N., Sarrobert, C., Nussaume, L. and Graham, I.A. (2000) Promoter trapping of a novel medium-chain acyl-CoA oxidase, which is induced transcriptionally during Arabidopsis seed germination. *J. Biol. Chem.* **275**: 34375-34381.

Eccleston, V.S. and Ohlrogge, J.B. (1998) Expression of lauroyl-acyl carrier protein thioesterase in *Brassica napus* seeds induces pathways for both fatty acid oxidation and biosynthesis and implies a set point for triacylglycerol accumulation. *Plant Cell.* **10**: 613-621.

Eccleston, V.S., Cranmer, A.M., Voelker, T.A. and Ohlrogge, J.B. (1996) Medium-chain fatty acid biosynthesis and utilization in *Brassica napus* plants expressing lauroyl-acyl carrier protein thioesterase. *Planta.* **198**: 46-53.

Faergeman, N.J. and Knudsen, J. (1997) Role of long-chain fatty-acyl CoA esters in the regulation of metabolism and in cell signalling. *Biochem J.* **323**: 1-12.

Fox, S.R., Rawsthorne, S. and Hills, M. J. (2000) Role of acyl-CoAs and acyl-CoA-binding protein in regulation of carbon supply for fatty acid biosynthesis. *Biochem. Soc. Trans.* **28**: 672-674.

Fox, S.R., Rawsthorne, S. and Hills, M.J. (2001) Fatty acid synthesis in pea root plastids is inhibited by the aciton of long-chain acyl-CoAs on metabolite transporters. *Plant Physiol.* **126**: 1259-1265.

Froman, B.E., Edwards, P.C., Bursch, A.G. and Dehesh, K. (2000) ACX3, a novel medium-chain acyl-CoA oxidase from Arabidopsis. *Plant Physiol.* **123**: 733-741.

Gerhardt, B. (1992) Fatty acid degradation in plants. *Prog. Lipid Res.* **31**: 417-446.

Germain, V., Rylott, E., Larson, T.R., Sherson, S.M., Bechtold, N., Carde, J-P., Bryce, J.H., Graham, I. and Smith, S.M. (2001) Requirement for 3-ketoacyl-CoA thiolase-2 in peroxisome development, fatty acid β-oxidation and breakdown of triacylglycerol in lipid bodies of Arabidopsis seedlings. *Plant J.* **27**(4): 1-13.

Girke, T., Todd, J., Ruuska, S., White, J., Benning, C. and Ohlrogge, J. (2000) Microarray analysis of developing Arabidopsis seeds. *Plant Physiol.* **124**: 1570-1581.

Gühnemann-Schäfer, K. and Kindl, H. (1995) Fatty acid β-oxidation in glyoxysomes. Characterisation of a new tetrafunctional protein (MFP III). *Biochim. Biophys. Acta.* **1256**: 181-186.

Harwood, J.L. (1988) Fatty acid metabolism. *Ann. Rev. Plant Physiol.* **39**: 101-138.

Hayashi, H., De Bellis, L., Ciurli, A., Kondo, M., Hayashi, M. and Nishimura, M. (1999) A novel acyl-CoA oxidase that can oxidise short-chain acyl-CoA in plant peroxisomes. *J. Biol. Chem.* **274**: 12715-12721.

Hayashi, M., Toriyama, K., Kondo, M. and Nishimura, M. (1998) 2,4-dichlorophenoxybutyric acid-resistant mutants of Arabidopsis have defects in glyoxysomal fatty acid β-oxidation. *Plant Cell* **10**: 183-195.

Hellyer, S.A., Chandler, I.C. and Bosley, J.A. (1999) Can the fatty acid selectivity of plant lipases be predicted from the composition of seed triglyceride? *Biochem. Biophys. Acta.* **1440**: 215-224.

Hills, M.J., Kiewitt, I. and Murkherjee, K.J. (1990) Lipase from *Brassica napus* L. discriminates against *cis*-4 and *cis*-6 unsaturated fatty acids and secondary and tertiary alcohols. *Biochim. Biophys. Acta.* **1042**: 237-240.

Hooks, M.A., Bode, K. and Couee, I (1996) Higher plant medium- and short-chain acyl-CoA oxidases: identification, purification and characterization of two novel enzymes of eukaryotic peroxisomal β-oxidation. *Biochem. J.* **320**: 607-614.

Hooks, M.A., Fleming, Y., Larson, T.R. and Graham, I.A. (1999b) No induction of β-oxidation in leaves of Arabidopsis that overproduce lauric acid. *Planta.* **207**: 385-392.

Hooks, M.A., Kellas, F. and Graham, I.A. (1999a) Long-chain acyl-CoA oxidases of Arabidopsis. *Plant J.* **19**: 1-13.

Kamiryo, T., Parthasarathy, S. and Numa, S. (1976) Evidence that acyl-CoA synthetase activity is required for repression of yeast acyl-CoA carboxylase by exogenous fatty acids. *Proc. Natl. Acad. Sci. USA.* **73**(2): 386-90.

Kato, A., Hayashi, M., Takeuchi, Y. and Nishimura, M. (1996) cDNA cloning and expression of a gene for 3-ketoacyl-CoA thiolase in pumpkin cotyledons. *Plant Mol. Biol.* **31**: 843-852.

Kinney A.J. (1998) Plants as industrial chemical factories – new oils from genetically engineered soybeans. *Fett. Lipid.* **100**: 173-176.

Kirsch, T., Loffler, H.G. and Kindl, H. (1986) Plant acyl-CoA oxidase: purification, characterization and monomeric apoprotein. *J. Biol. Chem.* **261**: 8570-8575.

Knutzon, D.S., Hayes, T.R, Wyrick, A., Xiong, H., Davies, H.M. and Voelker, T.A. (1999) Lysophosphatidic acid acyltransferase from coconut endosperm mediates the insertion of laurate at the sn-2 position of triacylglycerols in lauric rapeseed oil and can increase total laurate levels. *Plant Physiol.* **120** (3): 739-746.

Larson, T.R. and Graham, I.A. (2001) A novel technique for the sensitive quantification of acyl-CoA esters from plant tissues. *Plant J.* **25**: 115-125.

Lee, M., Lenman, M., Banaś, A., Bafor, M., Singh, S., Schweizer, M., Nilsson, R., Liljenberg, C., Dahlqvist, A., Gummeson, P-O, Sjödahl, Green, A. and Stymne, S. (1998) Identification of non-haem diiron proteins that catalyse triple bond and epoxy group formation. *Science.* **280**: 915-918.

Mekhedov, S., Martínez de Ilárduya, O. and Ohlrogge, J. (2000) Toward a functional catalog of the plant genome. A survey of genes for lipid biosynthesis. *Plant Phys.* **122**: 389-401.

Mittendorf, V., Bongcam, V., Allenbach, L., Coullerez, G., Martini, N. and Poirier, Y. (1999) Polyhydroxyalkanoate synthesis in transgenic plants as a new tool to study carbon flow through β-oxidation. *Plant J.* **20**(1): 45-55.

Ogiwara H., Tanabe, T., Nikawa, J. and Numa, S. (1978) Inhibition of rat-liver acetyl-coenzyme-A carboxylase by palmitoyl-CoA. Formation of equimolar enzyme-inhibitor complex. *Eur. J. Biochem.* **89**(1): 33-41.

Ohlrogge, J.B. (1999) Plant metabolic engineering: are we ready for phase two? *Curr. Opin. Plant Biol.* **2**: 121-122.

Ohlrogge, J.B. (1994) Design of new plant products – engineering of fatty-acid metabolism. *Plant Physiol.* **104** (3): 821-826.

Ohlrogge, J.B. and Jaworski, J.G. (1997) Regulation of fatty acid synthesis. *Ann. Rev. Plant Physiol. Plant Mol. Biol.* **48**: 109-136.

Pistelli, L., Perata, P. and Alpi, A. (1992) Effect of leaf senescence on glyoxylate cycle enzyme activities. *Aust. J. Plant. Physiol.* **19**: 723-729.

Poirier, Y., Ventre, G. and Caldelari, D. (1999) Increased flow of fatty acids toward β-oxidation in developing seeds of Arabidopsis deficient in diacylglycerol acyltransferase activity or synthesizing medium-chain-length fatty acids. *Plant Physiol.* **121**: 1359-1366.

Preisig-Müller, R. and Kindl, H. (1993) Thiolase mRNA translated *in vitro* yields a peptide with a putative N-terminal presequence. *Plant Mol. Biol.* **22**: 59-66.

Preisig-Müller, R., Gühnemann-Schäfer, K. and Kindl, H. (1994) Domains of the tetrafunctional protein acting in glyoxysomal fatty acid β-oxidation. *J. Biol. Chem.* **269**: 20475-20481.

Richmond, T.A. and Bleecker, A.B. (1999). A defect in β-oxidation causes abnormal inflorescence development in Arabidopsis. *Plant Cell* **11**: 1911-1923.

Scherling, C.K., Hummel, R., Hansen, J.K., Borsting, C., Mikkelsen, J.M., Kristiansen, K. and Knudsen, J. (1996) Disruption of the gene encoding the acyl-CoA-binding protein (ACB1) perturbs acyl-CoA metabolism in *Saccharomyces cerevisiae*. *J. Biol. Chem.* **271**: 22514-22521.

Voelker, T.A., Hayes, T.R., Cranmer, A.M., Turner, J.C. and Davies, H.M. (1996) Genetic engineering of a quantitative trait: Metabolic and genetic parameters influencing the accumulation of laurate in rapeseed. *Plant J.* **9** (2): 229-241.

Voelker, T.A., Worrell, A. C., Anderson. L., Bleibaum, J., Fan, C., Hawkins, D.J., Radke, S.E. and Davies, H.M. (1992) Fatty-acid Biosynthesis Redirected to medium chains in transgenic oilseed plants. *Science.* **257**: 72-74.

White, J. A., Todd, J., Newman, T., Fockes, N., Girke, T., Martínez de Ilárduya, O., Jaworski, J. G., Ohlrogge, J. B. and Benning, C. (2000) A new set of Arabidopsis-expressed sequence tags from developing seeds. The metabolic pathway from carbohydrates to seed oil. *Plant Physiol.* **124**: 1582-1594.

Wiberg, E., Banas, A. and Stymne, S. (1997) Fatty acid distribution and lipid metabolism in developing seeds of laurate-producing rape (*Brassica napus* L.). *Planta* **203**: 341-348.

Wiberg, E., Edwards, P., Byrne, J., Stymne, S. and Dehesh, K. (2000) The distribution of caprylate, caprate and laurate in lipids from developing and mature seeds of transgenic *Brassica napus* L. *Planta.* **212**(1): 33-40.

15

POLYHYDROXYALKANOATE SYNTHESIS IN PLANT PEROXISOMES

Yves Poirier

Université de Lausanne, Lausanne, Switzerland

KEYWORDS

Biodegradable plastics; novel biopolymers; polyhydroxybutyrate; futile cycling; unsaturated fatty acids.

INTRODUCTION

Polyhydroxyalkanoates (PHAs) are high-molecular-weight polyesters of hydroxyacids synthesized by a wide variety of bacteria (Anderson and Dawes, 1990; Braunegg *et al*, 1998; Madison and Huisman, 1999; Steinbüchel 1991; Sudesh and Doi, 2000). Poly(3-hydroxybutyrate) (PHB), one member of the family of PHAs, was first discovered in *Bacillus megaterium* by M. Lemoigne in 1926 (Lemoigne, 1926). It was only several decades later that the value of PHAs as polymers having properties of thermoplastics and elastomers was recognized. Since then, the potential application of PHAs as biodegradable, renewable and environmentally-friendly plastics has been a major driving force supporting research on its synthesis in bacteria, as well as its application in polymer chemistry. A wide spectrum of PHAs has now been characterized and numerous bacterial genes involved in their synthesis have been cloned (Steinbüchel and Valentin, 1995; Steinbüchel and Hein, 2001). However, despite recent advances in fermentation technologies and the basic attractiveness of PHA as a substitute for petroleum-derived polymers, the major hurdle facing commercial exploitation of PHA in consumer products is the high cost of bacterial fermentation, making bacterial PHA five to ten times more expensive then polypropylene and polyethylene (Poirier *et al*, 1995).

A. Baker and I.A. Graham (eds.), Plant Peroxisomes, 465–496.

It is in view of the inadequacy of bacterial fermentation as a cheap production system for PHA that synthesis of PHA in plants was seen as an attractive proposition (Poirier *et al*, 1995; Poirier, 1999; Poirier, 2001). The basic premise is that since crops plants can produce annually millions of tons of starch and oils at costs below $1.0/kg, synthesis of PHA in crops could potentially be the only way of producing PHA on a large scale and at a cost competitive to petroleum-derived plastics.

Synthesis of PHA in plants was first demonstrated in 1992 by the accumulation of PHB in the cytoplasm of cells of *Arabidopsis thaliana* (Poirier *et al*, 1992a). Since then, a range of different PHAs have been synthesized in various species through the creation of novel metabolic pathways either in the cytoplasm, plastid, or more recently in the peroxisome (Poirier, 2001). As with bacterial PHA, the initial driving force behind synthesis of PHA in plants has been for the biotechnological production of biodegradable polymers. However, more recently, PHA synthesis in plants has emerged as a useful and novel tool to study fundamental aspects of plant metabolism. This stems from the fact that PHA synthesized in plant acts as a final carbon sink that can be used either as an indicator or a modifier of the carbon flow through different pathways. So far, the potential of PHA to study plant metabolism has been mostly exploited for the study of fatty acid degradation in plant expressing the PHA biosynthetic pathway in the peroxisome.

This review will mainly focus on the synthesis of PHA in plant peroxisomes and how this system can be used both for the production of novel biopolymers for industry, as well as a novel tool to study fatty acid metabolism. However, since PHA is naturally synthesized in bacteria and most of our knowledge on PHA synthesis and degradation has been obtained from studies in bacteria, a brief review of the main pathways involved in PHA synthesis in bacteria will first be given. This will be followed by a description of the major milestones of PHA synthesis in the cytoplasm and plastids of plants before describing in more detail the synthesis of PHA in plant peroxisomes.

SYNTHESIS OF PHA IN BACTERIA

Introduction

PHAs have been shown to occur in nearly one hundred different genera of bacteria, encompassing Gram-positive and Gram-negative species, as well as some phototrophic bacteria and cyanobacteria (Steinbüchel, 1991; Kim and Lenz, 2001). Although the majority of PHAs are composed of R-(-)-3-hydroxyalkanoic acid monomers ranging from 3 to 16 carbons in length, some PHAs can also contain 4- or 5-hydroxy acids (Steinbüchel and Valentin, 1995). PHAs are synthesized as polymers containing typically between 1000-10 000 subunits linked through ester bonds. They are hydrophobic molecules and accumulate in bacteria as intracellular inclusions of 0.2-0.5 μm in diameter. The size of the granule is not fixed, but can be influenced by the presence of amphipathic proteins, named phasins, found on the surface of the inclusions (Wieczorek *et al*, 1995). The role of PHA in bacteria is mainly to act as a reserve of carbon and an electron sink. Typically, PHA accumulates to high levels in bacteria which are grown in media containing an excess supply of carbon, such as glucose, but limited in one essential nutrient, such as phosphate or nitrogen (Anderson and Dawes, 1990; Braunegg *et al*, 1998; Madison and Huisman, 1999). Upon addition of the limiting nutrient, the intracellular PHA is degraded to its monomers and the carbon released used for growth. Degradation of intracellular PHA is dependant on specific enzymes, the PHA depolymerases (Jendrossek, 2001).

Well over one hundred different hydroxyacids have been found to be incorporated in PHAs, with the major diversity being found in the length and the presence of functional groups in the side chain of the polymer (Steinbüchel and Valentin, 1995). Although many of the aliphatic monomers have been found in PHA produced by bacteria in their natural environment, a larger fraction of monomers having functional groups in their side chains, such as unsaturated bonds or halogenated groups, have been incorporated into PHA following growth of bacteria under laboratory conditions in media containing exotic sources of carbon.

Bacteria-synthesizing PHAs have been traditionally sub-divided into two groups. One group, including the bacterium *Ralstonia eutropha*, produces short-chain-length PHA (SCL-PHA) containing monomers ranging from 3 to 5 carbons in length, while a distinct group, including a number of Pseudomonads such as *Pseudomonas oleovorans,* synthesize medium-chain-length PHA (MCL-PHA) containing monomers ranging from 6 to 16 carbons in length (Anderson and Dawes, 1990; Braunegg *et al*, 1998; Madison and Huisman, 1999). This division between SCL- and MCL-PHA is largely determined by the substrate specificity of the PHA synthase responsible for the polymerization of the substrate R-3-hydroxyacyl-CoA. While there are some clear distinctions between some of the biochemical pathways involved in the synthesis of SCL-PHAs and MCL-PHAs, there can also be significant overlap between them. Furthermore, it has recently become clearer that this division between SCL- and MCL-PHA is not strict, since several bacteria have been found that can synthesize a hybrid PHA that can include monomers from 4 to 8 carbons (Clemente *et al*, 2000; Fukui and Doi, 1997; Hall *et al*, 1998; Matsusaki *et al*, 1998).

A number of enzymes and metabolic pathways have been shown to be implicated in the synthesis of a wide spectrum of PHAs in bacteria. We wish here only to give a brief overview of the main pathways in order to provide an adequate background for the discussion of PHA synthesis in plant peroxisomes. The readers are referred to several excellent recent reviews to learn more on various aspects of bacterial PHA, including biochemical synthesis (Anderson and Dawes, 1990; Braunegg *et al*, 1998; Madison and Huisman, 1999; Steinbüchel, 1991; Sudesh *et al*, 2000), degradation in the environment (Jendrossek *et al*, 1996; Jendrossek, 2001), fermentation technology (Kessler *et al*, 2001), and applications of PHAs (van der Walle *et al*, 2001).

Synthesis of short-chain-length polyhydroxyalkanoate

PHB is the most widespread and thoroughly characterized PHA found in bacteria. A large part of our knowledge on PHB biosynthesis has been obtained from *Ralstonia eutropha* (formerly *Alcaligenes eutrophus*) which can accumulate PHB up to 85% of the dry weight (dwt) (Steinbüchel and Schlegel, 1991). In this bacterium, PHB is synthesized from acetyl-coenzyme A (CoA) by the sequential action of three enzymes (Figure 1). The first enzyme of the pathway, 3-ketothiolase, catalyzes the reversible condensation of two acetyl-CoA moieties to form acetoacetyl-CoA. Acetoacetyl-CoA reductase subsequently reduces acetoacetyl-CoA to R-(-)-3-hydroxybutyryl-CoA, which is then polymerized by the action of a PHA synthase to form PHB. The PHA synthase of *R. eutropha* has been shown to accept the R-isomer 3-hydroxybutyryl-CoA but not the S-isomer (Haywood *et al*, 1989).

Figure 1: Pathway of PHB and P(HB-HV) synthesis in R.eutropha

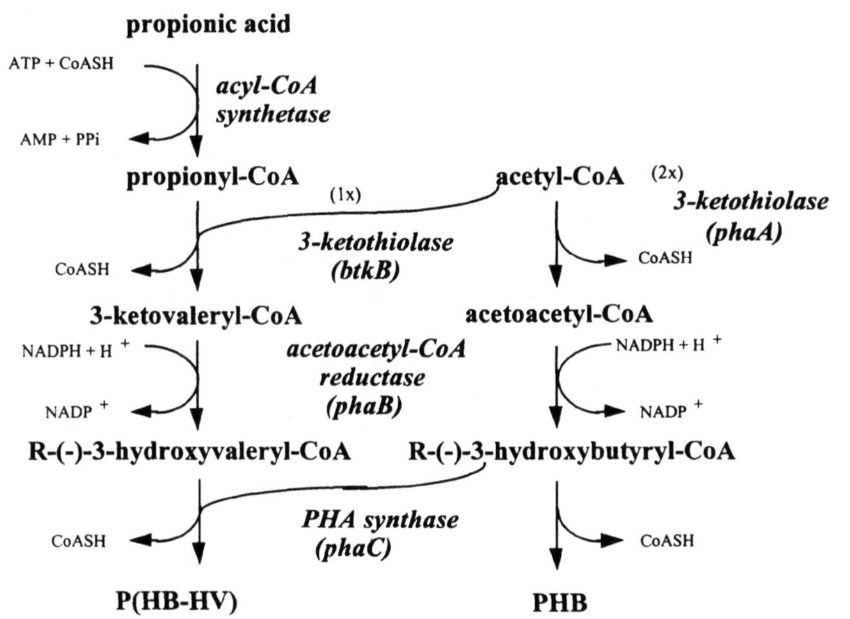

Figure 1. The genes encoding the main enzymes are indicated in brackets.

PHB is a water and air impermeable polymer that is relatively resistant to hydrolytic degradation, making it superior to starch-derived plastics, which are moisture sensitive. PHB is, however, a highly crystalline polymer, making it a relatively stiff and brittle thermoplastic (de Koning, 1995). The polymer is also difficult to process since its melting point (Tm = 175°C) is only slightly lower then the temperature at which it starts to degrade to crotonic acid (approximately 185°C). These properties seriously limit its use in a wide range of commodity products.

Because PHB homopolymer has relatively poor physical properties, extensive efforts have been invested on the synthesis of SCL-PHA co-polymers that have better properties. Incorporation of 3- or 5-carbon monomers into a polymer composed mainly of 3-hydroxybutyrate leads to a decrease in the crystallinity and melting point of the polymer compared to PHB homopolymer (de Koning, 1995). The co-polymer poly(3-hydroxybutyrate-*co*-3-hydroxyvalerate) (P[HB-HV]) is, thus, less stiff and tougher than PHB, as well as easier to process, making it a good target for commercial application (de Koning, 1995). A number of PHAs with different 3- to 5-carbon monomers have been produced in *R. eutropha*, the nature and proportion of these monomers being influenced by the type and relative quantity of the carbon sources supplied to the growth media (Steinbüchel and Schlegel, 1991). For example, addition of propionic acid or valeric acid to the growth media containing glucose leads to the production of a random copolymer composed of 3-hydroxybutyrate and 3-hydroxyvalerate P(HB-HV) (Steinbüchel and Schlegel, 1991). The biochemical pathway of P(HB-HV) synthesis from propionic acid is shown in Figure 1. In *R. eutropha*, condensation of propionyl-CoA with acetyl-CoA is mediated by a distinct 3-ketothiolase, named btkB, which has a higher specificity for propionyl-CoA then the 3-ketothiolase encoded by the *phaA* gene (Slater *et al*, 1998). Reduction of 3-ketovaleryl-CoA to R-3-hydroxyvaleryl-CoA and subsequent polymerization to form P(HB-HV) are catalyzed by the same enzymes involved in PHB synthesis, namely the acetoacetyl-CoA reductase and PHA synthase.

A novel pathway has more recently been elucidated for the synthesis of SCL-PHA containing 4- to 6-carbon monomers in *Aeromonas caviae* and *Rhodospirillum rubrum*. In these bacteria, the substrates R-3-hydroxybutyryl-CoA or R-3-hydroxyhexanoyl-CoA are synthesized through the action of an R-specific enoyl-CoA hydratase, converting crotonyl-CoA or hexenoyl-CoA to the corresponding 3-hydroxyacyl-CoA (Fukui *et al*, 1998; Reiser *et al*, 2000). Since these enoyl-CoAs are intermediates of the fatty acid β-oxidation pathway, synthesis of these SCL-PHAs is, therefore, linked to fatty acid degradation, a pathway that was first discovered for the synthesis of MCL-PHA (see below).

Synthesis of medium-chain polyhydroxyalkanoate

Although MCL-PHAs are typically described as elastomers, their actual physical properties are very diverse and can range from soft plastics to elastomers, rubbers and glues. The properties of MCL-PHAs are directly dependent on the monomer composition (de Koning *et al*, 1994). The range of monomers found in MCL-PHA is particularly broad, with numerous monomers containing functional groups, such as double or triple bonds, phenoxy and halogenated groups (Steinbüchel and Valentin, 1995). The presence of reactive groups in the side chain of MCL-PHA offers interesting opportunities to modify the physical properties of the polymer after extraction. For example, electron-beam irradiation of a MCL-PHA containing unsaturated monomers resulted in the conversion of a soft amorphous polymer into a cross-linked polymer with properties of a rubber (de Koning *et al*, 1994).

There are two major routes for the synthesis of bacterial MCL-PHA (Steinbüchel and Füchtenbusch, 1998). The first is the synthesis of MCL-PHA using intermediates of fatty acid β-oxidation (Figure 2). This pathway is found in several bacteria, such as *Pseudomonas fragii* and *Pseudomonas oleovorans*, which can synthesize MCL-PHA from fatty acids or alkanoic acids. In this pathway, the monomer composition of the PHA produced is directly influenced by the carbon source added to the growth media. Typically, the PHA is composed of monomers that are 2n ($n \geq 0$) carbons shorter than the substrates added to the growth media. For example, growth of *P. oleovorans* on octanoate generates a PHA co-polymer containing H8 and H6 monomers (the prefix H refers to a 3-hydroxyacid), whereas growth on dodecanoate generates a PHA containing H12, H10, H8 and H6 monomers (Lageveen *et al*, 1988). Alkanoic acids present in the media are directed to the β-oxidation pathway where a number of 3-hydroxyacyl-CoA intermediates can be generated. Since the PHA synthase accepts only the R isomer of 3-hydroxyacyl-CoA and the bacterial β-oxidation cycle normally only generates the S isomer of 3-hydroxyacyl-CoA (Gerhart, 1993; Schultz, 1991), bacteria must have enzymes capable of generating R-3-hydroxyacyl-CoA. Several enzymes have been identified which could produce R-3-hydroxyacyl-CoA. One enzyme is a 3-hydroxyacyl-CoA epimerase, mediating the reversible conversion of the S and R isomers of 3-hydroxyacyl-CoA. This activity is found as a part of the multi-functional protein (MFP), the second enzyme participating in the core β-oxidation cycle. The bacterial MFP possesses, in addition to the 3-hydroxyacyl-CoA epimerase activity, an enoyl-CoA hydratase I, S-3-hydroxyacyl-CoA dehydrogenase and a Δ^3-Δ^2-enoyl-CoA isomerase activity (Hiltunen *et al*, 1996). As described in previously for *A. aeromonas* and *R. rubrum*, *Pseudomonas aeruginosa* has been shown to have R-specific mono-functional enoyl-CoA hydratase II enzymes, converting directly enoyl-CoA to R-3-hydroxyacyl-CoA (Tsuge *et al*, 2000). Finally, it is speculated that some bacteria may also have a 3-ketoacyl-CoA reductase that could specifically generate R-3-hydroxyacyl-CoA, although such an enzyme has not yet been unambiguously identified. It has, however, been shown that the enzyme 3-ketoacyl-acyl carrier protein (ACP) reductase, participating normally in the fatty acid biosynthetic pathway, may also act on the 3-ketoacyl-CoA to generate R-3-hydroxyacyl-CoA (Taguchi *et al*, 1999), and thus contribute to MCL-PHA synthesis.

Figure 2: Synthesis of MCL-PHA using intermediates of fatty acid β-oxidation intermediates

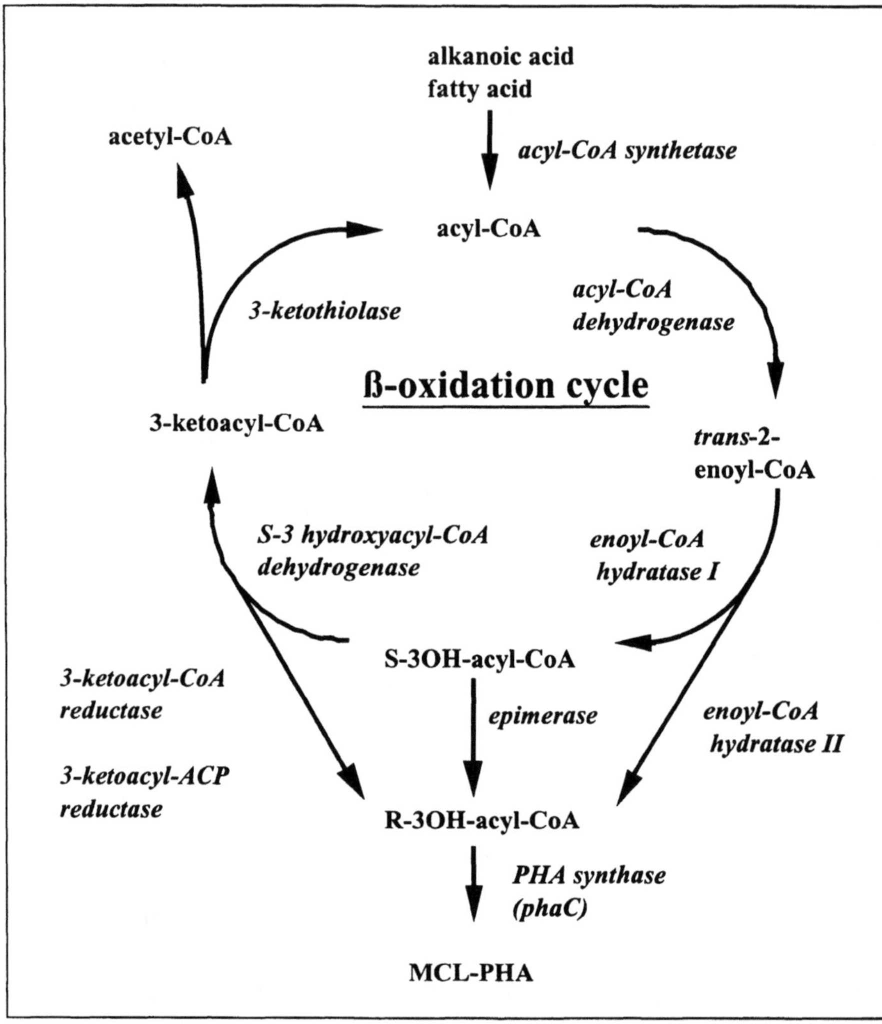

An alternative route for MCL-PHA in bacteria is through the use of intermediates of fatty acid biosynthesis (Steinbüchel and Füchtenbusch, 1998). This pathway is also found in numerous *Pseudomonads*. In contrast to *P. fragii* and *P. oleovorans*, which can only synthesize MCL-PHA from related alkanoic acids present in the growth media, *P. aeruginosa* and *Pseudomonas putida* can synthesize a similar type of MCL-PHA when grown on unrelated substrates, such as glucose (Haywood *et al*, 1990; Timm and Steinbüchel, 1990). Detailed analysis of the composition of PHA produced by *P. putida* grown on glucose revealed the presence of unsaturated monomers that were structurally identical to the acyl-moieties of the R-3-hydroxyacyl-ACP intermediates of *de novo* fatty acid biosynthesis (Huijberts *et al*, 1992). The implication that intermediates from the *de novo* fatty acid biosynthetic pathway could be used to form PHAs was further supported by studies using ^{13}C-labelled acetate (Huijberts *et al*, 1994; Saito and Doi, 1993). A key enzyme linking PHA synthesis and fatty acid biosynthesis has first been identified in *P. putida* and the corresponding gene cloned. This protein, named phaG, was shown to have a 3-hydroxyacyl-CoA-ACP transferase activity (Rehm *et al*, 1998). Expression of the corresponding gene in *P. fragii* and *P. oleovorans* confers to these bacteria the novel capacity to synthesize PHA from glucose (Fiedler *et al*, 2000; Hoffmann *et al*, 2000; Rehm *et al*, 1998).

SYNTHESIS OF POLYHYDROXYALKANOATE IN PLANTS

Introduction

Almost a decade has passed since the first report of the synthesis of a relatively small amount of PHA in the plant *A. thaliana*. Since then, synthesis of various SCL- and MCL-PHAs has been demonstrated in several plants, including *A. thaliana*, rapeseed, tobacco, corn and cotton. The PHA biosynthetic pathway has successfully been introduced either in the cytoplasm, plastid or peroxisome. In this section, the status of PHA synthesis in the cytoplasm and plastid will be briefly reviewed before describing the synthesis of PHA in peroxisomes in greater details.

Synthesis of PHA in the cytoplasm and plastid

Despite its relatively poor physical properties as a thermoplastic, PHB was initially targeted for production in plants because the first bacterial PHA biosynthetic genes that were cloned were for PHB synthesis in the bacterium *R. eutropha* (Peoples *et al*, 1989a, b; Slater *et al*, 1988). The *R. eutropha phaB* and *phbC* genes, encoding, respectively, the acetoacetyl-CoA reductase and PHA synthase, were expressed in *A. thaliana* under the control of the cauliflower mosaic virus (CaMV) 35S promoter. The highest amount of PHB measured in the shoots of these hybrid plants was approximately 0.1% dwt (Poirier *et al*, 1992a). Analysis of the transgenic plant cell by electron microscopy revealed that PHB accumulated in the form of inclusions that had a similar size and appearance compared to bacterial PHA inclusions. Plants expressing high level of acetoacetyl-CoA reductase in the cytoplasm have shown a strong reduction in growth (Poirier *et al*, 1992b). Although the reasons for the dwarf phenotype have not been unambiguously determined, it was hypothesized that the diversion of cytoplasmic acetyl-CoA and acetoacetyl-CoA away from the endogenous isoprenoid and flavonoid pathways might lead to a depletion of essential metabolites which may affect growth. Similar results have also been obtained with the expression of the PHB biosynthetic pathway in the cytoplasm of cells of *B. napus* (Poirier, 2001).

PHB production has also been demonstrated in the cytoplasm of cotton fibre cells (John, 1997; John and Keller, 1996). In this approach, PHA is not produced as a source of polyester to be extracted and used by the plastic industries, but rather as an intracellular agent that modifies the heat exchange properties of the fibre. The *phaA*, *phaB* and *phaC* genes from *R. eutropha* were expressed in transgenic cotton under the control of a fibre specific promoter (John, 1997; John and Keller, 1996). PHB accumulated in the cytoplasm up to 0.3% dwt of the mature fibre.

Since the limited supply of cytoplasmic acetyl-CoA was thought to be the main factor limiting PHB accumulation and causing reduction in plant growth in the first-generation transgenic plants, expression of the PHB pathway in a compartment with a higher flux through acetyl-CoA, such as the plastid, was thought to be a potential solution. The *R. eutropha* phaA protein, encoding the 3-ketothiolase, as well as phaB and phaC proteins, were thus modified for plastid targeting and the genes were individually expressed in *A. thaliana* under the control of the constitutive CaMV 35S promoter. Following combination of all three genes through cross-fertilization, hybrid transgenic plants were shown to produce PHB up to 14% of the shoot dwt (Nawrath *et al*, 1994). Transmission electron microscopy revealed that PHB inclusions accumulated exclusively in the plastids, with some of these organelles having a substantial portion of their volume filled with inclusions. Recently, it has been shown that transformation of *A. thaliana* with the same three *pha* genes, but combined on a single T-DNA vector, led to the isolation of plants producing up to 40% dwt PHB (Bohmert *et al*, 2000). While plants containing 3% dwt showed only a relatively small reduction in growth, plants accumulating between 30-40% dwt PHB were dwarf and produced no seeds (Bohmert *et al*, 2000). As previously observed by Nawrath *et al* (1994), all plants producing above 3% dwt PHB showed some chlorosis (Bohmert *et al*, 2000). These results indicated that, although the plastid can accommodate a higher production of PHB with reduced impact on plant growth compared to the cytoplasm, there was, nevertheless, a limit above which alteration in some of the chloroplast functions was evident (Bohmert *et al*, 2000; Nawrath *et al*, 1994). Similar results have been obtained by scientists at Monsanto who have demonstrated the production of PHB in the plastids of corn leaves and stalks up to 5.7% dwt (Mitsky *et al*, 2000).

In view of the potential limitation of producing PHB in the chloroplast of green tissues, such as leaves, it has been an important goal to evaluate the synthesis of PHB in the leukoplasts of seeds of an oil crop. This has been accomplished by the successful production of PHB in oilseed leukoplasts of *B. napus* transformed with the three modified bacterial genes *phaA*, *phaB* and *phaC* expressed under the control of the fatty acid hyroxylase promoter from *Lesquerella fendeleri* (Houmiel *et al*, 1999; Valentin *et al*, 1999). PHB levels up to 7.7% fwt of mature seeds was reported with no significant effect on seed viability or germination (Houmiel *et al*, 1999).

In view of the poor physical properties of PHB, synthesis of the more flexible co-polymer P(HB-HV) in plants was attractive. As described earlier, synthesis of P(HB-HV) in the bacterium *R. eutropha* relies on the production of propionyl-CoA. The strategy adopted by Slater *et al* (1999) to produce propionyl-CoA in the plastid was the conversion of 2-ketobutyrate to propionyl-CoA by the pyruvate dehydrogenase complex (PDC). Since PDC would have to compete for the 2-ketobutyrate with the acetolactate synthase, an enzyme involved in isoleucine biosynthesis, the quantity of 2-ketobutyrate present in the plastid was enhanced through the expression of the *E. coli ilvA* gene, which encodes a threonine deaminase (Slater *et al*, 1999).

The genes encoding the *E. coli ilvA*, the *R. eutropha phaB* and *phaC*, as well as the *bktB* gene from *R. eutropha* encoding a novel 3-thiolase having high affinity for both acetyl-CoA and propionyl-CoA (Slater *et al*, 1998), were all modified to add a plastid leader sequence to the enzymes. Constitutive expression of the ilvA protein along with bktB, phaB and phaC proteins in the plastids of *A. thaliana* lead to the synthesis of P(HB-HV) in the range of 0.1-1.6% dwt, with the fraction of 3-hydroxyvalerate units being between 2-17 mol% (Slater *et al*, 1999). Expression of the same pathway in the developing seeds of rape showed accumulation of P(HB-HV) in the range of 0.7-2.3 % dwt , with a 3-hydroxyvalerate content of 2.3-6.4 mol% (Slater *et al*, 1999).

Synthesis of polyhydroxyalkanoate in the peroxisome

Synthesis of medium-chain-length polyhydroxyalkanoate

As previously described, MCL-PHAs form a large group of polymers with a wide spectrum of physical properties, ranging from soft plastics to glues. The approach used to synthesize this group of polymer in plants was to divert the 3-hydroxyacyl-CoA intermediates of the peroxisomal β-oxidation of endogenous fatty acids for MCL-PHA production (Mittendorf *et al*, 1998). The phaC1 synthase from *P. aeruginosa* was thus modified at the carboxy-end by the addition of the last thirty four amino acids from the isocitrate lyase of *B. napus*, which harbour a type I peroxisomal targeting sequence. Addition of this thirty four amino acid peptide to the chloramphenicol acetyl transferase was previously shown to target the foreign protein to the peroxisome (Olsen *et al*, 1993). The modified *phaC1* gene was expressed under the control of the CaMV35S promoter and transformed into *A. thaliana* (Mittendorf *et al*, 1998). Targeting of the PHA synthase in plant peroxisomes was analyzed by immunolocalization using polyclonal anti-PHA synthase antibodies. In addition to confirming the correct localization of the bacterial enzyme to the peroxisome, TEM also showed the presence of electron-lucent inclusions within the peroxisomes (Figure 3) (Mittendorf *et al*, 1998). These inclusions had a size and a general appearance similar to bacterial PHA granules as well as PHA granules found in the plant cell cytoplasm and plastids. Furthermore, the peroxisomes of MCL-PHA-producing plants were significantly enlarged compared to peroxisomes of wild type plants. This situation is reminiscent of the enlargement of the leukoplasts of transgenic *B. napus* producing PHB in the plastids of developing embryos (Houmiel *et al*, 1999), indicating that these different organelles can adjust their size to accommodate the accumulation of inclusions.

Figure 3: Accumulation of MCL-PHA inclusions in the peroxisome

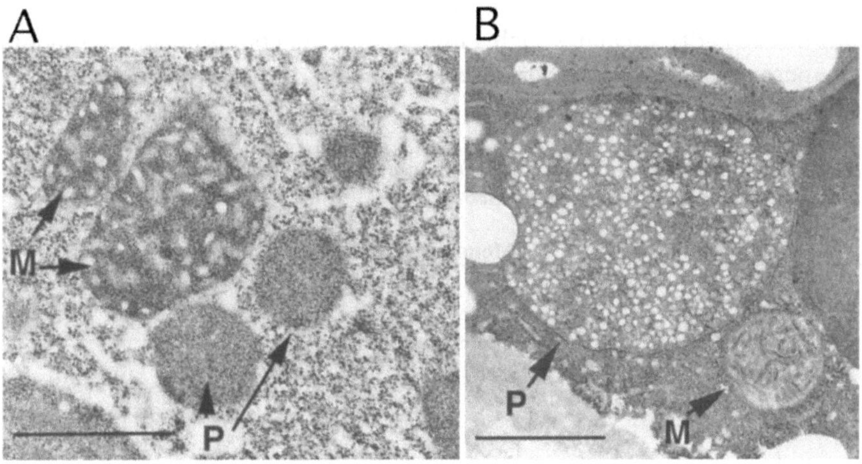

Figure 3. Peroxisomes (P) and mitochondria (M) are shown for a cotyledon cell of a wild type A. thaliana plant (A) and of a transgenic plant accumulating MCL-PHA inclusions in the peroxisomes (B). Bars represent 1 μm.

The monomer composition of the MCL-PHA produced in plants is complex and reflects both the broad substrate specificity of the PHA synthase of *P. aeruginosa* as well as the nature of fatty acids present in plant cells. Thus, peroxisomal PHA was composed of fourteen different monomers, including saturated and unsaturated monomers ranging from 6 to 16 carbons (Table 1) (Mittendorf *et al*, 1998). The 3-hydroxyacids found in plant MCL-PHA could be clearly linked to the corresponding 3-hydroxyacyl-CoA generated by the β-oxidation of saturated and unsaturated fatty acids (Figure 4). The production of MCL-PHA in *A. thaliana* was relatively low, with a maximal level of 0.4% dwt in seven-day-old germinating seedlings. In mature leaves, PHA levels decreased to approximately 0.01-0.02 % dwt. This decrease is not thought to reflect the degradation of PHA over time, since plants do not have PHA depolymerases, but rather the fact that in expanding green tissues the plant weight increases faster than the rate of PHA synthesis. Interestingly, a two- to three-fold increase in PHA was observed in senescing tissues. Together, these data support the link between β-oxidation and PHA synthesis, since the former pathway is most active during germination and senescence. In contrast to PHB synthesis in the cytoplasm and plastid, no negative effects of low level peroxisomal MCL-PHA synthesis on plant growth or seed germination were observed (Mittendorf *et al*, 1998). It remains possible, however, that adverse effects may arise if yields of MCL-PHA reaches such a high level as to compromise the function of the peroxisome or the carbon reserve available for use by the germinating seedlings.

Table 1: Monomer composition of PHA produced in transgenic plant expressing a PHA synthase with/without a caproyl-ACP thioesterase

		Monomer composition (mol%)[1]													
	PHA % dwt	**6**	**8**	**8:1**	**10**	**12**	**12:1**	**12:2**	**14**	**14:1**	**14:2**	**14:3**	**16**	**16:2**	**16:3**
PHAC3.3	0.01	3.1	13	21	4.7	4.8	2.8	7.5	3.8	4.6	3.7	13	3.0	5.3	10
TP 2.4	0.08	7.2	37	5.8	37	3.6	0.3	1.1	2.4	0.6	0.4	1.4	1.8	0.3	1.5

[1]Monomer composition of MCL-PHA isolated from 40-day-old leaves. Transgenic plant PHAC3.3 expresses only the P. aeruginosa PHA synthase in the peroxisome while line TP 2.4 expresses both the P. aeruginosa PHA synthase in the peroxisome and the C. lanceolata caproyl-ACP thioesterase in the plastid. Adapted from Mittendorf et al (1999).

Similar to the PHA synthase from *R. eutropha*, the PHA synthase of *P. aeruginosa* is thought to accept only the R-isomer of 3-hydroxyacyl-CoAs. The wide range of monomers found into plant MCL-PHA suggests that, as with bacteria, plants also have enzymes capable of converting the β-oxidation intermediates to R-3-hydroxyacyl-CoA (Figure 5). One source of R-3-hydroxyacyl-CoA could come from the hydration of *cis*-2-enoyl-CoA by the enoyl-CoA hydratase I activity of the plant MFP. Whereas the enoyl-CoA hydratase I of the plant MFP converts the usual β-oxidation intermediate *trans*-2-enoyl-CoA to S-3-hydroxyacyl-CoA, the same enzyme will produce R-3-hydroxyacyl-CoA from *cis*-2-enoyl-CoA (Hiltunen *et al*, 1996). This later substrate is generated after several rounds of the β-oxidation cycle for unsaturated fatty acids having a *cis* double bond at an even position (see later for further discussion). From this pathway, degradation of linoleic (18:1 *cis*Δ9,12) or related fatty acids would generate only the substrate R-3-hydroxyoctanoyl-CoA while degradation of linonelic acid (18:1 *cis*Δ9,12,15) and related fatty acids would generate the substrate

R-3-hydroxyoctenoyl-CoA. Although the corresponding saturated and unsaturated 8-carbon monomers are abundant in peroxisomal PHA, this pathway clearly cannot account for the twelve other monomers found in the PHA. One potential enzyme that could generate a broad range of R-3-hydroxycayl-CoAs would be the 3-hydroxyacyl-CoA epimerase present on the plant MFP (Hiltunen *et al*, 1996). Alternatively, a monofunctional enoyl-CoA hydratase II has been detected in cucumber (Engeland and Kindl, 1991) which could be the homologue of the enoyl-CoA hydratase II isolated from various bacteria producing PHA (Fukui *et al*, 1998, 1999; Reiser *et al*, 2000).

Figure 4: Spectrum of 3-hydroxyacyl-CoAs (denoted by the prefix H) generated by the degradation of the four major plant fatty acids through the β-oxidation cycle

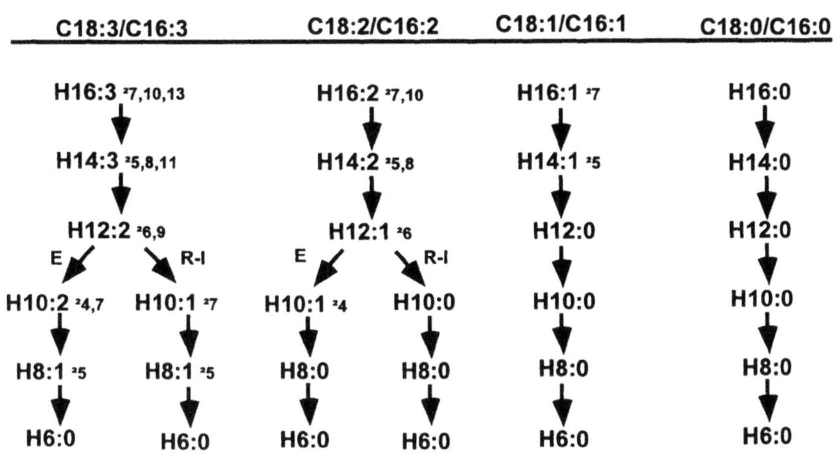

Figure 4. For fatty acids with a cis unsaturated bond at an even-carbon, two pathways of degradation are possible, namely the epimerase (E) and reductase-isomerase (R-I) pathways. All potential 3-hydroxyacids ranging from 6 to 16 carbons are incorporated in plant MCL-PHAs, except for H10:2Δ4,7 and H10:1Δ4.

Figure 5: Modification of fatty acid metabolism for the synthesis of MCL-PHA in plant peroxisomes

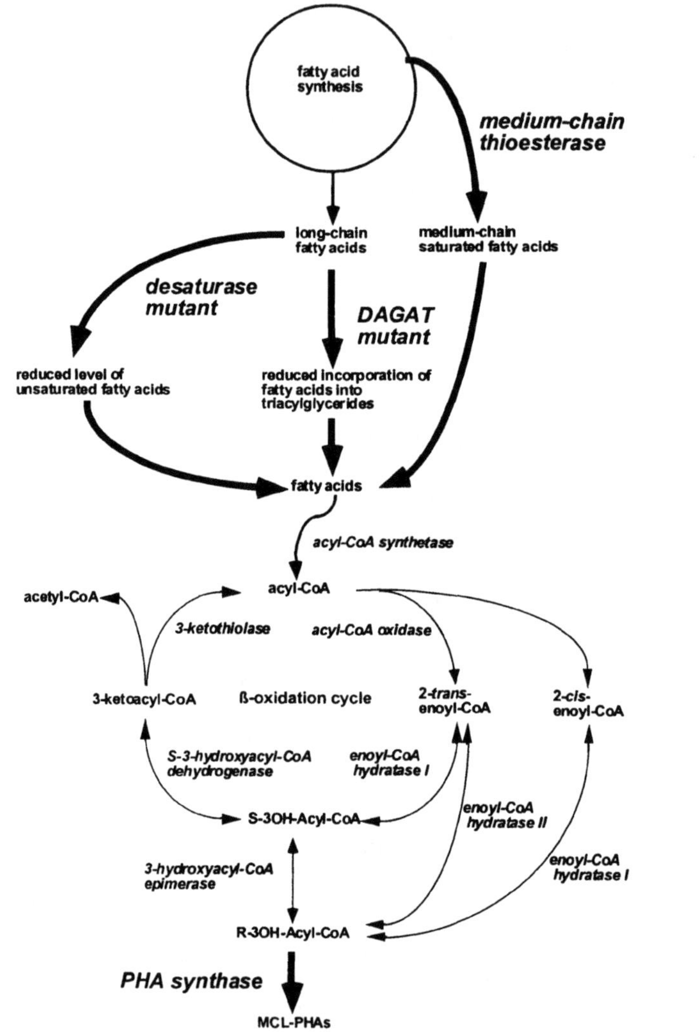

Figure 5. The pathways created or modified by the expression of transgenes from bacteria (PHA synthase) or plant (medium-chain thioesterase), or the use of mutant genes (DAGAT and fatty acid desaturases), are highlighted in bold.

Synthesis of polyhydroxybutyrate

Acetyl-CoA, the building block of PHB biosynthesis, is found not only in the cytoplasm and plastids, but also in peroxisomes, where it is generated by the β-oxidation cycle. Synthesis of PHB in the peroxisome has recently been demonstrated in transgenic Black Mexican Sweet corn suspension cell cultures (Hahn *et al*, 1999). In these experiments, the *phaA*, *phaB* and *phaC* genes from *R. eutropha* were modified in order to add the amino acids RAVARL at the carboxy terminal end of each protein. The terminal tripeptide ARL is a type I peroxisomal targeting signal and has previously been shown to localize the enzyme glycolate oxidase to the peroxisome of tobacco (Volokita *et al*, 1991). Biolistic transformation of maize suspension culture with a mixture of all three genes lead to the isolation of transformants expressing all three enzyme activities and accumulating up to 2% dwt PHB (Hahn *et al*, 1999). No detailed effects of peroxisomal PHB biosynthesis on plant metabolism has been reported. As no transgenic plants have been obtained from these transformed cells, it is also difficult at this point to evaluate the potential effects of peroxisomal PHB synthesis on growth and metabolism. It is, nevertheless, interesting to note that more PHB appears to be produced from acetyl-CoA in cells growing in media supplemented with sucrose, which are likely to have low levels in β-oxidation activity, compared to the synthesis of MCL-PHA from 3-hydroxyacyl-CoA intermediates generated during germination and break down of storage lipids. Although many factors could be involved in producing these differences, including the activity of the two different PHA synthases in the peroxisome, it is interesting to speculate that acetyl-CoA, the final product of the core β-oxidation cycle, may be more available for use by a competing pathway, than 3-hydroxyacyl-CoAs, which are intermediates from the same pathway.

Modulating the monomer composition of polyhydroxyalkanoate

The MCL-PHA synthesized in *A. thaliana* contained a relatively high proportion of monomers larger than ten carbons, as well as high proportion of unsaturated monomers, including di- and tri-unsaturated monomers (Table 1). From similar MCL-PHA produced in bacteria, it is expected that such a polymer would have a low melting point and behave like a glue at room temperature. In order to modify the properties of the MCL-PHA produced in plant peroxisomes to make them more similar to soft plastics or elastomers, it was necessary to decrease the proportion of unsaturated monomers and of longer-chain monomers (≥H12). Furthermore, commercial exploitation of plants for MCL-PHA synthesis would require amounts of PHA of the order of 10-15% dwt. It was, thus, important to find ways of modulating both the quantity and the monomer composition of the MCL-PHA synthesized in plant peroxisomes. This was first achieved by influencing the nature and quantity of fatty acids that were targeted to the β-oxidation cycle (Figure 5) (Mittendorf *et al*, 1999).

Modulation of MCL-PHA quantity and monomer composition was achieved by influencing the nature and quantity of fatty acids that were targeted to the β-oxidation cycle. This was first demonstrated by growing transgenic plants in liquid media supplemented with detergents containing various fatty acids (Mittendorf *et al*, 1999). Addition of free fatty acids or fatty acid esters to the growth media resulted in both an increased accumulation of MCL-PHA and a shift in the monomer composition that reflected the intermediates generated by the β-oxidation of the external fatty acids. For example, addition of the detergent polyoxyethylenesorbitan esterified to lauric acid (Tween-20) resulted in a eight- to ten-fold increase in the amount of PHA synthesized in fourteen day old plants, compared to plants growing in media without detergent. The monomer composition of the MCL-PHA synthesized in plants grown in media containing Tween-20 showed a large increase in the proportion of saturated even-chain monomers with ≤ 12 carbons, and a corresponding decrease in the proportion of all unsaturated monomers. This shift in PHA monomer composition is accounted for by the fact that β-oxidation of lauric acid, a 12 carbon saturated fatty acid, generates saturated 3-hydroxyacyl-CoA intermediates of 12 carbons and lower. Additional feeding experiments have shown that addition to the plant growth media of either tridecanoic acid, tridecenoic acid (C13:1 Δ12) or 8-methyl-nonanoic acid resulted in the production of MCL-PHA containing mainly saturated odd-chain monomers, unsaturated odd-chain monomers with the double bond at the terminal carbon, or branched-chain 3-hydroxyacid monomers, respectively (Mittendorf *et al*, 1999). These results were interesting in that they demonstrated that the plant β-oxidation cycle was capable of generating a large spectrum of monomers which can be included in MCL-PHA, even from fatty acids which are not present to significant quantities in a plant such as *A. thaliana*. Furthermore, experiments with these unusual fatty acids demonstrated that all 3-hydroxyacids between 6 and 16 carbons that could be generated by the β-oxidation cycle (via the 3-hydroxyacyl-CoA intermediate) were found in the MCL-PHA. These results support the concept that analysis of PHA monomer composition produced from unusual fatty acids can be used as a tool to study the degradation pathway of such fatty acids (see later).

Although the strategy of feeding external fatty acids to plant cells could bring about a dramatic shift in MCL-PHA composition, this approach was impractical in the context of agricultural production of PHA. Strategies were thus sought to improve PHA composition by modifying the endogenous fatty acid biosynthetic pathway (Mittendorf *et al*, 1999). The first example of this approach was the expression of the peroxisomal PHA synthase in two mutants of *A. thaliana* deficient in fatty acid desaturation. The *fad2-1* mutant is deficient in the extra-plastidial 18:1 desaturase, resulting in an increase in mono-unsaturated fatty acids and in a decrease in di-unsaturated and tri-unsaturated fatty acids present in leaf, root and seed lipids (Miquel and Browse, 1992). The triple mutant *fad3/fad7/fad8* shows a large reduction in the 16:2 and 18:2 desaturase, resulting in a large decrease in the synthesis of the tri-unsaturated fatty acids (McConn and Browse, 1996). Expression of the PHA synthase in the *fad2-1* mutant background resulted in the production of a PHA with an increased proportion of monomers derived from mono-unsaturated fatty acids (H16:1, H14:1) or both mono-unsaturated and saturated fatty acids (H12, H10, H6), and a decrease in monomers derived from tri-unsaturated fatty acids (H14:3, H12:2, H8:1) and di-unsaturated fatty acids (H16:2, H14:2, H12:1) (Mittendorf *et al*, 1999). Expression of the PHA synthase in the *fad3/fad7/fad8* background produced a PHA that was almost completely deficient in the monomers derived from tri-unsaturated fatty acids (H16:3, H14:3, H12:2 and H8:1) (Mittendorf *et al*, 1999). Since numerous fatty acid desaturases have now been cloned and expressed in transgenic plants to control the number and position of unsaturated bonds in fatty acids, this approach could be extended to further modulate the proportion of a number of 3-hydroxyacid monomers in PHAs.

Studies on futile cycling of fatty acids

The first transgenic plants that were created for the synthesis of a novel exotic fatty acid were *A. thaliana* and *B. napus* expressing the lauroyl-ACP thioesterase from the California bay (Voelker *et al*, 1992). Although expression of the lauroyl-ACP thioesterase under the control of the constitutive CaMV35S promoter led to lauric acid accumulation in the triacylglycerides of the seed, no lauric acid could be found in leaves, despite the fact that CaMV35S is known to be a strong promoter in this tissue. Surprisingly, studies on chloroplasts isolated from these transgenic plants demonstrated that a large fraction of the fatty acids synthesized was lauric acid (Eccleston *et al*, 1996). Interestingly, it was noted that the isocitrate lyase activity was increased in the leaves of plants expressing the thioesterase (Eccleston *et al*, 1996). Collectively, these data supported the hypothesis that in vegetative tissues, the newly synthesized lauric acid was degraded via the β-oxidation cycle instead of accumulating in membrane lipids, thus creating a futile carbon cycle (Eccleston *et al*, 1996). The presence of a futile cycle was, however, not restricted to vegetative tissues, but could also be shown for developing seeds accumulating lauric acid in triacylglycerides. Analysis of the fate of lauric acid in developing *B. napus* seeds expressing high levels of lauroyl-ACP thioesterase also revealed that a substantial portion of fatty acids were converted to water-soluble compounds (Eccleston and Ohlrogge, 1998). These data, combined with the observed increase in acyl-CoA oxidase activity in developing transgenic seeds, indicated that even in developing seeds accumulating lauric acid in triacylglycerides, a substantial portion of lauric acid could be recycled through the β-oxidation cycle.

These studies on lauric acid-producing rapeseed indicated that expression of a thioesterase might be used to increase the flux of saturated medium-chain fatty acids towards peroxisomal PHA biosynthesis (Figure 5). This hypothesis was tested in *A. thaliana* by combining the constitutive expression of the peroxisomal PHA synthase with the *fatB3* gene from *Cuphea lanceolata* encoding a plastidial caproyl-ACP thioesterase (Mittendorf *et al*, 1999). Expression of both enzymes lead to a seven- to eight-fold increase in the amount of MCL-PHA synthesized in mature plant shoots as compared to transgenics expressing only the PHA synthase (Table 1). Furthermore, the composition of the MCL-PHA produced in the thioesterase/PHA synthase double transgenic plant was strongly shifted towards saturated 3-hydroxyacid monomers containing 10 carbons and less. This shift is in agreement with an increase in the flux of decanoic acid towards β-oxidation triggered by the expression of the caproyl-ACP thioesterase (Mittendorf *et al*, 1999).

These results clearly indicate that futile cycling of fatty acids in transgenic *A. thaliana* expressing a medium-chain fatty acyl-ACP thioesterase could be detected by peroxisomal PHA. Interestingly, constitutive expression of the California bay lauroyl-ACP thioesterase in *A. thaliana* was shown not to lead to an increase in the genes or enzymes involved in β-oxidation in vegetative tissues (Hooks *et al*, 1999). It is hypothesized that the level of activity of the β-oxidation cycle found in vegetative tissues of wild type *A. thaliana* are sufficiently high to accommodate the degradation of lauric acid in these tissues. Together, these results showed that analysis of the quantity and monomer composition of PHA synthesized in the peroxisome could be a more sensitive indicator of the flow of fatty acids towards β-oxidation than the activity of genes or enzymes involved in β-oxidation.

The relation between fatty acid futile cycling in plants expressing a medium-chain acyl-ACP thioesterase and peroxisomal PHA synthesis was further extended to the developing seeds (Poirier *et al*, 1999). Synthesis of MCL-PHA has been demonstrated in seeds of *A. thaliana* by expressing the peroxisomal PHA synthase gene under the control of the seed-specific napin promoter of *B. napus*. In such transgenic plants, MCL-PHAs accumulated to 0.006% dwt in mature seeds and the monomer composition was similar to the PHA synthesized in germinating seedlings, indicating a minimal turn-over of fatty acids in developing seeds (Poirier *et al*, 1999). Expression of both the PHA synthase and FatB3 caproyl-ACP thioesterase in the leukoplasts of developing seeds resulted in a nearly twenty-fold increase in seed PHA, reaching 0.1% dwt in mature seeds, as well as a large increase in the proportion of 3-hydroxyacid monomers containing $\leq$ 10 carbons in PHA (Poirier *et al*, 1999). These data clearly indicate that even though expression of the caproyl-ACP thioesterase in seeds leads to the accumulation of medium-chain fatty acids in triacylglycerides to levels approaching 20 mol%, there is still a significant proportion of these fatty acids that are channeled towards β-oxidation. This flux towards the β-oxidation cycle in developing seeds is thought to be quite significant, considering that there is only a four-fold difference between the maximal amount of PHA synthesized in *A. thaliana* germinating seedlings (0.4% dwt), where β-oxidation is thought to be maximal, and the PHA synthesized in the developing seeds expressing the thioesterase (0.1% dwt), where metabolism should normally be mainly devoted to the synthesis of fatty acid instead of degradation. These data support the notion that more medium-chain fatty acids are synthesized than can be accumulated in the triacylglycerides, indicating a bottleneck in the incorporation of these fatty acids in storage lipids.

Synthesis of MCL-PHA in the peroxisomes of developing *A. thaliana* seeds has also revealed the presence of an increased cycling of fatty acids towards β-oxidation in a mutant deficient in the enzyme diacylglycerol acyltransferase (DAGAT) (Poirier *et al*, 1999). The *tag1* mutant of *A. thaliana* was shown to be deficient in DAGAT activity in developing seeds, resulting in a decreased accumulation of triacylglycerides and corresponding increase in diacylglycerides and free fatty acids in seeds (Katavic *et al*, 1995). It was hypothesized that the imbalance created between the capacity of the plastid to synthesize fatty acids and the ability of the lipid biosynthetic machinery of the endoplasmic reticulum to include these fatty acids into triacylglycerides might have two basic consequences. These would be that either fatty acid biosynthesis would be feedback-inhibited in order to match it with triacylglyceride biosynthesis, or that excess fatty acids that cannot be included in triacylglycerides would be directed towards peroxisomal β-oxidation. Expression of the peroxisomal PHA synthase in the *tag1* mutant resulted in a ten-fold increase in the amount of MCL-PHA accumulating in mature seeds compared to expression of the transgene in wild type plants (Poirier *et al*, 1999). Although these results do not directly address whether fatty acid biosynthesis is decreased in the *tag1* mutant, they clearly indicate that a decrease in triacylglyceride biosynthesis results in a large increase in the flux of fatty acids towards β-oxidation (Figure 5). Thus, carbon flux to the β-oxidation cycle can be modulated to a great extent and appears to play an important role in lipid homeostasis in plants, even in tissues which are primarily devoted to lipid biosynthesis, such as the developing seeds.

Studies on the degradation of unsaturated fatty acids

As discussed earlier, changes detected in the monomer composition of peroxisomal PHA produced in plants that are either fed with external fatty acids, deficient in the synthesis of particular unsaturated fatty acid, or expressing a medium-chain acyl-ACP thioesterase, clearly reflect both the nature of the fatty acid being targeted to the peroxisome and how this fatty acid is degraded by the β-oxidation pathway (Mittendorf *et al*, 1999). It is, thus, possible to use peroxisomal PHA as a tool to elucidate the pathways involved in the degradation of fatty acids, including unsaturated and unusual fatty acids. Knowledge on the biochemistry of fatty acid degradation could have an important impact in plant biotechnology projects aimed at producing transgenic plants accumulating valuable fatty acids, such as vernolic acid and ricinoleic acid (van de Loo *et al*, 1993). This is because normal germination of transgenic seeds accumulating an exotic fatty acid may require the presence of additional specialized enzymes required to handle the presence of novel groups on the fatty acids, such as epoxy or hydroxy groups, during fatty acid degradation (Gerhart, 1993). Furthermore, as revealed by the work with the *C. lanceolata* caproyl-ACP thioesterase, peroxisomal MCL-PHA may be used to measure the extend of the loss of exotic fatty acid in seeds through futile cycling.

The usefulness of using peroxisomal MCL-PHAs to dissect the biochemical pathway of fatty acid degradation in plants has recently been demonstrated for the β-oxidation of unsaturated fatty acids (Allenbach and Poirier, 2000). Degradation of fatty acids with *cis*-double bonds on even-numbered carbons requires the presence of auxiliary enzymes in addition to the enzymes of the core β-oxidation cycle. This is because hydration of *cis*-2-enoyl-CoA by the 2-enoyl-CoA hydratase I present in the MFP generates the R-isomer of 3-hydroxyacyl-CoA, which is not a substrate for the S-3-hydroxyacyl-CoA dehydrogenase. Two alternative pathways have been described to degrade these fatty acids (Figure 6) (Schulz, 1991). One pathway, which is referred as the reductase-isomerase pathway, involves the participation of the enzymes 2,4-dienoyl-CoA reductase and Δ^3-Δ^2-enoyl-CoA isomerase, resulting in the conversion of *trans*-2,*cis*-4-dienoyl-CoA to *trans*-2-enoyl-CoA, which can be degraded via the core β-oxidation cycle. The second pathway, which is referred as the epimerase pathway, involves epimerization of R-3-hydroxyacyl-CoA to the S isomer via the 3-hydroxyacyl-CoA epimerase present on the MFP or the action of two stereo-specific enoyl-CoA hydratases. Whereas degradation of fatty acids with *cis*-double bonds on even-numbered carbons in bacteria and mammalian peroxisomes was shown to mainly involve the reductase-isomerase pathway (Yang *et al*, 1986), analysis of the relative activity of the enoyl-CoA hydratase II and 2,4-dienoyl-CoA reductase in plants indicated that degradation occurred mainly, if not exclusively, through the epimerase pathway in plants (Engeland and Kindl, 1991).

Figure 6: Alternative pathways for the degradation of fatty acids with cis-double bonds on even-numbered carbons

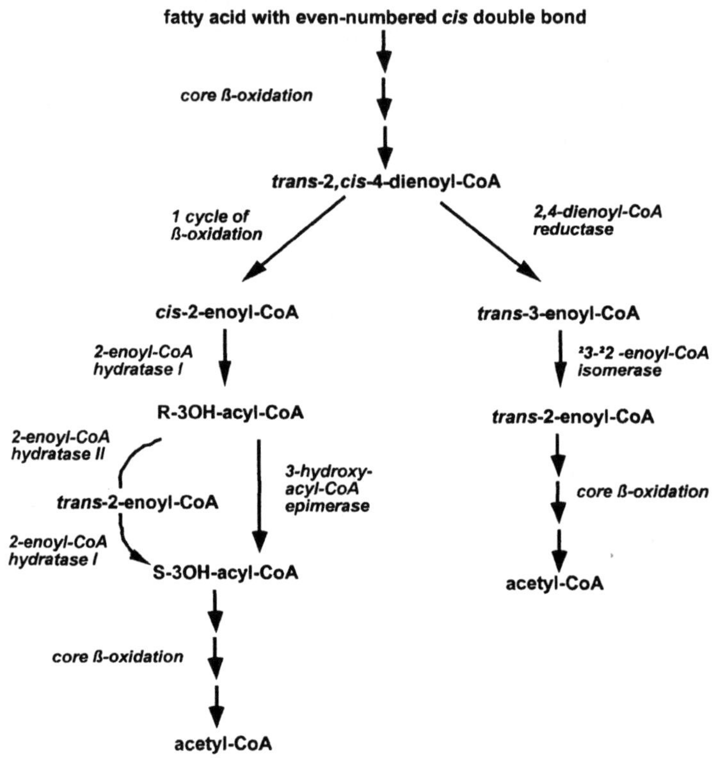

Figure 6. The right branch is referred to as the reductase-isomerase pathway, while the left branch is referred to as the epimerase pathway.

The relative contribution of the reductase-isomerase and epimerase pathways could be examined in transgenic plants synthesizing peroxisomal PHA since the degradation of *cis*-10-heptadecenoic or *cis*-10-pentadecenoic acids via these two different pathways results in the introduction of some distinctive 3-hydroxyacid monomers in PHA (Figure 7) (Allenbach and Poirier, 2000). For example, degradation of *cis*-10-pentadecenoic acid via the epimerase pathway would generate the unsaturated intermediate S-3-hydroxy-*cis*-4-nonenoyl-CoA and the saturated intermediate R-3-hydroxyheptanoyl-CoA. In contrast, degradation of the same fatty acid through the reductase-isomerase pathway would generate the distinctive intermediate S-3-hydroxynonanoyl-CoA as well as S-3-hydroxyheptanoyl-CoA. Thus, incorporation of 3-hydroxynonanoic acid in PHA from plants fed with *cis*-10-pentadecenoic acid would indicate contribution of the reductase-isomerase pathway. Although incorporation of the unsaturated 3-hydroxy-*cis*-4-nonenoic acid monomer in PHA could, in theory, be an indicator of the epimerase pathway, this monomer is undetectable in PHA, presumably because the presence of the double bond near the 3-hydroxy group creates a steric hindrance precluding the use of the monomer by the PHA synthase. However, the distinctive feature of the degradation of *cis*-10-pentadecenoic acid via the epimerase pathway is the direct synthesis of the R- isomer of 3-hydroxyheptanoyl-CoA, leading to an increase proportion of this monomer in PHA compared to PHA synthesized from *trans*-10-pentadecenoic acid. Analysis of PHA produced from transgenic plants fed with *cis*-10-pentadecenoic acid revealed the presence of both 3-hydroxynonanoic acid as well as an increased proportion of 3-hydroxyheptanoic acid compared to PHA from plants fed with *trans*-10-pentadecenoic acid (Allenbach and Poirier, 2000). These results indicated a significant contribution of both the reductase-isomerase and epimerase pathways in the degradation of fatty acids with *cis*-double bonds on even-numbered carbons. These results were further confirmed by comparing the PHA monomer composition in plants fed with either *cis*-10-heptadecenoic or *trans*-10-heptadecenoic acid (Allenbach and Poirier, 2000).

Figure 7: Spectrum of 3-hydroxyacyl-CoAs generated by the degradation of cis-10-pentadecenoic acid

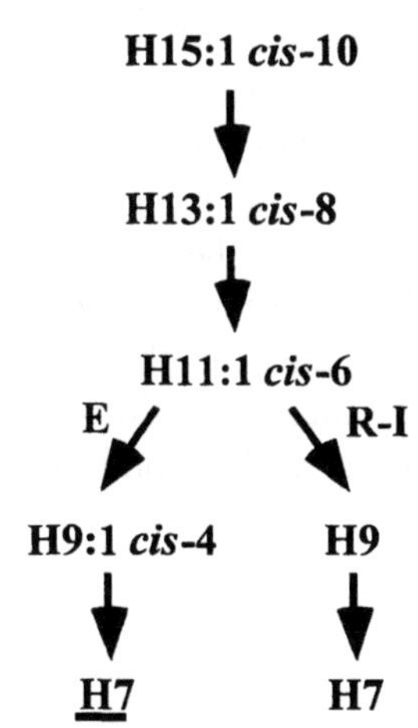

Figure 7. Only 3-hydroxyacyl-CoAs between 7 to 15 carbons are indicated. The epimerase (E) and reductase-isomerase (R-I) are indicated. All 3-hydroxyacyl-CoAs generated by the β-oxidation of unsaturated fatty acids are in the S-configuration, with the exception of R-3-hydroxyheptanoyl-CoA (underlined) generated by the epimerase pathway.

CONCLUSIONS

Beyond its direct impact in plant biotechnology as a biopolymer production system, synthesis of PHA in plants can also be exploited as a novel tool to study metabolic pathways. Because PHA represents a largely un-recyclable carbon sink in plants, it can be used to modify or monitor carbon flux to various pathways.

From the biotechnological aspect of biopolymer production, the main challenge for the synthesis of MCL-PHA in the peroxisome is to accomplish levels of production that will go beyond the present 0.1% dwt to reach 10-15% dwt. It is likely that further metabolic engineering of the β-oxidation cycle will be required in order to accomplish this goal. Nevertheless, peroxisomal MCL-PHA has so far proven to be a valuable tool to study a number of aspects of fatty acid metabolism, including the presence of fatty acid futile cycling and the pathways involved in the degradation of unsaturated fatty acids.

Although the potential of PHA as a novel analytical tool has been so far mostly exploited in the context of fatty acid degradation in the peroxisome, it is clear that its use could be expanded to the study of other biochemical pathways. For example, synthesis of PHB from acetyl-CoA present in the cytoplasm may provide valuable insight into how carbon flux to the flavonoid and isoprenoid pathways is regulated. Similarly, high level production of PHA in the plastid raises a number of interesting questions. For example, how does PHB synthesis in the plastids affect carbon flow to other compounds synthesized in the organelle, such as starch and fatty acids? How does the plant adjust, at the metabolic, enzymatic, and genetic levels, to accommodate for the synthesis of this new carbon sink? The novel tools of genomics, proteomics and metabolic profiling could be used to provide interesting answers to these questions and give general insights in plant biochemistry that would go well beyond the synthesis of PHA in plants as a plastic factory.

REFERENCES

Allenbach, L. and Poirier, Y. (2000) Analysis of the alternative pathways for the β-oxidation of unsaturated fatty acids using transgenic plants synthesizing polyhydroxyalkanoates in peroxisomes. *Plant Physiol.* **124**: 1159-1168.

Anderson, A.J. and Dawes, E.A. (1990) Occurrence, metabolism, metabolic role, and industrial uses of bacterial polyhydroxyalkanoates. *Microbiol. Rev.* **54**: 450-472.

Bohmert, K., Balbo, I., Kopka, J., Mittendorf V., Nawrath C., Poirier, Y., Tischendorf G., Trethewey R.N. and Willmitzer L. (2000) Transgenic Arabidopsis plants can accumulate polyhydroxybutyrate to up 4% of their fresh weight. *Planta.* **211**: 841-845.

Braunegg, G., Lefebvre, G. and Genser, K.F. (1998) Polyhydroxyalkanoates, biopolyesters from renewable resources: physiological and engineering aspects. *J. Biotechnol.* **65**: 127-161.

Clemente, T., Shah, D., Tran, M., Stark, D., Padgette, S., Dennis, D., Brückener, K., Steinbüchel, A. and Mitsky, T. (2000) Sequence of PHA synthase gene from two strains of *Rhodospirillum rubrum* and *in vivo* substrate specificity of four PHA synthases across two heterologous expression systems. *Appl. Microbiol. Biotechnol.* **53**: 420-429.

de Koning, G.J.M. (1995) Physical properties of bacterial poly((R)3-hydroxyalkanoates). *Can. J. Microbiol.* **41** (Suppl. 1): 303-309.

de Koning, G.J.M., Van Bilesen, H.M.M., Lemstra, P.J., Hazenberg, W., Withold, B., Preusting, H., Van der Galiën, J.G., Schirmer, A. and Jendrossek, D. (1994) A biodegradable rubber by crosslinking poly(hydroxyalkanoate) from *Pseudomonas oleovorans*. *Polymer.* **35**: 2090-2097.

Eccleston, V.S. and Ohlrogge, J.B. (1998) Expression of lauroyl-acyl carrier protein thioesterase in *Brassica napus* seeds induces pathways for both fatty acid oxidation and biosynthesis and implies a set point for triacylglycerol accumulation. *Plant Cell.* **10**: 613-621.

Eccleston, V.S., Cranmer, A.M., Voelker, T.A. and Ohlrogge, J.B. (1996) Medium-chain fatty acid biosynthesis and utilization in *Brassica napus* plants expressing lauroyl-acyl carrier protein thioesterase. *Planta.* **198**: 46-53.

Engeland, K. and Kindl, H. (1991) Evidence for a peroxisomal fatty acid β-oxidation involving D-3-hydroxyacyl-CoAs: characterisation of two forms of hydrolyase that convert D-(-)-3-hydroxyacyl-CoA. *Eur. J. Biochem.* **200**: 171-178.

Fiedler, S., Steinbuchel, A. and Rehm, B.H.A. (2000) PhaG-mediated synthesis of poly(3-hydroxyalkanoates) consisting of medium-chain-length constituents from non-related carbon sources in recombinant *Pseudomonas fragi*. *Appl. Environ. Microbiol.* **66**: 2117-2124.

Fukui, T. and Doi, Y. (1997) Cloning and analysis of the poly(3-hydroxybutyrate-*co*-3-hydroxyhexanoate) biosynthesis genes of *Aeromonas caviae*. *J. Bacteriol.* **179**: 4821-4830.

Fukui, T., Shiomi, N. and Doi, Y. (1998) Expression and characterization of (R)-specific enoyl coenzyme A hydratase involved in polyhydroxyalkanoate biosynthesis by *Aeromonas caviae*. *J. Bacteriol.* **180**: 667-673.

Fukui, T., Yokomizo, S., Kobayashi, G. and Doi, Y. (1999) Co-expression of polyhydroxyalkanoate synthase and (R)-enoyl-CoA hydratase genes of *Aeromonas caviae* establishes copolyester biosynthesis pathway in *Escherichia coli*. *FEMS Microbiol. Lett.* **170**: 69-75.

Gerhart, B. (1993) Catabolism of fatty acids (α- and β-oxidation). In: *Lipid metabolism in plants*. Moore, T.S. (ed.). pp. 527-565. Florida: CRC Press Inc.

Hahn, J.J., Eschenlauer, A,C., Sleytr, U.B., Somers, D.A. and Srienc, F. (1999) Peroxisomes as sites for synthesis of polyhydroxyalkanoates in transgenic plants. *Biotechnol. Progr.* **15**: 1053-1057.

Hall, B., Baldwin, J., Rhie, H.G. and Dennis, D. (1998) Cloning of the *Nocardia corallina* polyhydroxyalkanoate synthase gene and production of poly-(3-hydroxybutyrate-*co*-3-hydroxyhexanoate) and poly-(3-hydroxyvalerate-*co*-3-hydroxyheptanoate). *Can. J. Micro.* **44**: 687-691.

Haywood, G.W., Anderson, A.J. and Dawes, E.A. (1989) The importance of PHB-synthase substrate specificity in polyhydroxyalkanoate synthesis by *Alcaligenes eutrophus*. *FEMS Microbiol. Lett.* **57**: 1-6.

Haywood, G.W., Anderson, A.J., Ewing, D.F. and Dawes, E.A. (1990) Accumulation of a polyhydroxyalkanoate containing primarily 3-hydroxydecanoate from simple carbohydrate substrates by *Pseudomonas* sp. strain NCIMB 40135. *Appl. Environ. Microbiol.* **56**: 3354-3359.

Hiltunen, J.K., Filppula, S.A., Koivuranta, K.T., Siivari, K., Qin, Y.-M. and Häyrinen, H.-M. (1996) Peroxisomal β-oxidation and polyunsaturated fatty acids. *Ann. N. Y. Acad. Sci.* **804**: 116-128.

Hoffmann, N., Steinbüchel, A., Rehm, B.H. (2000) The *Pseudomonas aeruginosa phaG* gene product is involved in the synthesis of polyhydroxyalkanoic acid consisting of medium-chain-length constituents from non-related carbon sources. *FEMS Microbiol. Lett.* **184**: 253-259.

Hooks, M.A., Fleming, Y., Larson, T.R. and Graham, I.A. (1999) No induction of β-oxidation in leaves of Arabidopsis that over-produce lauric acid. *Planta.* **207**: 385-392.

Houmiel, K.L., Slater, S.. Broyles, D., Casagrande, L., Colburn, S., Gonzalez K., Mitsky, T.A., Reiser, S.E., Shah, D., Taylor, N.B., Tran, M., Valentin, H.E. and Gruys, K.J. (1999) Poly(beta-hydroxybutyrate) production in oilseed leukoplasts of *Brassica napus*. *Planta.* **209**: 547-550.

Huijberts, G.N.M., Eggink, G., de Waard, P., Huisman, G.W. and Witholt, B. (1992) *Pseudomonas putida* KT2442 cultivated on glucose accumulates poly(3-hydroxyalkanoates) consisting of saturated and unsaturated monomers. *Appl. Environ. Microbiol.* **58**: 536-544.

Huijberts, G.N.M., de Rijk, T.C., de Waard, P. and Eggink, G. (1994) ^{13}C nuclear magnetic resonance studies of *Pseudomonas putida* fatty acid metabolic routes involved in poly(3-hydroxyalkanoate) synthesis. *J. Bacteriol.* **176**: 1661-1666.

Jendrossek, D. (2001) Microbial degradation of polyesters. *Adv. Biochem. Engin. Biotechnol.* **71**: 293-325.

Jendrossek, D., Schirmer, A. and Schlegel, H.G. (1996) Biodegradation of polyhydroxyalkanoic acids. *Appl. Microbiol. Biotechnol.* **46**: 451-463.

John, M.E. (1997) Cotton crop improvement through genetic engineering. *Crit. Rev. Biotechnol.* **17**: 185-208.

John, M.E. and Keller, G. (1996) Metabolic pathway engineering in cotton: biosynthesis of polyhydroxybutyrate in fibre cells. *Proc. Natl. Acad. Sci. USA.* **93**: 12768-12773.

Katavic, V., Reed, D.W., Taylor, D.C., Giblin, E.M., Barton, D.L., Zou, J., MacKenzie, S.L., Covello, P.S. and Kunst, L. (1995) Alteration of seed fatty acid composition by an ethyl methanesulphonate-induced mutation in *Arabidopsis thaliana* affecting diacylglycerol acyltransferase activity. *Plant Physiol.* **108**: 399-409.

Kessler, B., Weusthuis, R.A., Witholt, B. and Eggink, G. (2001) Production of microbial polyesters: fermentation and downstream process. *Adv. Biochem. Engin. Biotechnol.* **71**: 159-182.

Kim, Y.B., Lenz, R.W. (2001) Polyesters from microorganisms. *Adv. Biochem. Engin. Biotechnol.* **71**: 51-79.

Lageveen, R.G., Huisman, G.W., Preusting, H., Ketelaar, P., Eggink, G. and Witholt, B. (1988) Formation of polyesters by *Pseudomonas oleovorans*: effect of substrates on formation and composition of poly-(R)-3-hydroxyalkanoates and poly-(R)-3-hydroxyalkenoates. *Appl. Environ. Microbiol.* **54**: 2924-2932.

Lemoigne, M (1926) Produits de déshydration et de polymérisation de l'acide β-oxobutyrique. *Bull. Soc. Chem. Biol. (Paris).* **8**: 770-782.

Madison, L.L. and Huisman, G.W. (1999) Metabolic engineering of poly(3-hydroxyalkanoates): from DNA to plastic. *Microbiol. Mol. Biol. Rev.* **63**: 21-53.

Matsusaki, H, Manji, S., Taguchi, K., Kato, M., Fukui, T. and Doi, Y. (1998) Cloning and molecular analysis of the poly(3-hydroxybutyrate) and poly(3-hydroxybutyrate-co-3-hydroxyalkanoate) biosynthesis genes in *Pseudomonas* sp. Strain 61-3. *J. Bacteriol.* **180**: 6459-6467.

McConn, M. and Browse, J. (1996) The critical requirement for linolenic acid is pollen development, not photosynthesis, in an *Arabidopsis* mutant. *Plant Cell.* **8**: 403-416.

Miquel, M. and Browse, J. (1992) Arabidopsis mutants deficient in polyunsaturated fatty acids synthesis: biochemical and genetic characterization of a plant oleoyl-phosphatidylcholine desaturase. *J. Biol. Chem.* **267**: 1502-1509.

Mitsky, T.A., Slater, S.C., reiser, S.E., Hao, M. and Houmiel, K.L. (2000) Multigene expression vectors for the biosynthesis of products via multi-enzyme biological pathways, PCT application WO 00/52183.

Mittendorf, V., Bongcam, V., Allenbach, L.,Coullerez, G., Martini, N. and Poirier, Y. (1999) Polyhydroxyalkanoate synthesis in transgenic plants as a new tool to study carbon flow through β-oxidation. *Plant J.* **20**: 45-55.

Mittendorf, V., Robertson, E.J., Leech, R.M., Krüger, N., Steinbüchel, A. and Poirier, Y. (1998) Synthesis of medium-chain-length polyhydroxyalkanoates in *Arabidopsis thaliana* using intermediates of peroxisomal fatty acid β-oxidation, *Proc. Natl. Acad. Sci. USA*. **95**: 13397-13402.

Nawrath, C., Poirier, Y. and Somerville, C.R. (1994) Targeting of the polyhydroxybutyrate biosynthetic pathway to the plastids of *Arabidopsis thaliana* results in high-levels of polymer accumulation. *Proc. Natl. Acad. Sci. USA*. **91**: 12760-12764.

Olsen, L.J., Ettinger, W.F., Damsz, B., Matsudaira, K., Webb, M.A. and Harada, J.J. (1993) Targeting of glyoxysomal proteins to peroxisomes in leaves and roots of a higher plant. *Plant Cell*. **5**: 941-952.

Peoples, O.P. and Sinskey, A.J. (1989a) Poly-β-hydroxybutyrate biosynthesis in *Alcaligenes eutrophus* H16. *J. Biol. Chem.* **264**: 15293-15297.

Peoples, O.P. and Sinskey, A.J. (1989b) Poly-β-hydroxybutyrate (Poly(3HB)) biosynthesis in *Alcaligenes eutrophus* H16: Identification and characterization of the Poly(3HB) polymerase gene (*Poly(3HB)C*). *J. Biol. Chem.* **264**: 15298-15303.

Poirier, Y. (2001) Production of poylesters in transgenic plants. *Adv. Biochem. Engin. Biotechnol.* **71**: 209- 240.

Poirier, Y. (1999) Production of new polymeric compounds in plants. *Curr. Opin. Biotechnol.* **10**: 181-185.

Poirier, Y., Dennis, D.E., Klomparens, K. and Somerville, C. (1992a) Polyhydroxybutyrate, a biodegradable thermoplastic, produced in transgenic plants. *Science*. **256**: 20-523.

Poirier, Y., Dennis. D., Klomparens, K., Nawrath, C. and Somerville, C. (1992b) Perspectives on the production of polyhydroxyalkanoates in plants. *FEMS Microbiol. Rev.* **103**: 237-246.

Poirier, Y, Nawrath, C. and Somerville, C. (1995) Production of polyhydroxyalkanoates, a family of biodegradable plastics and elastomers, in bacteria and plants. *Biotechnology*. **13**: 142-150.

Poirier, Y., Ventre, G. and Caldelari, D. (1999) Increased flow of fatty acids towards β-oxidation in developing seeds of *Arabidopsis thaliana* deficient in diacylglycerol acyltransferase activity or synthesizing medium-chain fatty acids. *Plant Physiol.* **121**: 1359-1366.

Rehm, B.H., Krüger, N. and Steinbüchel, A. (1998) A new metabolic link between fatty acid *de novo* synthesis and polyhydroxyalkanoic acid synthesis. The PHAG gene from *Pseudomonas putida* KT2440 encodes a 3-hydroxyacyl-acyl carrier protein-coenzyme a transferase. *J. Biol. Chem.* **273**: 24044-24051.

Reiser, S.E., Mitsky, T.A. and Gruys, K.J. (2000) Characterization and cloning of an (R)-specific trans-2,3-enoylacyl-CoA hydratase from *Rhodospirillum rubrum* and use of this enzyme for PHA production in *Escherichia coli*. *Appl. Microbiol. Biotechnol.* **53**: 209-218.

Saito, Y. and Doi, Y. (1993) Biosynthesis of poly(3-hydroxyalkanoates) in *Pseudomonas aeruginosa* AO-232 from ^{13}C-labelled acetate and propionate. *Int. J. Biol. Macromol.* **15**: 287-292.

Schubert, P., Steinbüchel, A. and Schlegel, H.G. (1988) Cloning of the *Alcaligenes eutrophus* genes for synthesis of poly-β-hydroxybutyric acid (Poly(3HB)) and synthesis of Poly(3HB) in *Escherichia coli*. *J. Bacteriol.* **170**: 5837-5847.

Schulz, H. (1991) Beta-oxidation of fatty acids. *Biochim. Biophys. Acta.* **1081**: 109-120.

Slater, S., Houmiel, K.L., Tran, M., Mitsky, T.A., Taylor, N.B., Padgette, S.R. and Gruys, K.J. (1998) Multiple β-ketothiolases mediate poly(β-hydroxyalkanoate) copolymer synthesis in *Ralstonia eutropha*. *J. Bact.* **180**: 1979-1987.

Slater, S., Mitsky, T.A., Houmiel, K.L., Hao, M., Reiser, S.E., Taylor, N.B. Tran, M., Valentin, H.E., Rodriguez, D.J., Stone, D.A., Padgette, S.R., Kishore, G. and Gruys, K.J. (1999) Metabolic engineering of *Arabidopsis* and *Brassica* for poly(3-hydroxybutyrate-co-3-hydroxyvalerate) copolymer production. *Nature Biotechnol.* **17**: 1011-1016.

Slater, S.C., Voige, W.H. and Dennis, D.E. (1988) Cloning and expression in *Escherichia coli* of the *Alcaligenes eutrophus* H16 poly-β-hydroxybutyrate biosynthetic pathway. *J. Bacteriol.* **170**: 4431-4436.

Steinbüchel, A. (1991) Polyhydroxyalkanoic acids. In *Novel biomaterials from biological sources.* (ed. Byrom, D). New York: MacMillan.

Steinbüchel, A. and Füchtenbusch, B. (1998) Bacterial and other biological systems for polyester production. *Trends Biotechnol.* **16**: 419-427

Steinbüchel, A. and Hein, S. (2001) Biochemical and molecular basis of microbial synthesis of polyhydroxyalkanoates in microorganisms. *Adv. Biochem. Engin. Biotechnol.* **71**: 81-123.

Steinbüchel, A. and Schlegel, H.G. (1991) Physiology and molecular genetics of poly(β-hydroxy-alkanoic acid) synthesis in *Alcaligenes eutrophus*. *Mol. Microbiol.* **5**: 535-542.

Steinbüchel, A. and Valentin, H.E. (1995) Diversity of bacterial polyhydroxyalkanoic acids. *FEMS Microbiol. Lett.* **128**: 219-228.

Sudesh, K. and Doi, Y. (2000) Synthesis, structure and properties of polyhydroxyalkanoates: biological polyesters. *Prog. Polym. Sci.* **25**: 1503-1555.

Taguchi, K., Aoyagi, Y., Matsusaki, H., Fukui, T. and Doi Y. (1999) Co-expression of 3-ketoacyl-ACP reductase and polyhydroxyalkanoate synthase genes induces PHA production in *Escherichia coli* HB101 strain. *FEMS Microbiol. Lett.* **176**: 183-190.

Timm, A. and Steinbüchel, A. (1990) Formation of polyesters consisting of medium-chain-length 3-hydroxyalkanoic acids from gluconate by *Pseudomonas aeruginosa* and other fluorescent pseudomonads. *Appl. Environ. Microbiol.* **56**: 3360-3367.

Tsuge, T., Fukui, T., Matsusaki ,H., Taguchi ,S., Kobayashi, G., Ishizaki, A. and Doi, Y. (2000) Molecular cloning of two (R)-specific enoyl-CoA hydratase genes from *Pseudomonas aeruginosa* and their use for polyhydroxyalkanoate synthesis. *FEMS Microbiol. Lett.* **184**: 193-198.

Valentin, H.E., Broyles, D.L., Casagrande, L.A., Colburn, S.M., Creely, W.L., DeLaquil, P.A., Felton, H.M., Gonzalez, K.A., Houmiel, K.L., Lutke, K.. Mahadeo, D.A., Mitsky, T.A., Padgette, S.R., Reiser, S.E., Slater, S., Stark, D.M., Stock, R.T., Stone, D.A., Taylor, N.B., Thorne, G.M., Tran, M. and Gruys, K.J. (1999) PHA production, from bacteria to plants. *Int. J. Biol. Macromol.* **25**: 303-306.

van de Loo, F.J., Fox, G.G. and Somerville, C. (1993) Unusual fatty acids. In: *Lipid metabolism in plants*. Moore, T.S. Jr., (ed.). pp. 91-126. Florida: CRC Press Inc.

van der Walle, G.A.M., de Koning, G.J.M., Weusthuis, R.A. and Eggink, G. (2001) Properties, modifications and applications of biopolyesters. *Adv. Biochem. Engin. Biotechnol.* **71**: 263-291.

Voelker, T.A., Worrell, A.C., Anderson, L., Bleibaum, J., Fan, C., Hawkins, D.J., Adke, S.E. and Davies, H.M. (1992) Fatty acid biosynthesis redirected to medium chains in transgenic oilseed plants. *Science.* **257**: 72-74.

Volokita, M. (1991) The carboxy-terminal end of glycolate oxidase directs a foreign protein into tobaco leaf peroxisomes. *Plant J.* **1**: 361-366.

Yang, S.Y., Cuebas, D. and Schultz, H. (1986) 3-Hydroxyacyl-CoA epimerase of rat liver peroxisomes and *Escherichia coli* function as auxiliary enzymes in the β-oxidation of polyunsaturated fatty acids. *J. Biol. Chem.* **261**: 12238-12243.

Wieczorek, R., Pries, A., Steinbüchel, A. and Mayer, F. (1995) Analysis of a 24-kilodalton protein associated with the polyhydroxyalkanoic acid granules in *Alcaligenes eutrophus. J. Bacteriol.* **177**: 2425-2435.

INDEX

MIX
Papier aus verantwortungsvollen Quellen
Paper from responsible sources
FSC® C105338
FSC
www.fsc.org

If you have any concerns about our products,
you can contact us on
ProductSafety@springernature.com

In case Publisher is established outside the EU,
the EU authorized representative is:
Springer Nature Customer Service Center GmbH
Europaplatz 3, 69115 Heidelberg, Germany

Printed by Libri Plureos GmbH
in Hamburg, Germany